Natural Enemies

An Introduction to Biological Control

This second edition of *Natural Enemies* will give students, professionals, and anyone wishing to learn the basics of biological control a fully updated and thorough introduction. The book discusses the huge diversity of organisms used in the control of pests, weeds, and plant pathogens and compares the many different strategies referred to as "biological control": the introduction of exotic natural enemies; application of predators, parasitoids, and microorganisms for shorter term control; and manipulation of the environment to enhance natural enemy populations.

The authors present the ecological concepts which form the bases of biological control and discuss recent changes to make biological control safe for the environment. Case studies are included throughout, providing in-depth examples of the use of different organisms and strategies in a variety of ecosystems. A new chapter covers the current challenges, the impact of climate change, the problem of invasive species, and how biological control can aid sustainability.

Ann E. Hajek is a professor in the Department of Entomology at Cornell University, New York, USA. Her career has focused on biological control and insect pathology, with research ranging from developing biopesticides to investigating why epizootics of insect diseases occur. The International Organization for Biological Control, Nearctic Regional Section, presented her with their Distinguished Scientist Award in 2011.

Jørgen Eilenberg is a professor in the Department of Plant and Environmental Sciences at the University of Copenhagen, Denmark, where he leads a research team studying insect pathogens and biological control. He was the president of the Society for Invertebrate Pathology from 2012 to 2014.

Natural Enemies

An Introduction to Biological Control

Second edition

ANN E. HAJEK
Cornell University

JØRGEN EILENBERG
University of Copenhagen

Shaftesbury Road, Cambridge CB2 8EA, United Kingdom

One Liberty Plaza, 20th Floor, New York, NY 10006, USA

477 Williamstown Road, Port Melbourne, VIC 3207, Australia

314–321, 3rd Floor, Plot 3, Splendor Forum, Jasola District Centre, New Delhi – 110025, India

103 Penang Road, #05–06/07, Visioncrest Commercial, Singapore 238467

Cambridge University Press is part of Cambridge University Press & Assessment, a department of the University of Cambridge.

We share the University's mission to contribute to society through the pursuit of education, learning and research at the highest international levels of excellence.

www.cambridge.org
Information on this title: www.cambridge.org/9781107668249

DOI: 10.1017/9781107280267

First published 2004
Second edition 2018

A catalogue record for this publication is available from the British Library

ISBN 978-1-107-66824-9 Paperback

Contents

Preface

This book provides a new and updated view on what was presented in the 2004 edition by the same name. The first edition was single-authored by Ann Hajek while in this version Jørgen Eilenberg has joined Ann as a coauthor. As in 2004, our goal has been to write an introductory book with broad coverage of the diverse uses of natural enemies for control of invertebrate and vertebrate pests, weeds, and plant pathogens. There are numerous excellent books on biological control that provide more advanced coverage, but our book is intended for people that do not already have extensive knowledge in these subject areas.

The field of biological control has grown and changed in many ways since 2004 and an updated and revised version is needed. Natural enemies are important for pest control in agriculture, in both greenhouses and fields, in forestry, and in managing and protecting natural areas as well as in controlling medical and veterinary pests. One reason for the growth in biological control, especially in agriculture and forestry, is that not as many new chemical pesticides are coming on the market and many older ones can no longer be used. In addition, the public is often more knowledgeable and less tolerant of the uses of pest controls that are potentially harmful to people and the environment. What has changed since 2004 that requires a new book? In particular, in 2004 classical (importation) biological control was considered by many people concerned about the environment and especially biodiversity as being a practice that had great potential for being dangerous to our environment. Since then, the amount that this strategy has been used has decreased significantly, but, with the need to control ever-increasing introductions of invasive species and the development of safer methods for classical biological control, use of this strategy with environmentally safe measures is now beginning to rebound. Another big change is the growth in the diversity and use of natural enemies not released for permanent establishment, a strategy that is generally called augmentation; in addition to arthropod predators and parasitoids applied augmentatively, products based on invertebrate pathogens, insect-pathogenic nematodes, antagonists of plant pathogens, and plant pathogens killing weeds are produced by commercial companies. A prime example of augmentative use of biological control agents today is the greenhouse industry, where natural enemies are often used extensively and very successfully for a variety of reasons including enhanced worker safety. Conservation biological control has also been growing, with practical and theoretical emphases on understanding how we can change environments in order to manipulate the preexisting biodiversity toward controlling pests.

Our book is arranged in the same way as in 2004, beginning with two introductory chapters about why to use biological control and defining biological control. Then we present sections describing the major strategies (classical biological control, augmentation, and conservation), followed by sections on biological control of invertebrates and vertebrates, biological control of weeds, and biological control of plant pathogens, and plant parasitic nematodes. We next discuss safe use of biological control and its integration with other control methods. Finally, we examine the future for this growing and changing field, including discussions on biological control in the context of invasive species, climate change, and the potential for increased use against medically important arthropods, and we discuss how biological control fits into the context of sustainability.

Throughout the book, we have included stories from around the world about how diverse natural enemies have been used successfully in the context of different strategies. As in the 2004 book, we do not provide references throughout the text, with the goal of making this introductory book more easily readable. However, the references that have been used in writing the book are listed at the end of the book in the general references section. We have emphasized examples of biological control that are in practical use, while describing some pest/natural enemy systems that are close to utilization, and only occasionally discussing systems that are simply tantalizing. Those readers interested in biological control agents that hold promise but are not yet being used or who want more in-depth information are referred to the further readings suggested at the end of each chapter, as well as to the large number of reviews and enormous number of primary papers in the scientific literature.

This book is intended as a basic presentation and readers should not need an extensive background in entomology, weed science, or plant pathology. We have attempted to use scientific jargon as little as possible and have provided a glossary at the end to help with specific terms used in the text. We have used English common names for pests and natural enemies when possible, but not all of these organisms have common names (e.g., among insects, actually relatively few have common names). Therefore, we have always provided Latin genus and species names regardless of whether the English common name is given. We assume this will also be helpful for readers who are not familiar with English common names.

Both authors of this book have taught biological control over decades to undergraduate and graduate students at their respective universities and this book grew from their interests in providing more accessible background readings for students taking their classes. The 2004 book has been used for classes around the world and we hope that this version, with many examples from different continents of use of natural enemies in a variety of contexts, finds similar use. In particular, we hope that in reading this book, you will find the interactions between natural enemies and their hosts as fascinating and interesting as we do and will feel that it makes sense to use these relationships to control pests toward increased safety for humans and the environment whenever possible.

Generalization pertaining to biology must always be followed by exceptions. In fact, making generalizations virtually means leaving out at least some of the fascinating variability found in biological systems. There are many tales of amazing interactions and relationships among natural enemies and their hosts or prey and only a small fraction of these could be included in this book. The diversity of manipulations of biological systems for pest control also made it difficult to decide which examples to include in a book such as this. There are really many good stories to be told. Our emphasis has been on providing a glimpse of the diversity of natural enemies used and the diversity of biological control approaches that have been applied. In summary, with this book we hope that we have shared our personal excitement about the field of biological control and that you will become as fascinated as we are with the practice and potential of using natural enemies to control pests.

Acknowledgments

This book would not have been possible without the help of many knowledgeable, kind, and helpful colleagues. First, special thanks go to Sana Gardescu for her amazing assistance with organizing figures and tables and copyright permissions as well as proofreading. David C. Harris provided help with editing early drafts and Scott Salom helped with proofreading the final version. Edwin Lewis wrote the boxed story about the use of entomopathogenic nematodes in pistachios. All chapters were sent out for external review and we thank external reviewers Raghavan Charudattan, Matthew Cock, Julia Crane, Jodi Gangloff-Kaufmann, Jeff Garnas, Jennifer Grant, Roma Gwynn, Kevin Heinz, Martin Hill, Mark Hoddle, Kim Hoelmer, Judith Hough-Goldstein, Trevor Jackson, Marc Kenis, Douglas Landis, Madoka Nakai, David Shapiro-Ilan, Anthony Shelton, Jennifer Thaler, and Saskya van Nouhuys. We also thank many scientists from different continents for information and opinions, including Arthur Agnello, Renato Bautista, Norbert Becker, Gary Bergstrom, David Biddinger, Eric Brennan, Stephen Danielson, Italo Delalibera Junior, Antonio DiTommaso, Sanford Eigenbrode, Mary Louise Flint, Jeffrey Garnas, Martin Hill, Hariet Hinz, Judith Hough-Goldstein, Tero Klemola, Thomas Kuhar, Lawrence Lacey, Ellen Lake, Edwin Lewis, Pamela Marrone, Russell Messing, Lindsey Milbrath, Michael Milgroom, Madoka Nakai, Steven Naranjo, Louis Nottingham, Katja Poveda, Paul Pratt, Mohsen Ramadan, Neil Reimer, George Roderick, John Sanderson, Mark Schwarzländer, David Shapiro-Ilan, Deborah Sharp, Lene Sigsgaard, Donald Steinkraus, Philip Tipping, Joop van Lenteren, Mark Whalon, Rachel Winston, and Mark Wright.

We thank Michael Raupp for allowing use of his lovely photo of a lacewing for our cover and Catherine Tauber and Oliver Flint for identifying the lacewing. We also thank others who helped with figures including Yuri Baranchikov, Roy Bateman, Charlotte Bering (the drawing of the virus-killed nun moth caterpillars by her late father, Claus Bering), Raghavan Charudattan, Marina Cheyushova, Matthew Cock, Regina Kleespies, Tero Klemola, David Mota-Sanchez, Olga Paschenko, Anthony Shelton, and Tomi Vanek (the painting of the predatory mite by his late father, Gašpar Vanek).

This book has not changed entirely from the 2004 version, although the text has definitely changed quite a lot, and more than we had expected. We are not going to list again the many people who helped in many ways and were acknowledged in the 2004 version, but we thank them again as parts of the 2004 book remain. In particular, many of the figures used in the 2004 book have been used again. We especially want to acknowledge again Alison Burke, who helped extensively with organizing the excellent figures for the 2004 version and even drew some of the illustrations herself.

We also want to thank our editors, Dominic Lewis and Lindsey Tate, for their steadfast faith and patience that this revised edition could and would be completed. Finally, we thank our spouses, James Liebherr and Betty Østergaard Jacobsen, for their encouragement, patience, and assistance throughout the long process of revising, writing, and putting this book together.

Part I

Introduction

1 Why Use Natural Enemies?

Humans share the planet Earth with about 8.7 million species of complex organisms, those whose cells have nuclei. If we add the number of species of microbes without nuclei, this number would be far greater. Each species eats, grows, and reproduces in different ways in different locations around the world, but virtually no species does this in isolation. All species are interconnected to some extent, with some organisms more dependent on others, especially those higher in the food chain. Tigers would not live long without their prey being available, just as rabbits would not survive for long without plants to eat. Humans have quite a dominant position in many ecosystems and we depend on many other species for food and shelter. Especially because the influence of humans is so pervasive throughout the world, humans also compete with many organisms. When other organisms compete for resources with humans or negatively affect humans in other ways, they are generally regarded as "pests."

Humans have been plagued by pests since before recorded history. A pest can be formally defined as any organism that competes with humans for resources used for food, fiber, or shelter. These pests eat crop plants used for food or trees used for lumber as well as plants, such as cotton, used for fiber. Pests can also disrupt human and animal health and well-being, making organisms directly affecting human and animal health, such as mosquitoes carrying pathogens that cause diseases like malaria or dengue, pests too. Thus, the definition of the term pest needs to be broad because of the great diversity in the ways that pests negatively impact humans. Pests are as diverse taxonomically as they are in the ways that they compete with humans, ranging from microorganisms to plants and to animals with or without backbones. With such variability comes a variety of adaptations making some organisms that compete with humans tough adversaries.

There are many different means for controlling pests (see Chapter 19) but this book is principally covering control methods using living organisms, a strategy called biological control. We will therefore not be covering all pests but only those specifically targeted by biological control. The major types of pests that are addressed by biological control include invertebrates (especially arthropods that often attack plants or animals) and vertebrates, weedy plants, and microorganisms, called plant pathogens, that attack plants (often crop plants or forest trees).

1.1 Historical Perspective on Chemical Pest Control

Humans have always needed to control pests affecting them directly, such as mosquitoes or bed bugs, or competing with them for a great diversity of resources. Through the ages, pest-control practices have changed dramatically. The earliest known record for the use of naturally occurring compounds for pest control was around 2000 BCE (BC) in a Hindu book written in India that referred to using poisonous plants to control pests. At the time of the pharaohs, the ancient Egyptians used compounds extracted from plants to help with insect control. Around 1000 BCE, Homer the Greek mentioned using sulfur as a fumigant to control pests, and in 77 BCE, Pliny the Roman reported that arsenic was insecticidal. Around 1100 CE (AD) soap was used as an insecticide in China. From the 1500s to 1600s, approaches to pest control seem to have changed. Plants having insecticidal compounds were used more extensively by Europeans, so that in the 1800s tobacco extracts and nicotine smoke were applied for insect control. Around the same period, the use of inorganic compounds also increased for pest control. In the 1800s, we see the first mention of a mixture concocted for pest control that became widely used: Paris green, an arsenic-based compound, was developed and applied against Colorado potato beetles, *Leptinotarsa decemlineata*, in the United States. Bordeaux mix, a combination of copper sulfate and hydrated lime, was developed in 1882 in Bordeaux, France, to control plant-pathogenic fungi on grapes and other fruit.

However, throughout these times, the overriding methods for pest control were cultural practices, such as destroying pest-infested fields, allowing them to lie fallow, and rotating crops. For example, when soybean crops are rotated with corn (maize), the populations of soil-dwelling nematodes that attack soybean roots decline when corn is growing, and after the corn crop has been harvested, soybeans can again be planted in that field. Other cultural controls included practices such as altering dates for planting and harvesting, using trap crops, planting mixtures of crops, managing drainage, and removing crop residues that harbor pests. Through use of cultural controls, growers were basically manipulating and augmenting the naturally occurring processes of pest suppression.

Several developments took place between World Wars I and II, setting the stage for major changes in pest control. Industries developed methods for large-scale production and chemists vastly improved their abilities to synthesize chemicals. In 1939, both DDT for control of insects and 2,4-D for control of weeds came on the scene. These extremely effective compounds revolutionized pest control. Since that time, a cascade of different compounds belonging to an increasing number of chemical classes have been synthesized for pest control. Most of the early compounds were effective against a broad spectrum of pests, killed pests very quickly, and were relatively easy to apply using spray equipment. Availability of these synthetic chemical pesticides vastly improved the potential for successful harvests and, consequently, use of these compounds skyrocketed.

Use of pesticides (such as insecticides, fungicides, herbicides, etc.) over time increased, but these changes are not easy to quantify. Figure 1.1 illustrates the increase in sales of different types of pesticides on the worldwide market between 1980 and 2011.

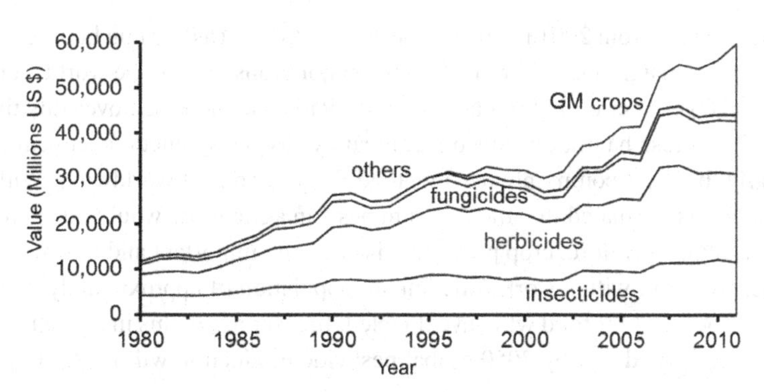

Figure 1.1 Worldwide pesticide markets from 1980 to 2011. GM crops = genetically modified crops. Data compiled from the annual reviews of the British Agrochemicals Association (updated by Roy P. Bateman from Bateman, 2000).

In 2011, while 55 percent of pesticide sales occurred in North America and Europe, a significant amount (45 percent) of pesticides were sold in Asia, South America, and the rest of the world. The total value of pesticide sales increased more than 3.5 times between 1980 and 2011. Looking at the value of pesticides can be a misleading statistic, because, over time, the potency of pesticides has increased, confounding comparisons through time (e.g., moving back through time, more of a compound had to be purchased for a similar effect). An alternate way to look at this could be to evaluate the area of land on which pesticides were applied, but unfortunately such data are not readily available. A major fact to be gleaned from Figure 1.1 is that among the numerous types of pesticides, from 1980 to 2011 the use of herbicides increased substantially, although the use of fungicides and insecticides also increased. Also during this period, genetically modified (GM) crops for insect and weed control entered the market (in 1996) and have become widely adopted in industrialized and developing countries. By 2011, GM corn, cotton, soybeans, and canola, the only large-acreage GM crops available at the time, accounted for 22.4 percent of the total sales ('GM crops' in this context include those containing insecticidal toxins and/or resistant to herbicides). Other crops designed to control insects, weeds, and viruses are entering the market and are predicted to become an increasingly larger component of the value of the global pesticide market.

The bottom line is that the global market for pesticides is projected to reach 3.2 million tons in 2019, valued at US $81.18 billion worldwide. In 2007 alone, greater than 2.4 billion kilograms of pesticides were applied worldwide and the United States, with a large area of agricultural land, accounted for 20 percent of this use. However, in the United States in 2013, it was estimated that 80 percent of pesticide use was agricultural while 20 percent was nonagricultural, including use for homes and gardens where more pesticide is often applied per square meter. Between 2005 and 2009, 2.2 kg of pesticide per hectare of arable land were applied in the United States, while rates of application were even higher in other countries such as China (10.3 kg/ha), an example of a country with extensive agricultural land use, and Colombia (15.3 kg/ha), where coffee is a valuable crop with heavy pressure from pests and therefore abundant use of pesticides can occur. Using data from 2001 to 2003, pests were estimated as causing between 26 and 40 percent losses in agricultural production for major crops around the world; surprisingly for the forty-year period before, although pesticide use increased over this time, these levels of crop loss had not changed significantly. Today, synthetic chemical pesticides are clearly the most commonly used method for pest control worldwide. Without crop protection, it is estimated that the losses to pests in agriculture would be approximately 60–86 percent. Therefore, crop protection is critically important and this need will only increase. By 2050, with a worldwide human population of approximately 9 billion, the worldwide demand for food will have doubled. In response to this increased demand for food, it is estimated that by 2050 global pesticide production will be 2.7 times greater than in 2000.

1.2 Why Consider Biological Alternatives for Pest Control?

Synthetic chemical pesticides are used so widely because they often work very well for controlling pests. However, pesticides are not always the optimal solution; sometimes they cannot control pests effectively for a variety of reasons. The major reasons that alternatives to synthetic chemical pesticides have been developed are presented below. In describing these scenarios, control of arthropods (e.g., insects and mites) will be used for examples although similar issues occur relative to the control of weeds and plant pathogens.

1.2.1 The Pesticide Treadmill

Although synthetic chemical pesticides are still the most widely used method for pest control, there are growing reasons to consider alternatives. Frequently, when pesticides are applied to control arthropods, naturally occurring controls are severely disrupted and natural enemies that normally live by consuming the pest are no longer abundant or even present. When this happens, if the target pest reinvades the treated area, there are no or few natural enemies present and the target pest population increases again unchecked, frequently to higher densities than were present initially (target pest resurgence) (Figure 1.2). Figure 1.3 shows the growth of an outbreak in a target pest, the California red scale, *Aonidiella aurantii*, as a result of regular spraying of low doses of DDT.

Since many natural enemies are often killed when broad-spectrum pesticides are applied, other organisms that had not been pests before the treatment can increase to densities that cause damage. This occurs because natural enemies that had previously maintained the nonpest populations at low densities are no longer present or abundant enough to provide control. This is known as a secondary pest outbreak (Figure 1.4). This scenario of a secondary pest outbreak can be demonstrated with increases in peach silver mites, *Aculus fockeui*, a species that was not a problem until the pyrethroid fluvalinate was applied to peach trees in Japan for control of other peach pests (including fruit borers, aphids, and spider mites). Before the application of fluvalinate, the peach silver

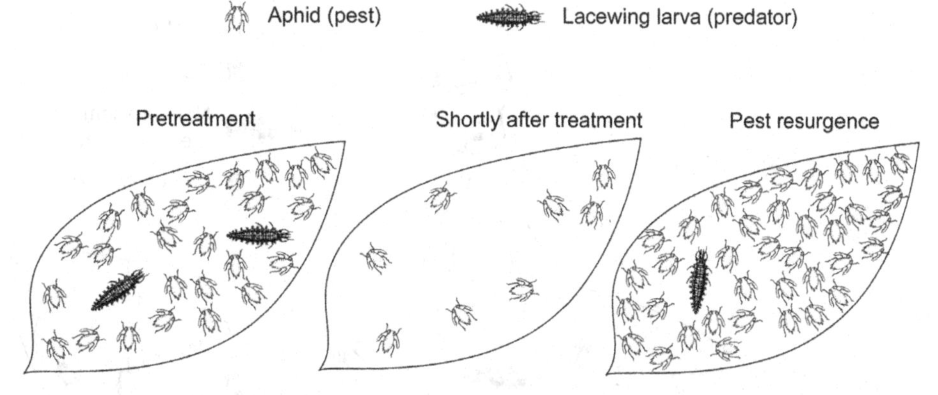

Figure 1.2 Target pest resurgence can occur when natural enemies are destroyed. Pesticides often kill a higher proportion of natural enemies than pests so that after application the pest can increase again rapidly (Flint & Gouveia, 2001).

mite population was naturally regulated, but after the pesticide treatment the predatory mite population was decimated and the peach silver mites were able to increase with little to stop them (Figure 1.5). New York State apple orchards provide an example of the diversity of secondary pests that can become problematic because of the application

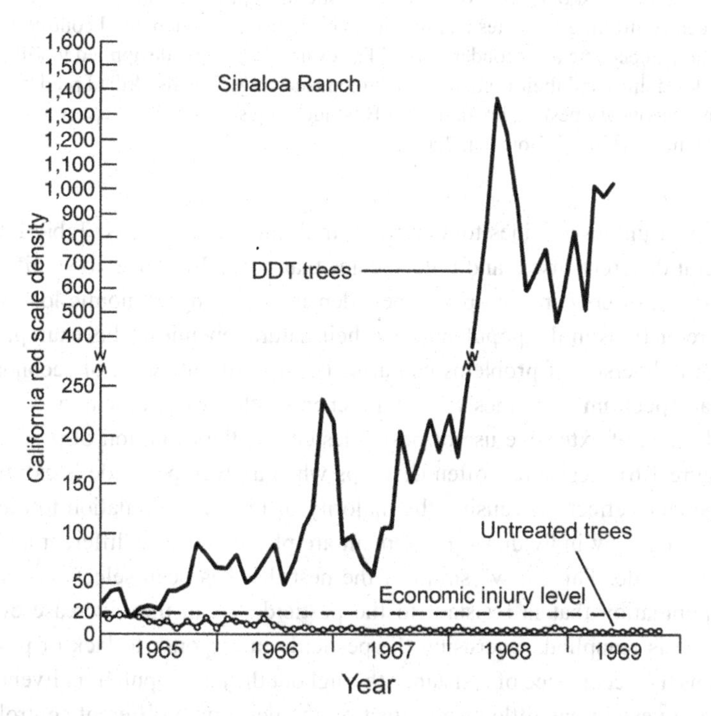

Figure 1.3 Increases in California red scale, *Aonidiella aurantii*, on citrus trees associated with monthly sprays of low doses of DDT, compared with nearby untreated trees under biological control (DeBach, 1974).

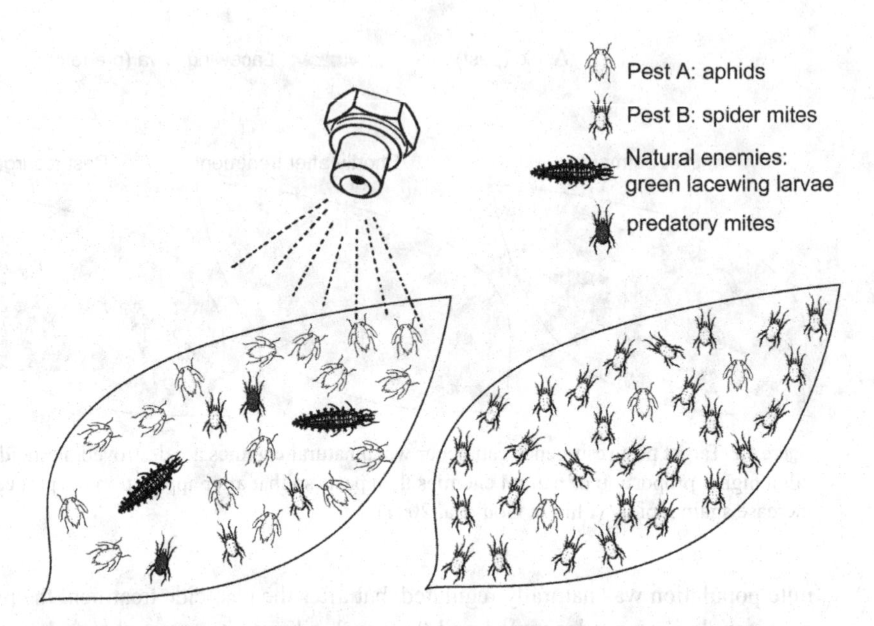

Figure 1.4 Secondary pest outbreaks occur when pesticide applications kill the natural enemies that have been controlling a species that has not been a pest. Without natural control, this species increases and can become a "secondary pest." For example, a pesticide applied to kill Pest A (aphids) killed aphids and their predators, the green lacewings, but also killed predatory mites, resulting in a secondary pest outbreak of Pest B (spider mites) previously kept at lower densities by predatory mites (Flint & Gouveia, 2001).

of broad-spectrum insecticides for control of multiple primary pests (Table 1.1). In this case, several different insect and mite species that normally cause no significant trouble for apple growers can multiply to pest densities causing economic losses because of severe reductions in the populations of their natural enemies. This example demonstrates that a diversity of problems can arise because of outbreaks of secondary pests when broad-spectrum pesticides kill natural enemies that are not the targets.

A third effect of extensive use of pesticides can be the evolution of pesticide resistance (Figure 1.6). Resistance often develops when a given pesticide is extremely, but not 100 percent, effective, causing the majority of the pest population to die after an application. The few individuals that remain are physiologically different and can tolerate the pesticide. This "new" strain of the pest that has been selected – that is, the survivor population that is resistant to the pesticide – can then increase even when the pesticide is reapplied. Overusing the pesticide in response to lack of pest control only hastens the occurrence of resistance throughout the pest population. Eventually, the pesticide in question has little or no effect on the pest and a different control strategy must be used. It is often assumed that when a new synthetic pesticide has been developed and is heavily applied, it will only be a matter of a few years before resistance

Table 1.1 Primary and secondary arthropod pests in apples in New York State.

Type of pest	Species	Type of damage
Primary pests	Codling moth (*Cydia pomonella*) Plum curculio (*Conotrachelus nenuphar*) Apple maggot (*Rhagoletis pomonella*) Obliquebanded leafroller (*Choristoneura rosaceana*)	For all primary pests, larvae (immature stages) damage or bore into developing apples
Secondary pests	San Jose scale (*Quadraspidiotus perniciosus*) European red mite (*Panonychus ulmi*) White apple leafhopper (*Typhlocyba pomaria*) Woolly apple aphid (*Eriosoma lanigerum*) Two-spotted spider mite (*Tetranychus urticae*)	For all secondary pests, apples are not directly damaged, but overall tree health can be affected

Source: A. Agnello (pers. comm.)

begins to develop in the target population. The length of time before resistance evolves depends on many factors, but resistance is always a threat to any pesticide.

While resistance to pesticides was first reported in 1914, it did not create major problems until the 1940s. Resistance to DDT was first seen in 1946 in house flies, *Musca domestica*, only 7 years after use began. By 1948, pesticide resistance was seen in 14 target species and by 2013 close to 600 species of arthropods displayed resistance to a variety of insecticidal active ingredients (Figure 1.7).

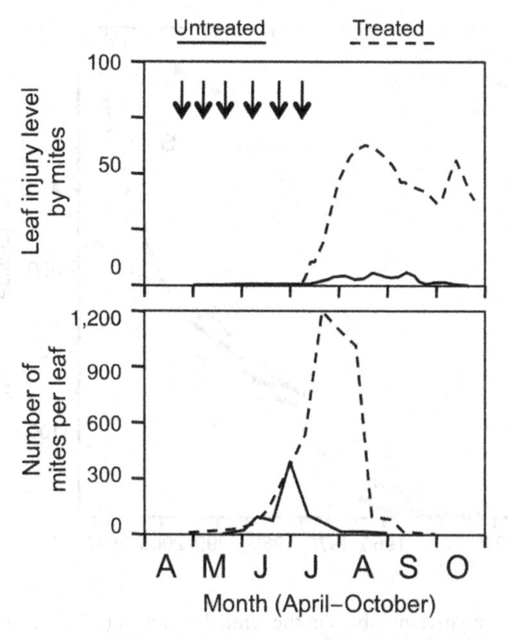

Figure 1.5 Effect of fluvalinate application on the population densities of peach silver mite in treated and untreated orchards with associated injury to peach leaves. Arrows indicate application times (based on Kondo & Hiramatsu, 1999).

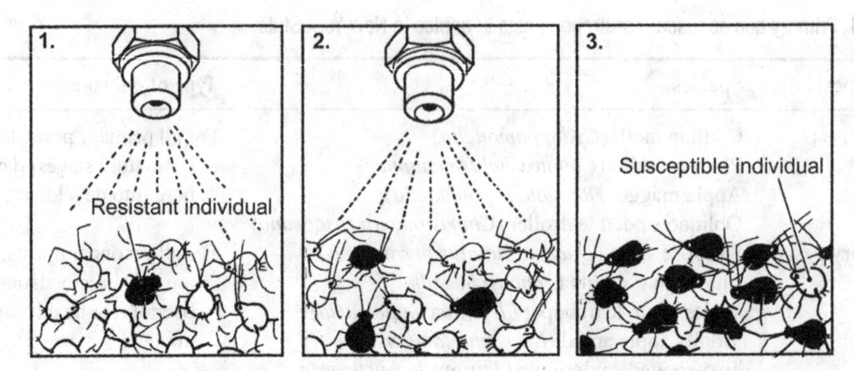

Figure 1.6 Pest populations can develop resistance to pesticides through natural selection. (1) When pesticides are applied, most individuals are killed, but a few are less susceptible and these remain. (2) The less susceptible individuals or their progeny are less likely to die with subsequent applications. (3) After repeated applications, the resistant or less susceptible individuals predominate and applying the same pesticide is no longer effective (Flint & Gouveia, 2001).

When resistance to a pesticide begins to develop, there is a characteristic series of events that often occur. First, growers may apply more of the pesticide, often not realizing that the lack of control is because of resistance. Next, growers might switch to a closely related pesticide, but once pests develop resistance to one pesticide in a pesticide class, they are often at least partially resistant to other similar pesticides. The

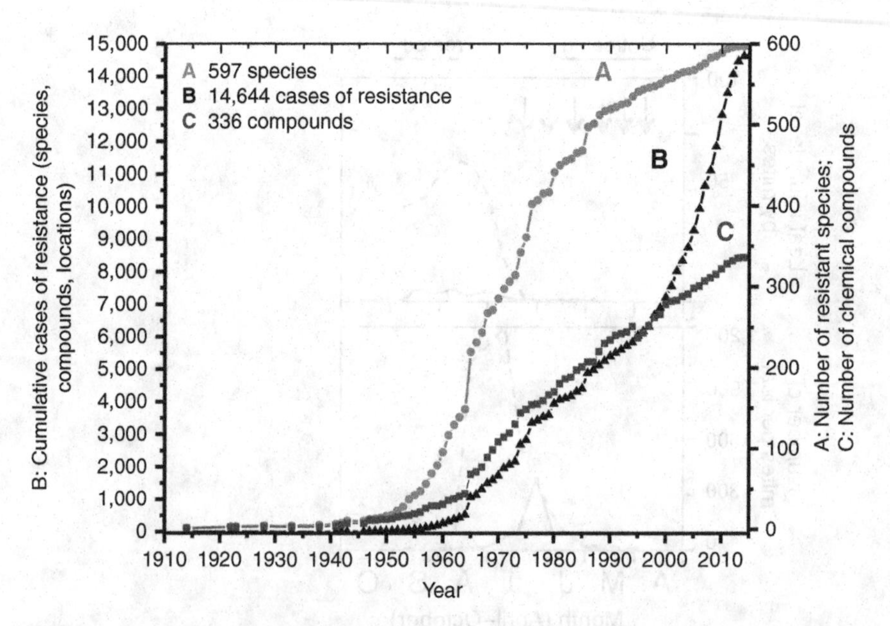

Figure 1.7 From 1945 to 2015, cumulative numbers in the United States of (A) arthropod species resistant to pesticides, (B) cases of resistance (by species, compound and location), and (C) chemical compounds with resistance documented (updated by D. Mota-Sanchez; data from Mota-Sanchez & Wise, 2017).

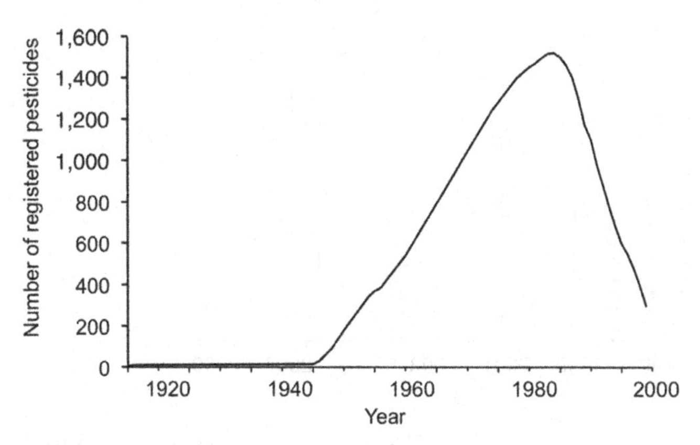

Figure 1.8 Total numbers of pesticides registered with the US Environmental Protection Agency from 1914 to 1999 (based on Mota-Sanchez et al., 2002).

grower might next choose a pesticide from another class of pesticides, for example switching from organophosphate insecticides to pyrethroids, under the assumption that the pest had acquired at least partial resistance to all organophosphates. However, pests can be resistant to several classes of pesticides at the same time so resistance can eventually develop to this second choice of control agent. To compound the troubles, frequently the alternative pesticide can be more costly. For example, with development of resistance to DDT, the organophosphate malathion was substituted at five times the cost. When resistance developed to malathion, fenitrothion, propoxur, or deltamethrin were often substituted by growers at 15–20 times the cost.

These three phenomena together (target pest resurgence, secondary pest outbreaks, and development of resistance in pest populations) have been termed the "pesticide treadmill." This can lead to increasing dependence on synthetic chemical pesticides, resulting in seemingly addictive use of this type of control.

1.2.2 Fewer Pesticides Are Available

As a result of the development of resistance to entire classes of pesticides, there is a constant demand for new types of pesticides, because the pesticides that previously were effective no longer provide adequate control. However, the costs of developing and registering new pesticides have increased over time. Since about 1970, there has been a significant slowdown in the rate of new pesticides being introduced to the market. It is estimated that 140,000 insecticidal compounds need to be screened to discover one successful compound and, once a compound has been identified, it can take about US$250 million and 8–12 years to develop and register a new material for application. In addition, owing to increasing regulation, some of the pesticides that have been available for many years are no longer legally available for application. For both of these reasons, in many countries there are fewer pesticides registered and thus available for use (see Figure 1.8 for the trend in the United States) and there are increasingly fewer new synthetic chemical pesticides for use (Figure 1.9). As one example, fumigation with

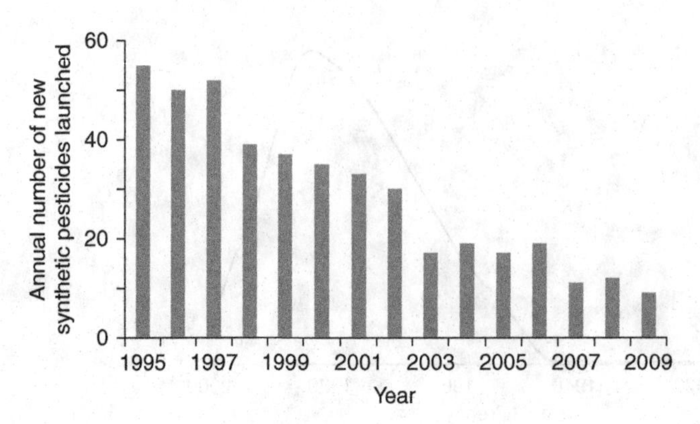

Figure 1.9 Annual numbers of new synthetic chemical pesticides for agriculture launched globally from 1995 to 2009 (based on Glare et al., 2012).

methyl bromide has been a mainstay for control of soil-borne and structural pathogens and pests as well as storage diseases of fruits and vegetables. However, use of this compound as a pesticide was phased out in most countries in the early 2000s, largely because of its role in ozone depletion. In its absence, alternative controls must be used. As another example, a moratorium was placed on use of three neonicotinoid insecticides by the European Union (EU) between 2013 and 2015 while the effects of these pesticides on wild bees were evaluated, with the future for use of these pesticides in the EU in question. In summary, there is a trend toward fewer synthetic chemical pesticide options as a result of increased resistance to existing insecticides and decreased development and registration of new compounds.

1.2.3 Synthetic Chemical Pesticides Aren't Always the Answer

There are some situations for which chemical pesticides are not the most appropriate choice for controlling pests. One example would be introduced exotic organisms that become pests; it has been estimated that 50,000 exotic organisms have been introduced to the United States alone and those having an impact of some kind are referred to as invasive species. In fact, invasive species are now considered a major problem worldwide as a result of the increasing human population frequently moving organisms around the globe and thereby altering ecosystems at an increasingly alarming rate. Many invaders can become pestiferous because of the fact that they are no longer associated with the natural enemies with which they coevolved. Among pests in agriculture, approximately 20–40 percent have been introduced from elsewhere. While most introduced organisms are accidental introductions, a small percentage of these were purposeful introductions such as crop plants and honey bees (*Apis mellifera*). Some were purposeful introductions with unexpected side effects, for example, the weed kudzu (*Pueraria* species) that was introduced to the southeastern United States to control erosion became established and then spread rampantly through much of the region, becoming a problematic weed. Introduced organisms are not always identified quickly, so they establish and become ubiquitous before it is possible to eradicate them. It is difficult to imagine how a synthetic chemical pesticide could easily solve a problem like kudzu, which has already spread extensively, without continual human intervention and its associated costs. Problems caused by such pests are therefore not readily addressed using synthetic chemical pesticides because more permanent control is what is needed. Fortunately, classical biological control has been used successfully against such pests (permanently introducing natural enemies from the land of origin of the pest; see Chapter 3). Unfortunately, by all predictions, accidental introductions of invasive species will only continue with the increased global movement of humans and materials (see Section 1.3.1).

Synthetic chemical pesticides, for a variety of reasons, simply cannot control some pests. Damaging stages of numerous arthropod pests live in the soil, especially those that feed on roots. Control of soil-dwelling arthropods is not as straightforward as control of externally feeding arthropods. It can be very difficult, if not impossible, to apply pesticides that will effectively reach soil-dwelling arthropods and plant-pathogenic nematodes. In the past, soil was sometimes fumigated because of this difficulty, but now many fumigants and two of the most effective nematicides, can no longer be applied in the United States. To add to this problem, the fumigants that are now available are very costly.

No chemicals are capable of controlling some pests. Just as there is no control for the common human cold by directly killing the causative viruses, there are no chemical options for directly killing viruses attacking plants. Therefore, general control tactics employed for control of plant-pathogenic viruses include cultural practices such as rotating crops and altering planting dates, use of insecticides to control the insects vectoring viruses, or use of virus-resistant strains of plants.

Sometimes a crop or habitat is just not amenable to the use of synthetic chemical pesticides. For rangeland weeds and insects, the areas impacted by these pests can be huge. However, rangeland is not high-value land and the yield from the land often cannot support the cost of spraying such huge acreages. For vegetable crops, small acreages are planted with a diversity of crops, each with its own pests. Pesticide manufacturers have to develop and register individual pesticides for particular crops and they have little monetary incentive to develop and register pesticides that will be used on such small areas. Therefore, there are often fewer chemical options for pest control in many crops planted on smaller acreages. As a third case where chemical pesticides are not optimal, aquatic weeds can be immense problems when they block waterways. These are usually controlled by manual removal and by applying herbicides but such solutions are only temporary and the problem then usually recurs. Because repeatedly applying herbicides to water can lead to the presence of herbicides throughout the environment, including in drinking water, such a control option is often frowned upon.

1.2.4 Concerns About Human and Environmental Health

The first general outcry by the public against use of synthetic chemical pesticides was championed by Rachel Carson who wrote *Silent Spring*, published in 1962 (Box 1.1).

Box 1.1 Rachel Carson

Rachel Carson was a quiet person who loved nature and whose writings had a profound influence on the creation of the level of environmental consciousness present today. She was born in 1907 in rural Pennsylvania, far from the ocean but, as her career developed, she followed her fascination with marine biology. She spent time at Wood's Hole Biological Laboratory on Cape Cod, Massachusetts, and her training was completed with a master's degree in zoology from Johns Hopkins University in 1932. In 1936, she began working as a scientist and editor for the US Fish and Wildlife Service, where she continued working for 15 years. She never married, caring for her mother and adopting her grandnephew after his parents died. During this time, she began writing about the natural history of the sea for the public. In 1952, the same year that she completed her prize-winning book *The Sea Around Us*, she resigned from her position as editor-in-chief of the US Fish and Wildlife Service to be able to concentrate on her writing.

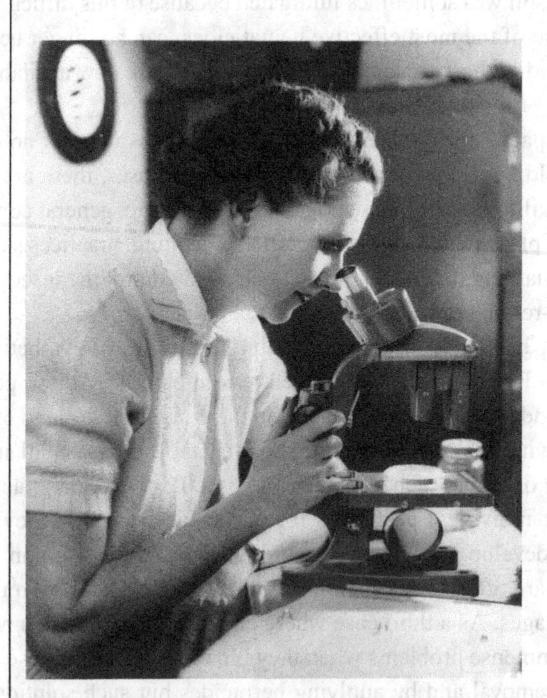

Figure Box 1.1 Rachel Carson (photo by Brooks Studios, courtesy of the Linda Lear Center for Special Collections & Archives, Connecticut College).

With some reluctance, after World War II Rachel turned her focus from the sea to the land. She was an avid birder and she was very aware of bird deaths linked with pesticide spraying. As she investigated further, she became disturbed by the misuse of synthetic chemical pesticides. She decided to take on the responsibility of informing the public about the side effects of pesticide use by writing a book.

Originally, she planned on using the title "Silent Spring" for a chapter on effects of pesticides on birds but eventually, in 1962, this title was used for the entire book.

By 1960, Rachel was already fighting breast cancer yet she persevered with publication of her book although she knew that unpleasantness would certainly follow publication of *Silent Spring*. As she had expected, the chemical industry and some members of the US government vehemently charged that she was an alarmist. Yet, Carson's message was unwavering: she proposed stopping the uncontrolled use of synthetic chemical insecticides that had long-lived activity. She demanded creation of new policies to protect humans and the environment. Her quarrel was with misuse of this technology for which the long-term effects were not known and she insisted on the fundamental rights of individuals to be free from contamination with toxic chemicals without their consent. Her book became a best seller and she lived long enough to see the issues she had raised discussed on television, in the US Congress, and in the British House of Lords. Many credit this quiet naturalist and excellent author with providing the sparks that initiated movements to protect the environment. Certainly, the increase in interest in biological control that began in the 1970s was spurred by desires to find alternative pest controls causing minimal impact to the environment.

The development of synthetic chemical pesticides provided unprecedented pest control, their use soared, and there were few regulations regarding their use. As an example, a Tennessee Game & Fish Commission biologist cited an application of 10 percent dieldrin granules (a compound more toxic than DDT) at 30 pounds per acre (= 33.6 kg/ha) for control of Japanese beetle, *Popillia japonica*, in a recreational area. The granules were so thickly applied that they covered picnic tables, and parents and children were told to brush them off from tables before eating. Excessive applications such as this resulted in extensive mortality of animals higher in the food chain than insects, for example, birds and fish. Rachel Carson, working as a wildlife biologist, became aware of these environmental side effects. She decided to write a book about this broad-scale, unregulated application of pesticides. In the book *Silent Spring*, she urged the government to investigate the effects of pesticide use and regulate pesticide application. President Kennedy read the book and was instrumental in initiating studies of the type Rachel had urged. The book generated extensive controversy at the time and, despite efforts by the chemical industry to suppress it, *Silent Spring* became a best seller. It is generally credited as the trigger that started the environmental movement. In 1970, President Nixon created the US Environmental Protection Agency as a direct repercussion from this controversy and today this organization regulates uses of insecticides, herbicides, and fungicides for pest control in the United States. Similar agencies in other countries were formed that now regulate pesticides in their respective countries.

Rachel Carson was correct that insecticides were having side effects on animals and the environment and, while more regulations were enacted in response to her efforts, levels of pesticide application since her time have only increased worldwide (although many pesticides used today are less persistent and are safer for the environment). This worrisome trend is a result of many factors including increased agricultural production and availability of less-persistent pesticides than DDT, which was the focus of much of Rachel Carson's writing. DDT is a chlorinated hydrocarbon and compounds in this group are known for their environmental persistence and movement through the food chain, so organochlorine pesticides have been banned in most areas, although they are still used for mosquito control in some developing countries where mosquito-vectored pathogens are otherwise difficult to control. Still, among 26 countries where pesticide application has been quantified, 46 percent applied more than 5 kg of pesticides per hectare of arable land between 2005 and 2009. Despite the likelihood that many of these pesticides were environmentally safer than those previously available, the naturally occurring flora and fauna have certainly been exposed to pesticides and this has resulted in environmental consequences.

After the development of organochlorine insecticides, the next groups of compounds to be developed were the organophosphate and carbamate insecticides. Although less persistent in the environment, they were much more toxic than chlorinated hydrocarbons and can lead to degeneration of nervous tissues. These groups have eventually also been largely phased out in the developed world, although their use in certain crops persists. Although these materials are more dangerous to humans, especially applicators and farm workers, they are cheaper than newer insecticides that have been developed and are often used in the developing world. The global pesticide industry continues to develop new compounds, especially in response to the lack of pesticides for specific pests, because of new regulations banning certain preexisting pesticides, and the evolution of pesticide resistance among many pests.

With years to study the nontarget impacts of pesticides, numerous unintended side effects from pesticide use have been identified. There are direct effects on nontarget animals and plants, some being lethal. However, some direct effects are sublethal and chronic, negatively affecting health and reproduction while not killing less susceptible species. A classic case was the reproductive failure in predatory birds, often attributed to eggshell thinning caused by DDT exposure through the consumption of contaminated prey. Another instance of unintended environmental effects is the reduction of sperm production caused by a commonly used soil fumigant, DBCP. Recent studies are showing that other commonly used pesticides negatively impact immune and reproductive systems, and these are linked with decreasing amphibian populations. There are also effects because of pesticide residues remaining in the soil or being leached into the water. Scientists who have been studying ecosystem-level impacts of pesticides suggest the persistence of pesticides in the environment can be a cause of widespread biodiversity loss, especially in aquatic systems and in soils. In recent years, huge declines in honey bee colonies have been linked to sublethal exposures to pesticides as one of several interacting causes of colony decline. We also know that chronic exposure of bumble bees to several insecticides reduced their pollen-collecting efficiency. The ecosystem-level impacts of the last two examples include decreased pollination, which certainly directly impacts human food production.

Even low rates of human exposure to some pesticides can cause irritation to skin or eyes, while exposure to high levels may cause mortality. Determining the effects of exposure to any given pesticide can be difficult, as it varies according to the exact compound, dose, method of exposure (inhalation, skin contact, eye contact, etc.), age, and stage of physiological development, and more. The effects of chronic exposure to lower levels of pesticides (exposures at doses lower than those causing acute effects and usually over a long period) are difficult to predict. With the relatively widespread use of pesticides worldwide, one can easily imagine that exposure to traces of pesticides occurs commonly during our daily lives, including in drinking water. It is difficult to determine the effects of such long-term exposure to trace amounts of pesticides. However, for families with children living near agricultural fields where pesticides are applied, some studies suggest that growing up with pesticide exposure can negatively impact the development of young human brains.

To their credit, the chemical companies that develop and market synthetic chemical pesticides are now producing compounds that are much safer for humans and the environment, but still effective for pest control. These are often referred to as "reduced-risk pesticides." Regulations are in place in developed countries to ensure safer, efficient, pesticide use. For example, as of 2013, 178 countries and the EU are part of an international treaty to phase out persistent organic pollutants (the Stockholm Convention on Persistent Organic Pollutants), particularly targeting organochlorine pesticides because of their persistence and bioaccumulation (although the United States and several other countries are not included in this agreement). In the United States, legislation has banned a number of chemical pesticides, use of pesticides is regulated, and the amounts of pesticides left on foods to be marketed for human consumption or animal feed must be below regulated limits. The United States and the EU have supported the use of alternative pest-control strategies and the Food and Agriculture Organization (FAO) of the United Nations (UN) has adopted a code on distribution and use of pesticides that promotes integrated pest management (Chapter 19) and natural pest-control strategies.

However, the pesticides that are banned in the United States and Europe are often still being produced and/or sold in developing nations where they are applied without regulation or with little enforcement of regulations. Short-term savings as a result of the low prices of older pesticides that are banned elsewhere can influence pest-control choices made by these farmers, although the pesticides they are using can damage human health and the environment.

It is very difficult to estimate the extent of effects of chemical pesticides on human health in developed countries and more difficult still in developing countries. In 1992, the World Health Organization estimated that 25 million cases of pesticide poisoning and 20,000 unintentional deaths occur each year, mostly among agricultural workers and members of rural communities. One survey from Nicaragua suggested that two-thirds of cases of pesticide poisoning are not reported. A summary stated that "50% of all pesticide-related illnesses and 72.5% of recorded fatal pesticide poisonings occur in developing countries, although these countries account for only 25% of the pesticides used world-wide" (Harris, 2000). While more than 80 percent of pesticides are applied in developed countries, 99 percent of poisonings occur in developing countries where regulation and education systems are not as well established.

1.2.5 Sustainable and Organic/Ecological Food Production

Last but not least, especially in developed countries, the public has become increasingly aware of the potential health and pollution hazards that accompany the use of many pesticides. To prevent and avoid such problems, many people choose to buy food that has been grown in an environmentally sustainable way and is often labeled "organic" or "ecological." Biological control plays a key role in organic food production (especially in high-value crops) and is recognized internationally, by the International Federation of Organic Agriculture Movements, as a means for pest control that fits into the overall scope of organic production. Different countries and regions may have different specific rules about the types of pest controls that can be used for a crop that is sold as "organic" but, as a general overriding rule, pest control in organic crops is based on "natural" compounds and agents, which include biological control organisms.

1.3 A Pest or Not?

The goal of biological control is to control pests. The status of a species as a pest at one time and place does not mean that species will always pose problems. The subjectivity of designation as a pest is illustrated by the fact that species that are pests to some people can be considered beneficial by others. A case in point would be the multicolored Asian lady beetle, *Harmonia axyridis*, introduced to the United States to control aphids. Unfortunately, this beetle species often forms large aggregations in sheltered locations to spend the winter. In the eastern United States, these beetles find their way into houses where they happily take up residence for the winter, most often being considered a nuisance. Therefore, these lady beetles are seen as beneficial biological control agents by some and pests by others.

Sometimes, an organism designated as a pest requiring control does not actually cause serious damage. As early as about 1915, in California citrus orchards, feeding by even a few tiny red scale insects or discolored spots because of feeding by other insects, decreased profits of citrus growers, even if not affecting the taste or nutritive value of oranges. As in many crops today, pests in California citrus are controlled to meet cosmetic standards that often require complete eradication of arthropods from a field. Luckmann and Metcalf (1994) provide a more ecologically based view regarding the presence of pests in crops. "Pest-management concepts dictate a tolerant approach to pest status. Indeed, it may be that not all pests are bad and that not all pest damage is intolerable."

Why some species become pests and others do not has been of great interest in biological control. When pest species become very abundant, understanding the cause for the perturbation allowing population increase can help toward developing methods for controlling the pest. Pests can be native species whose numbers have increased because new opportunities are offered as a result of human activity. For example, when crops are planted as monocultures, previously little-known native species that are able to feed on the crop plants can become important pests. Potatoes are not native to North

America but to South America. When they were first planted in North America, a previously poorly known beetle found these plants that it could eat and its populations increased phenomenally. This example is about the Colorado potato beetle, *Leptinotarsa decemlineata*, and is a case of a native insect attacking an introduced plant for which it was preadapted. Subsequently, the Colorado potato beetle was introduced to Europe where it maintained its status as a major pest of potatoes. However, in Europe, the Colorado potato beetle is an introduced insect attacking an introduced plant species grown as an important crop.

It has been estimated that 60–80 percent of all pests are native to the areas where they are pests. Yet, many examples of successful biological control involve pests that have been introduced from one area to another. We refer to organisms that are introduced to an area from somewhere else as introduced or exotic, while species that evolved in that area are called native or endemic. If the introduced organism becomes a pest, then we refer to it as invasive. As a result of recent controversy regarding the increasing importance of invasive species as pests and the extensive history of use of biological control against invasives, this group will be discussed in more depth.

1.3.1 Invasive Species

Movement of species to new locations around the world has been very common throughout human history. Many crop plants and domesticated animals were first moved by humans long ago and have been moved so extensively since then that it is difficult to trace exactly where the original species or strains came from. Such species of uncertain origin are called cryptogenic. The rate at which species could be moved long distances really only began to increase significantly once ocean-going sailing ships began to be built in the fifteenth century.

Wherever people traveled, they brought with them the plants they knew how to grow and use for food and, eventually, the animals that they knew how to raise or hunt. On oceanic islands, the waves of people following the explorers were often sealers and whalers and they purposefully brought and released goats and rabbits, while rodents and pet animals were released as well, albeit often unintentionally. The native flora and fauna on islands are often characterized as having few species and these are not well adapted to competition or predation. Thus, the relatively fragile ecology of many islands has been severely impacted by the multitudes of arrivals of non-native species that became established. Hawaii, in the middle of the Pacific Ocean, was a frequent stopover for ships with the result that the flora and fauna are dominated by introduced species; for example, about half of the terrestrial plants, 25 percent of the insects, most of the freshwater fish, and 40 percent of the birds in Hawaii are not native.

In the mid-1800s, European settlers, particularly in the Americas, Australia, and New Zealand, who were far from home, and Europeans curious about exotic species and potential commercial exploitation of new species, formed so-called "acclimatization" societies. The goal of such societies was specifically to foster importation and establishment of new species. Such acclimatization societies were an extreme cause for introductions, but even where such societies did not occur, new species were still accidentally introduced but just not purposefully. Since the time of acclimatization societies, global trade and global travel have only increased and the numbers of organisms that are inadvertently moved from one geographic area to another have increased in parallel.

What makes it so easy for introduced species to become established in new locations? Changes in land use, disturbance, and destruction of natural areas have often opened habitats for introduced species to become established. It has also been hypothesized that environmental changes such as warming oceans and land and changes in large-scale disturbance regimes (e.g., suppressing forest fires) leave natural systems in imbalance so that invaders can more easily become established. Today, growing numbers of introductions of introduced species are common around the world and few countries guard against them effectively enough to prevent new introductions.

A summary of six countries (the United States, the British Isles, Australia, South Africa, India, and Brazil) in 2001 estimated that by that time, already 120,000 introduced species had become established in these countries. In the United States alone it has been estimated that there are more than 50,000 species that are exotic. Estimates from 14 countries suggest that from 7–47 percent of the species of terrestrial plants present had been introduced and, of these, approximately 15 percent of the introduced species had become pests and are called invasives.

Ultimately, there is a cost from invasive species that become established. Invasive species have been estimated as costing the United States more than US$130 billion dollars per year in damage to agriculture, forests, rangelands, and fisheries. When pests attack a commercial product, it is relatively easy to ascribe a cost to their impact but it is much more difficult to calculate monetary values for the damage caused by invasives to native flora and fauna and the ecosystems in which they live. In more recent years, interest has grown regarding exotic species that become established and which then outcompete members of the native flora and fauna. Such introductions that affect the biodiversity of an area are sometimes referred to as "biological pollution." In extreme cases, competition can extend to localized extinctions of native species. The presence of invasives can also disrupt ecological processes, such as disrupting associations among community members that depend upon each other. For example, the European herbaceous plant, garlic mustard, *Alliaria petiolata*, invading North American forests produces antifungal compounds that disrupt mutualistic fungi associated with the roots of tree seedlings. This results in reduced growth of tree seedlings after garlic mustard invades an area. Unfortunately, being able to predict which introduced species will become invasive and then predict the ecosystem-wide impact of invaders has proven difficult. Certainly, the effects from invasions are difficult to document because frequently prior to introductions, in many cases, the standard patterns of activity and abundance of native species have not been documented. Without such information about the initial composition, abundances, and interactions of native communities, it is difficult to quantify which changes in ecosystems occur as a result of invasives and which are normal fluctuations in that ecosystem. However, with increasing efforts over time, scientists are working on answering such questions in order to pinpoint which invasives are causing disruptions in ecosystems and therefore need to be controlled.

Further Reading

Carson, R. (1962). *Silent Spring*. Boston, MA: Houghton-Mifflin.

Graham, F. (1970). *Since Silent Spring*. Boston, MA: Houghton-Mifflin.

Köhler, H.-R. & Triebskorn, R. (2013). Wildlife ecotoxicology of pesticides: Can we track effects to the population level and beyond? *Science*, **341**, 759–765.

Mascarelli, A. (2013). Growing up with pesticides. *Science*, **341**, 740–741.

Pimentel, D. (ed.) (2002). *Biological Invasions: Economic and Environmental Costs of Alien Plant, Animal, and Microbe Species*. Boca Raton, FL: CRC Press.

Pretty, J. (ed.) (2005). *The Pesticide Detox*. London: Earthscan.

Sexton, S. E., Lei, Z., & Zilberman, D. (2007). The economics of pesticides and pest control. *International Review of Environmental and Resource Economics*, **1**, 271–326.

Stokstad, E. (2013). Pesticide planet. *Science*, **341**, 730–731.

Thacker, J. R. M. (2002). *An Introduction to Arthropod Pest Control*. Cambridge: Cambridge University Press.

Whalon, M. E., Mota-Sanchez, D., & Hollingworth, R. M. (2008). *Global Pesticide Resistance in Arthropods*. Wallingford, UK: CABI Publishing.

2 Introduction to Biological Control

> The amount of food for each species of course gives the extreme limit to which each can increase; but very frequently it is not the obtaining food, but the serving as prey to other animals, which determines the average numbers of a species.
>
> Darwin, 1859

2.1 Defining Biological Control

Populations of all living organisms are, to some degree, reduced by the natural actions of their predators, parasites, pathogens, and antagonists (this last term is mostly used in biological control to describe microorganisms attacking plant-pathogenic microbes). These processes, which occur among all living organisms, have been referred to as "natural control." However, when organisms that are considered pests are controlled by natural enemies, it is often called biological control (sometimes shortened to biocontrol) and the agents that exert the control are frequently called natural enemies. Humans can exploit biological control in many ways to suppress pest populations. The varied approaches for manipulating the activity of natural enemies to control pests differ based on the types of effort that are required, what natural enemies are involved and including the level of human involvement, and sometimes the suitability of approaches for commercial development.

Biological control has been defined many times, but a commonly accepted definition is provided below.

The use of living organisms to suppress the population of a specific pest organism, making it less abundant or less damaging than it would otherwise be. (Eilenberg et al., 2001)

To understand the basis for this definition, we need to discuss why biological control is used. Of course, there are a multitude of reasons. Development of biological control methods expanded greatly after synthetic chemical pesticide application became the dominant method for pest control. Use of biological control grew as a result of practical needs to find a solution to pest problems when chemical pesticides did not work or were not appropriate for controlling specific pests. Another major impetus for using biological control has been the fact that chemical pesticides can cause serious side effects, leading to major concerns about human health and the health and preservation of global and local environments. Biological controls leave no chemical residues and are usually quite host specific, especially in comparison to many synthetic chemical pesticides.

As years have passed and scientific research has advanced, the types of approaches available for pest control have increased in number and complexity. Within the field of biological control, a diversity of natural enemies can be used in many different ways. Another advance in control methods has been the ability to synthesize the active compounds used by pests for communication (pheromones), which can then be used for detecting and controlling those same pests. More unique types of control associated with the biologies of organisms have also been developed more recently. One example is the bacterium *Chromobacterium subtsugae*, which produces multiple compounds that have a range of effects on numerous species of insects, including oral toxicity, repellency, reduced oviposition, and reduced fecundity. The bacterium is mass-produced and both the media in which the bacteria grew (which contains these compounds) as well as the bacterial cells (most of which are dead) are applied to prevent high densities of numerous types of pests.

Likewise, the genes responsible for producing compounds that deter or harm pests can be spliced into other organisms where they are expressed for production of pesticidal compounds in the specific tissues where they are needed. A well-known example is genetically modified, or transgenic, plants. Genes that are currently used extensively for expression in plants were originally derived from insect-pathogenic bacteria, *Bacillus thuringiensis* (Bt). These genes encode production in plants of the same Bt toxins that kill insects feeding on these bacteria (Chapter 10).

Based on our definition, use of only the compounds produced by natural enemies would not be called biological control. The use of these applications could instead be included in the larger categories of "biologically based pest management" or "biorational pest management". However, disagreements over use of this terminology are far from resolved. Controversy centers on whether the organisms used for "biological control" must be living or must be just the source of compounds and genes. There is also controversy over what is meant by the term "living" and whether viruses are considered living or not. Viruses do not have their own metabolism, but they are able to replicate themselves; for that reason they have traditionally been considered as natural enemies. In one of the original descriptions, biological control agents (including parasites, predators, and pathogens) should provide self-sustained control with density-dependent responses to host populations (see Chapter 6). As the use of biological control expanded, a more inclusive definition was drafted by DeBach (1964) to include the activities of all parasites, predators, and pathogens that decrease other organisms' populations and not only relationships based on density dependence.

A definition published by the US National Academy of Sciences in 1988 expanded DeBach's definition to "the use of natural or modified organisms, genes, or gene products to reduce the effects of undesirable organisms (pests), and to favor desirable organisms such as crops, trees, animals, and beneficial insects and microorganisms." Some biological control experts embrace this expanded definition as a means toward growth of biological control through adoption of new technologies. Others worry that the expanded definition, including genes and gene products, could tarnish the positive image of biological control embraced by a public who might not welcome genetic engineering. Yet others feel that the recent additions to the definition lose the original aspect of interactions among populations of organisms. A solution to this controversy appears to have come in the use of the terms "biologically based pest management," "bioprotectants," and "biorational pest management," which include products from living organisms (sometimes also produced synthetically) as well as the living organisms themselves. The use of these alternate terms to retain the emphasis of reliance on biological interactions while preserving the definition of the already otherwise-defined term "biological control" circumvents these problems.

2.2 Natural Control

The concept of a "balance of nature" has been traced to ancient times when it was believed that the numbers of each species were virtually constant. It was thought that each species had a role and place, and extinction did not occur because it would disrupt the balance and harmony of nature. Outbreaks of species were often considered aberrations, having something to do with gods punishing humans for wrongdoing. During the nineteenth century, especially after Darwin's contributions, biologists and early ecologists began trying to understand how this "balance of nature" was attained and maintained. This concept of a balanced condition maintained in nature is found in literature, such as a 1906–1907 book for school children, *The Wonderful Adventure of Nils Holgersson*, written by Selma Lagerlöf from Sweden, a winner of the Nobel Prize in literature. She described an outbreak of the nun moth (*Lymantria monacha*) in a spruce forest, which she said occurred as a result of humans' misunderstanding of nature. However, when the clever creatures in the forest took action, a disease (an insect virus) caused an epidemic in the nun moth population the next year, this ensured that the correct natural balance was reestablished.

Populations of the majority of species in nature are thought to be under naturally occurring regulation through complex interactions within food webs and they therefore do not increase to compete with humans. The problems historically addressed by biological control come from pestiferous species that have evaded the web of natural controls that could restrict their numbers. The goal of several different biological control strategies is to reestablish this level of self-sustained natural control, either by introducing a natural enemy for permanent establishment (classical biological control) or by altering the environment to conserve or enhance the activity of natural enemy populations (conservation biological control). While the third major strategy, augmentation, is often more immediate and is not aimed at establishing long-term regulation of pest populations, it relies on these same basic interactions of natural enemies living at the expense of hosts. In fact, classical biological control introductions have been likened to "ecological experiments on a grand scale" and illustrate both the "escape" of pest species relieved of natural enemies and their demise when enemies are restored to the system.

The extent to which successful biological control coordinated by humans compares with the natural control in regulating naturally occurring populations has been questioned. It seems that biological control often has succeeded by establishing a single strong association between one pest species and one natural enemy species, a more simplified type of regulation than normally would be seen in nature. Simple or not, the basic tenet behind the use of classical and conservation biological control is that "natural control" can be used to reduce the pest population. Augmentative releases have the goal of reestablishing control of pests, usually in simplified or manipulated systems. Ecologists have worked for many years investigating the interactions between pests and their natural enemies to understand what is necessary for establishing, reestablishing, or maintaining natural levels of control.

Control of pests by natural enemies is often thought of as a function of a healthy ecosystem and one that is beneficial for all living organisms. In this light, biological control is considered one of the most important "ecosystem services." This concept of ecosystem services, looking at aspects of ecosystems in relation to their benefits to all living organisms, is also much broader, including functions such as pollination and the cycles that clean our water.

2.3 Diversity in Biological Control

Biological control differs significantly depending on whether the pests are invertebrates, vertebrates, plants, or microorganisms. For biological control of invertebrates, hosts are usually small and sometimes mobile (at least in some life stages). Emphasis has been on plant-eating arthropods and arthropods of importance to public health. Virtually all natural enemies used for biological control of arthropods kill pests directly. Mortality of the pest is often very quick with predators, but there can be a time lag with parasitoids or pathogens because these natural enemies often must first develop using the hosts as food before killing them (although this is not always the case).

Pestiferous plants range from small herbs to large trees; these are stationary and, at times, dense. Biological control of weeds requires many individual natural enemies to damage a weed, unless the natural enemy attacks the so-called "Achilles heel" for that plant species (e.g., a part of the plant or a period of its life cycle that is especially vulnerable). When an Achilles heel is being targeted, fewer biological control agents could be necessary to control the weed. Mortality of the weed is always delayed, if the plant dies at all, although growth and seed production would be reduced more quickly. Also, in contrast to biological control of arthropods, weeds do not move except through seed dispersal or the movement of vegetative propagules (parts of plants that can themselves take root), so herbivorous natural enemies generally do not have as much difficulty locating their target pests as do natural enemies attacking animals. Weedy plants can escape from a natural enemy by dispersing and establishing a new population at some distance from the source. However, finding such new isolated plant populations is still considered less of a problem for weed-feeding natural enemies compared with the difficulty for arthropod-attacking natural enemies of finding and attacking mobile arthropod pests. Weeds are also different from arthropods as pests because competition with other plants can be important in mediating the outcome of biological control. If weeds can be partially suppressed by herbivory or disease, then the weed can more easily be outcompeted by other plants that are hopefully desirable.

For the microorganisms causing plant disease, biological control is caused by the multitudes of microbial antagonists that compete with multitudes of plant-pathogenic microbes. Both plant pathogens and their antagonists are usually tightly linked with specific habitats. For many programs, antagonists to plant diseases are successfully applied preventively, so the time before control is effective is not an issue.

Scientists working to control these diverse pests must adopt very different tactics in relation to the importance and immediacy of the pest problem, the type of impact on the pest that is needed, and the ability of both natural enemy and host/prey to disperse.

2.3.1 Is Biological Control Always Appropriate?

Biological control is principally used to combat arthropod pests, weeds, and plant pathogens and only in a few instances has it been used to control vertebrate pests (see Chapter 11). It has been used extensively for terrestrial systems, both above ground and in the soil, starting with uses in agriculture and forestry and later being applied to natural ecosystems as well as parks and gardens for control of invasive and native pests. Natural enemies have also been successful in controlling arthropods and weeds in freshwater ecosystems, principally when used in contained bodies of water. Natural enemies have only in a few cases been used in marine ecosystems but, given the growing number of invasive species in such ecosystems, this possibility may be revisited. However, there are certainly some types of pests and conditions for which biological control might not be the most appropriate type of control. These conditions include situations where pests must be totally eliminated very rapidly. Several such examples are described in the following three subsections.

Economic Injury Level of a Crop

The presence of an organism in association with humans or some valued resource does not always mean that the organism needs to be controlled. This is particularly true with agricultural pests or pests that can be present at low densities without causing problems. A concept that has been developed to determine whether an organism needs to be controlled is called the economic injury level (EIL). The economic injury level is defined as the lowest density of pests that will cause economic damage. An economic threshold is generally set below the economic injury level and once densities of a potential pest reach this threshold, control practices should begin (Figure 2.1). If managers wait until pest densities reach the economic injury level, the pests are sure to increase over that density and cause economic loss before being controlled.

Of course, economic injury levels are very dynamic and differ based on crop value, management costs, degree of injury, and crop susceptibility to injury. Among these factors, crop value is perhaps the most notorious for fluctuations and unpredictability. Not

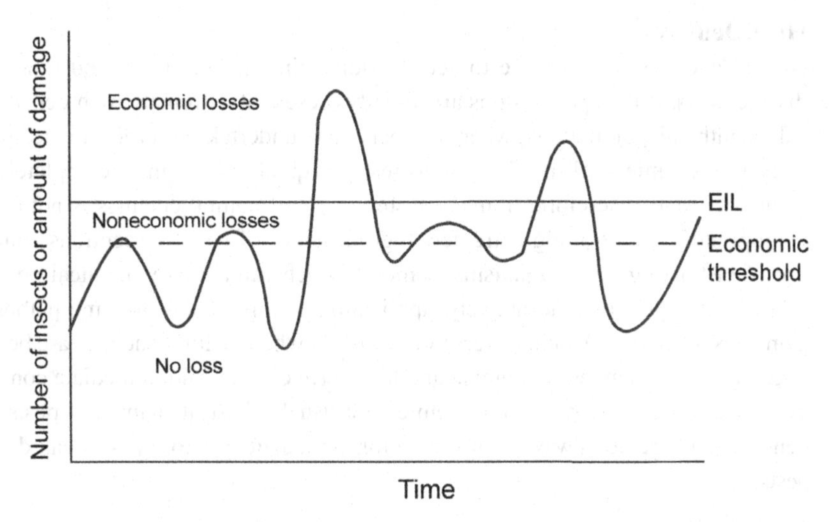

Figure 2.1 Relationship among the density of an insect population, the economic threshold used to trigger management decisions, and the economic injury level (EIL) above which economic loss occurs.

all pest problems are amenable to economic analysis so it can be difficult to determine how much money is lost because of damage caused by a pest. Therefore, use of an economic injury level is most suited for pests of agricultural crops. While the actual economic injury level can change each year to some extent, this relationship clearly demonstrates that control is not always necessary and that sampling the pest population should be conducted to determine when pests are dense enough that control is necessary.

The economic injury level of the crop being managed is important when considering use of biological control. Some crops, such as cut flowers, cannot withstand much, if any, damage before losing their economic value. The economic injury level of these crops is therefore set very low; few to no pests or damage can be present before economic losses are incurred. Biological control that could be used on this type of crop would need agents that act very quickly to kill all pests, such as fast-acting microorganisms applied inundatively. For crops that can tolerate the presence of some pests before economic injury occurs, there is more flexibility in the types of biological control that can be used. Natural enemies such as some parasitoids and pathogens that take some time before killing hosts can protect crops in systems with higher economic injury levels, such as when pests feed on the foliage of greenhouse vegetables; in this situation, the presence and activity of pests for some period before their death will not always ruin the product. For long-term biological control of established pests in the field, a system with a higher economic injury level is often thought to be necessary so that some pests remain present at low densities to maintain a natural enemy population in the area. Then, if the pest population increases, the natural enemies are already present and can respond more quickly, rather than if they were absent and needed to recolonize the site.

Host Density

Natural enemies require time to act, including time to find hosts and time to kill or disable hosts. If pest populations are already present at extremely high densities (often called outbreak populations) when the decision to undertake control is made, most types of natural enemies will not be able to respond quickly enough to completely prevent further damage. Therefore, in many systems, use of natural enemies is not considered appropriate for controlling extremely high densities of pests. Exceptions could be the use of fast-acting microorganisms, some of which can act like chemical pesticides in killing pests quickly. Alternatively, application of high doses of some pathogens can gain a fast response. Another exception would be where natural enemies are being introduced for permanent establishment and long-term control, and immediate control is not expected. Nevertheless, natural enemies are usually best at managing pests at lower densities and are not always appropriate for immediate control of outbreak densities of pests.

Eradication

When an invasive species is introduced to a new area, governmental bodies frequently decide to eradicate it, which means to totally eliminate it from that area. Successful eradication programs are often large undertakings. When eradication programs are focused on rapid action, slower-acting types of biological control agents are certainly not appropriate and fast-acting and lethal chemical pest-control agents are often used. However, in some cases, natural enemies have been used when pests targeted for eradication have occurred in urban areas. Repeated aerial sprays of the insect-pathogenic bacterium *B. thuringiensis* were used in British Columbia, Canada, to eradicate the Asian gypsy moth (*Lymantria dispar asiatica*) in 1992 and in Auckland, New Zealand, to eradicate the white-spotted tussock moth, *Orgyia thyellina*, in 1996–1998 and the painted apple moth, *Teia anartoides*, beginning in 2002.

2.4 History of Biological Control

The first records of biological control describe habitat manipulation to increase natural enemy populations. As early as 324 BCE (BC), growers in China encouraged populations of the weaver ant, *Oecophylla smaragdina*, in citrus trees to control caterpillars and large boring beetles. This species of ant builds large paper nests in trees, resulting in legions of ants inhabiting the trees. Colonies could be purchased or were moved from wild trees into orchards. In addition, to foster movement of ants within the orchard, bamboo runways were placed between trees. Surprisingly, these practices were still seen in the Shan States of North Burma in the 1950s. In 1775, a similar practice was reported from date growers in Yemen, who moved colonies of predatory ants (most probably *Formica animosa*) from the mountains to date groves to control pest insects.

These earliest uses of natural enemies to control pests involved manipulations of preexisting natural enemies visible to the naked eye that were generalist predators feeding on many types of prey. With scientific advances, other groups of natural enemies that were smaller and more specialized began to be investigated and then considered as control agents. The fact that smaller invertebrates live as parasites of larger invertebrates was first reported in the 1600s. With the invention of the microscope by Antonie van Leeuwenhoek in the late 1600s, it became possible to learn more about these ever-smaller natural enemies. Although microorganisms had been seen previously, it was not until 1835 that microorganisms were first shown to be the cause of an insect disease by Agostino Bassi, working with the fungal pathogen *Beauveria bassiana* infecting silkworm, *Bombyx mori*, larvae (caterpillars). In 1874, W. Roberts, working with the fungus *Penicillium* and bacteria, first demonstrated that microorganisms could inhibit one another and, in 1908, M. C. Potter first demonstrated such inhibition among plant-pathogenic microorganisms.

As explorers set out to discover new lands and establish trading colonies, relatively rapid movement of humans around the world became possible. Movement of plants that could be used as crops followed and pests were often accidentally introduced with the crops. In some cases, organisms that were familiar to colonists from Europe were purposefully introduced, only to become pests, as with rabbits introduced to Australia. It was frequently found that organisms virtually unknown in their areas of endemism could become major pests in areas where they had been introduced and it became commonly accepted that this was as a result of their release from control by natural enemies in their area of origin. This hypothesis states that a pest is able to increase to high densities in the absence of the natural enemies that regulate populations of that pest in its area of endemism.

As practices in agriculture and forestry improved, single cultivars were grown in ever-larger monocultures. These changes were accompanied by greater pest problems caused by both native and introduced pests. With such pest problems, the world was eager to accept synthetic chemical pesticides when they were developed. The synthetic chemical insecticide DDT and the synthetic chemical herbicide 2,4-D were first tested and used for pest control around 1942 after which development and use of a great diversity of pesticides followed. Although natural enemies had been discovered and described much earlier, developments in the use of natural enemies for control only seriously diversified and escalated after problems with DDT became evident.

Of course, scientists had been thinking of using natural enemies for pest control long before the advent of synthetic chemical pesticides. Even Linnaeus suggested using predatory insects to control insect pests in 1752. The term "biological control" was coined in relation to plant pathogens by C. F. von Tubeuf in 1914 and then applied to insects by H. S. Smith in 1919. While similar basic principles underlie much of biological control, control of different groups of pests evolved quite separately. Scientists working with these different groups of pests and different groups of natural enemies needed specific training. Scientists trained as entomologists generally specialized either in predators and parasitoids for controlling arthropods or, with backgrounds in plant science and entomology, in phytophagous arthropods for use against weeds. Knowledge of microbiology, plant science, and plant pathology is necessary for plant pathologists working to control plant pathogens or to control weeds with microbes, and knowledge of different types of microorganisms and entomology is required to work on pathogens for control of arthropods. Scientists working to control invertebrates or vertebrates, weeds, or plant pathogens historically had few opportunities for interchange although they certainly communicated results within each subdiscipline. The different subdisciplines thus partly developed their own definitions and practices.

Similarly, as biological control grew, different strategies (classical, augmentation, conservation; see Chapters 3, 4, and 5) were developed to address the diverse array of pest control problems. As different types of pests were given different priorities for control in different types of systems and geographic areas, different types of biological control were emphasized. For example, regions with abundant invasive species were among the first to use classical biological control. Augmentation has developed extensively in some areas with large greenhouse industries.

In more recent years, there has been an attempt toward fostering communication among practitioners working with different types of natural enemies and pests and with different strategies. Notably, several books published in the last decade are cross-disciplinary in scope. The goal of the scientific journal of the International Organization for Biological Control, named *BioControl*, is to publish scientific research from all of the different branches of biological control. In 1991, two new journals, *Biological Control: Theory and Application in Pest Management* and *Biocontrol Science and Technology*, were started specifically to publish research results from across all types of biological control research.

Because of the independent nature and growth of the different subdisciplines, short histories of each will be presented separately.

2.4.1 Controlling Arthropod Pests

Before the advent of restrictions on movement of organisms around the world, pest introductions were numerous and frequently caused dramatic outbreaks. The cottony cushion scale (*Icerya purchasi*), an insect attacking citrus, was introduced to southern California where it caused enough damage in the mid- to late 1800s to threaten the existence of the emerging California citrus industry. A predatory lady beetle (*Rodolia cardinalis*) and a parasitic fly (*Cryptochaetum iceryae*) were introduced from Australia, the original home of the scale insect. These introductions led to phenomenal success and brought public attention to biological control (see Box 3.1). For a period following this success, there were many introductions of predatory and parasitic insects around the world to control introduced pests, particularly lady beetles were introduced to control aphids and scale insects, but no programs were as successful as the California citrus example. This period of seemingly haphazard introductions following the cottony cushion scale success was considered a little too enthusiastic by some, who later called this period the "lady bird [lady beetle] fantasy."

Introduction of exotic natural enemies to control introduced pests has remained very active. After DDT first became available, classical biological control programs continued to be undertaken at an increasing rate by entomologists trying to repeat the success they had seen with cottony cushion scale. Although many introductions of natural enemies were made, especially in the 1950s–1970s, at this time this control strategy was not as dependable as with the cottony cushion scale (see Chapter 3). The high rate at which classical biological control organisms were released was possibly driven by scientists trying to compete with chemical pesticides. Without extensive background information, releases were often on a "try it and see" basis, hoping for quick success. In 1983, Howarth published his first article criticizing the nontarget effects of introductions of exotic natural enemies, especially regarding classical biological control of insects and weeds (see Chapter 18). The rate of introductions against exotics subsequently slowed, with increased emphasis on nontarget testing. However, use of classical biological control has not stopped and this strategy remains occasionally the best option for specific pest situations, when environmentally safe natural enemies exist.

As use of biological control grew, practitioners began investing more effort in using natural enemies in ways other than classical biological control. In England in 1895, the egg parasitoid *Trichogramma* became the first natural enemy to be mass-produced for release to control pest arthropods. The ability to mass-produce parasitoids and predators was developed further, with the parasitoid *Encarsia formosa* mass-produced in Europe before World War II to control the greenhouse whitefly, *Trialeurodes vaporariorum*. However, predators and parasitoids were not used extensively for augmentation until the 1970s, when use of mass-produced natural enemies in greenhouses escalated. This type of augmentative use of natural enemies has increased exponentially since then through the development of companies that mass-produce different types of natural enemies and then navigate the regulations and permits to sell and send them for biological control use (see Chapter 4).

Work with pathogens to control arthropods began in earnest later than work with predators and parasitoids, in part because scientific advances were necessary to be able to work easily with microorganisms. However, already in 1879, the Russian scientist Elie Metchnikoff conducted the first experiments to use the fungus *Metarhizium anisopliae* to control the wheat grain beetle, *Anisoplia austriaca*. It was, however, not until the twentieth century that scientists began to understand what insect viruses are and how they work. Pathogens began being developed to be used as formulated products, often referred to as biopesticides (see Chapter 4). In 1948, a bacterial pathogen for control of Japanese beetles (see Chapter 10) was the first insect pathogen registered for control in the United States. As will be described, the number of arthropod pathogens used has increased to fulfill the specific needs of different systems.

Today, use of natural enemies to control arthropods is usually part of integrated pest-management programs (IPM). The concept of IPM was proposed in 1959 (see Chapter 19), and its adoption has increased since then, both in response to systems in which pesticides are not effective or cannot be used and systems where use of natural enemies for control is preferred. Conservation or enhancement of the resident natural enemies of arthropods for control is also increasingly included as part of IPM programs.

2.4.2 Controlling Weeds

As stated by Goeden and Andrés (1999), "Like so many other aspects of science, [the study of biological control of weeds] began by accident." In 1795, a scale insect called cochineal that was cultured commercially as a source of carmine dye was introduced from Brazil to northern India for dye production. However, the species that was introduced was not the superb dye-producer *Dactylopius coccus*, but by accident, it was a related species, *Dactylopius ceylonicus*. Instead of reproducing well on the spineless prickly pear grown specifically for dye production, *D. ceylonicus* moved onto its natural host plant the prickly pear *Opuntia vulgaris* that had been introduced to northern India and had become a problematic weed. The value of *D. ceylonicus* as a control agent was realized and, from 1836 to 1838, this species was introduced to southern India and then, in the 1860s, to Sri Lanka. In both areas, *D. ceylonicus* provided successful control of the weedy *O. vulgaris*.

Classical biological control of weeds grew and was used in numerous countries with the principal emphasis being the use of herbivorous insect natural enemies to control introduced, perennial weeds in relatively undisturbed areas such as rangelands. Emphasis shifted in the late 1950s and early 1960s when programs were initiated against aquatic and semi-aquatic weeds, annuals, biennials, and weeds growing in croplands, along roadsides, and invading natural ecosystems. The diversity in types of weeds to control and types of natural enemies to use for control continues to increase today.

The second type of approach, use of plant pathogens for mass application and more immediate weed control without expecting permanent establishment, began around 1971. Research in the late 1960s through the 1980s began the development of bioherbicides for control of a diversity of weedy plants, and research and development of plant pathogens as bioherbicides continues today.

2.4.3 Controlling Plant Pathogens and Plant Parasitic Nematodes

Biological control of plant pathogens got its start much later. Because this field is based totally on microorganisms, more technically advanced techniques were required for its growth. The first biological control strategy that was used extensively against arthropods, classical biological control, was not appropriate against plant pathogens. Early in the 1900s, plant pathologists recognized that microorganisms could suppress plant diseases and this activity could be manipulated through cultural and management practices. The first trials attempting to suppress plant disease by adding beneficial microorganisms to soil occurred in the 1920s. It was not until the 1950s that the first biological control organism targeting plant pathogens, *Phlebiopsis gigantea*, was commercially used to control infection of cut tree stumps by *Heterobasidion annosum*, a fungal pathogen that has the potential to spread through root grafts to healthy trees nearby (Box 16.2). Biological control of plant pathogens and plant parasitic nematodes was destined to continue to grow through development of biopesticides, especially against pathogens in the soil environment. From 1995 to 2000, many biological control organisms became available as commercial formulations for suppression of plant diseases under 80 different product names.

Soils in some regions were identified as suppressive or conducive to plant pathogens attacking banana as early as 1922. Naturally suppressive soils have now been identified for numerous crops, and plant pathologists are working on understanding the mechanisms involved in suppression and thus developing ways to create suppressive soils.

There is a great need for biological control to control soil-dwelling plant parasitic nematodes and, at present, few biological control products are available.

Therefore, scientists from a diversity of backgrounds have worked to develop methods for biological control of animal, plant, and microbial pests and their findings have been reported and summarized in many books. Some of the recent books presenting broad coverages of biological control are listed at the end of this chapter. Many additional excellent books and reviews are narrower in scope and cover biological control of specific types of pests, specific types of natural enemies, different types of resources to protect, or biological control in different geographic regions and some of these will be referenced at the ends of the following chapters.

2.5 Studying Biological Control

Use of biological control generally requires much more background information about the biology and ecology of pests than use of chemical pesticides. For all types of biological control, it is necessary to demonstrate that natural enemies are effective at controlling pests. Methods have been developed in ecology for evaluating the importance of natural enemies throughout the life of a host or prey species. Life tables are used to document the effects of natural enemies on pest populations of different ages. This type of analysis is easier to use with insects that have discrete life stages, such as eggs, different larval instars (stages between molts), pupae, and adults. Mathematical models can then be used to explore interactions or suggest hypotheses about what regulates population densities. However, these types of analysis do not really demonstrate the efficacy of natural enemies in natural or managed systems and experimental methods are needed to prove this.

There are numerous ways the effects of natural enemies can be demonstrated experimentally. The following brief discussion will include methods used for evaluating biological control of arthropods but comparable methods are appropriate for documenting success of other types of natural enemies against pests. For all of these experimental methods, the emphasis is on documenting a difference in host/prey populations when natural enemies are absent versus when they are present.

2.5.1 Sampling

Quantification of pest densities before and after impacts of natural enemies has often been used to demonstrate that natural enemies have been effective at suppressing pest populations. Percent infection, parasitism, and level of predation have all been used to demonstrate that natural enemies have been responsible for pest control. However, the important information regarding whether control was achieved is the absolute number of surviving pests and not the percentage that survive. This is because if pest populations were very high and even if there was a high percentage of mortality, there could still be enough remaining individual pests to cause significant damage. Critics of this

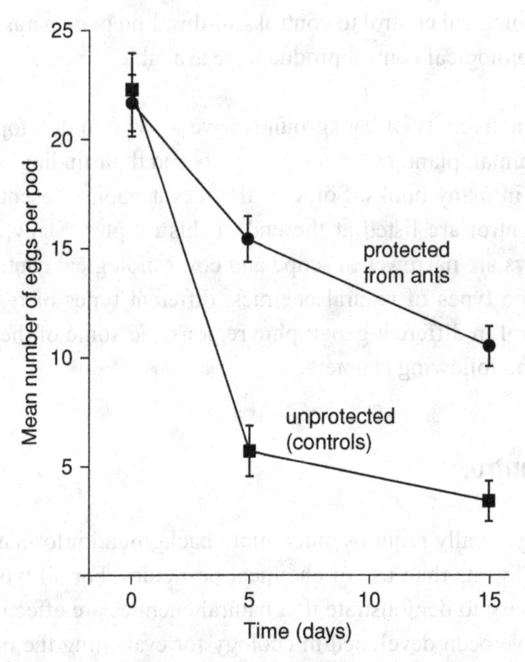

Figure 2.2 Mean bruchid beetle eggs per seed pod of *Acacia farnesiana* on control branches unprotected from ant predation, and on protected branches, at days 0, 5, and 15. Branches were protected from predatory ants by wrapping a 10-cm-wide band of tape around the base and applying sticky material to the tape (Traveset, 1990).

approach have pointed out that some new pests can decrease in numbers when no biological control introduction has been made. Therefore, after introductions of natural enemies, densities should be simultaneously quantified in areas with and without the natural enemies. These techniques are often considered standard for introductions of exotic natural enemies against introduced pests but are certainly appropriate for other types of biological control as well. With all of these methods, the bottom line is that finally the pest should have been controlled to the extent that there is no, or insignificant, damage.

2.5.2 Cages

Cages have been used to evaluate the effects of natural enemies more frequently than any other method. Cages can exclude natural enemies from a segment of the host or prey population and the subsequent differences in densities between the wild population and the protected population can be used to indicate the effect of the natural enemies. For example, such techniques were used to evaluate the impact of predators on cereal aphid populations. It was shown that the caged populations increased at a far greater rate than the uncaged populations. The caged population was protected from natural enemies while the uncaged population was exposed to natural enemies in the field. Cages do not have to enclose the prey completely but only exclude the natural enemies of concern, as in the case of exclusion of predatory ants using sticky bands (Figure 2.2). Alternatively,

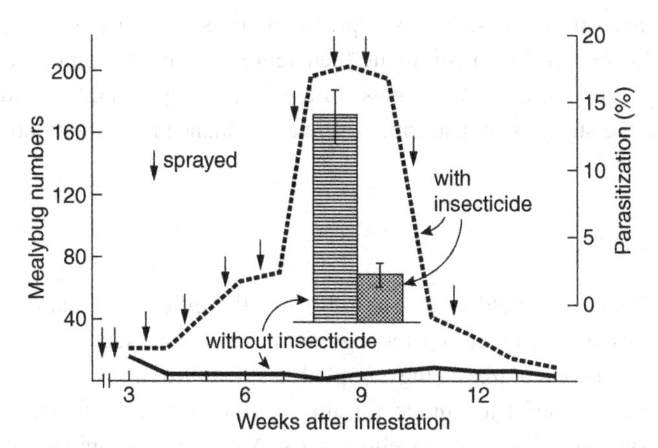

Figure 2.3 Cassava mealybug, *Phenacoccus manihoti*, population development in insecticide-treated and untreated plots, with the mean levels of parasitism shown in inset columns (Neuenschwander & Herren, 1988).

cages are used for including natural enemies with hosts so that both natural enemies and hosts can be sampled. This is especially useful for natural enemies with mobile hosts that might be difficult to locate in significant numbers once a study begins. For example, gypsy moth larvae (caterpillars) will frequently feed on foliage high in tree canopies but can be caged on lower branches during experiments so that groups of insects can be repeatedly observed and quantified.

Results from studies using cages must be interpreted with caution because it is extremely difficult to make conditions within cages completely realistic. Densities of hosts and natural enemies can be unrealistic because dispersal is not possible, encounter rate between natural enemy and host could be artificially elevated, behavior of the host or its natural enemy can be affected by the cage itself, and the cage can create an interior microclimate different from the area outside of the cage.

2.5.3 Removal Techniques

An effective method used for removal of natural enemies to evaluate their effect is the "insecticidal check method." This method has been used primarily to study the effect of insect natural enemies but would also be applicable for studying microbes that can be killed with pesticides. An area can be sprayed with selective insecticides to exclude insect natural enemies, and pest densities in this area (either arthropod pests or weeds) can later be compared with controls where no treatments were applied (Figure 2.3). This method assumes that the pest population will not be totally eradicated by the spray but the natural enemy population will be totally killed or will be so sparse that it will not increase rapidly. This is not a bad assumption in some cases, for example, when sprays are made early in the season and natural enemies are present but hosts/prey are in a more resistant stage. In such cases, this method is very appropriate. However, there can be problems interpreting results if the pesticide affects the treated area in ways other than killing the natural enemies. For example, spraying sublethal doses of pesticides on some phytophagous mites can stimulate their reproduction and sometimes this can lead to greater production of female mites. In these cases, spraying pesticides would bias results from the study, demonstrating artificially enhanced pest populations in the sprayed area.

2.5.4 Prey Enrichment

Adding prey or hosts to a field can be used to test the efficacy of natural enemies. This technique is especially useful for testing nonmobile stages, such as eggs or pupae because it ensures that each area being sampled has a well-defined initial population density. This is also a useful technique for studying hosts that would be difficult to find in the field, such as soil- or tree-dwelling pests. A common technique for studying entomopathogenic nematodes and fungi in the soil is to place larvae of a very susceptible species such as the wax moth, *Galleria mellonella*, in the soil and later retrieve them to evaluate the percentage of larvae found and killed by these natural enemies. As with studies using cages, interpretation of data from enrichment studies can be difficult because in most instances the hosts and prey placed in the field might not be found exactly at that time or place or in that density under natural conditions.

2.5.5 Direct Observation

This technique is useful for predators or parasitoids that can be observed attacking hosts. It can be especially difficult to determine to what extent predators have been the cause of declining pest populations. In many cases the predator eats the entire prey item and leaves no evidence behind of its meal, and thus, this event cannot be counted in the field. While direct observation is simple, it requires a huge time commitment and cannot be used if the predator or prey are cryptic or are easily disturbed, or if the predator consumes the prey very quickly. However, direct observation was used successfully to monitor the numbers of green rice leafhoppers (*Nephotettix cincticeps*) preyed upon by four species of spiders, with part of the observations taking place at night using flashlights because it turned out that predation was primarily nocturnal. Alternatively, video recording equipment can be appropriate toward reducing the time an observer spends in the field, and has been used to study predation in both diurnal and nocturnal situations.

2.5.6 Evidence of Natural Enemy Activity or Presence

This method involves having some way to evaluate whether natural enemies have attacked specific prey. In some systems, natural enemies leave evidence of killed hosts or prey and this can be used to quantify activity. For example, mice feeding on gypsy moth pupae characteristically leave behind an empty pupal case that is different in appearance from pupal cases that have been attacked by ants, another major predator. Parasitoids can leave behind their pupal cases next to cadavers of hosts and the different parasitoid puparia are characteristic of different species. Cadavers of arthropods killed by pathogens often remain in the field for some period of time, during which they can be sampled. However, in many systems such helpful evidence is lacking and no indication of prior presence of a pest is left.

Laboratory-bench assays have been developed for evaluating the activity of natural enemies. Predators can be collected in the field and their gut contents can be evaluated to determine what they have eaten. Some widely used techniques are based on developing vertebrate antibodies to specific prey and then testing predators caught in the field to evaluate whether the antibodies react to their gut contents. These general types of tests are called immunoassays and methodology is similar to methods used in medical laboratories. If the antibodies react, it indicates that the predator had eaten that prey species. Accuracy in detection depends on the size of the prey, the size of the meal, the time since the meal, the means of feeding (sucking versus chewing), occurrence of closely related prey (which might also cause a reaction), and the sensitivity of the test. Electrophoresis can also be used to detect the presence of prey protein in the guts of predators.

Instead of developing antibodies or using electrophoresis, prey tissues can be marked using a variety of materials, including radioactive isotopes, rare elements such as rubidium, or dyes (e.g., fluorescent dyes). Suspected predators can be collected and assayed for markers.

More recently, various PCR-based molecular methods have proven very useful to document whether a microbial natural enemy is present in the host, or if an infected host is present in a predator. Also, modern PCR-based methods can monitor the presence of natural enemies in the environment. A big advantage is that such methods can determine specific genotypes of the natural microbial enemy and prove whether the exact strain that had been released had caused a target population to collapse. A disadvantage is that additional equipment and skills are needed, and that specific primers in many cases must be developed before such methods can be applied.

If pests are sampled from the field and percent parasitism and infection are to be determined, insects must be collected and reared under controlled conditions. However, rearing can be tricky and introduces its own bias to the study if the hosts are stressed during the process. If a technique such as immunoassay, DNA-based molecular methods, or marking prey is used to detect the presence of a parasitoid or pathogen within a host, rearing the hosts before analysis is often not necessary, thus avoiding potential problems introduced when rearing insects.

All of these techniques require special training and equipment. However, there are distinct advantages because each study method provides a different type of information compared with others. Also, some techniques are time efficient, which can be a concern when working in biological systems where critical time periods can be brief. Field-collected samples need only to be processed to a limited extent at the time of collection and often can then be stored for evaluation later or shipped for evaluation elsewhere. Field work is generally very labor intensive for a defined period of time and the prospect of being able to process samples at a later date makes it possible to evaluate more samples and thus learn more about natural enemy activity.

Further Reading

Bellows, T. S. & Fisher, T. W. (eds.) (1999). *Handbook of Biological Control*. San Diego, CA: Academic Press.

Campbell, R. (1989). *Biological Control of Microbial Plant Pathogens*. Cambridge: Cambridge University Press.

DeBach, P. & Rosen, D. (1991). *Biological Control by Natural Enemies*, 2nd edn. Cambridge: Cambridge University Press.

Eilenberg, J., Hajek, A. E., & Lomer, C. (2001). Suggestions for unifying the terminology in biological control. *BioControl*, **46**, 387–400.

Eilenberg, J. & Hokkanen, H. M. T. (eds.) (2006). *An Ecological and Societal Approach to Biological Control*. Dordrecht, Netherlands: Springer.

Greathead, D. J. (1994). History of biological control. *Antenna*, **18**, 187–199.

Gurr, G. & Wratten, S. (eds.) (2000). *Biological Control: Measures of Success*. Dordrecht, Netherlands: Kluwer Academic.

Heimpel, G. E. & Mills, N. J. (2017). *Biological Control: Ecology and Applications*. Cambridge: Cambridge University Press.

Van Driesche, R. G., Hoddle, M., & Center, T. E. (2008). *Control of Pests and Weeds by Natural Enemies: An Introduction to Biological Control*. Chichester, UK: Wiley-Blackwell.

Part II

Strategies for Using Natural Enemies

3 Classical Biological Control

This strategy provided the first means developed on a large scale for using natural enemies for pest control, hence the name "classical" biological control.

Classical biological control is the intentional introduction of an exotic biological control agent for permanent establishment and long-term pest control. (Eilenberg et al., 2001)

Importantly, the goal is quite specific: to release an exotic natural enemy into a new environment so that it will become established and will regulate a pest population over the long term without further intervention. In referring to this strategy, sometimes the term "classical" has been replaced with "importation," "introduction," or "inoculation" – or "introduction of new natural enemies" has been used as the entire name for this strategy instead of "classical biological control." Classical biological control is often practiced by inoculating an environment with a small number of natural enemies that are expected to increase and the goal is permanent establishment. However, where permanent establishment is unlikely, another strategy can be used that involves releasing fewer natural enemies that are expected to increase, resulting in shorter-term persistence and control and not permanent establishment; this is called "inoculative augmentation" (see Chapter 4).

Classical biological control has been used extensively and, as we will discuss, some programs have been very successful (Figure 3.1). This strategy was initially developed to control exotic pests, based on the following scenario. Scientists noted that many introduced pests are not problematic in their areas of origin, where they are often controlled by a community of natural enemies. After introduction to a new area, in some cases the introduced species increased in numbers to become pests. It is thought that the pest is able to increase at least in part because in the new area the natural enemies that would naturally regulate this species where it is native were not present. This basic assumption underlying much of classical biological control has been called the "enemy release hypothesis" (see Chapter 6). The goal with classical biological control has often been to reestablish the natural balance (see Chapter 2) that controls the pest in its native habitat. This concept has been taken one step further by mentioning that the goal is establishing a "new natural balance" where the natural enemies being released would not themselves be killed by another tier of natural enemies (e.g., hyperparasitoids; see Chapter 8), as might occur in the area where the natural enemy is native.

The dramatically successful release of the vedalia beetle against the cottony cushion scale attacking citrus trees in California is often said to have launched the use of classical biological control (Box 3.1). Used against insect pests since the late 1800s, this

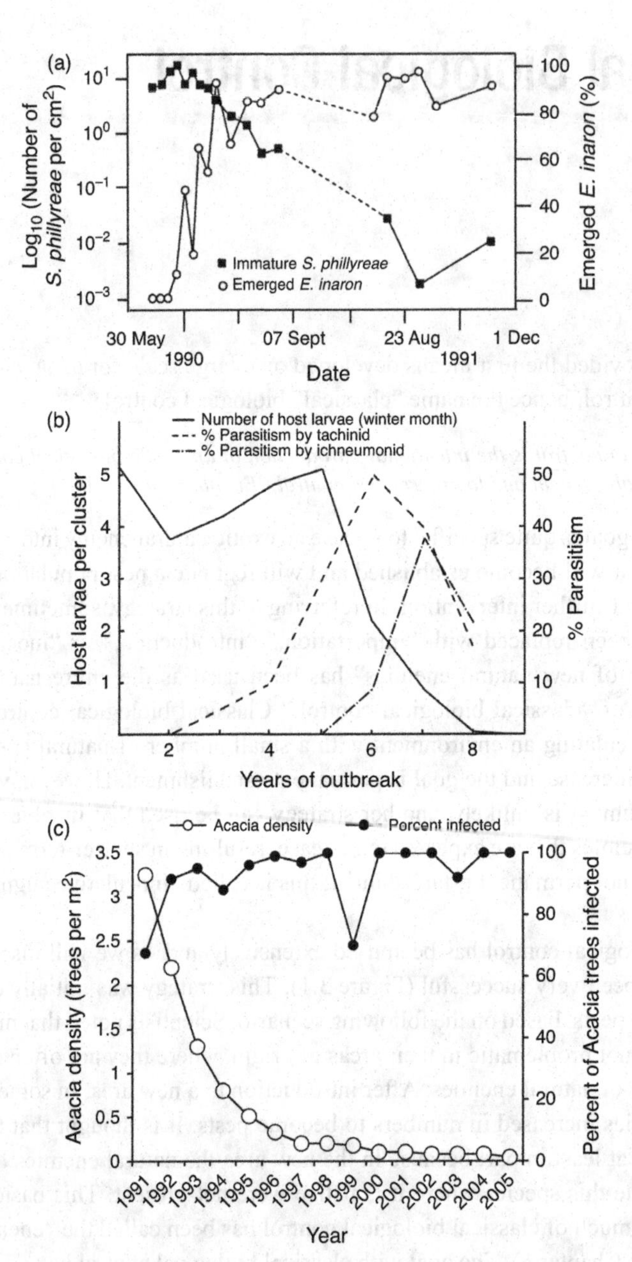

Figure 3.1 (a) After introduction of the parasitoid *Encarsia inaron*, changes in density of immature ash whitefly, *Siphoninus phillyreae*, and percentages of parasitized whitefly pupae on pomegranate in California, 1990–1991 (Bellows et al., 1992). (b) History of a winter moth, *Operophthera brumata*, infestation and parasitism by the tachinid *Cyzenis albicans* and the ichneumonid *Agrypon flaveolatum* in Nova Scotia. Winter moth was accidentally introduced and both parasitoids were introduced for classical biological control. Data are from seven different areas and time is the number of years the winter moth outbreak persisted, beginning one year before *C. albicans* appeared in the population (Embree, 1966). (c) Decrease in density of *Acacia saligna* trees over time at a site in South Africa and the percentage of trees infected with galls of the fungus *Uromycladium tepperianum*, which was introduced to the country as a biological control agent in 1987 (Wood, 2012).

Box 3.1 Introducing the Vedalia Beetle Against Cottony Cushion Scale

In southern California in 1868, the cottony cushion scale, *Icerya purchasi*, was a new pest attacking citrus, pear, and acacia. By 1880, it had spread all over California and was seriously damaging citrus orchards wherever it occurred. In 1886, frustrated growers were pulling out or burning citrus trees because they couldn't control this pest as it lay waste to their orchards and land values plummeted. Entomologists guessed that the scale was from Australia, the country from which much of the citrus had been imported. The head of entomology for the US government, Charles V. Riley, requested that someone be sent to Australia to search for natural enemies. However, this request was turned down because of a restriction on international travel for employees of the Division of Entomology. However, in 1888, Albert Koebele was sent to search for natural enemies in the guise of attending the International Exposition in Melbourne. In actuality, Koebele barely attended the meeting and instead traveled throughout Australia searching for natural enemies for this project. Even with assistance of Australian entomologists, it took a while for Koebele to locate cottony cushion scales. The most promising natural enemies Koebele found were a parasitic fly and a lady beetle (Coccinellidae), *Rodolia cardinalis* (although at the time that Koebele found it, this beetle was known as *Vedalia cardinalis*, hence the common name that has persisted).

Figure Box 3.1.1 Albert Koebele, the entomologist who collected the vedalia beetle from Australia to import to California for control of the cottony cushion scale (Swezey, 1943).

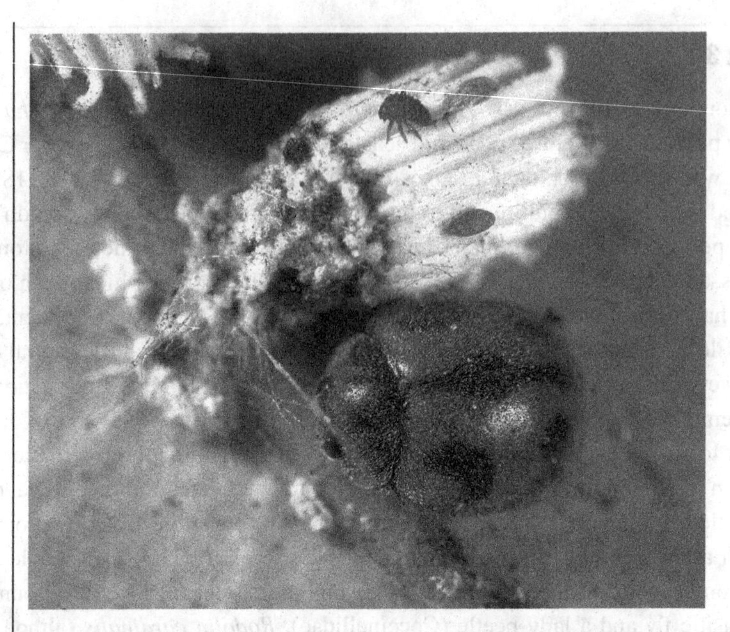

Figure Box 3.1.2 Vedalia beetle (*Rodolia cardinalis*) adult and larvae (immatures) with a wax-covered cottony cushion scale, *Icerya purchasi* (photo by Jack Kelly Clark, courtesy of the University of California Statewide IPM Program).

Koebele collected and sent both the flies and beetles in five shipments during which both scales and natural enemies had to be kept alive throughout the three-week boat trip from Australia to California. It is no small feat keeping organisms alive during transit even today, so the obstacles faced to keep citrus trees, scales, and natural enemies alive for three weeks on the open ocean in the late 1800s were substantial. During one voyage, the shipments were maintained in an ice house on board the ship but, during a gale, the parcels fell off their shelves and were crushed by cakes of ice falling on them. Despite these difficulties, by 1889, a total of 514 individuals of the lady beetles had arrived in California. These beetles were released and four months after the first release, adult vedalia beetles were swarming over a 3,200 tree orchard that had previously been heavily infested with cottony cushion scales. To hasten spread of the beetles, branches covered with scale-feeding beetles were transported to orchards where the beetles did not yet occur. By 1890, all infestations of the cottony cushion scale were completely controlled, the citrus industry was saved, and the total control program had cost less than $5,000, including salaries. The citrus industry has reaped benefits of millions of dollars annually ever since because of ongoing biological control of cottony cushion scale without further human intervention. The delighted Californians honored Mr. Koebele by giving him a gold watch and his wife received a pair of diamond earrings.

While Koebele was searching for natural enemies, for a while he thought that the parasitic fly he had collected would be the more important of the two natural enemies. The fly, *Cryptochaetum iceryae*, became established after releases of 1,200 individuals and, in fact, is the major agent controlling cottony cushion scale along the California coast. This example shows just how difficult it is for researchers to judge how successful a specific natural enemy will be; in this case, the vedalia beetle turned out to be astoundingly effective while the favorite, *C. iceryae*, was also successful but in a smaller area.

The vedalia beetle today continues controlling the cottony cushion scale in the interior of California. The scale can still be found and can even increase in abundance if pesticides are used in orchards so that beetles and flies are killed. In such cases, natural enemies are reintroduced and control is once more established. In addition, in the wake of such success these beetles and flies have been introduced in numerous other countries around the world where this pest occurred.

This example provided an early demonstration that biological control could be incredibly effective. This system was, in some ways, very unique and had many attributes that foretold success. *Rodolia cardinalis* is very specific, feeding only on scale insects and, even then, its host range is restricted. Most predators are not as specific as the vedalia beetle. In addition, this beetle had the ability to become established when only a few females were introduced. For example, only four females of this beetle were later introduced to Peru to control this same pest and the beetle became established.

strategy increased to almost 850 releases between 1960 and 1969 alone and classical biological control remains in use against arthropod pests and weeds today (Figure 3.2). However, it is very obvious that numbers of classical biological control introductions have declined since the 1970s for numerous reasons, especially emphasizing enhanced environmental safety, as will be discussed in Chapter 18.

3.1 Uses of Classical Biological Control

Classical biological control has predominantly been used for controlling insect pests and weeds and this strategy has been used in a few instances against vertebrate pests and plant pathogens (see Chapters 11 and 15, respectively). Some of the most untoward results, where natural enemies that were released attacked species other than the targets, have been obtained in the past using vertebrate natural enemies. Therefore, vertebrates are used only for classical biological control in a few very specific circumstances today (see *Gambusia* in Box 7.1 and grass carp in Section 14.4).

The principal types of natural enemies used for classical biological control have been insect parasitoids and predators for control of insect pests (Table 3.1). By 2010, some 2,171 species of insect biological control agents (predators and parasitoids) had been released for classical biological control in 6,158 introductions from different source countries, against 588 pests in 203 countries or islands around the world (Table 3.1). Classical biological control introductions for weed control have also been conducted extensively, with 1,555 attempted introductions of all kinds of natural enemies. Uses

Table 3.1 Statistics on classical biological control of insect pests using arthropod natural enemies and weeds using all natural enemies.

	Insect pests	Weeds
Attempted introductions	6,158	1,555
Establishments	2,084	982
Pest species	588	224
Agent species	2,171	468
Countries and islands	203	130
Satisfactory controls[1]	620	–[2]

Data on insect pests are through 2010 (BIOCAT database; Cock et al., 2016) and data on weeds were summarized by M. Schwarzländer, R. Winston, and H. Hinz in Winston et al. (2014). Both consider that each source country is counted as a separate introduction.

[1] Satisfactory control includes from partial to complete control (see Table 3.3).

[2] See Table 14.1.

of pathogens for classical biological control of arthropods trails far behind, with an estimate in 2016 of 144 introductions, including bacteria, fungi, viruses, nematodes, and protists as natural enemies. Obligate pathogens attacking weeds have been successfully introduced in some cases but the number of programs thus far is once again relatively few (Chapter 15).

Natural enemies generally considered appropriate for classical biological control have a considerable degree of host specificity so that natural enemy populations would increase when host populations increase and decrease when host populations decrease

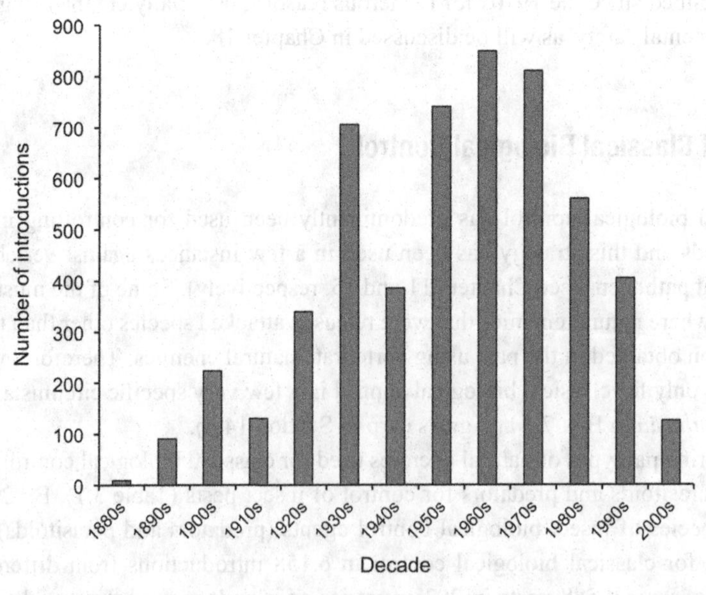

Figure 3.2 Numbers of classical biological control introductions of insect predators and parasitoids initiated against arthropod pests, by decade from 1880 to 2010 (Hajek et al., 2016).

(in a density-dependent relationship; see Chapter 6). Some microorganisms that are obligate pathogens can have such relations with hosts and these seem more appropriate for classical biological control. Why then have classical biological control programs used microbial natural enemies less frequently? Especially in the early years, it was more difficult to find, identify, and work with virulent microbial natural enemies for release, compared with macroscopic natural enemies. The majority of microbes used for biological control of plant pathogens are often ubiquitous microorganisms whose activity is not host specific and whose presence is more often related to the habitat than to presence of the plant pathogen. In addition, natural enemies used against plant pathogens are often thought to occur worldwide and, based on this assumption, there is no need to reunite introduced plant pathogens with their natural enemies.

Classical biological control programs are considered especially well suited to certain types of systems. Because the goal is to establish natural enemies permanently in a new environment where they will persist, this strategy has been applied more successfully to more permanent ecosystems, such as forests, natural areas, orchards, and perennial crops. This strategy has been used less frequently in shorter-term agricultural crops where simplified habitats are often disturbed and persistence of natural enemies that are released can be less certain. Classical biological control is not at all appropriate for highly manipulated systems such as greenhouse crops, where production facilities are emptied and cleaned at regular intervals, allowing no persistence of natural enemies. Classical biological control has often been used against pests introduced to relatively isolated regions, such as Australia and islands such as Hawaii or the central Californian agricultural area that is isolated by mountains and desert from other North American agricultural areas. For islands in particular, it has been hypothesized that since the fauna on islands is known to be less complex, introduced agents would have better chances of successful colonization. However, summaries of results show that natural enemies released on islands have not established more successfully than on continents, although for those natural enemies that become established, success in control efforts might be slightly greater on islands. Over time, the country with the most introductions of parasitoids and predators has been the United States, although this strategy is frequently used around the world, with the top 10 countries employing classical biological control located in North America, Oceania, Europe, Asia, and Africa.

Many different types of pests have been targeted by classical biological control programs but there are certain groups that seem to be controlled more successfully. Arthropod pests that are exposed and not hidden and are less mobile have been more successfully controlled because natural enemies have easier access to the pests. For this reason, use of parasitoids and predators against phloem-feeding insects such as aphids, scale insects, and mealybugs has been very successful, in part because these hosts are fairly sessile and usually feed externally on plants. These insects all belong to the order of insect pests that has been targeted most frequently, the Hemiptera. These small insects have frequently been accidentally introduced to new locations and have become pests, which is a common scenario for undertaking classical biological control. Important successes have also been made in controlling other groups of pests, such as caterpillars, beetles, and flies. Perhaps insect pests that live in concealed places have been more difficult targets because host ranges of their natural enemies tend to be ecologically determined and are not always based on the taxonomy of the host. For example, some parasitoids of wood borers are known to attack larvae of longhorned and death watch beetles as well as larval wood-boring bees. However, for classical biological control, a narrow host range is desirable so that nontargets will not be affected and host mortality has a greater chance of being density dependent, resulting in regulation of the pest population. Thus, trying to find natural enemies with a high degree of host specificity for pests in concealed locations can be problematic.

In classical biological control the natural enemy is often released in small quantities, with the expectation that the released organisms will multiply on their own to result in self-perpetuating permanent control. Historically, such release programs have been relatively inexpensive to conduct and can result in huge savings that are both short and long term. Thus, no self-sustaining profit is generated to support private industries producing the natural enemies. Therefore, governmental or international funding virtually always supports classical biological control programs and programs are carried out by international, national, or academic agencies. Because control is permanent and without cost to individuals, this type of strategy has been considered extremely appropriate for pests affecting resource-poor farmers who cannot pay for pest control. For example, large classical biological control programs against cassava pests in Africa, with funding from international aid organizations, have benefited subsistence farmers with no cost to them (see Chapters 7 and 8).

3.1.1 New Associations

Classical biological control was first developed to control pest species that had been introduced from other areas. However, classical biological control has not only been used in this scenario. In some instances, exotic natural enemies have been introduced against native pests, creating a new association because the natural enemy and the pest had not originated from the same area. Another type of new association has been used when either the area of origin of the pest is not known or effective natural enemies cannot be found associated with a pest in its area of origin. In these cases, researchers have searched for and introduced natural enemies attacking similar hosts in other areas. Once again, these are new associations because the natural enemy and pest are not assumed to have coevolved.

Some scientists have reasoned that interactions between a natural enemy and a pest that originated in the same location would have evolved toward a more benign association. Under this scenario, a natural enemy that had coevolved with its host would not be very virulent or efficacious or it would have driven the pest to extinction and would no longer be able to survive. The general concept is that a natural enemy that has not coevolved with a pest could be much more virulent and thus a more effective agent for control.

However, new associations have not been used as often as coevolved associations, although reviews of the literature demonstrated that they were used more frequently than had been assumed, and with greater success. Some uses of new associations have been

Table 3.2 Steps for a classical biological control program against an introduced pest.

1	Choose a target pest for which classical biological control would be appropriate and identify its origin. Increasing numbers of countries require that permission for foreign exploration be formally requested.
2	Acquire natural enemies, often through foreign exploration. The natural enemies must be sent to a quarantine to make certain they are without their own parasites or contaminants and for further evaluation.
3	Natural enemies for release will be chosen based on efficacy and safety testing in quarantine. Governmental approval for releases should be sought.
4	The natural enemy will be released in suitable habitats, using best estimates for how many individuals to release and how best to release them.
5	After establishment, distribution of the natural enemy throughout the distribution of the pest is frequently required, especially when the natural enemy does not spread quickly on its own.
6	Evaluation of the activity of the natural enemy. This step can sometimes take numerous years because establishment and increase of the natural enemy are not always immediate.

Source: based on Van Driesche & Bellows (1993).

extremely successful, such as the classic story of the coconut moth in Fiji (Box 18.3). The downside in the use of new associations is that for natural enemies to be successful when they are used in this way, a natural enemy must be less selective in the breadth of host species it will attack or it would not accept a "new association" host that it has never encountered before. While this has led to successful use of new associations, it also suggests that extra care must be taken to avoid potential nontarget effects, a serious consideration for classical biological control programs today (see Chapter 18).

3.2 Methods for Practicing and Evaluating Classical Biological Control

Methods for conducting classical biological control programs are relatively straightforward but not especially simple and require several stages (Table 3.2).

3.2.1 Determine the Identity of the Pest and its Area of Origin

For releases against introduced pests, the origin of the pest must first be determined. This sounds easy but has proven difficult in many situations, often because the pest is extremely widespread or the pest is not an outbreak species in its area of origin and therefore is not well known there. The coffee mealybug (*Planococcus kenyae*), a hemipteran that has been a serious pest of coffee plants in Africa, was first noticed causing serious infestations on coffee plants in Kenya, in 1923, after which populations spread. This mealybug also attacked some food crops. The fact that no parasitoids attacked it made it evident that this was a non-native insect. Plants in the genus *Coffea* are native to tropical Africa, Madagascar and neighboring islands, and tropical southeastern Asia. The search was on for parasitoids. However, misidentification of the host mealybugs led to searching for parasitoids in Sicily, the Philippines, and Java, from what turned out to be incorrect hosts. Success only occurred fifteen years later, when a mealybug expert searching in Uganda found nine species of parasitoids attacking the correct host. Among these, in 1938 one species, *Anagyrus* sp. near *kivuensis*, began being released and this species is credited as providing outstanding control.

The taxonomy of groups including introduced pests is frequently poorly understood and the true identity of a pest may not be known until adequate material has been collected from across the distribution of the pest and evaluated by specialists. In particular, because the natural enemies preferred for biological control are extremely host specific, help from systematists at this stage is critical for obtaining an accurate identification of the pest (Box 3.2), as the pest's identity is important information necessary for finding the correct host-specific natural enemies. In some cases, classical biological control programs have been unsuccessful until taxonomists reevaluated the identities of pests, only then collecting natural enemies that would yield successful control (see Box 8.4 and Box 13.1).

Box 3.2 Some Like It Hot

Whiteflies are one of the kinds of tiny insects that cause problems because they can increase to enormous numbers. These very small insects, like aphids and scale insects, feed on the phloem flowing through plants. For a short time after hatching from eggs, their first stage, called crawlers, move around to find the best place to feed and grow to the adult stage, which has white wings. The sweet potato whitefly (*Bemisia tabaci*; sometimes called the silverleaf whitefly) was first reported in Australia in 1994. It is a pest in numerous types of vegetable crops but also causes problems in melons, cotton, and soybeans. These whiteflies are damaging in three ways: their feeding directly damages plants, they excrete sugary sticky honeydew on which sooty molds grow that can cover plants, and they can vector viral plant pathogens. The strain of sweetpotato whitefly that had been introduced in Australia, referred to as *B. tabaci* biotype B, was the same biotype that was causing US$200–500 million of damage to agricultural crops in four southwestern US states in 1991 and 1992 and which had earned this whitefly a reputation as a "super-pest" and one of the hundred most important invasive species worldwide.

The classical biological control program that was developed to focus on this pest was a huge collaboration that took advantage of many different kinds of expertise that had not been available in the past. Whiteflies do not have many morphological features so systematists used molecular methods to discover that what looked like one whitefly species under the microscope was actually a species complex, with 24 very closely related species. As is typical of many invasive species, it was not certain exactly where this whitefly had been introduced from, especially as these 24 species occurred across many tropical and subtropical areas. Major problems caused by biotype B had been occurring in Texas, Arizona, and southern California and a climate matching program was used so that foreign exploration for natural enemies would concentrate on source areas with climates similar to the hot and dry climates experiencing outbreaks of biotype B. Pakistan, southern Spain, Ethiopia, and the United Arab Emirates were selected as places to concentrate searches. An enormous foreign exploration program began and eventually more than 235 field collections from 28 countries were made which yielded 13 species of parasitoids, several predators, and an entomopathogenic fungus. Once again, systematists using molecular methods were central to identifying these new natural enemies as well as differentiating between different strains of the same natural enemy species, as different natural enemy strains had different impacts on sweetpotato whiteflies.

The climate matching software was then used to determine which parasitoid release site in the United States was most similar to the areas in Australia where problems caused by biotype B were occurring. The Lower Rio Grande Valley was identified as the best fit, so the parasitoid with best establishment there, the tiny wasp, *Eretmocerus hayati*, was chosen for introduction in Australia. In 2004, it was released at 12 sites along the east coast of Queensland and within 3.3 years it had spread widely throughout the infested areas. It has been suggested that in part this parasitoid was transported accidentally by the nursery industry, which was the way that the pest had initially spread. This one tiny parasitoid species had a significant impact in controlling biotype B in Australia. The introduction program is considered as very successful, with from 30 to 85 percent parasitism recorded in vegetable crops. Growers are cautioned about only using insecticides (either to add to control of biotype B or for controlling other pests) that will not kill these parasitoids, which are now available for augmentative releases.

3.2.2 Foreign Exploration

Taxonomists are also critically important for identification of the species of natural enemies being considered for release. A species of natural enemy can occur across a broad distribution and we have learned that natural enemies of the same species but from different locations can be adapted to the climate in their area of endemism. Such locally adapted strains within a species have sometimes been referred to as biotypes or ecotypes. Some classical biological control programs have only attained success once natural enemies were collected from areas similar in climate to the area for introduction (Box 3.2). Today, climate modeling aids classical biological control in helping provide guesstimates on optimal regions for natural enemy searches. If targeted pests are native (which is not as common for classical biological control programs today), once again more knowledge of the species and strains of both pests and natural enemies is necessary to help pinpoint natural enemies attacking closely related hosts or prey that might be effective against the target pest needing control. In addition to climate, abiotic habitat characteristics can also be important to natural enemy establishment and success. For example, leafy spurges are native to Eurasia but invasive on rangelands in the Great Plains of North America where they cause great problems, occurring as many forms, species, and hybrids, usually referred to as *Euphorbia esula*. To find the best natural enemies for control of this mosaic of variability in leafy spurges across different areas of the Great Plains, soil characteristics were quantified to match natural enemy source habitats with release site habitats.

Within host or natural enemy species that have large geographic distributions, strains in different areas can differ genetically, leading to different relations with the target pest and this can provide important information for foreign exploration. Molecular-level techniques are now critically important to our being able to detect the variability within species, but biological studies in the lab are also still necessary for understanding how this translates to the potential for pest control. As an example of host and parasitoid variability, the alfalfa weevil (*Hypera postica*) is from Eurasia and three different strains of this pest were introduced accidentally to different areas of North America. Foreign exploration was undertaken and a parasitoid wasp, *Bathyplectes curculionis*, was introduced to control *H. postica*; while this parasitoid provided control against one weevil (Curculionidae) strain, for the other strains this was not the case, as parasitoid eggs were killed by the weevils' immune response. Foreign exploration began again and strains of *B. curculionis* that were not killed by the host were found and introduced.

When undertaking foreign exploration, once the area of origin has been identified, permission for foreign exploration must be sought from the appropriate countries. Historically, this was not required, but today countries where exploration is being conducted must be contacted regarding the necessity for collection permits. Today, in some countries, this step can be difficult as legislation defines natural enemies part of the "genetic resources" of a country. Permit requests for foreign exploration are now sometimes evaluated based on the thought that a profit could be made from the natural enemy in the country where it will be introduced and that this should be shared with the country of origin (see Section 20.3 for further explanation). Needless to say, such legislation was originally developed for different types of profit-driven programs. The classical biological control community, where no or little profit would be gained by the implementing agency, is arguing for the need to facilitate their activities through the continued free multilateral exchange of biological control agents.

With permission, a foreign expedition would then be undertaken. Foreign exploration can be difficult and time consuming because, in the area of origin of the pest, the pest itself can be difficult to find and natural enemies can be at very low densities. As well as collecting the natural enemies, it is important to gain as much information as possible about the pest and its natural enemies in their area of origin. It can become difficult to decide which natural enemies to emphasize for collection and eventual release, but information recorded as to natural enemy prevalence as well as host range in the area of endemism can help with such decisions. Any natural enemies that are collected must be cared for properly, so that they remain vigorous as they are subsequently sent to a receiving quarantine.

3.2.3 Prerelease Studies

Depending on the situation, prerelease studies may be done in the area of origin, the area of importation (in quarantine) or usually a combination of the two. The main goals are to evaluate efficacy of the natural enemy, test its host range, and make sure that there are no contaminants or pathogens associated with the natural enemy. However, several steps are necessary along the way to these goals. First, methods for rearing the natural enemies must be developed so that, hopefully, they will survive and increase in number. This usually means that quarantine personnel must also grow the pest in order to propagate the natural enemy. During rearing, any diseases or parasites of the natural enemies must be detected and eliminated. Early researchers found out the hard way that this is important. Potential problems were identified early when a parasitoid attacking another parasitoid (the hyperparasitoid *Quaylea whittieri*) was not recognized as such and was introduced for control of citrus black scale (*Saissetia oleae*), thus decreasing the effectiveness of primary parasitoids that had been introduced although, thankfully, with time this hyperparasitoid essentially disappeared.

Foreign explorers might find and send several strains of the same species, for example, small parasitoids that look the same but were collected from different locations. One conundrum has been the occurrence of morphologically identical natural enemies that can have very different host ranges. Such groups can be called species complexes. The gypsy moth fungal pathogen *Entomophaga maimaiga* belongs to just such a species complex (the *Entomophaga aulicae* species complex), which includes members with different specificities that can only be differentiated using molecular means or bioassays. Whitefly-attacking parasitic wasps that are virtually morphologically identical can also differ in host specificity. As molecular techniques are now used extensively, we can tell whether natural enemies collected in different areas differ significantly and should be maintained and evaluated separately. However, bioassays are still the best way to determine efficacy against different hosts.

Once there is a clean culture of the identified natural enemy, studies in the quarantine testing efficacy against the pest will require colonies of the pest and of the natural enemies and appropriate conditions for conducting bioassays, which might test subjects like efficacy of the natural enemy in controlling the pest under differing environmental conditions. Extensive host-range testing has also been a major goal in the quarantine so that there are no surprises regarding nontarget impacts if the natural enemy is chosen for release (discussed in more detail in Chapter 18). This now also starts earlier in the process as field studies in the region of origin of a natural enemy are increasingly being used to complement quarantine studies of host range, both to reduce time in quarantine for natural enemies and to provide more realistic host-range data from the area where the natural enemy is native.

To maintain the most effective natural enemies, time in the quarantine should be minimized so that the genetic variability in the natural enemy population is maintained. Also, time in quarantine should be limited to avoid selecting for optimal laboratory growth (see Box 9.1), trying instead to maintain maximal effectiveness of the natural enemy under field conditions.

3.2.4 Planning Releases

After a decision is made regarding which natural enemies should be released, governmental permits are required before release. A major requirement for such permits is general knowledge of the biology and host range of the natural enemies so that the risk of nontarget effects can be assessed. To protect the environment from any unexpected impacts of releases, the International Plant Protection Convention at the Food and Agriculture Organization (FAO) has developed an international standard regarding classical biological control releases and different countries have additional regulatory steps that must be taken (see Chapter 18). Classical biological control releases of microorganisms often undergo more scrutiny compared with release of parasitoids, predators, and herbivorous arthropods.

3.2.5 Releasing Natural Enemies

As with other stages of classical biological control programs, detailed knowledge of both the host and the natural enemy is necessary to optimize releases. No fail-proof system has been developed for the numbers of individuals that should be released in order for the natural enemy to become established. Not all natural enemy species are easy to grow in the laboratory and, in fact, some can be especially difficult to increase in numbers in a quarantine. Because of such difficulties, it is not always possible to release large numbers of individuals. In such instances, to ensure chances for mating, practitioners usually release many individuals at fewer sites instead of few individuals at many sites.

Certainly, it makes most sense to release natural enemies where there are large, healthy populations of the pest. For parasitoids, predators, and phytophagous insects, adults are usually released because they are ready to reproduce and will be less exposed to predation and other types of mortality before reproducing. Nevertheless, in some programs, parasitized hosts have been successfully released. For natural enemies like parasitoids, when adults are being released, it is also important to release them in areas where there are nectar sources, so that they will be able to feed, which is needed for field longevity of some parasitoids. Releasing natural enemies can be fine-tuned even further so that they are released under proper weather conditions and at an optimal time of day to promote establishment. For example, adults of tiny parasitic wasps might be released in shady locations at midday on a day with little wind. Insect-pathogenic fungi could also be released in the shade and in the evening so that when dew occurs overnight, the fungus can take advantage of the higher humidity and infect hosts. Alternative strategies would be to release arthropod hosts that have been parasitized or infected, introduce plants already infested with phytophagous natural enemies, or even introduce resting stages of the natural enemy, such as parasitoid pupae. Initial releases into field cages which are then opened is also practiced.

After releasing a natural enemy that becomes established, if the agent would then spread slowly on its own, classical biological control programs are generally extended to introduce the agent throughout the pest populations. In Australia, redistribution of phytophagous agents following establishment is now sometimes conducted through community groups interested in remediation of environmental problems. This community involvement helps to achieve a more systematic redistribution effort with a more rapid delivery of biological control results for the public.

Table 3.3 Terminology specific to classical biological control programs.

Establishment	Permanent occurrence of an imported natural enemy in a new environment.
Complete control	When no other control method is required or used, at least where the agent is established.
Substantial control	Other control methods are needed but the efforts required have been reduced as a result of the activity of the natural enemy.
Partial control	The natural enemy has some effect, but other means of control are still necessary (also called "negligible" control).

Sources: based on van den Bosch et al. (1982) and McFadyen (1998).

The program for control of cassava mealybug (*Phenacoccus manihoti*) faced a real problem with releasing natural enemies because cassava is a subsistence crop grown over a huge area of central Africa where transportation was difficult or nonexistent. To reach the most isolated areas most efficiently, adult parasitic wasps were placed in small vials that were dropped from airplanes. Wasps were able to escape from vials once they reached the ground. This strategy worked because the mealybug host populations were abundant throughout the release area and the parasitoids attacked all of the stages of the host. Predatory mites attacking cassava green mite (*Mononychellus tanajoa*) were released in a similar way, this time in vials with strings attached to them that would catch on crop plants and provide a route for the predators to crawl onto the foliage.

3.3 Success in Classical Biological Control

It is difficult to summarize success across the diversity of introduction programs. Purposefully releasing natural enemies for classical biological control is much more successful toward establishment and control than accidental introductions would be. However, experience has shown that the percentage of species that are released and which provide substantial control often falls short of expectations. It is important to be aware that after release, natural enemies might or might not continue living in the release location and if they do persist, effects in the new area can range from not providing control to providing excellent control. To evaluate classical biological control, a set of terms with specific meanings, such as establishment, substantial control, and complete control, are used (Table 3.3). Only about 34 percent of parasitoids and predators released against insects become established. It is more difficult to evaluate the percentages of releases that result in various degrees of control because scientists vary in their summarizations of results, especially if evaluations after releases were only subjective, as frequently occurred in earlier years. Of those programs where a predator or parasitoid became established to control insects, 29.7 percent yielded complete or substantial control of the pest. Thus, only 10.1 percent of the total attempted introductions of parasitoids and predators against insect pests resulted in satisfactory control from the 1880s to 2010s (Table 3.1). Looking at this trend over time, we can see that between the 1950s and

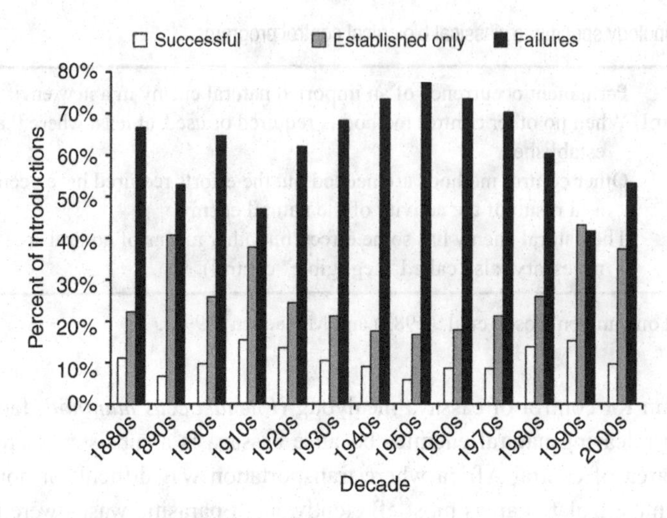

Figure 3.3 Results of classical biological control introductions of predators and parasites to control insect pests, showing the percentages of natural enemies introduced each decade which became established and contributed to successful biological control (white); which became established but have not been shown to contribute to control (gray); or which failed to become established (black) (based on Cock et al., 2016).

1990s, the levels of establishment and success increased each year (Figure 3.3). The small, relatively recent drop in successes afterward could in part be owing to the length of time after releases before final results are available (see Section 3.3.5).

Success rates are not greater because results from releases are often unpredictable. Although there are methods that are followed to try to achieve complete control through classical biological control programs, there are many unknowns, especially if working with a poorly understood pest or natural enemy. Van Lenteren (1980) very eloquently stated that many decisions while working on a classical biological control project can seem more like art than science, often relying on subjective intuition of researchers because adequate and complete objective information about a system is not available. All of the interactions in the environment that could affect the success of a natural enemy cannot be known before a release and even exhaustive laboratory studies of agents to be released do not always help us to predict the outcome of releases once the natural enemies are confronted with the pest under field conditions.

How do we improve the success rate of classical biological control? The thousands of classical biological control releases have been analyzed by scientists to discover trends in factors associated with successes and failures, ultimately to work toward improving the success rate. Such analyses have identified numerous factors associated with success at different stages during programs. Unfortunately, analyses of classical biological control of arthropods virtually never include weeds, and vice versa. Here, however, we will merge the findings from these two distinct types of classical biological control programs (targeting arthropods or weeds), especially concerning factors that are similar for both, to derive an idea of what factors are associated with a successful program.

3.3.1 Success in Establishing the Natural Enemy

Studies have demonstrated that establishment is improved if the climatic adaptations of the host and natural enemy are similar. Yet, so-called "climate matching" is not always the answer. A study of 178 projects with parasitoids and predators demonstrated that if a species of natural enemy did not become established after the first release, there was a greater chance of successful establishment if a different species altogether was released next instead of releasing further strains of the first species. For phytophagous natural enemies, a faster population growth rate was associated with successful establishment and establishment was improved when more individuals were released or when there were multiple smaller releases. For phytophagous species laying eggs in batches, there was a higher risk of mortality owing to predators or parasitoids, and species laying their eggs singly had a greater chance of establishment. Fortunately, it is considered rare for a natural enemy released for classical biological control not to persist in an area after becoming established in the first place. Therefore, although efforts may need to be repeated to establish a species of natural enemy, once that agent becomes established, it rarely goes extinct as long as habitat and hosts are present.

3.3.2 Habitats and Hosts Associated with Success

Classical biological control programs targeting pests in more stable habitats, such as perennial crops and orchards, forests, and wildlands, have been more successful because natural enemy populations can persist and therefore the natural enemy is better prepared to respond rapidly to increases in pest density. Often, successful control can occur in systems that are simpler, as with an exotic natural enemy in a managed system where the food web associated with the pest is simple. With a simpler food web, introduced natural enemies might escape the types of enemies that attack them in their native area, such as the hyperparasites that can attack parasites or the predators attacking predators.

Biological attributes of hosts can yield clues regarding probabilities of success. Biological control with parasitoids and predators has been more successful against more specialized, instead of generalized, pests. Success is also greater when pests live in exposed locations, such as external feeders on plant leaves versus stem borers. Analyses have been extended to evaluate specific groups of pests and one such study looked at the data from 150 biological control programs using parasitoids and predators against caterpillars. Successful parasitoids were most highly associated with two types of hosts: (1) hosts whose larvae were gregarious so that when a group of hosts was located, many parasitoid progeny could be produced, and (2) hosts that were plant feeders specializing on only a few host plant species, perhaps so that natural enemies could more easily locate hosts. Predators attacking caterpillars that were successful were often those attacking smooth-bodied caterpillars and caterpillars protectively colored to blend in with their surroundings (species that are not cryptically colored often contain defensive chemicals that deter predation). For biological control of weeds, there have been more successes in controlling plants with asexual reproduction rather than sexual because the plants are

Table 3.4 Comparison of use of parasitoids versus predators for classical biological control of insect pests.

	Introductions	Species	Successes	Successful species
Parasitoids[1]	4,759	1,512	505 (10.6%)	262 (17.3%)
Predators	1,399	468	115 (8.2%)	47 (10.0%)

Source: compiled 2016 by Matthew J. W. Cock, from the BIOCAT database.

[1] Introduced agents were categorized as parasitoid (predominantly wasps or flies) or predator (beetles) based on the major life strategy of their insect family.

then less diverse and the phytophagous natural enemies can specialize more easily (see Chapter 13 for additional weed attributes associated with success).

3.3.3 Successful Natural Enemies

There have been numerous lists of attributes of natural enemies associated with successful classical biological control. Successful parasitoids and predators often display (1) good searching ability for finding both host habitats and hosts, (2) a high degree of host specificity leading to a density-dependent relationship with the host, and (3) high fecundity. As suggested relative to establishment, the similarity between the climate where parasitoids originated and the climate of the release site can strongly influence whether the pest is controlled. Computer-driven climate matching was shown to be critically important with choosing species of parasitoids to release against the sweetpotato whitefly (Box 3.2). In addition, for parasitoids (1) lack of predators and parasites attacking them in the native fauna (i.e., showing that they have some defenses) and (2) use of alternate hosts or food have both been associated with success. For control of arthropod pests, although parasitoids have been used approximately three times as often as predators (Table 3.4), the success rates of these two groups are very similar. The greatest single predictor of success for phytophagous natural enemies of weeds was host specificity; agents that were more host specific were more useful for control.

Nick Mills asked the question "why classical biological control using parasitoids and predators has been more successful against Hemiptera (specifically phloem-feeding aphids, scale insects, etc.) and less successful against Lepidoptera (caterpillars)?" He investigated whether numerous biological attributes were associated with success. Success occurred more frequently when there was a low generation time ratio (development of parasitoids or predators was fast enough that they could keep up with development of the pest) and another study added that success more frequently occurred when there were more generations of the host/prey per year. A broad window of opportunity in which to attack the host or prey is beneficial; for example, natural enemies attacking long-lived life stages could be more successful than natural enemies attacking only a short-lived life stages. Finally, parasitoids with numerous individuals emerging from each host (see Chapter 6) were associated with more success in control.

Pests can be easily moved around the world and in such cases the same classical biological control agents are frequently released in multiple countries; these programs have been referred to as "me too" projects. As you might expect, releasing natural enemy species that were successful elsewhere often leads to success in a subsequent location. In one summary, among 2,677 predators and parasitoids released for classical biological control in numerous countries, 22 species had been released in 22 to 58 countries, with the top two being the mealybug destroyer, *Cryptolaemus montrouzieri*, and the vedalia beetle, *Rodolia cardinalis*.

3.3.4 Numbers and Diversity of Releases

As the practice of classical biological control was being expanded, papers written by practitioners questioned how to improve establishment of natural enemies and control of the pest. There have been heated discussions regarding whether only one or many natural enemies should be introduced; one worry was whether natural enemies might compete with each other and thereby decrease the overall control if numerous species were introduced. However, in practice there are not many examples of this happening. Instead, in Mexico lady beetles and parasitoids work in tandem to provide control for higher versus lower density pink hibiscus mealybug populations (Box 3.3).

Box 3.3 A Tropical Pest Spreads to Mexico

Mealybugs are hemipterans that are sessile for most of their lives, similar in many ways to scale insects. This makes them good targets for classical biological control. The pink hibiscus mealybug, *Maconellicoccus hirsutus*, originally came from India and has been spread around the world to many tropical and subtropical areas. This mealybug causes big problems as it feeds on many types of both agricultural and horticultural plants (not only hibiscus), releasing a toxic saliva as it feeds that causes symptoms ranging from curling and distortion of leaves to killing entire plants when densities are high. These mealybugs reproduce quickly with up to 15 generations per year in laboratories and they don't always need males to reproduce, a biological trick called parthenogenesis; this means that they have great chances for reproduction wherever they are. Although in later stages females are immobile, for both sexes the nymphal stages emerging from eggs, crawlers, are very mobile and can even be blown on the wind to new locations. These mealybugs are covered with a waxy layer, preventing many natural enemies from eating them and insecticides are generally not effective for controlling them.

Pink hibiscus mealybug was first found in Mexico in 1999, but by 2004 heavy populations of the pink hibiscus mealybug were infesting teak and acacia plantations in the state of Nayarit on the western coast of Mexico. Methods for biological control had been developed, especially in the nearby Caribbean in the 1990s. A small lady beetle, *Cryptolaemus montrouzieri*, called the mealybug destroyer, that was originally from Australia, had been used extensively around the world for classical and augmentative biological control of different mealybugs. While this lady beetle isn't specialized, it predominantly eats mealybugs and soft scales, and it eats a lot of them at outbreak densities. In Mexico, it wasn't difficult to mass-produce the mealybug destroyers, so more than 6.5 million individuals were eventually released. Around the same time a small parasitoid that had provided effective control elsewhere, *Anagyrus kamali*, was found to be able to establish so it was mass-produced at new production facilities and 25 million individuals were released in Nayarit and Jalisco.

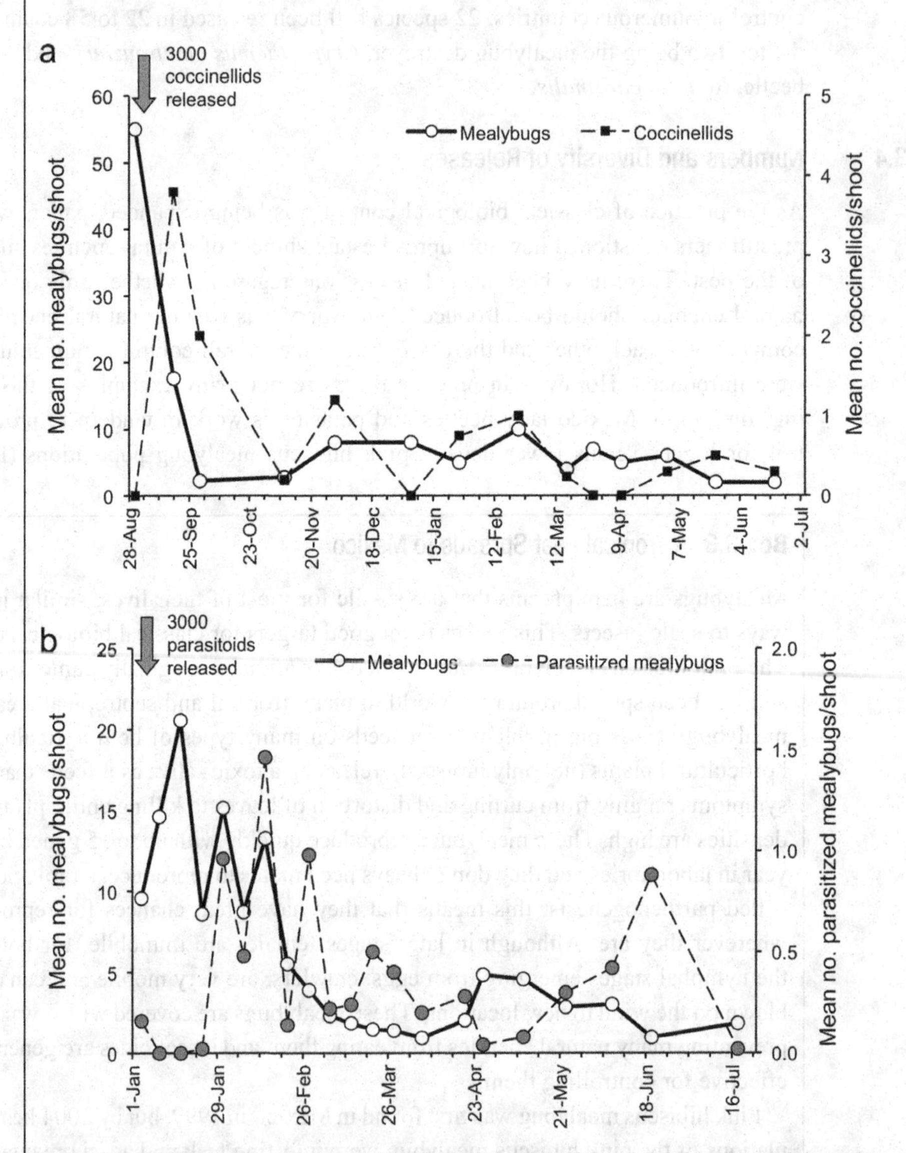

Figure Box 3.3.1 Examples of classical biological control of the pink hibiscus mealybug at two sites in Mexico: (a) a population of mealybugs at high density was rapidly controlled by releasing the "mealybug destroyer" lady beetle, an exotic coccinellid (2004–2005); (b) at lower mealybug densities, the exotic parasitoid *Anagyrus kamali* was released, providing stable control within a few months in 2005 (based on data provided by Trevor Williams & Hugo C. Arredondo-Bernal).

The two major natural enemies worked well in tandem. The lady beetles attacked high density mealybug populations but were not as effective once population levels decreased. On the other hand, the parasitoids were effective at reducing the population of mealybugs that were already at lower densities and keeping it low. The only places where these mealybugs remain as a problem are where insecticides are applied to control other pests. As for nontarget issues, the damage that can be caused by this pest was considered great and there are not many alternatives for control, so the potential nontarget effects predominantly to native soft scales and mealybugs were not considered a high enough risk to oppose use of this lady beetle or parasitoid.

In fact, in biological control of weeds, the "multiple stress" hypothesis suggests that numerous stressors acting in a complementary manner, often on different life stages, will lead to greater efficacy (see Chapter 13). Today, one release approach is for numerous species of natural enemies to be released against an arthropod pest or weed, in part owing to the unpredictability in results after releases. As an alternative, some practitioners think that it is better to introduce the natural enemy that is considered the best prospect for control (but which is also environmentally safe) and, after assessment, then proceed to introduce the next candidate if necessary, and so on; so, in this case, releases would be sequential. Usually, once an effective biological control agent has been released, more species are not introduced.

An analysis has shown that for programs where more natural enemies are introduced the probability for success increases, although at least against insect pests, this relationship seemed to reach a plateau when approximately eight natural enemy species were released against one pest (Figure 3.4). A study of biological control of weeds found a similar trend, with the success rate plateauing when about five agents had been released. Of course, these numbers reflect the numbers of natural enemy species released for each region; researchers have found over and over again that because an agent is successful in one climatic release area, this does not always mean that it will be successful in another. However, this analysis of the number of species released versus successful

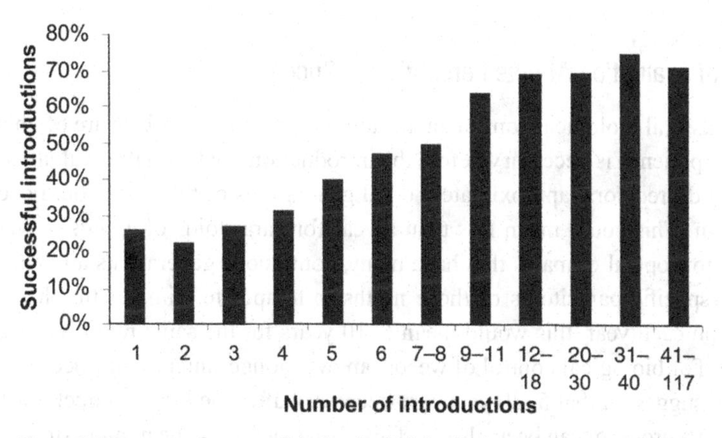

Figure 3.4 Percent of successful classical biological control programs using predators and parasitoids against insect pests, associated with various numbers of species introduced per program (compiled from 1992 BIOCAT database by Bradford A. Hawkins).

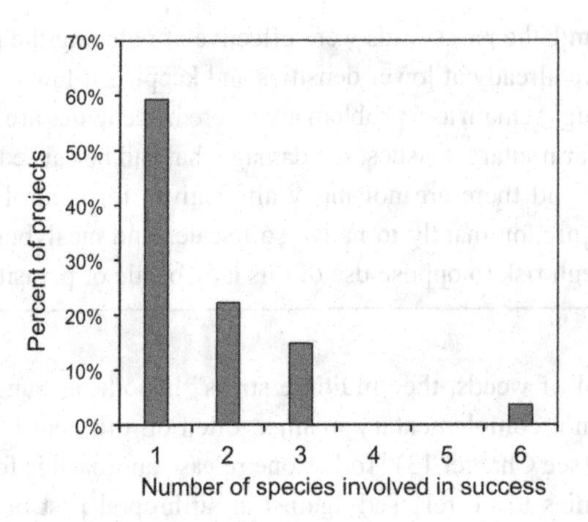

Figure 3.5 For successful classical biological control projects where multiple arthropod agents were released against arthropod pests, numbers of arthropod species involved in success for each project (based on Denoth et al., 2002).

control suggests that with increased effort in classical biological control programs there is an increased chance for success; if only one or two species are released, there is less chance of successful control compared with programs releasing more natural enemy species.

Curiously, one study demonstrated that for both biological control projects against weeds and against insect pests where multiple agents were released, more than 50 percent of the time only one of the multiple agents was considered the cause for successful control (Figure 3.5, with similar pattern with data analyzed from biological control of weeds projects). Therefore, it seems that when multiple agents are released they are often not acting together to enact control and, instead – Judy Myers calls this the "lottery ticket" to success – introducing more species is therefore often like buying more lottery tickets and thus increasing the chance of buying a winning ticket.

3.3.5 Length of Evaluation Affects Perception of Success

When classical biological control agents are introduced, the effects are often not immediate and patience is necessary before the introduction can be finally evaluated. For parasitoids and predators, approximately 6–10 generations of the pest should occur before evaluation. This could mean less than a year for parasitoids of aphids or scale insects adapted to tropical climates that have many, continuous generations all year. However, for host-specific parasitoids of those moths in temperate climates that have only one generation each year, this would mean 6–10 years for the same number of generations to occur. For biological control of weeds, an even longer interval has been suggested; it has been suggested that 5–20 years are necessary after the last introductions before the success of a program can be analyzed. This also depends on the number of weed seeds in the soil, as the presence of a seed bank that must be exhausted can mean that programs require even more time before control by a natural enemy can be finally evaluated.

3.3.6 Are Expectations Realistic?

What of the historically high failure rate of classical biological control programs? Using natural enemies and pests with characteristics leading to greater chances of success certainly will increase the rate of establishment and successful control. However, we could perhaps also change our expectations about classical biological control programs. Although one beauty of some of the examples of complete control has been their low cost, studies have shown that in systems where natural enemies are not successful, increased efforts can improve results. This increased effort can take the form of (1) introducing more agents, within bounds, (2) evaluating reasons for failures to gain a better understanding of the system, and (3) trying to find a point of weakness or "Achilles heel" of the pest as a focus for further efforts that would not fail as a result of this same issue. The low success rates also are, in part, a perception issue. When we calculate success rates using the proper lag times suggested above, we find greater success. This can be seen with programs in New Zealand for biological control of environmental weeds that have a success rate of 83 percent.

Rachel McFadyen made the point that classical biological control successes should not be evaluated by individual natural enemy species that have been released, as is usually done, but rather, according to whether a program against a specific pest was successful or not. She suggested we should be asking whether a weed being targeted was controlled, no matter how many agents were released or which agent(s) controlled it. Looking at results in this way, in one survey of results from South Africa, 6 weeds out of 23 targeted were under complete control and 13 more were under substantial control for a total success rate of 83 percent. In Hawaii, 7 weeds of 21 were under complete control with 3 more under substantial control for a success rate of 50 percent. Partial (i.e., negligible) successes fall somewhere in between success and failure but are seldom considered in analyses of successes although the natural enemies released are often helping to control the weed to some extent. This same situation applies to other types of classical biological control as well (see Chapter 19).

Some practitioners of classical biological control feel that the expectations for classical biological control are unrealistically high. Historically, programs were considered failures unless permanent, complete control was instituted. Practitioners of all types of classical biological control feel that natural enemies that have been released but which only cause moderate damage to targets are still valuable and should be considered as part of integrated control programs. As an example where such an intermediate effect was put to good use, in South Africa, leguminous mesquite trees (*Prosopis* spp.) were purposefully introduced for shade, firewood, and timber in desert areas but they then began to invade rangelands. Spread of mesquite was enhanced because livestock love to eat the seed pods and the livestock then spread the undigested seeds so that mesquite began taking over as a rangeland weed. There was a conflict of interest among farmers who used the trees to provide food for livestock and those who did not and wanted the rangeland back. A compromise was reached so that the spread of mesquite could be controlled but the established trees would not be affected. This was done by releasing a natural enemy that only affected reproduction of the trees, a bruchid beetle

(*Algarobius prosopis*) with larval stages feeding on seeds. Although this beetle showed great promise, at first control was not apparent because the seed pods were still being devoured by livestock before they could be invaded by beetles. With this knowledge, farmers protected the pods within fenced areas while the beetles were laying eggs and larvae were developing, but they could still feed the pods to livestock after the beetles had emerged. Using these procedures, the invasiveness of mesquite has been curtailed by the beetles stopping mesquite reproduction and thus arresting further spread of these weedy trees in rangeland, yet established trees were still present in areas where they were wanted.

3.4 Economics of Classical Biological Control

Although classical biological control programs cannot always be depended upon to successfully control a pest, they are still widely used because they require a relatively low investment and, with success, control is permanent. In addition, for situations such as controlling invasives in wildlands, where other types of control are difficult or impossible, natural enemies can disperse and find the pests on their own. Profits are often not apparent for a long time after establishment of the natural enemy, but with success, profits can far outweigh the costs of most classical biological control programs.

Costs for a classical biological control program include finding the natural enemies, identifying them, collecting and rearing them, and releasing them. Often few individuals are released, sometimes because they are difficult to rear. Yet it can be possible to release few individuals because they are expected to increase on their own in relation to the host population and then provide permanent control. Therefore, in general, costs have often been low over the short term for successful programs and, after releases, yearly costs for controlling that target pest are obliterated.

Cost:benefit ratios from classical biological control programs have been calculated for far too few successful programs, but the benefits always far outweigh the costs and often by considerable amounts. In Australia, an average cost:benefit ratio of 1:10.6 was developed by averaging several programs. This means that for every one dollar spent on the program, 10.6 dollars are saved because the crop was not lost and other controls did not have to be implemented. Frequently, cost:benefit ratios from individual programs are much higher, often exceeding 1:100. Benefits have been recorded from a program releasing a parasitoid against cassava mealybug (*Phenacoccus manihoti*) in Africa, yielding a cost:benefit ratio of 1:200 and releases of a virus against the rhinoceros beetle (*Oryctes rhinoceros*) attacking palms in east Asia and Oceania yielded a cost:benefit ratio of 1:120. A recent summary of cost:benefit ratios from classical biological control of weeds programs reported from 1:2.3 to 1:4,000. In fact, all of these cost:benefit estimates are based on some limited interval for the period over which control takes place and money is saved; so, in calculating the cost:benefit ratio for introduction of the vedalia beetle, normally benefits would be calculated for only 10 years after the success in 1890. However, today, more than 100 years later, we are still reaping benefits from the activity of the vedalia beetle so, in fact, the cost:benefit ratio should instead be calculated based on the savings between 1890 and today for a more accurate figure. Of course, cost:benefit analyses only portray economic benefits and cannot indicate benefits to the environment and human health and welfare as a result of classical biological control successes.

Further Reading

Cock, M. J. W., Murphy, S. T., Kairo, M. T. K., Thompson, E., Murphy, R. J., & Francis, A. W. (2016). Trends in the classical biological control of insect pests by insects: An update of the BIOCAT database. *BioControl*, **61**, 349–363.

Cock, M. J. W., van Lenteren, J. C., Brodeur, J., Barratt, B. I. P., Bigler, F., Bolckmans, K., Cônsoli, F. L., Haas, F., Mason, P. G., & Parra, J. R. P. (2010). Do new Access and Benefit Sharing procedures under the Convention on Biological Diversity threaten the future of biological control? *BioControl*, **55**, 199–218.

Gould, J., Hoelmer, K., & Goolsby, J. (eds.) (2008). *Classical Biological Control of* Bemisia tabaci *in the United States: A Review of Interagency Research and Implementation*. Dordrecht, Netherlands: Springer.

Hajek, A. E., McManus, M. L., & Delalibera, I., Jr. (2007). A review of introductions of pathogens and nematodes for classical biological control of insects and mites. *Biological Control*, **41**, 1–13.

Pratt, C. F., Shaw, R. H., Tanner, R. A., Djeddour, D. H., & Vos, J. G. M. (2013). Biological control of invasive non-native weeds: An opportunity not to be ignored. *Entomologische Berichten*, **73**, 144–154.

4 Augmentation: Inundative and Inoculative Biological Control

The second and third major ways to use biological control, inoculative and inundative biological control, both involve releasing biological control agents without the goal of permanent establishment. Although these two strategies have different goals and ways in which they work, there are strong commonalities and thus they both fall under the strategy of augmentation. These strategies are used to control pests when natural enemies occurring in an area are absent, when the control owing to natural enemies would naturally occur too late to prevent damage, or when natural enemies occur naturally in numbers too low to provide effective control. The term augmentation is used because natural enemies are being augmented.

4.1 Inundative Biological Control

The use of living organisms to control pests when control is achieved exclusively by the organisms themselves that have been released. (Eilenberg et al., 2001)

This strategy is directed toward rapid control of pests over the short term. In all cases, no significant reproduction by the natural enemy is expected in order to impact the pest population. Because control is only a result of the released individuals, inundative releases would have to be repeated if pest populations increase again after natural enemies are released. In practice, releases are often repeated if pest populations were not all present in a susceptible stage during the previous application, if new pests disperse into the crop, or if the crop is long lived, increasing the length of time it could become infested. The released agents must contact and kill a sufficiently high proportion of the pest population, or by other means reduce the damage level, to provide control. Of course, to achieve sufficient control rapidly it is important to release a large number of organisms to inundate the pest population.

Inundative control is often used for short-term crops grown in monocultures because viable populations of the natural enemies with the potential to increase do not occur in the habitats provided by temporary monocultures. Alternatively, inundative releases are appropriate where damage thresholds are very low, meaning that control needs to be achieved within a short time and rapid control is required at early stages of pest infestations.

In many ways, the goals and expectations of this strategy are similar to those for use of synthetic chemical pesticides. Perhaps the similarity of inundative biological control with the pesticide paradigm and the straightforward story about use of biological control agents to solve pest problems here and now have helped to account for the popularity of this approach compared with inoculative releases. Microbial natural enemies applied inundatively against arthropods, weeds, or plant pathogens have sometimes been referred to as microbial pesticides or biopesticides. However, the term biopesticide has different meanings in different instances and sometimes it also includes natural compounds produced by living organisms or pesticidal substances produced by plants that can contain added genetic material (sometimes also called plant-incorporated protectants). An example of inundative use of microorganisms would be applications of the bacterial pathogen *Bacillus thuringiensis*, used to control numerous species of insects (see Chapter 10). Inundative release is also the strategy used to apply a fungal pathogen against sugarcane spittlebugs in Brazil (Box 12.2), a viral pathogen against codling moth (*Cydia pomonella*) in European orchards (Box 11.2), and a fungal pathogen for control of weedy trees and bushes (Box 15.1). This last-mentioned natural enemy can also be called a bioherbicide or, because this is a fungus, a mycoherbicide.

Strengthening the view that microorganisms for biological control are similar to chemical pesticides, microorganisms for inundative release are often sold in forms similar to synthetic chemical pesticides. For example, they can be formulated as flowable concentrates or wettable powders and can be applied repeatedly, often with the same spray equipment that could be used to apply chemical pesticides. Also, product names for biopesticides are often similar to trade names of chemical pesticides and the name does not provide reference to the organism involved, for example, some products based on insect-killing bacteria have the trade name Dipel. Although there are some similarities between microorganisms and chemical pesticides, we should not think of using microorganisms in the same way as chemical pesticides. These are living organisms and care must be taken to store and transport them so that they remain alive and are released in an appropriate way. Owing to the large numbers of natural enemies that must be released when using an inundative approach, methods for cost-efficient and successful mass production, formulation, storage, transport, and release are critical for development and use of this strategy. As a caveat, because the term biopesticide makes it sound like the living organisms will be equivalent to synthetic chemical pesticides, many biological control practitioners and supporters prefer not to use this term, although it seems to be receiving greater acceptance in recent years.

A few of the many examples of macro-beneficials, usually not indicated with the term biopesticides but used inundatively, are lady beetles to control aphids, the predatory mite *Neoseiulus cucumeris* to control thrips, and nematodes to control fungus gnats. The term biopesticide applies well to these uses of macro-beneficials, since they are applied in huge numbers to achieve rapid control. When macro-beneficials are released for inundative biological control, in general, species that are chosen are those that kill quickly, as with natural enemies that are predators.

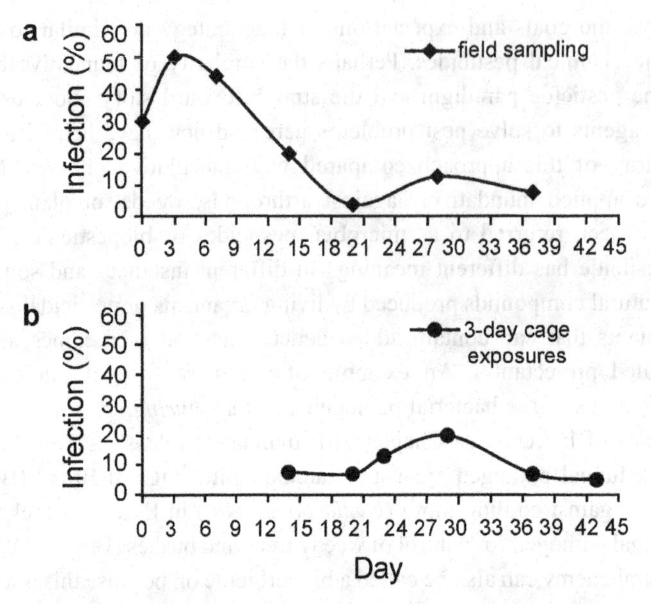

Figure 4.1 Primary infections initiated after release of *Metarhizium acridum* against African rice grasshoppers, followed by a second cycle of infections. (a) Infection of *Hieroglyphus daganensis* collected at varying times after fungal application on day 0. At each sampling date, 50 grasshoppers were collected and subsequently reared in the laboratory. (b) To detect the second cycle of infections, 14 days after the application, healthy grasshoppers were caged in the field for 3-day periods and then reared in the laboratory to detect whether infective fungal inoculum was still present in the environment (based on Lomer et al., 1997).

4.2 Inoculative Biological Control

The intentional release of a living organism as a biological control agent with the expectation that it will multiply and control the pest for an extended period, but not that it will do so permanently. (*Eilenberg et al., 2001*)

For this strategy, control is due not only to the released organisms themselves but also to their progeny. This strategy provides more long-term and self-sustained control than inundative releases. It is used in systems where, for example, a natural enemy can respond to and control a pest population, often in a density-dependent manner, but the natural enemy does not persist, or where a natural enemy provides density-dependent control but is difficult to mass-produce in large enough quantities for inundative releases. If an inoculative release is intended for predators, parasitoids, or pathogens, sufficient pest numbers (or other means for growth of the biological control agent) must be present following the initial release to support a second or third generation of the released agent and conditions that allow multiplication of the natural enemy must occur. Studies with the fungal pathogen *Metarhizium acridum* in central Africa have demonstrated this secondary cycling of infection where spores produced from the first cohort of grasshoppers that were killed in the field infect a second cohort (Figure 4.1). Although, for inoculative approaches, fewer natural enemies need to be released than with the inundative approach, these programs still usually require some aspects of mass-production and formulation, as well as storage and transport to supply enough agents at appropriate times and locations for release.

For biological control of plant pathogens, in general microorganisms that are released are intended to increase in the microhabitat where they are released, so they fall into the inoculation category of augmentation. This is especially true for those antagonists of plant pathogens that are used as bioprotectants so that plant pathogens will not be able to colonize a plant. Clearly, not only the original microorganisms released but later generations also are needed to colonize roots and wounds to antagonize disease organisms or to protect potential sites of infection.

When persistence in the release area is shortened as a result of seasonal effects and the natural enemy is released inoculatively each season, this strategy is called seasonal inoculative release. Seasonal inoculative release has been used in greenhouses, where a crop occupies an individual greenhouse for a finite period of time until harvest. Then, the greenhouse is emptied and cleaned in preparation for another crop. The goal of such natural enemy releases in greenhouses is usually to establish populations of natural enemies that will control the pests that are present and then these natural enemies will persist to respond to pest upsurges or new invasions while that same crop is present. Under standard greenhouse practices, the natural enemy populations are destroyed during greenhouse clean-out and new beneficials must be introduced into the next crop when new pests are detected. Such use of seasonal inoculative release in commercial greenhouses has been widely practiced (Box 4.1).

Box 4.1 Augmentative Releases of Macro-Biological Control Agents in Greenhouse-Grown Vegetables in Europe and Canada

In 1970, augmentation biological control was used in only 200 ha of greenhouses worldwide, while in 2017 biological control was much more commonly used, especially in vegetable production in Europe and Canada. Biological control has proven especially effective against pests that do not directly attack the parts of plants that are later harvested as vegetables. As a result, pesticide use for control of such pests has declined significantly in greenhouses where biological control is used. The steep increase in the demand for organically grown (or at least pesticide-free) vegetables for consumption has helped in creating a much larger market for biological control in greenhouses. In addition, the use of bumble bees for efficient pollination and the bees' susceptibility to insecticides has necessitated the use of noninsecticidal means for control. To make this increase in use of biological control agents possible, the number of companies producing biological control agents for sale has increased accordingly. Around 1970, only a few small companies produced natural enemies for sale but the use of biological control in greenhouses in Europe was just beginning. By 2017, companies producing biological control agents were numerous worldwide and had organized themselves to increase communication so that they could share methods and problems. Use of natural enemies for controlling pests on greenhouse ornamentals (e.g., poinsettias, chrysanthemums) has not been as common but more recently has been increasing. Also in nonproduction greenhouses like enclosed butterfly conservatories, the use of biological control has increased.

Some of the most important natural enemies used in greenhouses are the predatory mites, especially including *Phytoseiulus persimilis* against spider mites, and *Neoseiulus cucumeris* and *Amblyseius swirskii* against thrips. The small parasitoid wasp *Encarsia formosa* is used extensively against greenhouse whiteflies. Owing to the fragility of these agents, specialized methods for release have been developed. For the winged *E. formosa*, parasitized hosts are glued onto cards that are then placed throughout greenhouses (Figure 4.2). When adult *E. formosa* emerge from within the parasitized whitefly hosts, they are ready to fly and find hosts in which they lay their own eggs. Mites are released either by broadcasting them onto crop foliage or by deploying numerous perforated paper sachets in the crop canopy that contain predatory mites that are reproducing inside the sachet by feeding on nonpest prey mites. This strategy provides a slow and sustained release of the predatory mites.

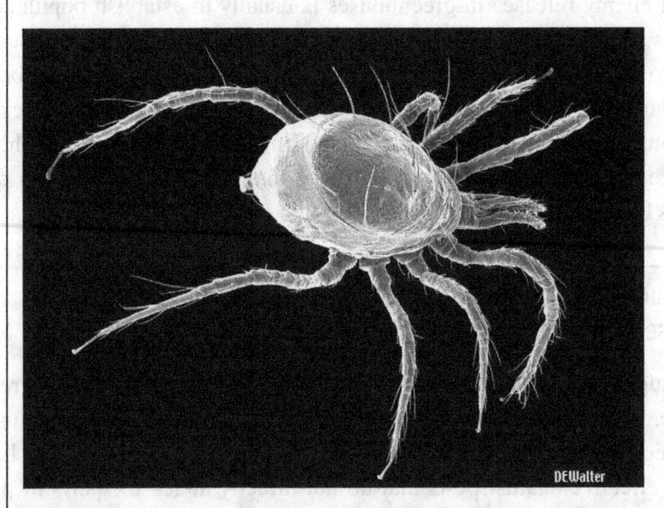

Figure Box 4.1.1 The mite *Phytoseiulus persimilis* (Family Phytoseiidae), an excellent predator of spider mites (photo by David Evans Walter).

Greenhouses are usually scouted to monitor the presence and densities of both pests and beneficials. Parasitoids and predators may be released when pests have been seen or they may be released preventively based on past histories of pest problems. For seasonal inoculative releases, determining release ratios can be critical. If too few natural enemies are released, control will not be obtained in time to protect the crop. Releases can sometimes be done using so-called "banker plants" that harbor nonpest insects serving as alternate hosts for the natural enemy. In this way a population of natural enemies can build so that they are present when the pest species increases in number.

Preventive releases, also called "blind releases," have become very popular. Such preventive releases are especially useful for control of small species (like thrips) or pests that are difficult to find when low in numbers but which can increase very quickly. The availability and ease of application of the most commonly used macro-beneficials make the preventive approach simple to use for many pests.

While both seasonal inoculative releases and preventive releases can be effective, the former requires more attention and knowledge of the system, while the latter provides excellent protection but depends on knowledge and action by the grower before any problem is observed. Therefore, a factor which should not be underestimated is the need for education of staff and growers who need to learn about the main pests and pathogens, how to monitor for them, and which biological control products to use. The great success of biological control in greenhouses in Europe and Canada has been attributed, at least in part, to the excellent information available to growers about how to most profitably use biological control in greenhouses.

Seasonal inoculative release is also appropriate to use outdoors for effective natural enemies that cannot persist in an area after release. The parasitoid *Pediobius foveolatus* attacking Mexican bean beetle, *Epilachna varivestis*, larvae in snap beans as well as soybeans has little tolerance to cold weather. In the mid-Atlantic and southern Appalachian Mountains in the United States, the hosts overwinter as adults so there is no overwintering refuge for the parasitoids. Therefore, this parasitoid species is mass-produced and released each year in mid-spring just after first instar larvae are found. Seasonal inoculative releases are also used for the fungus *Beauveria brongniartii* to control beetle grubs in soil in pastures in central Europe.

4.3 Inundative versus Inoculative Strategies

In practice, the distinction between inoculative and inundative releases is not always so precise. An important feature of inundative augmentation is that although the biological control agent is applied without the expectation that it will reproduce, it is a living organism capable of reproduction. So, when predators are released, feeding by the released agents would satisfy the expectation for inundative release and there is no expectation of reproduction by the predator or any residual effect as a result of the predator's offspring. In greenhouse systems the type of biological control might be categorized as inundative, not solely because there is no production of offspring, as any offspring produced might not find hosts to attack (and thus the offspring have no impact on the biological control). In practice, inundative biological control is probably often followed by residual effects if hosts are present and the released organisms can multiply after release. Conversely, with inoculative release, the majority of control can sometimes be caused by the released organisms themselves and the effects from progeny can be minimal, if reproduction is limited or because few prey/hosts remain.

Whether a natural enemy species is considered for inoculative or inundative biological control is determined in part by the difficulty and cost of producing adequate quantities of that agent for release. For example, while large volumes of the insect-pathogenic bacterium *B. thuringiensis* can be mass-produced in fermenters at a reasonable cost, producing large numbers of parasitic wasps for release is vastly more difficult and costly. In addition, the ability to store, transport, and release a certain natural enemy species will help determine whether it can be developed for inundative versus inoculative application.

The ability of those organisms being applied to reproduce after application and for their offspring subsequently to attack hosts influences which augmentative strategy is appropriate. For example, when *B. thuringiensis* is sprayed, although more bacterial cells are produced in infected insects after they die, subsequent bacterial generations virtually never go on to infect a significant proportion of the surviving hosts, owing to the nonepidemic nature of this bacterium. Therefore, use of *B. thuringiensis* is never considered an inoculative application. Conversely, parasitoids are adept at searching for hosts and the progeny of initially introduced parasitoids can have a significant impact on host populations. Therefore, whether parasitoids are intended for inundative or inoculative release, their progeny frequently continue to parasitize hosts, providing the benefits of an inoculative release.

4.4 Production of Natural Enemies by Industry

Use of an inundative release strategy, in particular, requires an industry to produce and distribute natural enemies. In general, companies producing natural enemies specialize in either macro-beneficials or microorganisms owing to the different types of equipment, methodology, and expertise needed. For both types of natural enemies, although some companies have been in existence for a long time, in this growing field many companies have not been in this business for long.

4.4.1 The Need for a Market

Augmentative biological control is not seriously adopted by growers unless a steady, reliable product is available to them, while for industry to invest the effort to develop such a natural enemy product, there must be reliable customers that will regularly purchase the product. A difficulty in developing natural enemies as products can be that some natural enemies, especially parasitoids and predators, cannot be stored for very long when they are not needed. If natural enemies are not needed and they do not survive well in storage, the producer or distributor will take a loss. Unfortunately, the need to control pests is volatile and not easy to predict, which can present difficulties for companies producing most beneficials. For a product to be developed, it is critical that the actual market is large enough to support production of that beneficial. Is the area in need of control sufficiently big? Are the pests, plant diseases, or weeds a continuous problem so that customers will reliably continue purchasing the natural enemy? Will the consumers be willing to cover the amount that the natural enemy will cost when they purchase products? In general, growers must be educated about biologicals and, for growers to use a biological, it must usually be simple to use and have some clear and significant advantage over pesticides. Also, growers frequently need some assurance that the biological control approach will work and, if it doesn't, they want to know what alternative approaches are available. Examples of advantages of using biological controls are increased yields when pesticide phytotoxicity no longer causes blossom drop in greenhouse vegetables, increased yields with the use of bumble bees for efficient pollination in greenhouse vegetables (and the supporting avoidance of pesticides, so that the pollinators survive), and preference for avoiding use of synthetic chemical pesticides by those who would be exposed to sprays or those purchasing the end-product. Also, the trend in many countries to phase out many synthetic chemical pesticides will be advantageous for development of new biological control products.

4.4.2 The Double-Edged Sword: Host Specificity

Host specificity of a natural enemy is often critical to its development for augmentation. Natural enemies with limited host ranges are considered safer for the environment. Also, with a limited host range the natural enemy should respond more closely to increases and decreases in host population density, searching harder when hosts are scarce instead of switching to another host species. Such a density-dependent response is, of course, a more critical feature for inoculative rather than inundative releases because reproduction of the natural enemies released is not expected with inundative releases.

However, host specificity can also influence the size of the market for the mass-produced natural enemy and thus the final cost of the product. If the natural enemy attacks a greater breadth of hosts, sales of that natural enemy may be greater depending upon the number of control alternatives for the pests in question. In contrast, a highly host-specific agent would not always generate enough sales to justify mass production; there must be a market large enough to support the costs of mass production. In practice, host-specific natural enemies used for augmentation have often been most viable if they fill a pest control niche in a high-value crop.

4.5 Products for Use

Biological control agents range in complexity from viruses to higher eukaryotes and methods for mass production, storage, and transport are equally divergent. The species of natural enemies that are chosen for inoculative or inundative releases are chosen in part through trial and error. They are known to attack a pest in nature but that is only the start and does not ensure that the species will be appropriate for augmentative biological control. Cost-effective methods for mass production and handling must be possible for any natural enemy to be mass-produced for augmentative release. For any organism, the process involves several stages, from choosing an efficacious strain for mass production to developing methods for releasing the natural enemy (Table 4.1). In addition, safety testing and registration of natural enemies can be required by governments (see Chapter 18).

Table 4.1 Steps necessary for developing a natural enemy for augmentative release.

1	Identification of a market searching for a pest control solution
2	Identification of a strain of a natural enemy for mass production that is effective against the target and has no or acceptable effects on nontargets
3	Development of a cost-effective method for mass production and quality control assessment
4	Development of storage methods and appropriate methods for transport
5	Development of methods for release and knowledge about quantities needed for release in different situations
6	Characterization of fit within agronomic practices, including integrated pest management approaches

4.5.1 Macroorganisms

Relative to use for augmentation, arthropod parasitoids and predators can be referred to as macroorganisms or macro-beneficials. Although some are small or even very small, they can usually be seen with the naked eye. Parasitoids and predators differ significantly from microorganisms in the ways they can be handled and released and the types of control they provide. Insect parasitic nematodes are generally included with macroorganisms when discussing augmentation. In many countries and regions in the world, parasitoids, predators, and nematodes do not have to be registered with governmental agencies so their development for control is not as difficult as with microorganisms, which must be registered. Issues for arthropods for biological control of weeds are similar to those for parasitoids and predators, although industries for these organisms are small at present.

Natural Enemy Strains/Species

With parasitoids and predators, biotypes or even species adapted to specific hosts and climates can be extremely important for achieving successful control. In the case of the parasitoid *Trichogramma ostriniae* attacking eggs of the European corn borer, *Ostrinia nubilalis*, this species has been shown to be so effective that only inoculative releases are needed, whereas inundative releases of other species of *Trichogramma* against European corn borers have repeatedly not provided acceptable control in the United States. In fact, augmentative releases of *Trichogramma* against European corn borer in the United States were not considered economically feasible until *T. ostriniae* began being developed.

Biological control researchers are aware that the strain of a natural enemy species can be critically important for efficacy but they are often limited by the strains that are available. Consequently, they may collect new strains worldwide to find unique and useful biological features. Finding new strains is accomplished through exploration but frequently also through obtaining new strains via mail from colleagues in other areas. Of course, when new strains are obtained, efficacy must be tested both in the laboratory and field as well as evaluating the potential for nontarget effects (see Chapter 18).

In recent years, the characteristics of natural enemies have sometimes focused on genetic engineering. However, because of the complexity of macroorganisms, this technology has not yet progressed far beyond inserting neutral markers into a few beneficial macroorganisms.

Mass Production

While efficacy against the target pest is critical, there will be no product unless the natural enemy can be grown in a cost-effective way. Macro-natural enemies are mass-produced in a variety of different ways depending on the organisms. The vast majority of macroorganisms need living hosts so they must be produced in insectaries where colonies of their hosts/prey can also be grown. The prey can be an alternative prey that might be easier for mass production; for example, some predatory mites for biological control are produced using flour mites, *Acarus siro*, for prey. The fact that two arthropod cultures (the host and the natural enemy) are often required to produce one macro-beneficial product, and the resulting product might have only a short shelf life, helps explain why macro-natural enemies often necessarily cost a bit more than insecticides. Further difficulty is found with cannibalistic predators, such as green lacewing larvae, that require considerable space for rearing so that they do not eat each other. In the unique case of the convergent lady beetle, *Hippodamia convergens* (Box 7.4), insects are field-collected and stockpiled, but they become unavailable seasonally and the quality of field-collected individuals is questionable if they are out of synchrony with the seasons at the release site and must be stored for long periods.

One way to make mass production more cost effective would be to produce the natural enemies on artificial diets. However, artificial diets are not used extensively for producing macro-beneficials because methods that have been developed are often not as successful as using living hosts. There is also concern that macro-beneficials reared in association with artificial diets will not learn the cues needed to locate hosts or host plants. It is critical not to alter the behaviors of the beneficial that make that species effective for control.

Quality control is a critically important issue for all agents, no matter how they are reared. There is the potential for inbreeding depression and adaptation to the methods used at the mass-production facility so that the natural enemies will no longer respond appropriately when encountering the pest. The best advice for avoiding such problems is to rear the natural enemy on the target pest on the type of plant or substrate that will be encountered in the field and under normal climate conditions, at least when beginning mass production. To ensure quality control, guidelines have been developed by the International Organization for Biological Control (IOBC) for production of the species of macro-natural enemies that are most widely used in greenhouses as well as for the egg parasitoid *Trichogramma*. To monitor quality over time, population attributes followed include emergence rates, sex ratio, length of the lifespan, fecundity, adult size, and predation/parasitism rate. In addition, there are regional standards for ensuring the quality of natural enemies when these reach the consumer (for example, the Association of Natural Bio-Control Producers in the United States have developed quality control standards).

Storage and Transport

Most macro-natural enemies cannot be stored very long so large numbers must be produced seasonally. Demand for natural enemies is usually not constant so the ability to mass-produce and then store macro-beneficials would be very helpful for maintaining availability. When storage is possible, this allows continuous production of smaller quantities of natural enemies instead of massive seasonal production. The possibilities for storing macro-beneficials differ by species. For example, diapausing predatory insidious flower bugs (*Orius insidiosus*) can be stored for up to eight weeks. Methods have been developed for long-term storage of green lacewing adults that can subsequently be brought into a reproductive state quickly and synchronously after storage. In contrast, green lacewing eggs, the stage that is usually released in the field, can only be stored in the cold for about three weeks. The entomopathogenic nematode *Steinernema carpocapsae* can be stored for up to five months at room temperature but up to a year if refrigerated.

Care must be taken with these living organisms to make sure that they arrive at the release site in excellent condition and are not crushed, asphyxiated, overheated, frozen, or released in transit. It is also often important to maintain humidity within packing containers so that predators and parasitoids do not die of desiccation en route, although some insect-pathogenic nematodes survive longer if they are dry. When transit requires several days, food can be packed along with the agents (e.g., honey for parasitoids and pollen or prey for predators). Because predators are often generalists they can be cannibalistic when hungry and at high densities will eat each other with the result that fewer individuals arrive at their destination than were packed initially. Packing cannibalistic species with paper, buckwheat hulls, or vermiculite helps to provide hiding places and reduces cannibalism. The excellent courier services available worldwide today make the rapid shipping that is needed possible; in previous years, shipping services often were not fast enough and macroorganisms that were ordered sometimes arrived in poor condition.

Release

Successful releases rely on a combination of factors that are surprisingly similar to the factors that influence successful pesticide use: application rate, timing (including time of day), synchrony with the pest's susceptible stages, coverage, and severity of rainfall after application. Repeated applications are often needed both for beneficials and for pesticides to assure synchrony of the application with the susceptible stages of the pest. One major difference between biological control and pesticide application is that release rates for beneficials should be adjusted to the density of pests, whereas pesticide application rates are geared for thorough coverage of surface areas. Thus, natural enemy:pest ratios are more important than the active ingredient per acre used for pesticides. Another major difference between releasing macro-beneficials inundatively and applying pesticides is that many beneficial arthropods are still released by hand.

The stage that is released is often determined by ease of transport and manipulation. Depending on the system, releases can be made by hand, as when cards bearing immature stages of the parasitoid *Encarsia formosa*, still within the skins of their

Figure 4.2 Card containing whitefly pupae parasitized by *Encarsia formosa* to be hung from plants in greenhouses and interior plantscapes where these parasitoids emerge and find whiteflies to parasitize (photo by Jack Kelly Clark, University of California).

dead whitefly pupal hosts, are placed in greenhouses (Figure 4.2). Predatory mites can be released by dispersing bran containing the mites by hand with a granular-pesticide dispenser or by tractor-mounted applicators. Releases of *Trichogramma* parasitoids in forests and fields have been made by placing cartons containing parasitized hosts on branches or by releasing cartons containing parasitized host eggs from helicopters or airplanes. To avoid predation before emergence of parasitic wasps, host eggs parasitized by *Trichogramma* can be released in fields within capsules that predators cannot penetrate but from which wasps can disperse. Beneficial nematodes are most often applied using conventional pesticide spray equipment.

4.5.2 Microorganisms

Augmentative release is the major strategy used for controlling insects, weeds, and plant pathogens with microbial natural enemies. In contrast with macroorganisms, commercial microorganisms can be easier to mass-produce, store, and apply (although this depends on natural enemy species). For industries in many developed countries to sell microorganisms for augmentation, the microorganism must be registered with appropriate governmental agencies. Costs of registration can be high so, for an industry to proceed with registration, a microbial product usually must be assured of sustained profits or industries often will not proceed with developing that microorganism for control.

Microbial Strain/Species

Searching for the "best" strain or isolate has been a major occupation of researchers working with microorganisms for biological control. With microorganisms, virulence can vary so much from strain to strain that major emphasis is placed on comparing virulence of multiple strains within a species of microorganism. While virulence studies are at first always done in the laboratory under controlled conditions, plant pathologists, especially, strongly suggest that such strain comparisons should also be conducted under field conditions. In the case of soil-dwelling microbes, results from the simplified systems in the laboratory can be very different from results in the complex microbial environment of field soil. For example, the fungal pathogen *Pochonia chlamydosporia* readily infects egg masses of nematodes protruding through galled roots and can provide effective control. However, through field studies it was found that different isolates of the fungus had differential survival in the root area, related to the ability of the fungus to colonize the surface of the root. Even different cultivars of tomato affected colonization of the roots by this fungus. Thus, testing microorganisms in the field to determine efficacy can be critical before proceeding further with developing mass-production methods.

The latest development for microorganisms to be used for biological control has been in the exploration of manipulations to enhance activity. Many microorganisms have much smaller and simpler genomes than macroorganisms and thus have been targets for use of genetic engineering techniques. Genes have been inserted into viruses and species of bacteria and fungi used for biological control to (1) enhance virulence, (2) confer resistance to pesticides, and (3) alter host range. Field trials have been conducted by releasing genetically engineered microbial agents in limited areas with the first studies only releasing microorganisms expressing genetic markers. Initially, these field trials raised a distracting media furor but now field studies can be conducted with more focus on results, at least in some countries. While few products on the market are genetically engineered, engineering microorganisms for biological control has been an area of research, although a major hindrance in several regions of the world has been that genetically manipulated organisms cannot be released in the field.

Mass Production

The microorganisms that have been used the most to date for biological control, for example, the bacterial pathogen *Bacillus thuringiensis*, require simple media and are relatively cheap to mass-produce in larger fermenters. For all microorganism species, it has been important to spend time optimizing nutrients (carbon sources, carbon to nitrogen ratios, etc.) and the fermentation environment (temperature, pH, aeration, rotation speed, etc.). Some microbial natural enemies do not readily or abundantly produce propagules, such as spores, in culture and strain selection as well as growth conditions have been manipulated during attempts to overcome this obstacle. In fact, spore production or the lack of it during mass production of fungi has in many cases determined the success or failure of species being considered for development. Newer production and formulation technologies, such as solid state fermentation or formulation in gelatin capsules, may pave the way for new products based on bacteria and fungi.

Among the fungi and bacteria, products are dominated by species and strains that are easy to grow in culture without live hosts. Other microorganisms, such as viruses (at present), microsporidia, and some obligate bacterial pathogens, can only be grown in living hosts. Therefore, both a stock culture of healthy hosts and a culture of infected hosts must be produced. The production system must avoid all risks of contamination of the stock insect culture, so reentry of a person from a microbial production unit into the insect stock production unit must be avoided in order to maintain consistent production of the microbe. The need for living hosts can seriously influence the extent to which microbial natural enemies are mass-produced. However, just because the natural enemy cannot be mass-produced outside of hosts has not stopped large-scale production of some insect-pathogenic viruses (see Chapter 11).

Quality control issues with mass-produced microorganisms include (1) assurance that cultures have not become contaminated, especially by microorganisms pathogenic to humans, (2) assurance that cultures are still sufficiently virulent to target species, and (3) assurance that active unit numbers, such as fungal spores, are as stated for the product.

Storage and Transport

Some microorganisms can be stored for months or years at room temperature, for example, *Bacillus thuringiensis* spores are thought to survive decades, if not longer. For these species, storage, shelf life, and shipping are similar to synthetic chemical pesticides. This is an important advantage because it allows year-round production and easy storage until the product is needed. Some fungi are more fragile and can be stored only for several months, often with refrigeration. The insect-pathogenic fungus *Lecanicillium lecanii*, sold to control aphids in greenhouses, is viable for a few months when kept cold.

Release

An advantage of microorganisms is that they can often be applied with preexisting equipment used for synthetic chemical pesticides. This can simply require adjustments to use bigger nozzle sizes that allow spores to pass. For materials that can be produced in large quantities and that need to be applied over extensive areas such as large fields, forests, or rangeland, application is even possible from the air (Figure 4.3). In contrast with macro-natural enemies, microorganisms must be deposited closer to the correct location of the pest because these species frequently have no ability to disperse and locate the pest, compared with mobile parasitoids and predators.

Microorganisms are almost always mixed with other materials to facilitate application; this is called formulation. Many microorganisms are sensitive to desiccation and solar radiation and cannot protect themselves after application. Therefore, formulations can include aids for extending the lives of microorganisms, such as protectants against ultraviolet radiation and materials to retain moisture. Formulations can also include materials added for delivery to the target pest, such as materials that help the microorganism stick to a plant leaf instead of being washed off. Formulations can help improve the activity of a microorganism; for example, optical brighteners added to insect-pathogenic viruses can help to compromise the gut wall of a caterpillar so that an insect-pathogenic virus can infect more readily. Fungal spores that require free water for germination can be formulated in vegetable oils that retain water; these formulations are then applied using ultra-low volume application equipment. Microorganisms can also be formulated to improve storage, ease of use, and compatibility with application equipment used for spraying pesticides.

Figure 4.3 Using an airplane to spray a microbial natural enemy such as *Bacillus thuringiensis* or nucleopolyhedrovirus for arthropod control in forests (photo by J. Podgwaite, USDA, Forest Service).

4.6 Natural Enemies Commercially Available for Augmentative Releases

While augmentative biological control strategies have not yet been used extensively for control of major pests in major crops, they can provide excellent control in more specialized systems, sometimes referred to as "niche" markets, often protecting higher value resources. Macro- and micro-natural enemies are mass-produced for augmentative release by a diversity of organizations, from large or small companies to farmers' cooperatives and national research organizations. Joop van Lenteren estimated in 2017 that there were 500 companies of varying sizes producing invertebrate natural enemies worldwide. Regional companies facilitate distribution from producers. Information about available products can be found in several different sources. There are published reviews listing available organisms and compilations of specific types of products are also produced. Much of this information can be found on the Internet. Within chapters of this book on each group of natural enemies, tables are included listing examples of products available for inundation or inoculation.

The availability of products changes over time: some organisms have been on the market for decades, new organisms are being introduced each year, and some organisms disappear from the market occasionally. There must be continuity in production of the natural enemies by the smaller industries producing them for these products to come into regular use because they are reliably available and effective. Of course, not all natural enemies are available in all countries and permits are needed for moving natural enemies from one country to another. As mentioned, microbial natural enemies must be registered for use with the government in each country (see Chapter 18). Suppliers

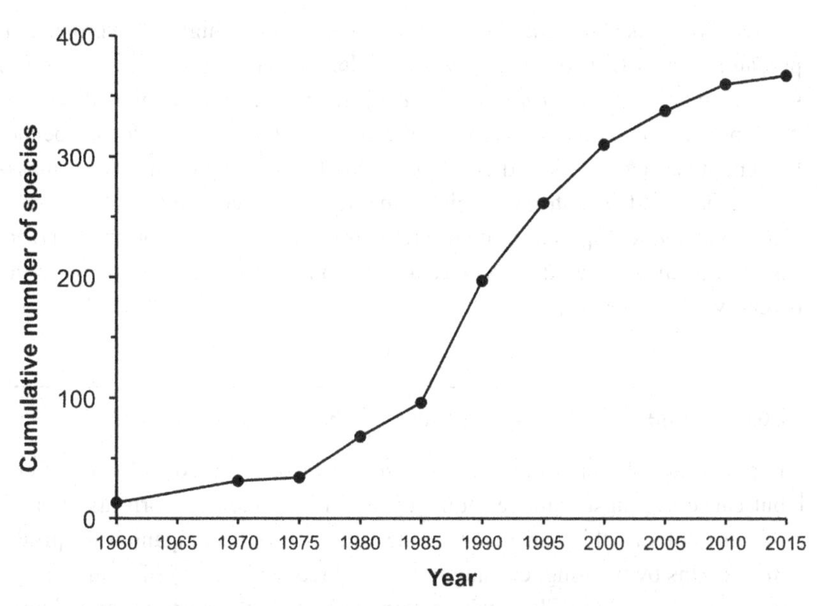

Figure 4.4 Cumulative numbers of invertebrate natural enemies commercially available for augmentative biological control worldwide, 1960 to 2015 (based on data from van Lenteren et al., 2018).

mostly provide detailed information about quantities to order, timing, and procedures for releasing natural enemies. Some suppliers have their web page split into two parts: one part is for growers who buy large quantities of a product, while another part is for private gardeners, who want to buy enough for treating a small area.

By 2016, approximately 350 species of invertebrate natural enemies (mostly insects, mites, and entomopathogenic nematodes) were produced commercially for augmentation (Figure 4.4). These natural enemies were applied on approximately 30 million ha worldwide in 2015. However, about 30 of these species make up more than 90 percent of worldwide use. For microorganisms, 209 strains of 94 species are being produced and sold for augmentation. The market for these biological control agents has been especially strong for control of invertebrate pests in greenhouses, with important contributions by Joop van Lenteren in the Netherlands (Boxes 4.1 and 19.2). Intensively managed greenhouse crops are especially well suited for augmentative biological control because many available pesticides kill bumble bees that are needed to pollinate greenhouse vegetables or produce phytotoxic effects in plants, including premature abortion of fruit and flowers. Releasing natural enemies requires less time than spraying chemical pesticides as there is no period after application when workers cannot reenter (for chemical pesticides there is usually a period after application during which people cannot reenter the sprayed area) and some key pests can only be controlled with natural enemies.

Greatly increased yields associated with the use of natural enemies in high-value greenhouse crops have justified their use instead of pesticides for decades. Augmentative uses of macro-natural enemies in open field agriculture has not been as extensive as greenhouse use but natural enemies are used in some high-value crops. For example, predatory mites (*Phytoseiulus persimilis*) have been widely used for control of two-spotted spider mites (*Tetranychus urticae*) in strawberry fields in California. The most widespread application of macro-beneficials in the field worldwide may be the use of the hymenopteran egg parasitoid *Trichogramma*. These egg parasitoids are mass-produced around the world to control caterpillars in a variety of ecosystems; for example, in China *Trichogramma* are applied in many field crops (Box 4.2) and on fruit and forest trees. Insect-attacking nematodes are used against numerous soil-dwelling pests in turf and a diversity of smaller crops.

Box 4.2 Use of *Trichogramma* against Asian corn borers in China

Larvae of the Asian corn borer, *Ostrinia furnacalis*, feed on all parts of corn plants but cause the most damage indirectly when they feed on corn tassels and kernels of corn cobs as this can result in the ears rotting accompanied by production of mycotoxins by the fungi causing the rot and reduced quality of grain. This borer is a major pest of corn in China where corn is planted over a greater area than any other grain crop. Since the 1970s, one way to prevent feeding damage by *O. furnacalis* has been to release tiny wasps in the genus *Trichogramma* that parasitize *O. furnacalis* eggs.

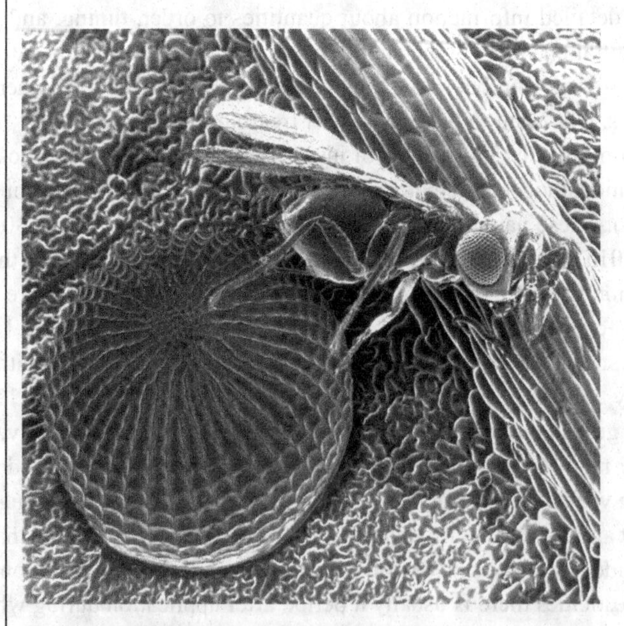

Figure Box 4.2.1 An adult female of the chalcidoid egg endoparasitoid *Trichogramma* ovipositing in an egg of the cabbage looper, *Trichoplusia ni* (photo by Gerald Carner).

Trichogramma develop in the eggs of moths and, while some "artificial" eggs were tested at one time, to produce *Trichogramma* you must either maintain a moth colony or harvest moth eggs in some other way. Chinese labs and businesses have investigated how to best accomplish mass production using three species of *Trichogramma*, with different attributes. *Trichogramma ostriniae* is native to China and is the dominant naturally occurring species, but it is not easy to mass-produce because it is particular about its hosts and therefore production efficiency is low. On the other hand, *T. dendrolimi* is easier to rear but it is less effective at controlling Asian corn borer compared with *T. ostriniae*. To use *Trichogramma* for control, host eggs are glued onto cards, exposed to parasitoids, and, at times when *O. furnacalis* eggs will be present in the field, the cards filled with parasitized eggs are placed in the field. When large host eggs are used, as those from native Asian silk moths, one host egg can produce from 50 to 260 *Trichogramma*. However, if the wasps are too crowded within host eggs, female body size and reproductive capacity are reduced and more males emerge than females. Researchers have learned the optimal amount of time to allow female *Trichogramma* to lay their eggs within host eggs in order to obtain the best oviposition once the parasitoids are released in the field. Also, mass producers need to stockpile cards full of *Trichogramma* that have not yet emerged so optimal conditions for cold storage have been studied and these egg parasitoids can then be available when they are needed.

Both inoculative and inundative release strategies have been tested at different times in different areas, with success demonstrated for both methods. In 2014 it was reported that nearly 4 million ha of corn were treated annually with *Trichogramma*, mainly in the northeastern corn-growing area of China. In line with the "Green Plant Protection" program that began in China in 2006, no insecticides are applied in many areas where *Trichogramma* are released; this allows resident natural enemies to return to add to control. For lower density pest populations, moth-killing lamps and trapping are used in addition to *Trichogramma* releases. If populations are higher, the bacterial biopesticide *Bacillus thuringiensis* is applied to plants and the fungal biopesticide *Beauveria bassiana* is applied to piles of infested corn stalks to kill overwintering larvae.

In addition, phytophagous arthropods, predominantly beetles and caterpillars, are available for augmentative release against weeds. In 2017, only 40 invertebrate biological control agents for weed control were commercially available for augmentation. However, these are mostly intended for inoculative release in areas where these species have not yet been released or did not persist.

The principal use of microorganisms in biological control is augmentative release. Numerous products based on bacteria, viruses, and fungi are available for control of arthropod pests and plant pathogens and a few species of fungi have been mass-produced for control of specific weeds (see appropriate chapters). However, among all of the microbial products used against invertebrate pests, the greatest use has been made of the bacterium *Bacillus thuringiensis*, which was estimated at more than 50 percent of the market share in 2015. Also, forty commercial products for control of plant pathogens were registered in the United States and the EU in 2005 and 2008, respectively.

Further Reading

Bailey, A., Chandler, D., Grant, W. P., Greaves, J., Prince, G., Tatchell, M. (2010). *Biopesticides: Pest Management and Regulation*. Wallingford, UK: CABI Publishing.

Glare, T., Caradus, J., Gelernter, W., Jackson, T., Keyhani, N., Köhl, J., Marrone, P., Morin, L., & Stewart, A. (2012). Have biopesticides come of age? *Trends in Biotechnology*, **30**, 250–258.

Glare, T. R. & Moran-Diez, M. E. (eds.) (2016). *Microbial-based Biopesticides: Methods and Protocols*. Dordrecht, Netherlands: Springer.

Gwynn, R. L. (2014). *The Manual of Biocontrol Agents: A World Compendium*, 5th edn. Alton, UK: British Crop Production Council.

Kabaluk, T. & Gadzik, K. (2005). *Directory of Microbial Pesticides for Agricultural Crops in OECD Countries*. Ottawa: Agriculture and Agri-Food Canada.

Lacey, L. A., Grzywacz, D., Shapiro-Ilan, D. I., Frutos, R., Brownbridge, M., & Goettel, M. S. (2015). Insect pathogens as biological control agents: Back to the future. *Journal of Invertebrate Pathology*, **132**, 1–41.

Marrone, P. G. (2014). The market and potential for biopesticides. In *Biopesticides: State of the Art and Future Opportunities*, A. D. Gross, J. R. Coats, S. O. Duke, J. N. Seiber (eds.), pp. 245–258. Washington, D.C., American Chemical Society.

Morales-Ramos, J., Guadalupe Rojas, M., & Shapiro-Ilan, D. (eds.) (2013). *Mass Production of Beneficial Organisms: Invertebrates and Entomopathogens*. New York, NY: Academic Press.

Paulitz, T. C. & Bélanger, R. R. (2001). Biological control in greenhouse systems. *Annual Review of Phytopathology*, **39**, 103–133.

Ravensberg, W. J. (2011). *A Roadmap to the Successful Development and Commercialization of Microbial Pest Control Products for Control of Arthropods*. Dordrecht, Netherlands: Springer.

van Lenteren, J. C. (ed.) (2003). *Quality Control and Production of Biological Control Agents: Theory and Testing Procedures*. Wallingford, UK: CABI Publishing.

van Lenteren, J. C. (2012). The state of commercial augmentative biological control: Plenty of natural enemies, but a frustrating lack of uptake. *BioControl*, **57**, 1–20.

van Lenteren, J. C., Bolckmans, K., Köhl, J., Ravensberg, W. J., & Urbaneja, J. (2018). Biological control using invertebrates and microorganisms: Plenty of new opportunities. *BioControl*, **63**, 39–59. DOI: 10.1007/s10526-017-9801-4.

Vurro, M., Gressel, J. (eds.) (2007). *Novel Biotechnologies for Biocontrol Agent Enhancement and Management*. Dordrecht, NL: Springer.

5 Conservation and Enhancement of Natural Enemies

This strategy for biological control differs from classical biological control and augmentation because natural enemies are not released. Instead, the resident populations of natural enemies are conserved or enhanced. This strategy is generally called conservation biological control and is defined as:

Modification of the environment or existing practices to protect and enhance specific natural enemies or other organisms to reduce the effect of pests. (Eilenberg et al., 2001)

In fact, this strategy was first principally developed to conserve natural enemies that were being decimated through use of synthetic chemical insecticides. Conserving natural enemies only later also began to be linked with enhancing them. For many years, our knowledge of how to conserve and enhance natural enemies grew only haltingly. These are more passive approaches and are usually directed toward longer-term control of pests. Conservation methods are usually not suitable for control of pests in high-value crops that can withstand little damage (i.e., crops having a low economic injury level), although there are exceptions (see Box 5.2). To develop effective conservation and enhancement of natural enemies we need to understand what factors are depressing natural enemy populations or otherwise inhibiting their ability to control pests, and these limitations must be alleviated. Alternatively, those factors limiting natural enemy populations must be identified so that they can be manipulated to enhance population levels of natural enemies or facilitate interactions between natural enemies and pests.

In more recent years, investigations of conservation and enhancement have blossomed for control of insects as well as plant diseases and plant parasitic nematodes. Taking advantage of these natural enemies for controlling pests is considered central to the services provided by ecosystems to benefit all living organisms (i.e., ecosystem services). Conservation biological control often is focused on agricultural systems. For arthropod control, a diversity of types of methods has been developed and these will be briefly described in this chapter (Table 5.1). As a type of conservation and enhancement, plant pathologists and nematologists use communities of organisms that develop within suppressive soils to control pests and this is described in Chapter 17.

5.1 Biodiversity Leading to Biological Control

Conservation biological control is largely based on the idea that increased biodiversity will lead to more stable systems in which there will be no outbreak of pests.

Table 5.1 Diversity in methods used for preserving and increasing arthropod natural enemy numbers and activity.

Conservation	Altering pesticide use
Enhancement	Providing food, often nectar and pollen sources Providing permanent habitats, shelter, and favorable microclimate Providing alternate prey or hosts (often present naturally in more diverse habitats)

However, is this correct? With greater biodiversity are pest outbreaks prevented? We know that introducing multiple species for classical biological control can be more successful than introducing fewer, although there are definitely cases where natural enemies that have been introduced interfere with each other either directly or in indirect ways. Direct interactions can include simply one predator eating another predator (intraguild predation, see Chapter 6) with an overall effect of increase in the prey population. There are many different possibilities for indirect interactions but one scenario would include two predators sharing the same prey species and one outcompeting the other. The big question is whether interference between natural enemies dampens their overall impact on pest, so that the effects are less than additive.

It is difficult to make generalizations across many different systems in order to answer this question of whether more natural enemies results in better control. So, scientists have enlisted meta-analysis, a method that involves finding a number of studies focusing on a subject and then using statistics to look for patterns. When Deborah Letourneau analyzed 266 sets of data on abundances of predators and parasitoids and their herbivorous prey or hosts, she found that increasing the diversity of natural enemies (in this case increasing the parasitoid species richness) generally was associated with greater suppression of herbivores in both temperate and tropical agriculture. Researchers commonly find lots of interconnected links in food webs, with what looks like numerous organisms needing very similar resources, and this can be thought of as functional redundancy. However, as scientists look closer they usually find that, in fact, the diverse organisms are using the ecosystem at least slightly differently from each other, acting in a complementary instead of redundant way. Therefore, it makes sense to try to conserve and help out the natural enemies that are already present and in return they provide an ecosystem service and control pest populations which can help to stabilize ecosystems.

It has also been suggested that biological control methods relying on biodiversity for providing natural enemies to control pests are best undertaken in landscapes (natural or managed) that are more complex and not areas with huge monocultures (growth of one type of plant), which can be thought of as simplified landscapes. In particular, relying on resident natural enemies includes specialist natural enemies that might be present but also largely depends on generalists. Whether specialists or generalists, natural enemies often do not find food and persist in large monocultures maintained with conventional agricultural methods including pesticide use. This knowledge has helped with developing methods toward conservation and enhancement of natural enemies to assist with pest control as part of integrated pest management.

Figure 5.1 Dead cotton aphid, *Aphis gossypii*, on a microscope slide. Presence of the fungal pathogen *Neozygites fresenii* is indicated by fungal spores on the aphid's wings and fungal cells growing throughout the body (photo by Donald Steinkraus).

5.2 Conserving Natural Enemies: Reducing Effects of Pesticides on Natural Enemies

If broad-spectrum pesticides are used as the principal method for controlling a pest, natural enemies are usually disrupted too much to be effective. To gain advantages from the resident natural enemies, most growers must change their goals and integrate use of pesticides that minimally disrupt natural enemies, although this is not always straightforward. As a first step, growers should collect data about the status of the pest in the crop system and then follow decision guidelines that provide thresholds for when pesticide treatments should begin in order to avoid economic loss (see Figure 2.1). Thus, unnecessary spraying that might deplete those natural enemies negatively impacted by spraying can be avoided. A specialized program to take advantage of a fungal pathogen of cotton aphids, *Aphis gossypii*, was successfully used to drastically reduce pesticide use for control of cotton aphids in the southeastern United States for 15 years (1993–2008) (Figure 5.1). In this case, the pesticide does not kill the fungal natural enemy, *Neozygites fresenii*, but growers could use information about predicted natural occurrence of epizootics caused by this fungal pathogen to avoid spraying the pests if enough of the pests were going to die from fungal infections to prevent problems. More recently, in Australia, a system for preserving natural enemies of *A. gossypii* in cotton has been implemented.

In a worst case scenario, use of agricultural chemicals directly kills natural enemies of the target pest or, by killing natural enemies of other organisms in the environment, secondary pests are created, thus starting the pesticide treadmill (see Chapter 1). Chemical pesticides also often reduce pest numbers to such low levels that natural enemies cannot persist. Because pests are often great dispersers and faster colonizers than natural enemies, the pests recolonize areas more quickly than natural enemies, leading to pest problems in habitats devoid of predators, parasitoids, and pathogens. Synthetic chemical pesticides do not always kill natural enemies but sublethal doses can decrease their longevity and fecundity, thereby decreasing their fitness and their effect on pests. Thankfully, not every natural enemy is killed by each agricultural chemical; some species are tolerant and some have developed resistance. With knowledge of specificity of action and the susceptibility of the most important natural enemies in a system, effective chemicals can sometimes be used that will harm pests while not harming natural enemies; these are sometimes called "selective pesticides." In addition, only certain pesticides could be used at specific times. In apple orchards in the northeastern United States, growers integrating types of control and not using specific pesticides at specific times can rely on predatory mites to control pest mites (Box 5.1).

Box 5.1 Conserving Predatory Mites in Apple Orchards (or Mighty Mites Conquer Mite Pests)

The best targets for biological control in apple orchards in Pennsylvania are the secondary foliage-feeding pests that do not harm fruit directly and for which low population densities can be present without impacting apple yield or quality. Previously, the primary apple pests were several species of caterpillars and when these were sprayed with many kinds of harsh insecticides, populations of leaf-feeding mites and other pests such as scale insects and aphids could erupt to damaging densities as secondary pests. In the 1970s to 1990s pest-control systems were developed to take advantage of a native small black lady beetle, *Stethorus punctum*, which feeds on pest mites. The primary caterpillar pests could often be controlled using pheromones to disrupt mating and this approach meant that the lady beetles were not killed by pesticides. Relying on *S. punctum* for mite control reduced acaricide use by 50 percent and saved growers approximately US$20 million over a 15-year period.

However, the situation changed, especially once a new invasive, the brown marmorated stink bug, *Halyomorpha halys*, began seriously damaging apples. Also, by the late 1990s a number of new insecticides, including neonicotinoids, began being used. These were pretty toxic to *S. punctum*, but they allowed survival of mite predators which provided effective control. The king among the several species of predaceous mites in these orchards has been *Typhlodromus pyri*, which is a long-lived generalist feeder, persisting by feeding on pollen, other mites and fungal spores when the leaf-damaging mites are absent or at low densities. These mite predators also never leave trees. Other predatory mites are also present in many orchards, but do not give reliable pest mite control because they are only predators (not feeding on alternate foods) and they leave the tree when prey is scarce or they reproduce too slowly to control the pests. Conserving populations of *T. pyri* has been shown to be even more effective than *S. punctum* with over 90 percent reduction in acaricide use that can save on average of around $1 million per year for Pennsylvania apple growers. The use of mite predators rather than acaricides has also stopped the rapid development of acaricide resistance among the pest mites, which often limited the effective field use of new acaricides to only 2–3 seasons of use before they stopped working.

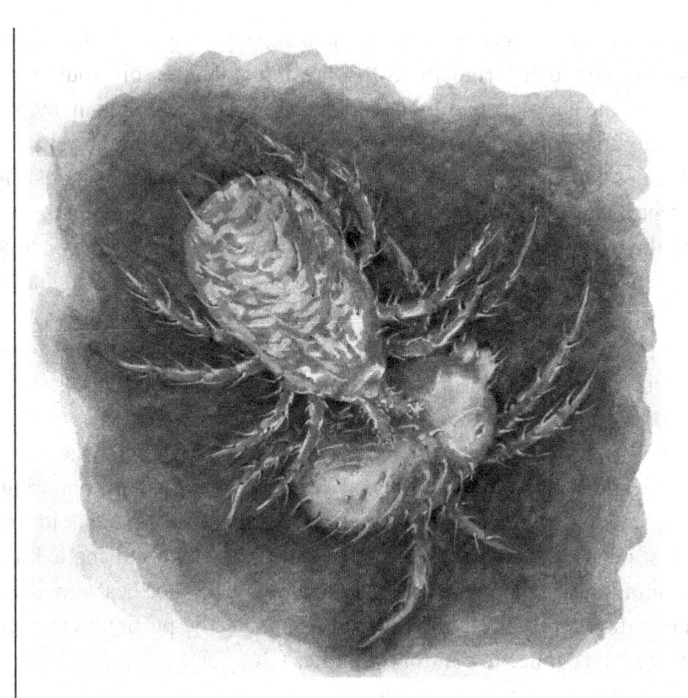

Figure Box 5.1.1 Predatory *Typhlodromus pyri* attacking a two-spotted spider mite, *Tetranychus urticae* (from a painting by Gašpar Vanek, courtesy of Tomi Vanek).

Apple growers, however, could not count on the efficacy of *T. pyri* without integrating other pest controls in their orchards. First, with help from Pennsylvania State University, growers must determine whether *T. pyri* is present and if it is not present, it can be introduced using infested shoots from "donor" sites either at the university orchards or from select grower sites. Second, once *T. pyri* is found or introduced, growers cannot use broad-spectrum pyrethroid or carbamate insecticides after apple trees bloom because these kill the predatory mites; instead selective insecticides and acaricides must be used. Pyrethroids and a few other insecticides are especially important to avoid because not only are they toxic to some natural enemies, they have a nasty side effect of mimicking hormones of phytophagous mites to make these pests develop faster and reproduce more. Apple growers who follow the guidelines for conserving *T. pyri* can save $50–100 per acre (0.4 ha) in acaricide costs alone. In addition, growers do not experience the losses in fruit quality and yield that they could have incurred from either not controlling the pest mites or having the control measures fail owing to acaricide resistance. This very successful biological control program has been adopted by most apple growers in Pennsylvania and many growers have not sprayed susceptible apple varieties for 10 years or more.

Pesticides can also be applied so that susceptible natural enemies are not exposed to them. Treating only part of a crop with a pesticide, for example, applications to alternate rows of apple trees, provides a habitat where predatory lady beetles feeding on phytophagous mites can survive. Some methods for applying pesticides can also protect natural enemies. Granular formulations that are applied to the soil would not affect natural enemies on the aboveground foliage. Systemic pesticides taken up by plants would not affect natural enemies that do not feed on the foliage, although systemics can also accumulate in pollen and nectar which are eaten by some natural enemies. However, as a warning regarding systemic pesticides, if the pests that are feeding on the plant are not affected by the systemic pesticide and it accumulates in their bodies, natural enemies that are sensitive and then attack these pests can be negatively affected via this indirect route.

In addition to avoiding natural enemy exposure in space, altering timing of treatments can also result in reduced exposure of natural enemies. Nonpersistent pesticides can be applied or applications can be made infrequently. Better yet, pesticides could be applied when natural enemies are not present during the season or when they are in a protected stage such as when they are still in protected overwintering locations or during pupation. For example, models were used to predict development of a parasitoid and its host, the cereal leaf beetle, *Oulema melanopus*, so that pesticides were applied in the spring before parasitoids had emerged. In apple orchards in the northeastern United States, growers integrating types of control and not using specific pesticides at specific times can rely on predatory mites to control pest mites (Box 5.1).

5.3 Enhancing Natural Enemy Populations

Habitats must provide resources needed by natural enemies, such as food, hosts or prey, shelter, and acceptable abiotic conditions to retain or attract active natural enemy populations. These are important provisions and a good start on how to increase the presence and build on the activity of natural enemies. In addition to maintaining provisions, it is also important to minimize disturbing the physical environments when possible, as with avoiding deep plowing that disrupts the habitats for natural enemies persisting in an area in the soil. Often crop habitats fail to provide constancy or necessary resources, or resources are not provided where or when natural enemies occur. To enhance natural enemy populations, we must learn what is necessary for retaining populations of important natural enemy species and devise methods to provide these limitations at the correct time and place. Therefore, understanding the biology and ecology of organisms in a system is critical. However, owing to the variability among species occurring in different systems, methods for natural enemy enhancement that are effective in one system are often not directly transferable to other systems.

Almost every enhancement strategy can be seen as some method for habitat manipulation to increase densities of natural enemy populations or to increase natural enemy effectiveness in controlling pests. In Chapter 2, we described constructions between trees in Chinese citrus orchards for dispersal of predaceous ants that were devised many centuries ago. Farmers have known for centuries that some habitats are more amenable to naturally occurring biological control than others and, when possible, have adjusted practices to take advantage of this.

A great diversity of methods have been investigated for enhancing natural enemy occurrence and effectiveness. However, enhancements as a result of many of these practices have not been effective enough for large-scale adoption or continued use, especially with large-scale agriculture. Today, newer and better methods for assessing effects of enhancements on natural enemy and pest populations are being tested and, in some cases, used. At present, the focus seems to be more on specialized and smaller-scale uses.

In fact, in recent years conservation biological control strategies have been receiving abundant interest. In theory, these strategies certainly are well suited to pest-management approaches integrating different control methods, such as integrated pest management, organic agriculture, and other types of sustainable agriculture (see Chapter 19). While enhancement strategies are more difficult to adapt to large monocultures in developed countries, they can be appropriate for smaller plantings. Certainly, organic growers and growers working toward decreasing pesticide use are very interested in methods for habitat management that will help them take advantage of the activity of resident or introduced natural enemies. However, methods must be tailored so they are still affordable because increasing the complexity of manipulations for production and control can also lead to additional costs, especially in increased manual labor, and this must be acceptable to growers.

5.3.1 Vegetational Diversity and Biological Control

In conservation biological control, manipulation of vegetational diversity to enhance natural enemies has been a major focus, based in part on the following findings. In 1973, Dick Root planted collard greens either in pure stands or in single rows surrounded by meadow vegetation. He found fewer herbivores and more natural enemies in the single rows. Based on his findings, he proposed a "resource concentration hypothesis" related to the dynamics of phytophagous arthropods which stated that "herbivores are more likely to find and remain on hosts that are growing in dense or nearly pure stands." This, of course helps explain associations between large acreages in monoculture and associated extensive pest problems. From practice we know that herbivorous pests are very able to find the crop plants that they need, especially when crops of their host plants are grown in large monocultures.

Associated with this same theme of vegetational diversity affecting plant–arthropod relations, based on his results Root proposed an "enemies hypothesis" stating that natural enemies would be more abundant in diverse plant communities. Thus, the extension from greater arthropod diversity in communities with mixed plant species was a resulting greater pressure by natural enemies on herbivores. The main message from these ideas was that increasing plant diversity was associated with decreased herbivore populations. This sounds encouragingly straightforward, but, when investigated in other systems, the same patterns are not always found.

While no one rule seems to explain the relations between vegetational diversity and natural enemy abundance, some studies have supported Root's 1973 proposals. A review of the literature showed that, among studies on field crops comparing monocultures with fields where several plant species were grown in the same field, parasitoids were more abundant in 72 percent of the cases with diverse plantings. One study clearly demonstrated that in systems with more vegetational diversity, by providing noncrop areas adjacent to fields of oilseed rape, parasitism of the pestiferous rape pollen beetles, *Meligethes aeneus*, was higher and crop damage was lower.

As a general rule, crop pests are less abundant as the biodiversity of a system increases. A review of 209 studies of 287 herbivorous arthropod species found that approximately 52 percent of the species had lower population densities when several crops were grown in a field (often called polyculture) instead of a single crop, while 15 percent of the herbivore species had higher densities in polycultures. Thus, while arthropod species had about a 50 percent chance of being less abundant in polycultures, there was also a chance, although lower, of more species of herbivorous arthropods feeding on plants in polycultures than in monocultures. What was causing this trend of the cases where herbivore numbers were lower in more diverse habitats? It is not the plant diversity per se that leads to lower pest numbers through increased natural enemies, but the resources that habitat diversity provides so that natural enemies can persist in these habitats.

In fact, sometimes vegetational diversity provides needed resources for natural enemies and other times it does not, as shown by the occasional occurrence of higher densities of herbivores in polycultures. In more diverse systems, parasitoids and predators might have more difficulty finding prey and hosts and then having a diversity of plants could benefit the herbivore. Alternatively, diverse systems could support greater populations of the natural enemies that kill the parasitoids and predators by providing shelter and alternate food. Unfortunately, systems can be idiosyncratic and must be considered separately to understand whether specific manipulations hinder or favor pests. While it is clear that vegetational diversity can have a profound effect on herbivores and natural enemies, at present there seems to be no overarching theory that consistently explains the relative importance to density of pests and natural enemies in simple versus diverse systems. Because relations seem to be system specific, the trick is to discover what resources are limiting natural enemies and determine how these can be added to the system.

Polyculture in Use

Natural enemy populations can be enhanced within a crop by planting several totally different types of crop plants within the same field. This practice of polyculture, which includes intercropping, has been used in over 2.3 million ha in northern China to reduce damage in cotton from cotton aphids. When cotton and wheat are interplanted, natural enemies are maintained in wheat fields. Wheat grows first and natural enemies feed on prey in wheat, but, as the cotton plants subsequently grow, the natural enemies then move into cotton when prey are present. Without wheat as a nearby alternative habitat, once cotton begins to grow predators will eventually arrive but usually they arrive too late to control aphid populations adequately.

Traditional crops for resource-poor Mexican and Central American farmers are frequently corn, beans, and squash, interplanted in the same small fields for subsistence.

With this diverse vegetation, predatory ants feeding on a broad diversity of prey have been reported maintaining control of a variety of pest species. Studies in Nicaragua and Mexico documented that several species of predatory ants were responsible for controlling fall armyworm, *Spodoptera frugiperda*, and corn leafhopper, *Dalbulus maidis*, on corn foliage and corn rootworm (*Diabrotica* spp.) eggs in the soil. When ants were excluded from plots, there was much higher crop damage.

5.3.2 Enhancing Habitats for Natural Enemies Within a Crop

When crops are grown as simple monocultures, as is common practice in large-scale agriculture, the resources required for persistence of natural enemies are often lacking. Food and shelter especially can be minimal after harvest and before a field is replanted. This situation can be altered in various ways, from providing shelter with favorable microclimatic conditions to providing alternate food, including nectar, pollen, and alternate hosts or prey. When crop habitats are transient, populations of natural enemies cannot be retained from year to year, much less increase over time. Management of crops can be altered in a great diversity of ways to preserve and enhance natural enemies. Only some of the methods that have been investigated are actually in use, perhaps because few methods have been shown to reliably suppress pest populations with an adequate decrease in pest damage. However, we will describe some of the diversity of manipulations that have been tested, including a few examples where methods are presently in use.

Providing Refuges Within a Crop

Construction of natural areas that occupy limited areas within fields provides an excellent example of successful control that has been accepted by growers. One of the most successful applications of conservation biological control has been the establishment of permanent strips of natural vegetation within crop, so-called "beetle banks." Beetle banks were developed as a cost-effective way to create favorable habitats for predatory invertebrates within fields by providing "islands" of diversity. Under standard agricultural practices in the United Kingdom, edges of fields often provided little overwintering habitat for predatory invertebrates. With very large agricultural fields, it could take a long time before the predators that survived the winter at the field edge or in some permanent vegetation in the general area invaded crops and they often would not travel to the centers of large fields, let alone feed there for prolonged periods. Beetle banks are constructed by creating a ridge or bank of earth, about 0.5 m high and 1.5–2 m wide, the majority of the length of a field, using two-directional plowing (Figure 5.2a). After establishing a ridge, it is planted in the autumn or the following spring with tussock- or mat-forming perennial grasses, such as cock's-foot (Figure 5.2b). The dense structure of this type of plant provides habitat for predatory invertebrates and, once established, excludes most weeds. In south-central England, densities of predators increased to more than 1,600 per m^2 after two years for some beetle banks; the most common predators were ground beetles and spiders. Studies in the United Kingdom showed that one beetle bank in a 20 ha field facilitated radiation of predatory ground beetles from beetle banks

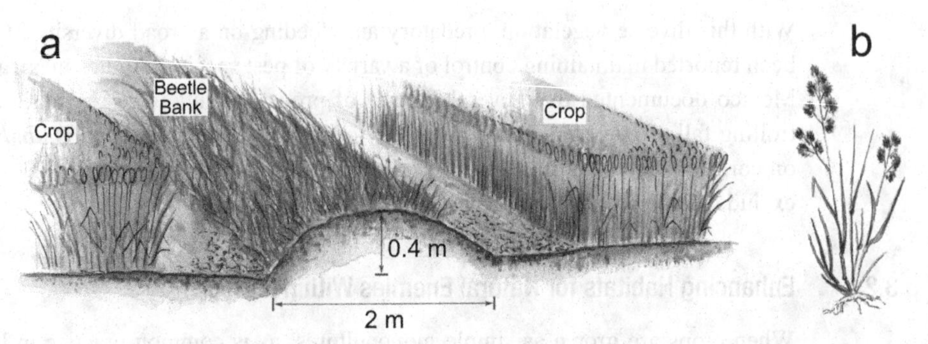

Figure 5.2 (a) Dimensions of grass ridges called "beetle banks" that create a permanent habitat for natural enemies of crop pests in cereal fields. (b) Cock's-foot grass, *Dactylis glomerata*, a tussock- and mat-forming perennial grass recommended for use in beetle banks because it provides shelter for natural enemies (courtesy of the Game & Wildlife Conservation Trust, UK).

in spring, resulting in a uniform distribution within cereal crops after the bank had been established for three years. Beetle banks have been tried in several crops but seem most effective for controlling aphids in cereals. This is an example in northern Europe of widespread grower adoption of habitat manipulation and this practice has become so well known that the term "beetle bank" is included in dictionaries. These man-made habitats strengthen natural controls in the fragmented and unstable environments created by intensive farming, along with reducing pesticide use. Decreased pesticide use is thought to outweigh costs of lost agricultural production and as an added bonus, beetle banks can provide nesting sites for vertebrates. However, a test of beetle banks in the northwestern United States did not find increased biological control and further study of this method in other regions should be conducted before adoption.

Providing Alternate Food

Augmentation biological control has been well developed for use in greenhouses (see Chapter 4) and, as one part, use of banker plants facilitates survival and persistence of natural enemies in greenhouses. Greenhouses are purposefully empty of insects when a new crop is started so if natural enemies are released when pests are absent or at low densities, it can be difficult for them to survive. Banker plants hosting alternate nonpest hosts and natural enemies are often placed in greenhouses before pests are present to avoid increases in pests once they arrive. The systems are usually very specific to certain major pests, with specific numbers of plants recommended per area of greenhouse, hosting alternate prey or hosts that will not infest the main crop. Banker plants can also provide nectar and pollen to natural enemies that can use these alternate foods.

Many studies have been conducted investigating how growing flowers within crops can help to feed natural enemies. In particular, this makes sense for holometabolous natural enemies, where the immature stages are predators or parasitoids and not as well adapted for dispersal while winged adults disperse and are not predatory but instead feed on nectar or pollen. However, hemimetabolous species, especially predatory bugs (Hemiptera) like mirids and anthocorids, also feed on pollen, nectar, and plant sap.

We now can analyze gut contents of predators and parasitoids and it seems that when flowering plants are not present, predators and parasitoids are often starving. When buckwheat (*Fagopyrum esculentum*) flowers were present in field cages, the parasitoid *Diadegma semiclausum* lived 28 days and actively parasitized hosts while without buckwheat flowers, these parasitoids only lived for one day. However, flowers from different plant species differ widely regarding their use by natural enemies and the flowers preferred by pollinators are often not those preferred by natural enemies.

When possible, plant species planted to provide flowers must be carefully chosen so that they are used by the primary natural enemies and are not used by pests or parasitoids of the natural enemies. Thus, judiciously adding flowers within fields can increase populations of predators and pollinators, and this is actively being done by organic lettuce growers in California (Box 5.2).

Box 5.2 Flower Power

Conservation biological control experts have tested many methods for supplying food for natural enemies but there appears to be a breakthrough. In organic lettuce crops in California, Eric Brennan has been experimenting with planting sweet alyssum, *Lobularia maritima*, along with lettuce to provide food for hoverfly (Syrphidae) adults.

Adult hoverflies often have coloring so that they mimic bees, and they gain some protection through this mimicry. The adults must eat to keep up their energy for all of the hovering, mating, and then laying eggs, and they usually eat nectar and pollen. The small larvae we are interested in are voracious insectivores, eating aphids, thrips, and other phloem-feeding insects, but the larvae don't travel very far, so we need the mobile adults to lay eggs on plants where there is food for the larvae. The lettuce aphid, *Nasonovia ribisnigri*, can develop to high population densities within heads of romaine lettuce in California. Because of their protected habitat within the lettuce

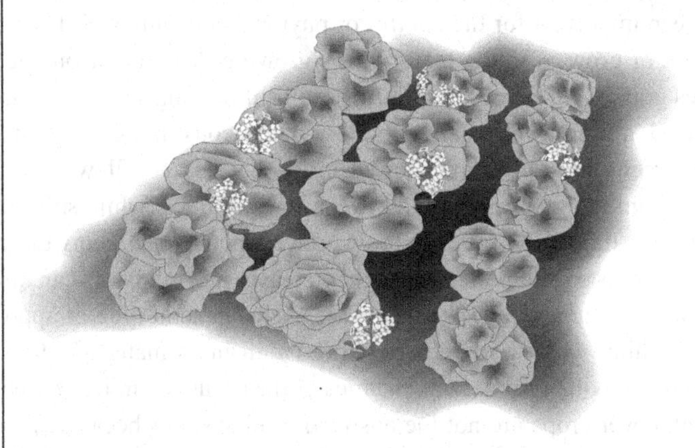

Figure Box 5.2.1 Lettuce interplanted with sweet alyssum (*Lobularia maritima*) to provide food for natural enemies in organic production fields in central California (drawn by Margaret C. Nelson).

heads, these aphids are difficult to control for conventional growers but especially for organic growers. However, larvae of numerous species of hoverflies make their way down inside of the lettuce heads and are excellent at finding and eating lettuce aphids. Methods have been developed for helping out the hoverflies in organic lettuce fields.

To provide food so that adult hoverflies will remain and lay eggs in romaine lettuce crops, researchers first compared different species of alyssum, a genus of flower preferred by hoverflies. Sweet alyssum plants produced the most flowers and these were transplanted so that entire rows were composed of sweet alyssum. This was called the replacement strategy because a significant portion of fields that could be producing lettuce were instead replaced by flowers. So, after it was evident that the replacement strategy was working, an additive strategy was tried instead, where sweet alyssum plants were added to normal rows of lettuce. This worked surprisingly well and meant that fields were still maximally planted with lettuce. This strategy works well for lettuce that has smaller heads; the issue is whether the lettuce plants with larger heads will compete with the sweet alyssum plants and whether this will be problematic. Therefore, intercropping with sweet alyssum is now being tested for potential use in fields with types of lettuce that grow larger heads, to optimize both use of sweet alyssum plants and production of organic lettuce.

Unfortunately, planting the flowering plants preferred by natural enemies does not seem to always work well within beetle banks as the main tussock-forming plants used for beetle banks outcompete the flowering plants. However, flowering plants could always be added elsewhere within a field containing beetle banks.

Cover Crops

A dense plant canopy can also improve natural enemy populations by providing a sheltered microhabitat within the crop. Cover crops in citrus orchards in Queensland, Australia, have been important for the control of phytophagous mites. Between 80 percent and 95 percent of growers in the major citrus-growing districts encourage the flowering of Rhodes grass (*Chloris gayana*) during the fruit-bearing season because the grass pollen produced is used as alternative food by predaceous mites. To do this, alternate inter-rows between citrus trees are mowed every 3 weeks to allow time for production of pollen from the grass growing between rows of trees, while still maintaining a neat orchard. In addition, 30–50 percent of growers plant *Eucalyptus* trees with hairy leaves in wind breaks so that pollen is caught on leaves and predators can build up long-term populations in these refuges near orchards. This general type of approach is also widely used in China, where cover crops are present in an estimated 135,000 ha of citrus orchards to provide pollen for natural enemies of the citrus red mite, *Panonychus citri*.

As a caveat, cover crops are not the answer for all systems because, in some cases, cover crop plants can compete with the crop plants and decrease yield. Cover crops can also provide resources for pests. In peach orchards, ground covers are often eliminated because they provide resources used by true bugs (Hemiptera) that feed on the peach flowers and fruit, resulting in scarring on peaches known as "cat-facing" (the surface of the fruit at harvest resembles the face of a cat).

Crop Residue Management

Many parasitoids and predators inhabit crop residues after harvest and burning or removing these residues can decimate natural enemy populations. Residues can be left in the fields, at least in part, to help conserve natural enemy populations. Several parasitoids attacking the sugarcane leafhopper, *Pyrilla perpusilla*, in India can effectively control pests if crop residues are not burned but are spread back onto fields.

Crop Management

In California, both alfalfa (i.e., lucerne) and cotton are commonly grown and the pestiferous lygus bugs, *Lygus hesperus*, feed on both, although they prefer alfalfa. Alfalfa is harvested several times each year and when an entire field of alfalfa is mowed during hot weather the lygus bugs leave within 24 hours. Often, they leave alfalfa and move to cotton where they can cause substantial damage leading to pesticide applications. This problem is clearly a result of harvesting practices so new harvesting practices were devised. If alfalfa is cut in alternating strips, then lygus will migrate not to cotton but to the noncut alfalfa strips that are nearby. Thus, chemical pesticides did not need to be sprayed on cotton. This practice also preserved the resident natural enemies in alfalfa because they moved to the nonharvested strips along with the lygus bugs. Unfortunately, this strategy was never widely adopted by growers because it was more expensive than standard practices. A strategy to interplant alfalfa with cotton was also proposed but this posed difficulties because these two crops have different water requirements and modifications of the water system and extra labor to cut alfalfa did not compensate for the reduced pesticide use. However, due to pesticide resistance and the fact that more insecticides are applied to cotton than any other crop, it has been suggested that these practices should be reconsidered.

During a monumental study in the hot Central Valley of California, Schlinger and Dietrick investigated whether there really were more natural enemies when alfalfa was strip-harvested. They had chosen a crop with an incredible biodiversity, there being more than a thousand different insect species in an average unsprayed field of alfalfa in California. They sampled 4.2 m^2 of alfalfa every two weeks and showed that all natural enemies except green lacewings were more abundant in strip-harvested fields (Table 5.2). In summary, strip-harvested alfalfa had four times as many natural enemies as regularly harvested fields. Insecticides were not needed on strip-harvested alfalfa but had to be applied twice to the regularly harvested field.

Plant Characteristics

If monocultures are grown, care can be taken to use cultivars of plants that enhance natural enemies. Extensive studies in greenhouses showed that the whitefly parasitoid *Encarsia formosa* (see Figure Box 8.5.1, p. 186) is very effective on numerous vegetable crops, but it consistently did poorly against whiteflies on cucumbers. A main

Table 5.2 Natural enemies associated with strip- versus regularly harvested alfalfa throughout the field season, Kern County, California.

	Strip-harvested	Regularly harvested
Spiders	1,094,000	105,000
Parasitic wasps	287,000	70,000
Big-eyed bugs (predators)	401,000	199,000
Lady beetles	437,000	57,000
Green lacewing larvae	206,000	195,000

Source: from Schlinger & Dietrick (1960); © Regents, The University of California)

reason behind the poor activity was that cucumbers are a good host plant for greenhouse whitefly (*Trialeurodes vaporariorum*) so the pest grows very fast when feeding on it. However, of equal importance, the cultivar of cucumber that was regularly planted had relatively large leaf hairs (trichomes) that reduced the walking speed of the parasitoid. The hairs also caught the sticky honeydew from the whiteflies (the sugar-rich liquid excreta produced from whiteflies, aphids, scale insects, and mealybugs) and if the parasitoids contacted the honeydew, they would become stuck to it. This situation was solved by breeding cucumbers to develop a cultivar with half of the leaf hairs compared with the commercial cultivar.

In contrast to the negative impact of leaf hairs described above, leaf hairs can enhance the abundance of beneficial phytoseiid and tydeid mites. Phytoseiid mites are predaceous and are used as agents for control of plant-feeding mites. In perennial cropping systems, persistence of these predators is a key to successful biological control, and abundance and persistence of phytoseiids is often greater on plant species and plant cultivars that have many leaf trichomes. These trichomes provide shelter that protects the beneficial mites from other predators, especially when trichomes are aggregated at junctions of leaf veins where they create hiding places called domatia (Figure 5.3). Trichomes also enhance the capture of pollen that often is eaten as an alternative food by predators if pests are not present. Furthermore, when on plants with leaf trichomes, predatory mites are less prone to leave the plant. On some crops such as plums and pears that lack leaf trichomes, mite biological control has generally not been successful. In contrast, on apples, where most cultivars have abundant leaf trichomes, mite biological control can be very successful (see Box 5.1). Among grape cultivars, there is wide variability in leaf trichomes and data suggest that the success of mite biological control is at least in part associated with variation in trichome density.

Other types of mites, tydeid mites, eat fungi (i.e., are fungivorous) and have been shown to suppress the plant diseases called powdery mildews, which are fungal pathogens that predominantly live on leaf surfaces. The effectiveness of tydeid mites in feeding on powdery mildews often hinges on the presence of tufts of leaf trichomes at leaf vein junctions that are used as refuges (Figure 5.3b). On some plants lacking these domatia, tydeid mites are scarce and mildew is not suppressed.

The waxiness of plant leaves can vary by plant cultivar and this surface characteristic has been shown to affect natural enemies attacking herbivores. Predatory insects

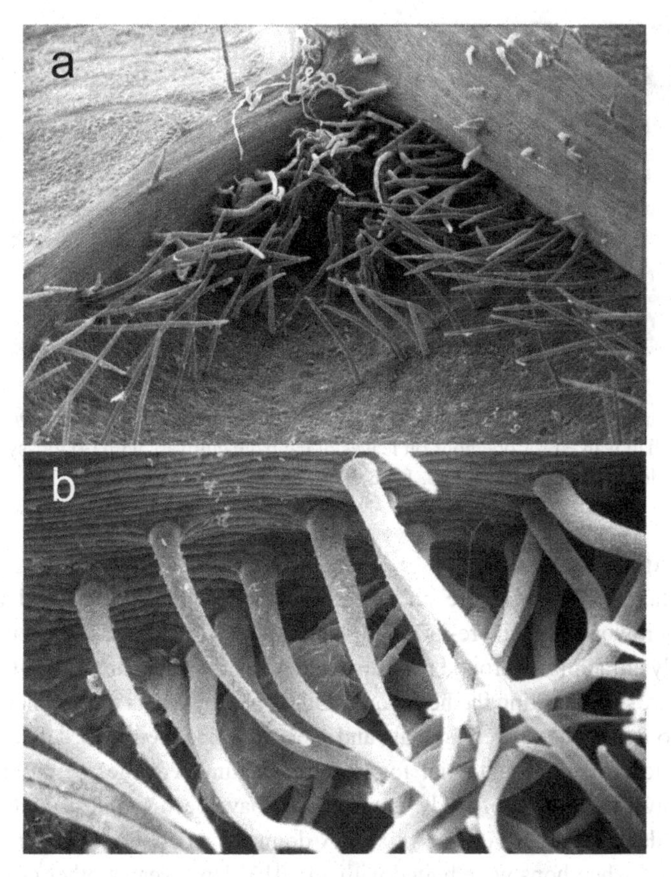

Figure 5.3 (a) Domatium at the intersection of veins on a leaf of wild river grape, *Vitus riparia*. (b) Tydeid mite, *Orthotydeus caudatus*, among leaf hairs within the domatium. The mite is approximately 0.2 mm long (from Agrawal, 2000; photos by Andrew Norton and Harvey Hoch).

were released on cabbage cultivars with reduced amounts of wax covering the leaves. Adult lady beetles, insidious flower bugs (*Orius insidiosus*), and larval green lacewings ate more diamondback moth (*Plutella xylostella*) larvae when hunting on low-wax cabbage leaves. These predators required much more time walking on waxy leaves because wax particles attached to their feet and they spent time either scrambling for attachment or grooming (Figure 5.4). Waxy coverings on leaves also affected pea aphids (*Acyrthosiphon pisum*), as more pea aphids were infected with the fungal pathogen *Pandora neoaphidis* on low-wax pea cultivars; companion studies suggested a mechanism with fewer of the spores of this entomopathogenic fungus adhering to waxy leaves so there was less inoculum to infect aphids on waxy-leaved plants.

At present, the plant cultivars favoring natural enemies described above, grape cultivars with increased hairs on leaves and cabbage and pea cultivars with glossy leaves, are not being used by growers. The next research steps should include field trials to demonstrate whether these cultivars that are beneficial to natural enemies will not be more susceptible to other pests and will be equally productive under field conditions.

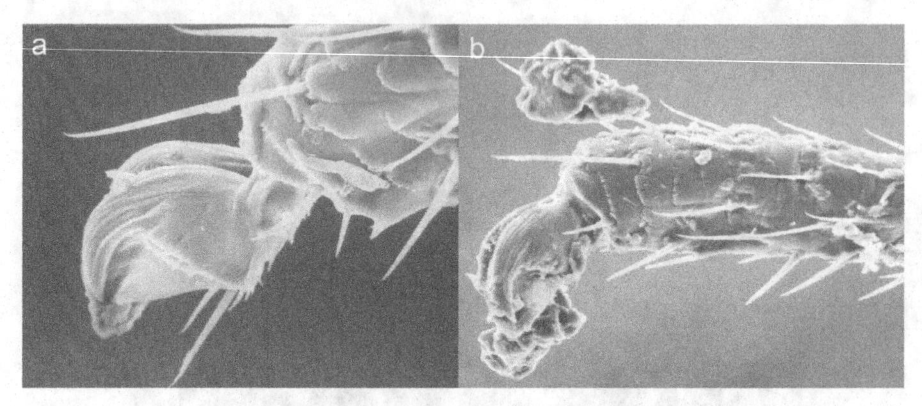

Figure 5.4 Scanning electron micrograph of the tarsi of adult insidious flower bugs, *Orius insidiosus*, that had walked for one hour on leaves of cabbage cultivars with (a) glossy leaves or (b) normal wax leaves (Eigenbrode et al., 1996).

Using Messaging by Plants

We are constantly learning more about the volatile chemicals produced by plants attacked by herbivores; herbivore-induced plant volatiles (HIPVs) act as plant distress signals, mainly attracting arthropod predators and parasitoids that arrive and function as plant bodyguards by attacking the herbivores. Other, nearby plants that are not yet being damaged, also detect the distress signals and then mobilize their own defenses. These HIPVs seem to have great potential for attracting natural enemies to protect plants. This is a relatively new field of study. Some HIPVs have been synthesized and used in crops and, to date, results have been mixed. A dramatic reduction in spider mites and aphids occurred when hops were baited with an HIPV. However, in other crops, sometimes additional herbivores or natural enemies killing the primary predators and parasitoids have been attracted, which is directly opposite to the desired result.

It seemed like sometimes HIPVs were attracting natural enemies which then arrived but they needed to have something to eat in order to remain in the area. Researchers have been investigating releasing HIPVs as well as planting flowering plants like buckwheat that are often preferred by natural enemies feeding on nectar and pollen. This strategy has been called "attract and reward" and sometimes results in higher densities of parasitoids compared with HIPVs alone.

Soil

Soil can frequently function as a reservoir for natural enemies but we know much less about what goes on underground than above ground. Suppressive soils that control plant pathogens and plant parasitic nematodes are well known and are now being investigated more fully (see Chapter 17) and soils can sometimes be manipulated to build suppressiveness in nonsuppressive soils, for example, by planting the same crop in the same fields for numerous years.

The soil can act as a reservoir for microorganisms, including pathogens of invertebrates. Research has shown that tillage of the soil can alter the ability of these soil-borne insect pathogens to infect pests. When fields are not tilled, spores of fungi infecting insects are at much greater abundance at the soil surface where they will contact pests but, with tillage, the spores can become buried where few hosts are present. In a similar example, velvetbean caterpillars (*Anticarsia gemmatalis*) feeding on grass in pastures are more abundant when fields are tilled because the viruses infecting and killing them are buried during tillage. Soil also serves as a reservoir for arthropod natural enemies. Numerous parasitoids utilize the soil for pupation and their populations are lower in tilled crops than untilled. For example, densities of parasitoids of rape pollen beetles (*Meligethes aeneus*) were 50–100 percent lower when rape fields were plowed.

Physical Environment

The physical environment strongly influences the activity of natural enemies. Application of water has been used to improve the microclimate within crops and enhance pathogens of insect pests. In greenhouses, activity of the fungus *Lecanicillium lecanii* can be enhanced by watering and providing nighttime temperatures that yield the high humidities necessary for optimal infection of insect hosts. Altering plant density also can be used to increase humidity in the microclimate occupied by pests and pathogens. Infections by the fungus *Metarhizium* (formerly *Nomuraea*) *rileyi* in three species of caterpillars were greatest when soybeans were planted early, in narrow rows with high seeding rates so that the plant canopy closed early, thus increasing the relative humidity in the microhabitat occupied by both host and pathogen. Another fungal pathogen *Zoophthora phytonomi* and its alfalfa weevil host (*Hypera postica*) have been manipulated to enhance disease epizootics; if alfalfa is cut early but left in rows, alfalfa weevils aggregate within the rows of harvested alfalfa. The microclimate within the windrows sitting in the sun is warm and humid and under these conditions this fungus causes high levels of infection among the crowded beetles. Models demonstrated that early-season insecticide decision thresholds, early harvesting, and relying on this fungus to decimate weevil populations instead of applying insecticides could increase profits by as much as 20 percent. This program was validated during field trials and recommended for alfalfa weevil management in the US state of Kentucky.

Providing Artificial Food for Natural Enemies

As we've discussed, within crops or alongside crops, vegetation can be manipulated to foster natural enemy populations through planting specific plants to provide nectar and pollen or alternate hosts or prey for natural enemies. However, food can also be directly supplied to bolster natural enemy populations. Natural enemies require carbohydrates for energy and protein for growth and reproduction and these nutrients can be limiting in simplified landscapes, like monocultures. Artificial food supplements or artificial honeydews have been developed with the goals of increasing immigration and reducing emigration of natural enemies, as well as increasing their fitness when they are present. For example, to bolster populations of green lacewings eating aphids in cotton crops, Ken Hagen applied a mixture of protein hydrolysate, water, and sugar to enhance green lacewing reproduction. In addition, studies have suggested that artificial application of pollen can increase population densities of predatory mites and thereby increase the impact of these predators on pestiferous mite species. Other studies have looked at applying artificial honeydew to provide food for parasitoids and to aggregate lady beetles. However, an analysis of results from 234 trials testing artificial food sprays demonstrated that inconsistent results were probably responsible for the minimal actual use of this strategy for control. In fact, in 87 percent of trials, natural enemy populations increased and in almost half of the trials (47 percent) pest populations decreased, but overall results suggested that consistency in results and integration with other methods need to be achieved before greater adoption is likely.

5.3.3 Enhancing Habitat for Natural Enemies: Using the Area Around the Crop

One of the best-known strategies for conserving and enhancing natural enemies is to provide "wild insectary" areas at the edges of fields of cultivated plants. These areas of "companion plants" can serve to provide food and shelter when there is no crop in the field or if the crop does not provide the resources needed by the natural enemy. These areas are much more effective if they are present over the long term so that natural enemy populations can build and persist there. One well-known example is the use of flowering plants along the edges of agricultural fields to provide nectar for parasitoids and pollen for predators. To this purpose, in Switzerland, "weed strips" of native flowering plants are frequently maintained in and around fields. Densities of numerous predators (ground beetles, predatory flies, damsel bugs, and spiders) increase when weed strips are present and this has been documented as promoting pest control and crop yield. As another example, so-called "insectary plants" are planted as cover crops between rows in vineyards in Europe, California, and New Zealand. Among the plants used for this purpose is lacy phacelia, *Phacelia tanacetifolia* (Figure 5.5), a plant from dry areas in the southwestern US and northern Mexico that grows quickly and flowers for a long time. These flowers are very attractive to natural enemies as well as pollinators. However, as mentioned earlier, all flowering plants providing food are not equivalent resources for all natural enemies. To test this, *Meteorus autographae*, a parasitoid wasp attacking numerous species of caterpillars was collected from cotton crops bordered by botanically diverse plants for bird conservation and these individuals were almost starving. In contrast, individuals of this parasitoid species collected from cotton fields bordered by flowering vetch (*Vicia*) had high energy reserves. Therefore, weedy areas at the sides of fields can assist with maintaining or increasing natural enemies but, once again, the types of weeds make a difference.

Several species in the plant genus *Euphorbia* naturally grow as weeds around sugarcane fields in Hawaii. These plants provided nectar and mating sites for adults of a tachinid fly (*Lixophaga sphenophori*) that parasitizes the sugarcane weevil, *Rhabdoscelis obscurus*. When herbicides were applied to ditch banks and field edges, these plants were all killed and there was a correspondingly great decline in the parasitic fly populations. Once it was recognized that these flowering weeds were important to naturally occurring control of these weevils, growers altered herbicide applications to spare these weeds.

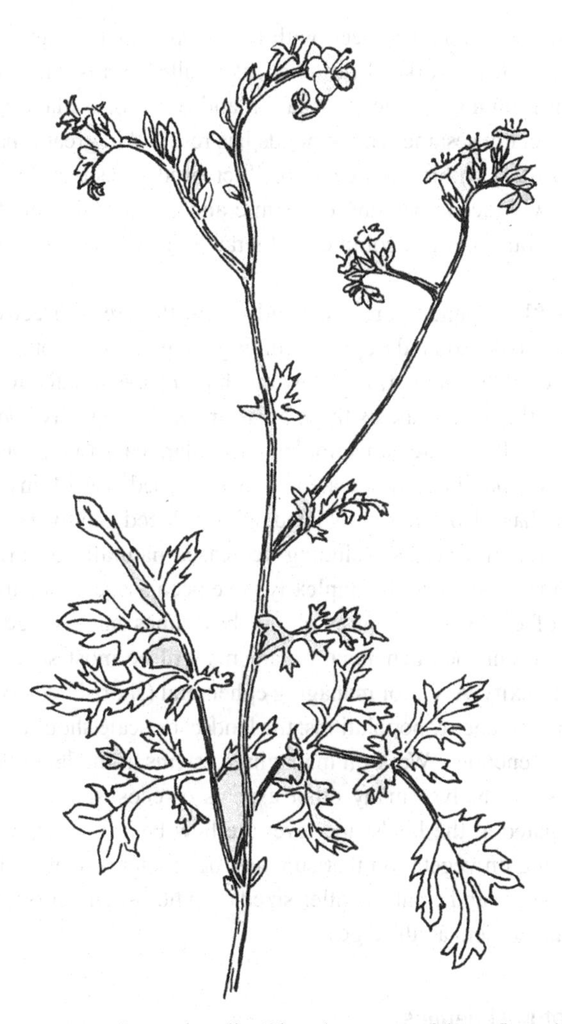

Figure 5.5 Lacy phacelia, *Phacelia tanacetifolia*, a flowering plant used to provide nectar and pollen alongside fields of crops (Leake et al., 1993; © John Benjamin Leake and Marcelotte Leake Roeder, reprinted by permission of the University of Oklahoma Press).

In California vineyards, the pestiferous grape leafhopper, *Erythroneura elegantula*, can be successfully controlled by the tiny egg parasitoid *Anagrus epos*. However, this parasitoid cannot overwinter in grape leafhoppers because the grape leafhoppers spend the winter as adults. Finding high levels of parasitism near streams, researchers discovered that this egg parasitoid can overwinter in the eggs of blackberry leafhoppers, *Dikrella californica*, occurring in this habitat. Because streams did not run through all vineyards, a clever alternative was found. This parasitoid could also overwinter in eggs of the prune leafhopper (*Edwardsiana prunicola*) that feeds on leaves of French prune trees. Thus, when French prune trees occurred alongside vineyards, the parasitoids could overwinter in the immediate vicinity of vineyards and lower leafhopper populations resulted.

Perhaps the greatest success story has been with rice crops in China and Vietnam. Rice paddies are surrounded by networks of earthen banks, called bunds. The bunds provide conduits between surrounding nonrice vegetation and rice crops. Planting nectar-producing plants, but especially sesame, on the bunds has resulted in greater parasitism of planthoppers and leafrollers within the rice crops. Increased parasitism has resulted in reduced pesticide use with accompanying economic and ecological benefits. As of 2012, planting sesame on bunds is a recommended strategy in China for suppressing planthopper populations.

Researchers have recently expanded areas of study from the areas directly around individual crops (the local scale) to landscapes, including the greater surrounding area. While local and landscape scales interact, we know much more specifically about complexity and functioning at the local scale as these areas are easier to study. On a much larger landscape scale, agricultural intensification, in particular, leads to increased landscape simplification, which has been associated with increased use of insecticides. Landscape simplification has also been associated with reduced biodiversity, which results in decreased ecosystem services including decreased naturally occurring pest control. Landscapes with more structural complexity have been associated with greater pest suppression. Tests of ecological theories along these lines have suggested that manipulation of habitats to enhance beneficial organisms will be most successful for habitats of moderate complexity. Also, for managing ephemeral (short-lived) systems to improve persistence of natural enemy populations, the landscape scale should be considered as a source of natural enemies. Work on managing systems at the landscape scale to enhance pest suppression involves many other aspects of ecosystems. At present, some ideas to be investigated at the landscape level are how best to manage noncrop habitats, especially focusing on vegetation that supports beneficial insects and reduces pests, and how best to preserve and create smaller-sized crop fields. Of course this work will require engaging stakeholders at all stages.

5.3.4 Providing Shelter for Natural Enemies

Protective habitats have been tested in numerous systems to enhance natural enemy populations. These can serve as sheltered resting locations when the natural enemies are active but can also provide longer-term shelter during the winter. Some examples of such structures are polyethylene bags provided as nesting sites for predatory ants in cacao plantations in Malaysia, boxes for wasps and overwintering lacewings, empty cans in fruit trees for earwigs, straw bundles for spiders in early-planted rice and cotton, or leaf litter around tree trunks as overwintering sites for predatory mites in apple orchards.

5.4 Conservation Biological Control Today

Conservation biological control is the biological control strategy that has seen the least practical use, although over the last decade interest and research has increased significantly. For classical biological control, numerous very useful databases have been maintained documenting releases and successes. For augmentative biological control, sales figures can provide some objective measure of use. It is more difficult to document the results of conservation biological control programs. As discussed, there are two major types of control used within this strategy: conservation of natural enemies and enhancement of natural enemies. In fact, decreasing pesticide use to preserve natural enemies can be readily quantified as savings in pesticides not purchased and applied. Using data from rice and apple crops, we know that reduced pesticide-induced mortality of natural enemies has aided control programs integrating different methods (see Chapter 19). Results from the diverse methods for enhancing natural enemy populations are often not as easily quantified yet habitat manipulations to enhance natural enemies are being implemented operationally in some systems. In greenhouses using biological control, banker plants providing alternate food for natural enemies are used. A current example of large field usage of conservation biological control and success is use of the ground-cover chick weed, *Ageratum conyzoides*, as well as other weedy ground covers, in 135,000 ha of citrus orchards in China. These various ground covers are beneficial for natural enemies of citrus red mite, *Panonychus citri*, in part by providing pollen, an alternate food that maintains populations of predatory mites.

Conservation biological control is in general not considered a "stand alone" type of control strategy. Thus, emphasis on use of conservation and enhancement of existing natural enemy populations as part of integrated pest-management programs or as ecologically based, sustainable approaches to agriculture seem the most appropriate implementation for this strategy. In one recent example, in Spain, the South American tomato pinworm or leafminer, *Tuta absoluta*, was introduced and during its first year in one area, it decimated tomato crops, sometimes resulting in zero marketable tomatoes. However, over the next 2–3 years, in some areas the damage from *T. absoluta* declined significantly. Growers had learned about the biology and behavior of this new invasive pest as well as integrating pest-management strategies but scientists had also monitored for natural enemies and found that native species had moved in and were controlling this pest in some areas. The impact of native natural enemies, which was named "fortuitous biological control" (but which we would also call natural biological control), was considered a key factor in the decline of this new invasive pest.

Further Reading

Cloyd, R. A. (2012). Indirect effects of pesticides on natural enemies. In *Pesticides – Advances in Chemical and Botanical Pesticides*, ed. R. P. Soundararajan, pp. 127–150. Rijeka, Croatia: InTech.

Gurr, G. M., Wratten, S. D., & Altieri, M. A. (2004). *Ecological Engineering: Advances in Habitat Manipulation for Arthropods*. Ithaca, NY: Cornell University Press.

Gurr, G. M., Wratten, S. D., Landis, D. A., & You, M. (2017). Habitat management to suppress pest populations: Progress and prospects. *Annual Review of Entomology*, **62**, 91–109.

Gurr, G. M., Wratten, S. D., Snyder, W. E., & Read, D. M. Y. (eds.) (2012). *Biodiversity and Insect Pests: Key Issues for Sustainable Management*. Chichester, UK: Wiley-Blackwell.

Jonsson, M., Wratten, S. D., Landis, D. A., & Gurr, G. M. (eds.) (2008). Conservation biological control. *Biological Control*, **45**, 172–280.

Landis, D. A. (2017). Designing agricultural landscapes for biodiversity-based ecosystem services. *Basic and Applied Ecology*, **18**, 1–12.

Landis, D. A., Wratten, S. D., & Gurr, G. M. (2000). Habitat management to conserve natural enemies of arthropod pests in agriculture. *Annual Review of Entomology*, **45**, 175–201.

Letourneau, D. K., Jedlicka, J. A., Bothwell, S. G., & Moreno, C. R. (2009). Effects of natural enemy biodiversity on the suppression of arthropod herbivores in terrestrial ecosystems. *Annual Review of Ecology, Evolution and Systematics*, **40**, 573–592.

Lu, Z. X., Zhu, P. Y., Gurr, G. M., Zheng, X. S., Read, D. M. Y., Heong, K. L., Yang, Y. J., & Xu, H. X. (2014). Mechanisms for flowering plants to benefit arthropod natural enemies of insect pests: Prospects for enhanced use in agriculture. *Insect Science*, **21**(1), 1–12.

Nilsson, U., Porcel, M., Świergel, W. & Wivstad, M. (2016). Habitat manipulation – as a pest management tool in vegetable and fruit cropping systems, with the focus on insects and mites. SLU (Swedish University of Agricultural Sciences, Uppsala, Sweden), EPOK – Centre for Organic Food & Farming. Retrieved from http://orgprints.org/30032/1/biokontrollsyntes_web.pdf.

Woltz, J. M., Werling, B. P., & Landis, D. A. (2012). Natural enemies and insect outbreaks in agriculture: A landscape perspective. In *Insect Outbreaks Revisited*, ed. P. Barbosa, D. K. Letourneau & A. A. Agrawal, pp. 355–370. Chichester, UK: Wiley-Blackwell.

Part III

Biological Control of Invertebrate and Vertebrate Pests

6 Ecological Basis for Use of Predators, Parasitoids, and Pathogens to Control Pests

Ecology may be the most intractable legitimate science ever developed.

(Slobodkin, 1988)

The ultimate goal of biological control is to manipulate systems to maintain populations of potential pests at low densities and thus prevent problems that they might cause. It follows that biological control in ecosystems can be thought of as a type of applied population ecology. In fact, natural enemies and their hosts have been the basis for many ecological studies investigating how populations are regulated. The field of ecology has gained from such studies which have been the basis of ecological theories. There is also a strong movement to use insights gained from ecological theory to help increase the success of biological control.

Different natural enemies to be used for biological control, as well as the diverse pests they attack, can have very different biologies. A basis for evaluating the spectrum of differences in species can be how fast they increase in numbers versus their long-term persistence in a competitive environment. The r-selected species increase very quickly but are not well known for persisting in an area. A great example would be an invasive weed species that is good at rapidly colonizing new habitats. At the other end of the spectrum, K-selected species are better competitors in communities and persist well in established ecosystems (Table 6.1). Looking at organisms along the r–K continuum provides an example of the kinds of characteristics that must be evaluated when considering how to control a pest, as well as when evaluating natural enemies to be used for pest control. For example, those lady beetles developing faster than their scale insect prey provide better control as opposed to lady beetles eating aphid prey that develop much faster than the lady beetles (see Section 7.2.1).

Information from the majority of population dynamics studies focused on natural enemies, and their interactions with their hosts can have relevance to classical biological control. Information from these studies may also be relevant to conservation biological control in providing information about correct conditions for optimization of activity of resident natural enemies. Inoculative releases are also dependent upon interactions between host and natural enemy. However, basic theory on long-term dynamics of natural enemies and hosts may have little application to inundative releases because the initial and often abundant releases of natural enemies are expected to control the pest. Because the vast majority of biological control of animals is focused on arthropods, and the majority of ecological studies have focused on predators and parasitoids, we will

Table 6.1 General characteristics of r- versus K-selected species that provide a way to compare pests based on general biologies.

r-selected organisms	Smaller-sized organisms Occupy less crowded ecological niches Many smaller reproductive units (offspring, seeds, spores) Fast development and short life expectancy, many offspring will not reach adult stage
K-selected organisms	Larger-sized organisms Occupy crowded ecological niches Fewer larger reproductive units Slower development and longer lives, high probability of reaching the adult stage

mostly discuss the ecological basis for biological control using these pests and natural enemies as examples.

6.1 Types of Invertebrate Pests

Invertebrates are ubiquitous and critically important members of ecosystems but, unavoidably, numerous diverse invertebrate life histories lead to competition with humans. Invertebrates eat plants grown by humans for food, supplies, or amenities and destroy many types of items built or stored by humans. They can carry plant pathogens from one plant to another. Invertebrates can destroy shrubs and trees in natural areas such as wetlands or forests and can thus change habitats and nutrient cycling in these areas. Invertebrates infest bodies of domestic animals as well as humans, either externally or internally, or they can simply bother us. Invertebrates can transmit diseases from one animal to another, with some pathogens spread only by these agents. Some invertebrates living in fresh water have been targeted by biological control programs but the majority of programs deal with terrestrial invertebrate pests that feed on plants.

When invertebrate pests are attacked by predators, they are referred to as prey and when attacked by parasitoids or infected by pathogens, they are referred to as hosts. For simplicity, throughout the following discussion, the terms host and prey will both be used to describe ecological relations of natural enemies.

6.2 Types of Natural Enemies

The natural enemies used to control invertebrates are taxonomically as well as functionally diverse. They include the functional groups of parasitoids, predators, and pathogens. Taxonomically, groups of natural enemies that are used for biological control range from fish to insects, mites, nematodes, and microorganisms, including bacteria, viruses, fungi, and single-celled organisms (referred to as protists). Different groups of natural enemies are emphasized for different control strategies. Classical and conservation biological control have predominantly been focused on insect parasitoids and predators and sometimes mites, while rarely insect pathogens. All types of natural enemies have been used inundatively. It would be far easier as well as more efficient to always use the same type of natural enemies for different situations requiring control, but not all groups of natural enemies have members that will provide effective control for each pest. Therefore, biological control practitioners must have training to be able to work with different types of natural enemies.

There is one main goal in biological control relative to interactions between natural enemies and their invertebrate pest prey and hosts. This is killing or disabling the individual pests as quickly as possible, while preventing further damage or injury by the pest to the greatest extent possible. Death of the pest can be rapid, as with an attack by a predator, or can be slower, when time is required as the natural enemy keeps the pest alive as a source of food while the natural enemy develops. Of course, for classical and conservation biological control as well as for inoculative releases, it is important for natural enemies to reproduce before they die, usually using the bodies of pests as food. In contrast, for inundative releases, reproduction of natural enemies is not expected for control.

6.3 Interactions between Natural Enemies and Hosts

Biological control does not occur when a few hosts are killed but rather when groups of individual hosts are killed and their population densities decline and remain low. Therefore, it is a phenomenon occurring at the population level. Studying populations that vary in space and time is typically more difficult than studying individual organisms. Progress has usually been made by starting with studying individuals under controlled situations, followed by controlled studies (often experimental) of combinations of natural enemy and host individuals in the laboratory and then moving to field studies. Information on outcomes of studies has been used to derive mathematical models, created to help provide answers about the interactions that cannot be directly gleaned from data collected in the field. Models can also be used practically to try to predict pest control by natural enemies. This type of approach (controlled studies providing data for developing models) is required because data from the field are typically influenced by many biological and abiotic factors and the complex interactions between these, and looking at field data from populations, one cannot readily see which are the key factors driving the observed population dynamics in a system. Experiments using mathematical models have been used extensively to investigate the emergent properties of groups of factors acting together. However, models are very sensitive to the assumptions used when building them, so starting from a good understanding of a system is critical.

Early work in developing ecological theory of population dynamics centered around interactions between predators and prey. Important interactions to be outlined were the responses by predators to changes in prey density. In classic studies, Buzz Holling was instrumental in investigating changes in predator behavior in response to changes in

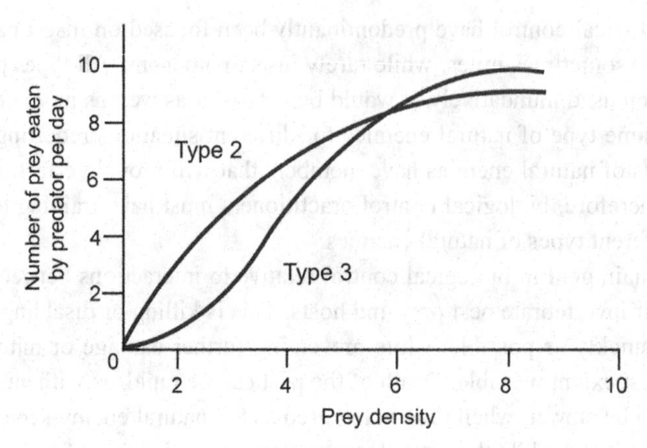

Figure 6.1 Two types of functional responses by predators to changes in prey abundance with satiation at high prey densities.

prey density that he called functional responses. A functional response is the behavioral response of an individual predator, or number of prey eaten per unit time, in relation to host density. It should be differentiated from a numerical response, which involves increasing reproduction or immigration in response to increasing prey density. Holling found that as prey density increased the number of prey eaten often increased quickly at first but then slowed to eventually reach a plateau at satiation (Figure 6.1, Type 2). Creating models for this response helped to identify the important components: (1) the rate of successful search (or rate of discovering prey), (2) the time available for searching, (3) the handling time (the time it takes the predator to eat the prey item and then be ready to search for another), and (4) predator hunger.

Functional responses were subsequently found to be characteristic of many invertebrate predators and also parasitoids. Responses by vertebrate predators were characterized better by a sigmoid response (Figure 6.1, Type 3). Through frequent contact with prey, as would occur at higher prey densities, vertebrates often learn how to find more successfully, catch, and handle prey and thus respond more quickly; so, at intermediate prey densities, the slope of the response is steeper although still reaching a plateau. Further studies showed that some invertebrates could also display sigmoid responses, especially those displaying more active searching in areas where more prey occurred.

Changing behavior when prey are more or less dense is only one component of a predator's response. Numerical response refers to the changes in numbers of predators when prey density changes. One can imagine an immediate increase in numbers of natural enemies as they gather at an aggregation of prey once it was discovered. For invertebrates we also commonly see more delayed responses, with increases in numbers of offspring following finding an abundance of prey or hosts, as a result of increased reproduction. These concepts of functional and numerical responses are central to development of models describing interactions between natural enemies and prey.

6.3.1 Natural Enemy Attributes

Early models suggested a number of general attributes characterizing successful biological control agents: (1) host specificity, (2) synchrony with the pest, (3) high rate of increase, (4) ability to survive periods with few to no prey, and (5) good searching ability. Such properties are more important for classical biological control or conservation strategies where natural enemies respond to prey and host populations across generations and tend to be more characteristic of parasitoids than predators or pathogens. Based on these general attributes, generalist predators would be less suited for classical biological control, because they can have lower rates of increase, are frequently not synchronized with specific pests, and many predators are less host specific. However, from experience we know that predators can be successful in classical biological control. In fact, even parasitoids that have been successfully used for biological control do not possess some of these attributes.

These attributes are quite general and research in particular systems has shown that seemingly minor differences in biologies can make big differences in efficacy for control. The parasitoids of the California red scale, *Aonidiella aurantii*, introduced to southern California, provide an excellent demonstration of the variability in attributes of natural enemies associated with successful control. This scale insect occurs worldwide on citrus, feeding through the bark of trees and can be a major pest. It was introduced to southern California sometime between 1868 and 1875 and the first attempts to introduce natural enemies to control this scale were made as early as 1889. Several different species of parasitoids were involved over time. The first species, *Aphytis chrysomphali*, was probably originally introduced from the Mediterranean around 1900 and it then spread across the distribution of citrus in southern California, but did not provide adequate control. It was not until *Aphytis lingnanensis* was introduced from southern China in 1948 that significant control was seen. By 1959, *A. lingnanensis* had spread throughout the citrus-growing area while the distribution of *A. chrysomphali* had become very limited in specific areas (Figure 6.2a). However, red scale control was still inadequate in the hot, interior valleys of southern California. Therefore, the introduction program continued and *Aphytis melinus* was introduced from India and Pakistan, successfully providing control of red scale populations in the hot interior valleys in California. By this time, *A. lingnanensis* could no longer be found in interior valleys; *A. melinus* had completely displaced *A. lingnanensis* in the interior regions where *A. lingnanensis* had not been effective (Figure 6.2b). It became clear that *A. lingnanensis* was dominant in more humid climates closer to the Pacific Coast with more even temperatures and *A. melinus* was dominant in interior, drier climates with more extreme temperatures. Although under control, red scale remained present throughout its distribution at low and fairly constant densities so that natural enemy populations could persist.

Ecologists have been very interested in this case of competitive displacement, asking what attributes were critical to the success of *A. melinus*. In the interior valleys, *A. melinus* displaced *A. lingnanensis* very quickly, within 1–3 years. Curiously, these two species are extremely closely related and can only be distinguished morphologically in the pupal stage or using molecular studies. Laboratory studies showed that

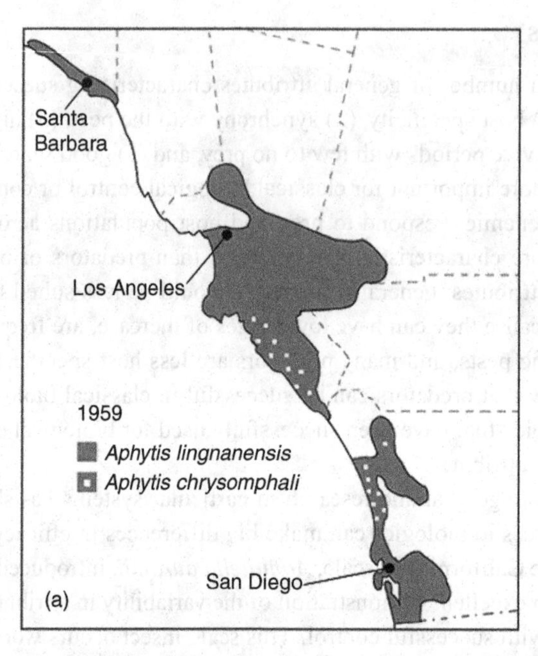

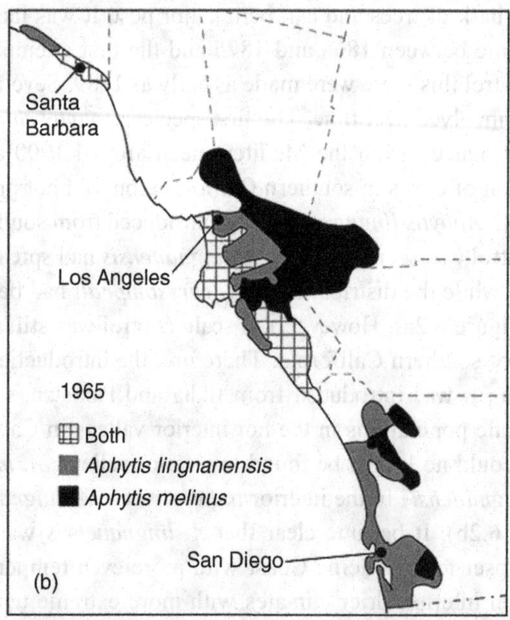

Figure 6.2 Distribution and relative abundances of three species of parasitoids attacking California red scale in southern California. (a) 1959: *Aphytis chrysomphali* had previously occurred throughout this area but by 1959 had been almost completely displaced by *Aphytis lingnanensis* (based on DeBach & Sundby, 1963). (b) 1965: *Aphytis melinus* had spread and displaced *A. lingnanensis* in interior areas, while *A. lingnanensis* was predominant in more coastal areas and *A. chrysomphali* was only found in one location (not shown) (based on DeBach et al., 1971).

Table 6.2 Exogenous (extrinsic) and endogenous (intrinsic) factors interacting to regulate populations.

Exogenous	Natural enemies (predators, parasites, pathogens)
	Food supply
	Weather
	Shelter
Endogenous	Sex and age
	Physiology
	Behavior
	Genetics

Source: based on Krebs (2001)

A. lingnanensis was a better searcher and, when an individual red scale was parasitized by both species, *A. lingnanensis* larvae outcompeted *A. melinus*. *Aphytis lingnanensis* might suffer higher mortality in the more extreme climates of the interior valleys but this finding was not enough to satisfy biologists in explaining the observed patterns. One detailed difference was eventually noticed and used in a model to investigate its effect on competition between these two species. As with other parasitic wasps, *Aphytis* females can control the sex of offspring when eggs are laid (see Chapter 8). Generally for *Aphytis*, male eggs are laid in smaller red scales and female eggs are laid in larger red scales. *Aphytis melinus* had an advantage because female eggs are laid in smaller red scales than would be acceptable for female eggs of *A. lingnanensis*. In the largest red scales, sometimes two female eggs are successfully laid by *A. melinus* when *A. lingnanensis* would lay only one. Therefore, *A. melinus* was able to produce more offspring than *A. lingnanensis* using the same red scale population. Adding these subtle differences in biology to a mathematical model of this system was enough to account for the rapid displacement of *A. lingnanensis* by *A. melinus*. These results also fit standard competition theory, which would predict that the winner of the competition would be the species that most reduces the equilibrium abundance of the common limiting resource, in this case the California red scale.

6.4 Population Regulation

When natural enemies control populations of prey or hosts, this has been called population regulation. Populations are generally thought to be controlled by some combination of extrinsic (or exogenous) factors: these are factors external to the host or prey population, such as the effects of natural enemies or climate, and intrinsic (or endogenous) factors, such as genetic changes in a population or intraspecific (within that species) competition (Table 6.2). Population regulation has been the subject of many studies and much discussion has focused on understanding how densities of populations in communities are maintained. For our purposes, it is important to understand how pest populations are controlled by natural enemies to try to improve success rates

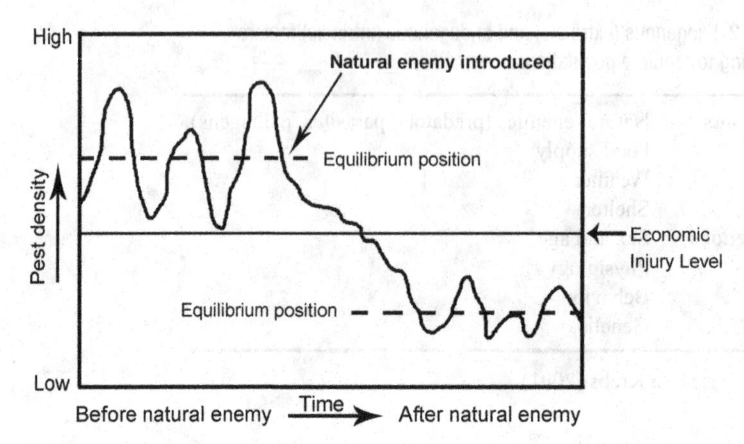

Figure 6.3 Hypothetical results of a classical biological control introduction in which the average abundance of a pest is reduced after introduction of a natural enemy, demonstrating stable equilibria both before and after the natural enemy is introduced (based on Flint & Dreistadt, 1998).

for environmentally safe biological control. A key question concerns what governs the interactions between natural enemies and hosts to allow their coexistence yet result in suppression of host or prey populations by natural enemies. Several issues, including the behavior of natural enemies and pests, responses of natural enemies to pest densities, and actions of natural enemies and hosts on a spatial scale are central to developing theories regarding how natural enemies persist and coexist with their prey or hosts. Also, more recent emphasis by scientists has been toward understanding how communities play parts in determining population densities of pests and natural enemies.

6.4.1 System Equilibria and Stability

For biological control to be successful (at least classical and conservation), it has long been thought that the natural enemy–host relationships should be stable. This means that populations of the pest would constantly be present and would fluctuate in density around some equilibrium density. After introduction of a natural enemy, that equilibrium density would decline to a new stable level at which natural enemy populations would track host populations (Figure 6.3). In contrast, in an unstable system, fluctuations could occur with resulting extinctions of pests or natural enemies; problems would be that with reinvasion of the pest, the natural enemy would probably not be present for some time and the pest would therefore have escaped control.

An early predator–prey model by Nicholson and Bailey used discrete generations with one generation of host and parasitoid per year but the results from this model were unstable and fluctuated wildly through time until both the host and natural enemy became extinct. Something needed to be changed as this very simple model was not representing what was occurring in nature. This initial model was based on encounters between host and parasitoid that occurred randomly, so changes were made. Once

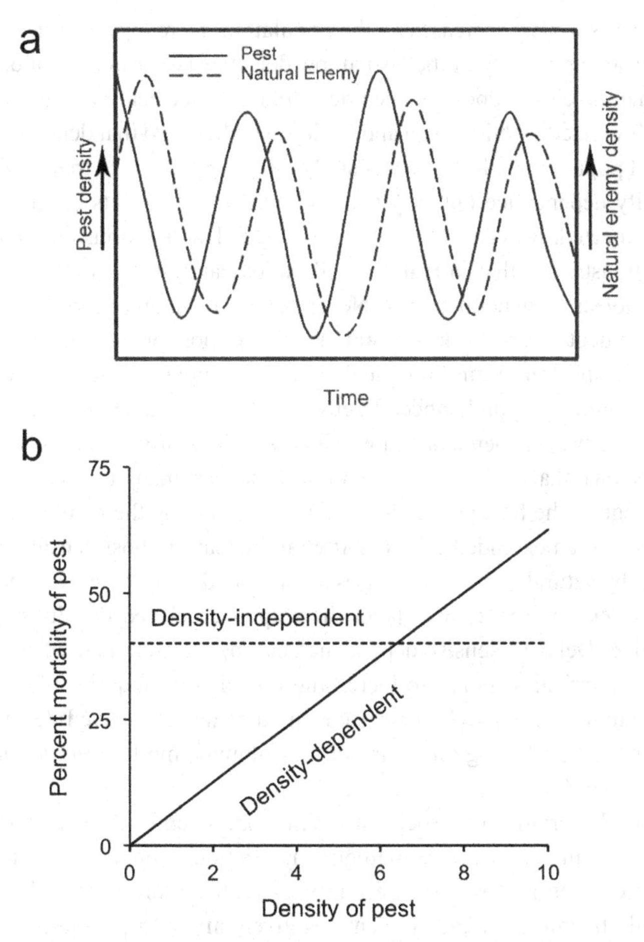

Figure 6.4 (a) Hypothetical density-dependent relations in a predator–prey (or natural enemy–pest) system with discrete generations. (b) Contrast in pest mortality rates between density-dependent and density-independent associations.

natural enemies in the model could aggregate in locations with high densities of the host, the model became stable. However, this theoretical model was still very simple and when it was changed by introducing parasitoid movement, the model again lost stability as populations went extinct. Researchers began to question whether this idea of a stable equilibrium for populations was real. If interactions are in fact unstable, how is coexistence of natural enemy and host then achieved and maintained?

6.4.2 Density Dependence

Central to the issue of regulation by natural enemies is the concept of density-dependent mortality, the relationship where the level of mortality inflicted on members of a host or prey population changes in relation to the density of the host or prey population (Figure 6.4). While this type of positively density-dependent mortality would increase as the population increases it also decreases as the population decreases, as a negative feedback. The decrease in mortality of the host at low densities is a critical attribute because in this way the natural enemy does not become extinct (but see below). This concept was central to models created by Nicholson and Bailey, who believed that density-dependent factors regulated populations. Researchers studying natural enemies tried to fit them to models of density dependence but they soon found that data points often did not fall directly where expected. Instead of being density dependent, relationships can be "density vague," demonstrating that in reality in all biological systems responses are often variable but demonstrate general trends. Nevertheless, for many years scientists held that density-dependent responses by natural enemies to host or prey populations were required for successful long-term biological control. In support of this, a recent meta-analysis demonstrated a preponderance of density dependence in almost 1,200 systems.

We can look at density dependence more closely and classify it into different types. Scientists have shown that looking at population responses, many times a delay occurs between the change in the host population and the response by the natural enemy population; this is due to an extended time lag after an increase in host density and before mortality caused by natural enemies increases and is called delayed density dependence. How much of a delay is required to label the response delayed density dependence remains subjective. Delayed density dependence can be characteristic of insect populations where a numerical response to increasing host density requires time for a new generation of natural enemies to be produced. In an example, delayed density dependence is hypothesized as driving outbreaks of the autumnal moth, *Epirrita autumnata*, in Fennoscandia (Box 6.1).

Of course not all mortality is associated with natural enemies or with host density. Density-independent mortality occurs without any relation to host or prey density and this has sometimes been referred to as a limiting factor (Figure 6.4b). Classic examples of this would be when a weather event negatively affects a population regardless of density, such as an early freeze causing extensive mortality among noncold-hardy species, or when herbivorous insects die because host plant quality declines as a result of drought. Among early theorists, some felt that density-independent processes were extremely important and, for a time, the relative importance of density independence versus density dependence in determining host densities was a matter of great debate.

6.4.3 Host Metapopulations

We have been discussing changes through time but we now know that spatial variability can also have a strong impact on pest population densities and on predator–prey or host–parasite dynamics. Zooming outward from natural enemies attacking one localized population of a pest, we can gain insights looking at a larger spatial scale, the metapopulation level. A metapopulation is a set of local populations connected to each other through dispersal. In reality, many species, unless at outbreak densities, have aggregated distributions. While host densities might decrease at one location, it is highly likely that they would by chance not decrease throughout all patches of a metapopulation. A pest population could then all die in a localized area where there would then be

Box 6.1 Autumnal Moth Cycles Explored: Specialist versus Generalist Natural Enemies

Every nine to eleven years somewhere in northern Fennoscandia, caterpillars of a small moth that often feeds on birch trees, the autumnal moth (*Epirrita autumnata*), increase to high densities that cause serious defoliation after which they decline in numbers. While these cyclic outbreaks occur in northern areas, autumnal moth populations remain at consistently lower densities in the south without cyclic increases and decreases. This scenario provided an excellent system for scientists to study why cyclical outbreaks of the same species occurred in one area but not another.

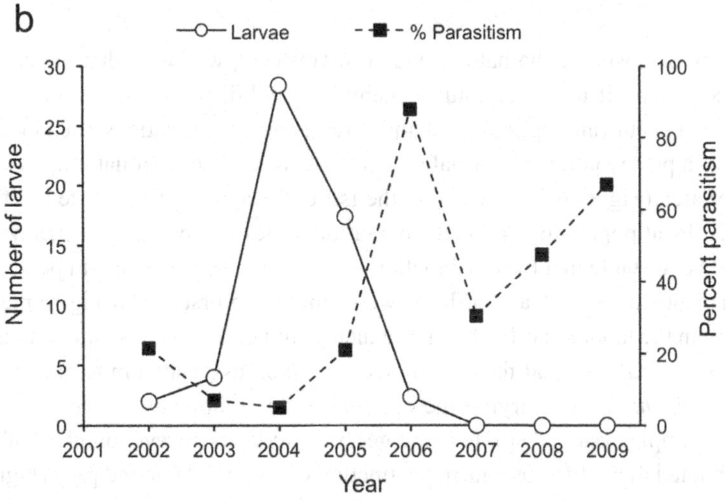

Figure Box 6.1.1 (a) A parasitoid wasp (*Agrypon flaveolatum*) and a caterpillar of the autumnal moth (*Epirrita autumnata*). (b) Densities of healthy autumnal moth larvae (open circles and solid lines) and percent larval parasitism (squares and dashed lines) from 2002 to 2009 in northern Finland (Klemola et al., 2010; photo by T. Klemola).

Researchers investigated many factors: plant responses induced by caterpillar feeding, food shortage after all of the leaves had been eaten by high density populations of autumnal moth caterpillars, or competitors, parasitoids, predators, and maternal effects (for example, if mothers were smaller or had poor nutrition, they could lay fewer viable eggs or their eggs might provide suboptimal nutrients for hatchlings). Over many years of study, researchers hypothesized that in the south there are more generalist natural enemies than in the north. In the south, birds and other generalist predators and parasitoids are thought to keep populations low, although these natural enemies mostly are not responding to the moth density. In the south, only predation of pupae was shown to depend on moth density, suggesting that the generalists maintain populations at low densities but then pupal predators further regulate populations so that outbreaks never start. In contrast, in the northern autumnal moth populations where outbreaks occur, the situation is quite different. There is a shorter growing season and fewer natural enemies attack autumnal moth eggs, larvae, and pupae. In particular, specialist parasitoids are the dominant natural enemies and it is hypothesized that these parasitoids are helped in finding host caterpillars by host plant-induced volatiles released as trees are defoliated. Although these parasitoids respond to moth density, this response occurs in a delayed fashion over time. Populations of the parasitoids only increase when the autumnal moth population is already decreasing and therefore they are not preventing the development of outbreak populations or the abundant damage that occurs at the peak of an outbreak. Thus, the specialist parasitoids respond in a delayed density-dependent manner that helps to control outbreaks, but only after a delay, which helps to drive the outbreak cycles of autumnal moths in northern Finland.

no prey or hosts for the natural enemies. However, we know that under such circumstances, many if not most natural enemies would disperse to find more prey/hosts. If dispersal occurs among local pest and natural enemy populations, then localized extinction of a pest would not especially result in extinction of that natural enemy across the larger area (Figure 6.5). Therefore, the issue of stability of the system differs between single, local population and metapopulation scales. A host/natural enemy combination could be unstable in a local population but on a regional scale could persist stably.

Elegant studies by Carl Huffaker were among the first to investigate metapopulation theory in the laboratory. Huffaker conducted studies using six-spotted mites, *Eotetranychus sexmaculatus*, that feed on oranges and their associated mite predator, *Galendromus occidentalis*, by varying the environmental complexity of the system. He found that in simple systems with few oranges, the predators always found all of the prey and annihilated them, thereby causing extinction of both predator and prey (Figure 6.6a). He added more trays of oranges with petroleum jelly barriers, creating metapopulations in this universe, but allowing some movement among oranges. In this way, prey and predator populations were maintained for 70 weeks (Figure 6.6b) after which time the study was terminated. This study demonstrated that having a heterogeneous environment in which prey occurred as metapopulations could lead to system stability.

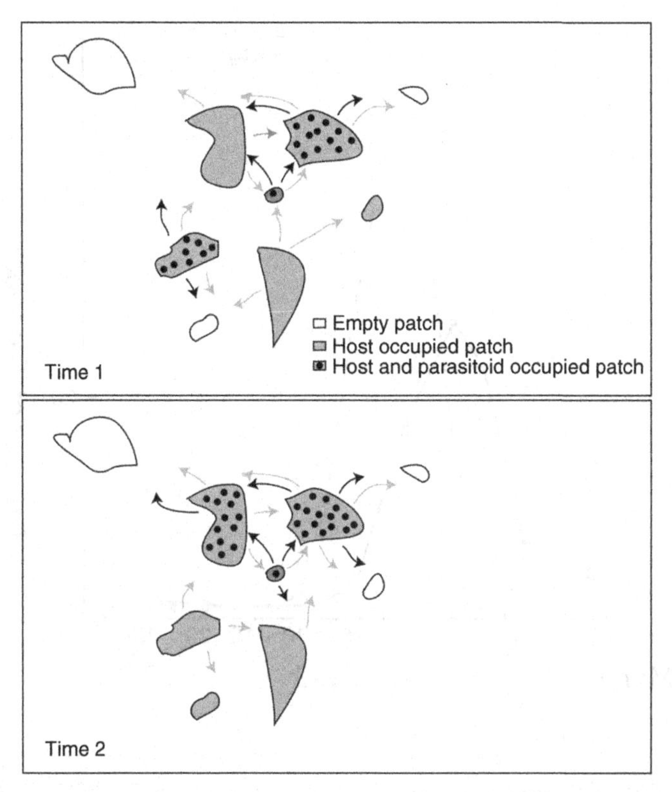

Figure 6.5 Hypothetical metapopulation dynamics showing patches that hosts could occupy and spatial distributions of host and parasitoid populations as they change through time (here shown as Time 1 and Time 2 for the same area) by colonizing patches and disappearing from patches (drawn by Saskya van Nouhuys).

In systems where pests are phytophagous, the host plant, pest, and natural enemies often all occur in aggregated distributions as a result of resource concentration. In the field, populations of herbivores have been shown to be interconnected into metapopulations by dispersal. In work with parasitoids, specialized natural enemies often only occur in a subset of host populations at any one time. Natural enemies usually must arrive at sites after the host; therefore they are constrained in their ability to colonize a new area successfully when dispersing, that is, a dispersing herbivore only needs to find a patch of host plant but the parasitoid must find a patch of host plant where the herbivore already occurs. Certainly, this concept of metapopulations that account for localized extinctions with reinvasions by pests and natural enemies is important to our understanding of the mechanisms by which natural enemies respond to and control pests in the field. Metapopulation dynamics have been reported in biological control for the lady beetle introduced against the cottony cushion scale (see Box 3.1). These beetles can eat all of the scales on individual citrus trees but movements among trees allow both scales and beetles to persist. Similar patterns have also been reported for mites in apple orchards and some greenhouse pests.

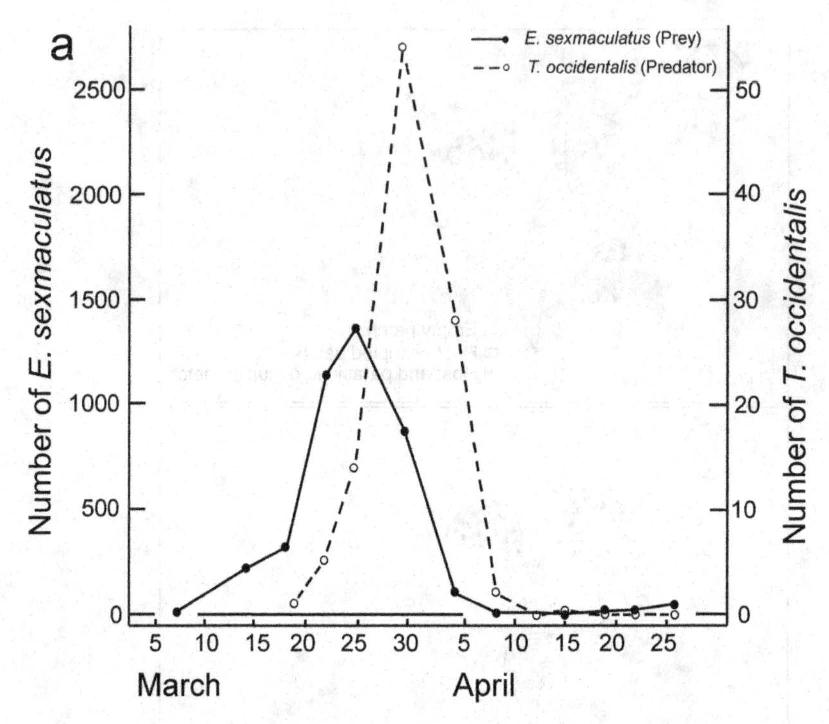

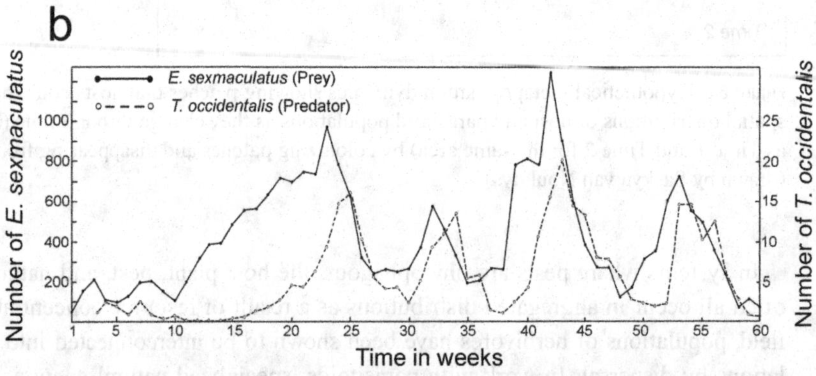

Figure 6.6 (a) Densities per unit area of orange for the prey mite *Eotetranychus sexmaculatus* and the predatory mite *Galendromus occidentalis* in a universe of 40 oranges, only 20 of which provided food for the prey (based on Huffaker, 1958; © Regents, University of California). (b) Predator–prey interactions between *E. sexmaculatus* and *G. occidentalis* in a complex 252-orange system in which one-twentieth of each orange was exposed for possible feeding by the prey (based on Huffaker et al., 1963; © Regents, University of California).

6.4.4 Refuges for Hosts

As stated above, to achieve stability between natural enemy and host populations, both players must be present in an area. This can be accomplished by recolonization of spatially isolated patches after host extinction (as explained above) or, alternatively, by persistence of the host population in some way within the same area. It is often thought that an optimal situation in biological control is for both the pest and the natural enemy to remain at very low densities, that is, at low, stable equilibria (Figure 6.3). If hosts have some sort of refuge where they cannot be killed, they can persist in that area. In a simple case, this could be a space in which the natural enemy could not reach the host. Another type of refuge can be based on timing, as when vulnerable stages of hosts or prey avoid natural enemies by occurring very early or late in the season when the natural enemies are not active. Alternatively, if the natural enemy is omnivorous and switches to other food when the host is scarce, low-density host populations would be relieved from predator pressure because if prey were scarce, predators could feed on other prey or even switch to feeding on plants (see Section 7.3.4).

6.5 Exploring Factors Impacting Pest Regulation and Its Stability

A major question arising about interactions between natural enemies and hosts has become: "To what extent does stability occur in these relationships?" By stability we mean a relationship where numbers of the natural enemy and host fluctuate around some equilibrium density, preferably at low densities of pests, with neither going extinct but also neither going into outbreak densities. The goal of biological control, of course, is for the equilibrium density of the pest to drop and remain below the economic injury level (Figure 6.3; see Chapter 2). Early researchers considered that stability in systems was necessary for classical biological control to be successful.

In part investigating questions about system stability, William W. Murdoch and colleagues studied relations between red scales and the parasitoid *Aphytis melinus* (see Section 6.3.1) on citrus in southern California for 20 years. In this system of citrus orchards, there are few other natural enemies or pests but the system they studied is surprisingly stable. They created outbreak populations of the scales in caged trees, with other trees serving as controls, and then monitored changes. They also compared experimental results with models that they had developed based on the two main species (i.e., host and parasitoid) so that they could test the importance of the different interactions in this system. The outbreaks that had been experimentally created returned to low level pest scale populations very rapidly (Figure 6.7). The major processes that they found were driving the system are attributes that are not uncommon. First, the scales are not vulnerable to the parasitoid when they are adults so this created a developmental refuge for the scale population; in fact, it is not uncommon for natural enemies to attack only certain stages of hosts/prey. Second, these specialized parasitoids developed much faster than the scales and this gave the natural enemies a numerical advantage against the scales, an attribute that is also not uncommon for effective natural enemies. An additional feature that plays a less dominant part is the fact that the parasitoids will indiscriminately attack already-parasitized scales, which was unsuccessful and acted to decrease parasitoid efficiency at high parasitoid densities so that the host population was not overexploited. In this system, it seemed that spatial processes or greater community diversity does not always play a significant part in system stability.

System stability is only achieved under what has been called the "paradox of enrichment" or the "paradox of biological control." For natural enemies, as their efficiency increases, it becomes a problem that with fewer and fewer remaining prey the natural

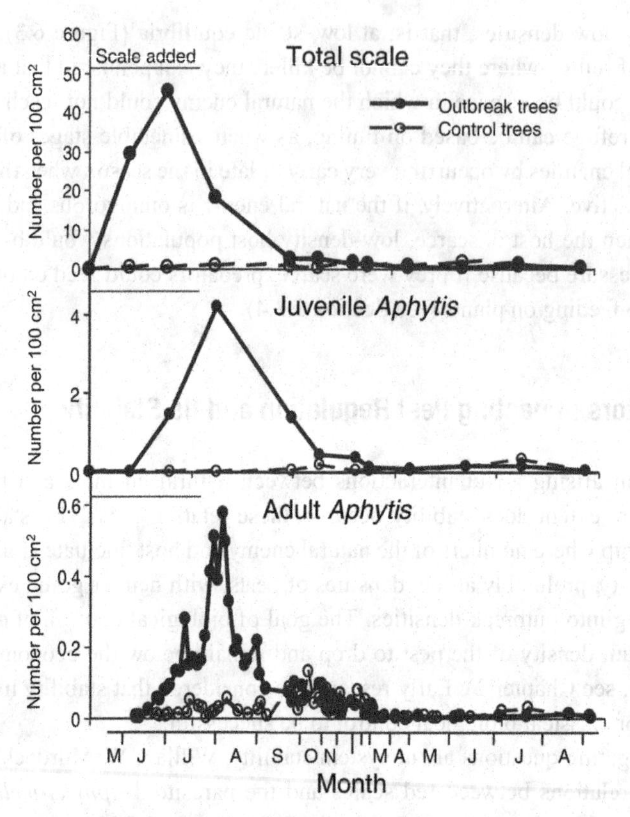

Figure 6.7 Mean densities of scale insects and *Aphytis* parasitoid wasps in four outbreak and ten control trees over a period of 16 months, from May through August of the following year (Murdoch et al., 2005; reprinted with permission from AAAS).

enemies' chances for survival decrease (because there are not enough prey). So, with increasingly effective natural enemies, it seems that predator–prey cycles will be of increasing amplitude, resulting in more system volatility and less pest control over the longer term. As a result, it has been hypothesized that highly virulent (efficient) natural enemies might not be the best for maintaining a persistent system where the pest is stably under control.

It is uncertain how unique the results from the *Aphytis*/red scale system are and there are few such experiments for drawing comparisons, perhaps because studies along these lines are frequently conducted under the duress of trying to control pests and not to evaluate concepts. As a comparison, let's investigate a locally nonpersistent system, such as a seasonal agricultural crop. Because the natural enemies cannot rely on the pest as a resource year round, it is assumed that generalist predators would be important. The pest must be mobile, to invade the crop, and the enemies must also be mobile to follow the pest; this kind of system is completely different from the red scale/*Aphytis*/citrus system described above. In this type of system, local extinction of the pest could occur, a nonsusceptible stage of the host/prey (a refuge) is not necessary and natural enemies cannot persist locally (see Section 6.4.3). For systems like this with aphids as hosts/prey there are a few examples demonstrating successful biological control. Pea aphids, *Acyrthosiphon pisum*, appear to be controlled each year in just such a situation in alfalfa fields in Wisconsin. In this system there is an introduced specialist parasitoid, *Aphidius ervi*, as well as a diversity of generalist predators, most of which are true bugs or beetles. The generalist predators move into alfalfa after the aphids and they feed on the aphids and reproduce. Unfortunately, part of the predation kills the parasitoids but this parasitoid develops much more slowly than the aphids and is relatively ineffective for control, which is predominantly caused by the generalist predators in the system.

6.5.1 Allee Effects

Lack of an equilibrium would mean that natural enemies and hosts would undergo local extinction, but how common is this? In the field, based on introductions of exotic species that might become invasive, it is estimated that approximately 10 percent of introduced organisms become established. In fact, many organisms purposefully introduced for classical biological control (which in reality can be thought of as human-aided introduction of an exotic species) do not become established after introduction. In some instances, as with accidental introductions of exotic species that will potentially cause extensive damage, eradication is attempted by public agencies, to drive populations of invasives to extinction in newly colonized areas. The goal of such an eradication program is for the pest population to decrease to such a low density that the Allee effect is seen. The Allee effect is a phenomenon whereby fitness (ability to successfully reproduce) is positively correlated with population size. In particular, the Allee effect is seen when animal or plant species decrease to low densities and their rate of increase declines. This can also be thought of as inverse density dependence (an increased effect at low densities) that can drive populations to extinction. This effect can often occur as a result of (1) failure to find mates in low-density populations; (2) failure to thrive at low densities if cooperation among individuals is needed, as among gregarious feeders; or (3) inbreeding depression. Therefore, natural enemies do not have to find all hosts for the host population to become locally extinct because low-density host populations could decline to extinction solely as a result of Allee effects. If hosts become extinct in an area, this does not especially mean that biological control will be ineffective, because metapopulation theory would suggest that a localized area could be recolonized, both by the pest and the natural enemy. Allee effects can also help to explain why colonization or recolonization of an area by natural enemies can fail, if the number of colonizing individuals is too low for the population to establish and increase.

6.5.2 Responding to Pest Population Increase

Parasitoids and predators are usually considered important for preventing pest outbreaks, but how do they respond quickly enough if pests have such patchy distributions and systems are not stable, with local extinctions occurring? Two very different life history strategies of parasitoids and predators, both thought to allow natural enemies to respond to increasing host populations, have been called "search and destroy" and "lying in wait." "Search and destroy" is employed by natural enemies that are highly host specific and are also good at searching for and finding their hosts. Spatial patchiness of the pest allows the pest to survive but these natural enemies eventually find and destroy individual pest populations, after which the natural enemies disperse and search for other populations of the host. This response works even better if the natural enemy develops faster than the host. Characteristics of natural enemies using this strategy, such as narrow host range, excellent ability to find hosts, and high rate of numerical increase, have long been considered the goals for successful classical biological control agents such as *Aphytis* parasitoid species controlling California red scale or vedalia beetle controlling cottony cushion scale.

The "lying in wait" strategy is quite different, and is characteristic of populations of polyphagous natural enemies that are continuously present in local areas subject to pest infestation. When the pest is not present, these natural enemies survive for a time without food or by eating alternate food (sometimes including each other or plant material; see Chapter 7). These natural enemies thus persist in areas whether hosts are present or not and are present and ready to respond when the pest is once more present and/or beginning to increase. This type of response is characteristic of predaceous mites that keep phytophagous mites under control in greenhouses, orchards, and vineyards; the effectiveness of these predators is evident once these predators are eliminated by pesticides and populations of phytophagous mites erupt.

6.5.3 The Enemy Release Hypothesis

Most organisms are considered as being under natural control, with their populations being regulated by a combination of the availability of required resources and activity of natural enemies as both of these can be limiting. However, under conditions such as simplified agricultural monocultures or in cases of invasive species we know that pest populations can increase to densities that need to be controlled as these limitations are not having an impact. Especially with invasive species, it can be quite evident that there are insufficient natural enemies attacking at a new introduction site compared with in the region of origin. In support of this, many times these invasives are not pests in their area of origin and can even be poorly known and rare.

Over many years, as introductions of alien species that become invasive have become increasingly common occurrences, an assumption grew that the population sizes of invasive species were high because the invasive had arrived without the natural enemies regulating its populations in its area of origin; this hypothesis was aptly named the "enemy release hypothesis." The follow-up assumption has been that going to the geographic area that is the origin of the pest, natural enemies can be collected for introduction to areas where the pest has invaded; this, of course, has been a basis for classical biological control. This has especially applied to situations where an invasive species arrived and became established in a foreign continent, for example, arriving as a contaminant with shipped goods. However, ecologists and invasion biologists began studying this phenomenon and they soon realized that it is difficult to prove the enemy release hypothesis. More recently, a larger analysis looking across many studies determined that in

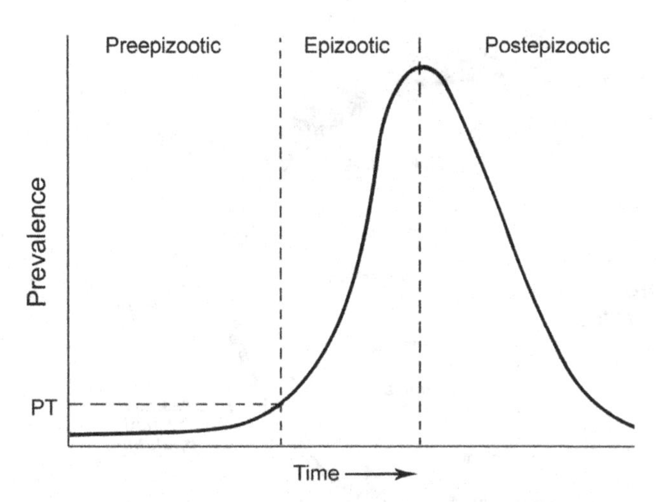

Figure 6.8 The three general stages of an epizootic cycle demonstrating the increase and decrease in disease prevalence. The preepizootic phase occurs as long as prevalence is below a perception threshold (PT), followed by a dramatic increase in prevalence that is often short-lived relative to the other phases. The postepizootic phase occurs as disease prevalence decreases (based on Brown, 1987).

about 50 percent of the cases populations of an invasive are out of control because they have escaped from natural enemies while in the remaining 50 percent of cases a diversity of other interactions are probably helping the invasive to reach and remain at high densities.

6.6 Microbial Natural Enemies Attacking Invertebrates

Much of the theory regarding population regulation relative to biological control has principally been developed with parasitoids and predators in mind. However, we know that pathogens causing infectious diseases can be important natural enemies. Given free rein, many pathogens are known to be effective natural enemies and because of their rapid reproductive rate, they can cause epizootics (a term used to describe unusually high levels of disease in animal populations, that is similar to the term epidemic) in host populations (Figure 6.8). Attributes of the biology and ecology of microbes are mostly different enough from parasitoids and predators that these should be mentioned separately. Microbes causing infectious diseases are highly variable in many biological attributes, so, for example, viruses, bacteria, and fungi cannot be treated as being the same. Host/pathogen models, based on invertebrate pathogens, were developed with separate models for different sets of characteristics.

Most insect pathogens do not have mobile propagules that actively search for hosts. Therefore, susceptible hosts often must contact pathogens to become infected. However, some invertebrate pathogens have mobile free-living stages or actively discharged

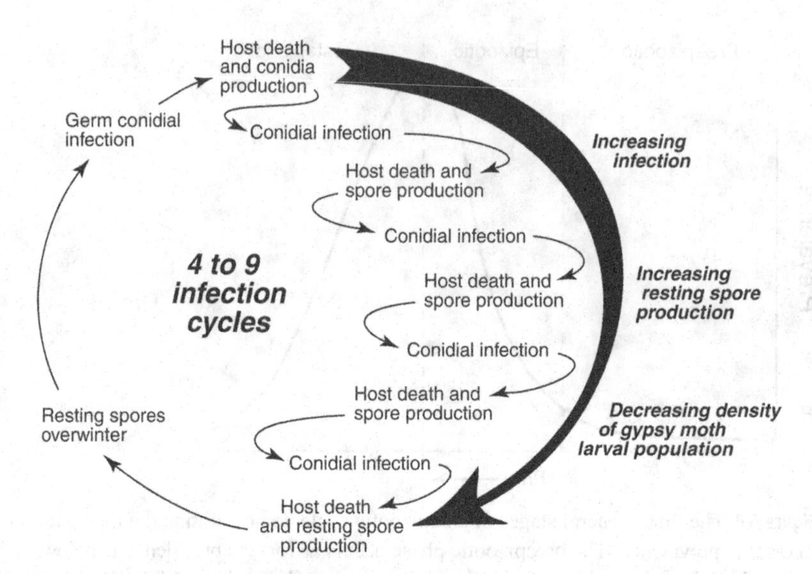

Figure 6.9 Multiple cycles of *Entomophaga maimaiga* infection during one gypsy moth field season that can result in epizootics. Models have indicated that during one gypsy moth generation (from egg to adult) four to nine infection cycles can occur (drawn by Frances Fawcett).

infective units, which help the pathogen to contact susceptible hosts. However, different pathogens may require vastly different numbers of propagules to achieve infection. Biologies of pathogens are diverse and can vary from causing fatal diseases that can increase in prevalence, killing many hosts during epizootics, to causing chronic infections that decrease host reproduction without causing mortality. Even different strains of the same pathogen species can vary in virulence, the disease-producing power of a pathogenic microorganism. For a pathogen causing fatal disease, the speed with which the pathogen strain kills a host can be used as a measure of virulence.

Some pathogenic microbes have long-lived stages that persist in the environment. Such stages are well known from various bacteria, baculoviruses and fungi that infect insects; these persistent stages often accumulate in the soil or the bottoms of bodies of water and these locations then act as reservoirs where the pathogen persists in the environment. When pathogen propagules from such reservoirs infect hosts, this can be thought of as a first cycle of infection, often called primary infection. Primary infection can be followed by multiple cycles of infection, called secondary infection cycles, when pathogen propagules produced from cadavers of recently killed hosts infect healthy hosts. Under optimal conditions, in this way pathogens can develop and kill hosts quickly, and such cycles of infection (from infection to microbial reproduction to infection of another host) can occur numerous times during one season (Figure 6.9). Rapid increases in the numbers of hosts infected as a result of multiple cycles of infection (secondary cycling) can result in levels of infection characteristic of an epizootic.

Simple models have suggested that pathogens of intermediate pathogenicity are more effective as naturally occurring biological control agents, while highly pathogenic microbes may drive cycles of extreme decreases (due to epizootics) in host populations followed by outbreaks. When adding heterogeneity in virulence in the pathogen population to the model, the tendency of the host population to repeatedly increase to outbreak densities that then decline during disease epizootics dampens. However, data on dynamics of host populations have shown that a diversity of factors affecting host mortality (such as all natural enemies, abiotic conditions, and host food) are linked with the occurrence of cycles of pest outbreaks and subsequent population crashes, so disease epizootics alone are not responsible for these fluctuations in host density.

6.7 Food Webs and Community Ecology

Much of our discussion thus far has dealt with systems composed of one natural enemy and one host or, at most, a few natural enemies and one host. However, in nature the interactions between these participants are only part of greater webs of interactions among many different organisms living within the same environment (Figure 6.10). In recent years scientists have increasingly investigated the interactions between host or prey species and the activity of natural enemies in biological communities. To accomplish this requires investigating interactions within food webs.

Bob Paine (1996) described food webs as "the ecologically flexible scaffolding around which communities are assembled and structured." In fact, it has become evident that models and studies isolating one host and one natural enemy often do not replicate dynamics seen under field conditions. Thus, in more recent years, scientists have increasingly investigated the impacts of hosts and natural enemies within the context of larger biological communities to understand what exactly is driving population densities and dynamics and, for our interests, biological control.

Systems can indeed become extremely complex and scientists have tried to classify them based on the major factors organizing the composition of the system. Systems where natural enemies seem to provide the major control of organisms feeding at lower trophic levels, such as lady beetles that control populations of aphids that feed on plants, are said to be "top-down." Conversely, systems where the primary producers such as plants seem to organize the dynamics, such that the herbivore populations are limited by the plant populations, are termed "bottom-up." Unfortunately, although this categorization helps us think about factors driving interactions, as communities become more diverse (include more interacting species) such simplified and directional control of dynamics is often not so evident.

For either type of system, with top-down or bottom-up dominance, there are often numerous species at each trophic (or feeding) level. The species at the different trophic levels can influence each other in what is called a trophic cascade, with direct interactions for levels next to each other and indirect impacts on trophic levels with more distant connections. A biological control example could be that the population of a predator increases, while the population of the herbivorous insect that it feeds on decreases and the plant the herbivore feeds on in turn increases. Therefore, the impact of the predator on the plant is positive but indirect. The different members of food webs can also be

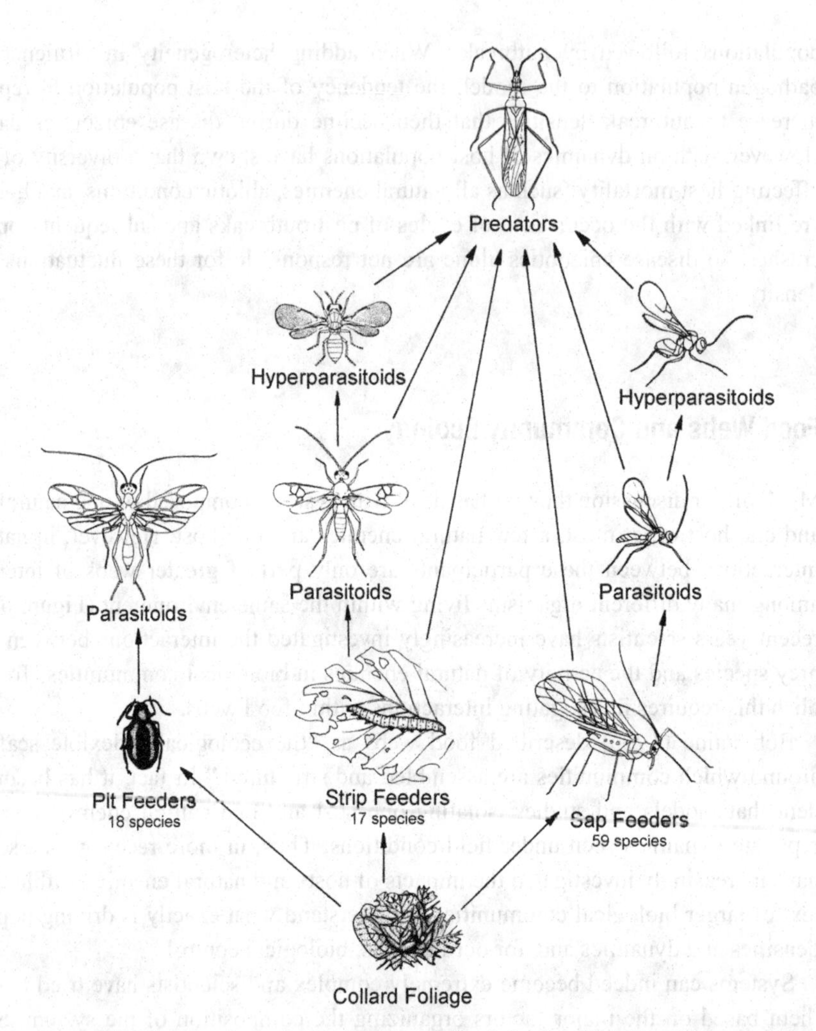

Figure 6.10 The simple food web found on collards by Root (1973). This community contains three guilds of herbivores, the pit feeders, strip feeders, and sap feeders, with parasitoids, hyperparasitoids, and predators feeding on the herbivores and, to some extent, feeding on each other (Price, 1984).

thought of as resources (basically, what is eaten) and consumers (organisms doing the eating) and how they interface would then be called consumer–resource interactions.

Within a food web, organisms that utilize common resources in the same manner can be thought of as a guild (as in Figure 6.10). These coexisting species often compete with other members of the same species, with other species in their trophic level, or with species in other trophic levels. Sometimes members of the same guild compete for the same resources and this can be called exploitative (or resource) competition (Figure 6.11a). In this case, there does not have to be direct interaction between the competitors but they share a common and limiting resource. Competition can also occur by competitors killing each other and this decreases competition for the limited resource.

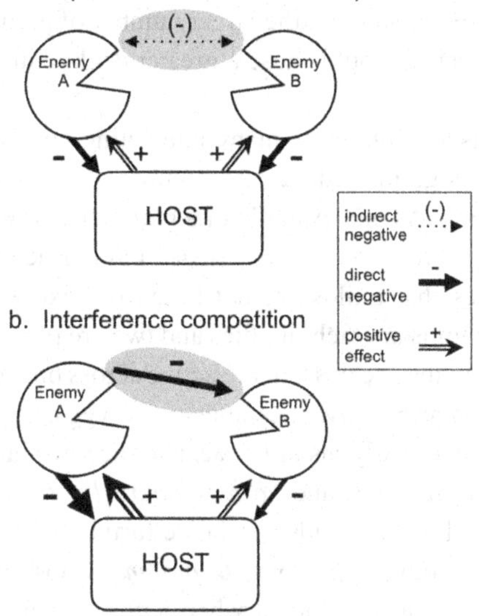

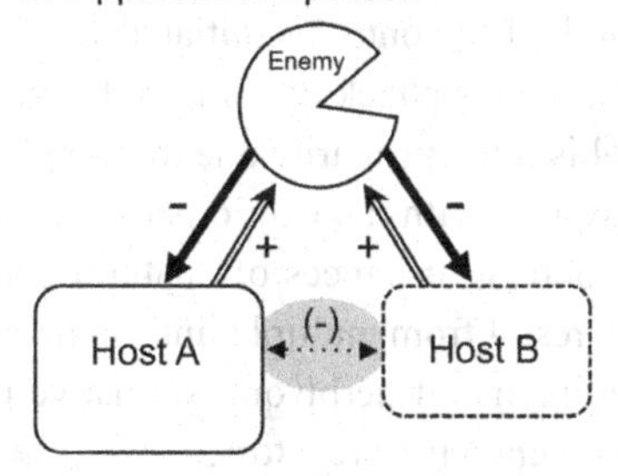

Figure 6.11 Diagrams of the three major types of competitive interactions (highlighted in gray): (a) exploitative or resource competition; (b) interference competition; and (c) apparent competition. Single solid lines indicate direct negative interactions. Double lines indicate direct positive interactions. The widths of the lines indicate the strength of the interactions. Dashed lines indicate indirect negative interactions (drawn by Sana Gardescu).

This so-called interference competition can occur within the same species or across different species in the same or different trophic levels (Figure 6.11b). When coexisting predators eat each other, this has been called intraguild predation. Research has shown that this can relax the control of the shared prey, as predators are partially satiated by feeding on other predators, which also decreases the number of predators feeding on the prey. This can be seen with green lacewings and true bug predators feeding on cotton aphids (*Aphis gossypii*). In experiments, when only green lacewings were present, cotton aphid densities were low. However, when predatory true bugs were also added to the system, they spent part of their efforts eating green lacewings and aphid populations rebounded. A third type of competition that is important to biological control is called apparent competition (Figure 6.11c). In this case, a natural enemy feeds on several host or prey species and this can indirectly negatively impact the prey. Populations of the natural enemy become more abundant because there is a lot of food. However, when one of the prey species was initially less abundant, the larger number of predators (supported by feeding on several prey species) would cause a greater decline in the species that was initially less common.

Communities can be thought of as assemblies of many interacting guilds. Thinking of groups of similar-acting species as guilds reduces the number of components in a community, thus facilitating their study. As one example of a guild, there are five major species of parasitoids that attack gypsy moth (*Lymantria dispar*) in northeastern North America. This example demonstrates that guilds are not based on taxonomy but on ecological roles; three of these parasitoids are tachinid flies and two are parasitic wasps. Within this guild, there is succession in the species that are active across one gypsy moth generation; one parasitic wasp (*Ooencyrtus kuvanae*) attacks the eggs, another parasitic wasp (*Cotesia melanoscelus*) attacks early instar larvae, and the three tachinid flies (*Compsilura concinnata, Blepharipa pratensis*, and *Parasetigena silvestris*) attack and kill late instar larvae and pupae. The three tachinids specialize further, with *C. concinnata* prevalent in low-density gypsy moth populations, *B. pratensis* most abundant at intermediate densities, and *P. silvestris* most abundant when outbreak populations are declining.

In the simplest situation when classical biological control is initiated, the food web of concern is simple, with only a few natural enemies attacking an introduced herbivore. In such a system, effective top-down control is often primarily due to a single specialist parasitoid species, usually in a simplified system with the exotic herbivore feeding on an exotic plant in a cultivated habitat. In contrast, instances of "natural control" (see Chapter 2) that have been documented often result from multiple links in more complex food webs. For example, populations of native insect herbivores on native plants in a natural habitat can be regulated by a guild of generalist predators.

Frequently, species of several different trophic levels influence each other. This has been called tritrophic interactions when three trophic levels are involved. In plant–herbivore–natural enemy tritrophic interactions, the host plant of a pest can affect a natural enemy, thus influencing the resulting population levels of the host. For example, if a specific plant provides resources that benefit the natural enemy such as protected refuges among leaf hairs (see Section 5.3.2), more natural enemies will be present, resulting in fewer hosts. As a simple example, experiments were conducted with pea aphids in microcosms containing either one or three host plants with either one or three predators. More aphids were consumed when there were three predator species so there was a top-down effect. However, predators were less efficient on one of the host plant species, *Vicia faba*, which also contributed a bottom-up effect to the study (Figure 6.12). Interactions between gypsy moth, mice, and acorns also demonstrate how several very different trophic levels can affect each other across time (Box 6.2).

An active area of research, looking at another kind of tritrophic interaction, is based on herbivore-induced plant volatiles. In situations where natural enemies respond to these volatiles in order to find hosts/prey, this indirect plant defense has been described as the plant calling for bodyguards to help provide protection. Natural enemies that are both above ground and below ground respond to these volatiles. However, most of

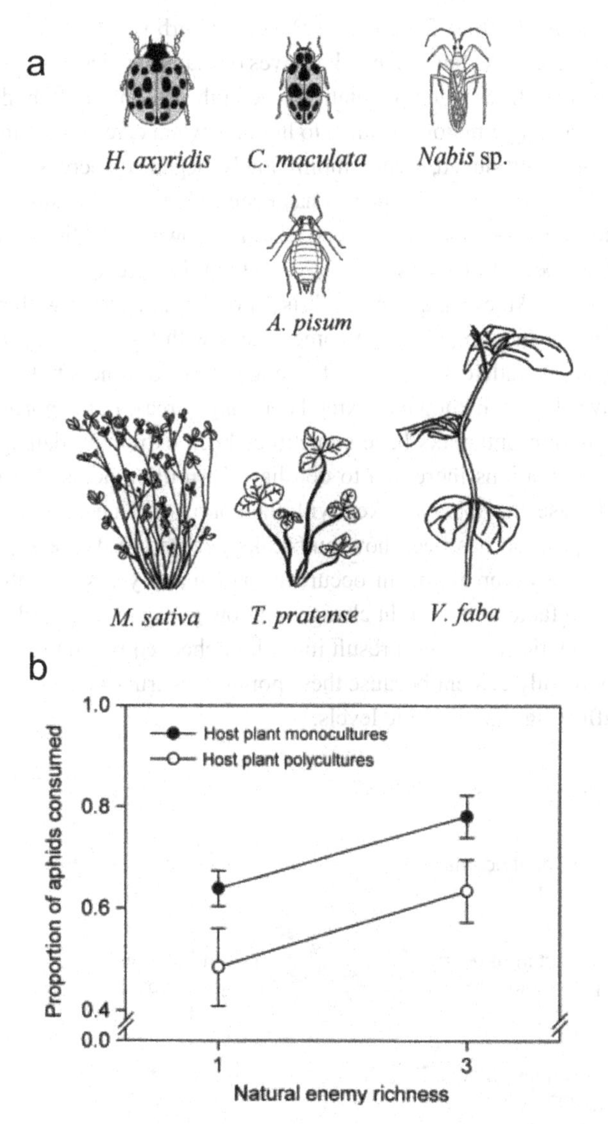

Figure 6.12 (a) The organisms within the tritrophic system tested in this microcosm experiment. (b) Results demonstrating increased aphid consumption when all versus only one predator species (of any of the three predators) were present and, in this system, decreased aphid consumption occurred in more diverse habitats (polycultures) versus monocultures (Aquilino et al., 2005; illustration by B. Feeny).

Box 6.2 Of Mice, Moths, and Acorns

The gypsy moth (*Lymantria dispar*) and white-footed mice (*Peromyscus leucopus*) and the oak trees (*Quercus* spp.) they both depend on provide an excellent example of how interconnections within food webs can have far-reaching effects. The gypsy moth was introduced from Europe to a Boston suburb in 1869. Gypsy moth caterpillars (larvae) prefer oaks but will eat the leaves of many species of trees during early spring, killing some trees when populations are high, but, more often, decreasing tree growth and causing a major nuisance to humans when caterpillars are abundant. After it was first introduced, gypsy moth slowly began to increase in abundance and spread. The spread by this species has been rather slow because females are flightless so this species cannot move very fast on its own. In addition, massive federal programs have been aimed at stopping or slowing the spread.

In northeastern North America, gypsy moth is an outbreak species with populations able to increase to huge numbers, at some places with 8–10-year cycles, and then decrease to virtually undetectable levels for long periods of time. Outbreaks can be localized but have also been known to extend over large areas. In the northeastern United States, population outbreaks have sometimes been extremely damaging; in 1981, gypsy moth populations increased to defoliate 13 million acres (5.3 million ha). Many times, diseases have been linked with the abrupt declines in gypsy moth outbreaks. The big question has been how outbreaks get started. If we know that, perhaps we can prevent outbreaks from occurring. For many years, scientists and land managers studied factors that might change to allow gypsy moth populations to increase from low densities and would result in such unchecked population growth. The answer was not readily evident because these population eruptions were actually driven by factors affecting other trophic levels.

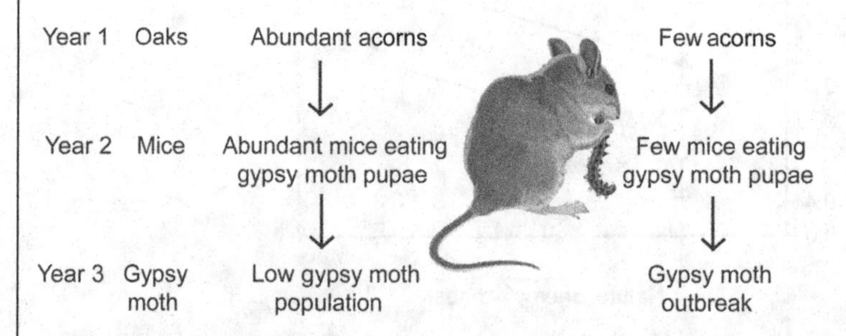

Figure Box 6.2.1 The generalized relations among the levels of acorn production by oak trees, *Quercus* spp. (Year 1) and subsequent white-footed mouse, *Peromyscus leucopus* (Year 2), and gypsy moth, *Lymantria dispar* (Year 3) population densities (illustration from USDA, Forest Service).

Researchers knew that white-footed mice were important predators because they eat gypsy moth pupae occurring near the ground in early summer (gypsy moth has one generation per year). However, acorns produced by oak trees are a dominant food for these mice over the winter. Factors determining the abundance of acorns produced in any one year are complex, including both tree genetics and weather. Overall, oak trees produce large crops of acorns only every 2–5 years. During the years that many acorns are produced, abundant food is available for the mice and more mice survive the winter. In the spring following a fall with a great abundance of acorns, the mouse population will have increased and predation on gypsy moth pupae will be high. When fewer gypsy moth pupae survive, fewer moths emerge and fewer gypsy moth eggs are laid. The gypsy moth eggs then overwinter and the following year there are few gypsy moth caterpillars. Alternatively, when few acorns are produced, the next year, mouse populations will be low (as there was little food over the winter) and few gypsy moth pupae will be eaten so lots of gypsy moth females will survive to lay eggs. Thus, the third year after a low acorn crop, gypsy moth populations will increase. With the high fecundity of gypsy moth, it does not take many years of decreased pupal predation by mice before gypsy moth populations begin climbing to outbreak numbers. While other natural enemies of gypsy moth certainly play a part, it seems that predation by mice is a key factor keeping gypsy moth populations at low densities. Therefore, those factors affecting mouse populations are indirectly setting the stage for outbreaks of gypsy moth to occur.

the research on this has been based on responses of one natural enemy species to one host species on one kind of plant in simplified systems. But of course, naturally occurring systems are much more complex, even for monocultures in simplified landscapes. Therefore, researchers have investigated what happens when there are both foliar- and root-feeding herbivores (caterpillars of *Spodoptera littoralis* and larvae of *Diabrotica virgifera virgifera*, respectively) on the same corn plants. In this case, the responses by natural enemies still occurred but were reduced compared with experiments with one host/one natural enemy systems where damage was only above ground or only below ground. Interestingly, the parasitoid attacking these caterpillars above ground (*Cotesia marginiventris*) was able to learn to recognize the volatiles produced by plants signaling aboveground feeding but they could not recognize odors signaling belowground feeding. On the other hand, the entomopathogenic nematodes (*Heterorhabditis megidis*) attacking the beetle larvae below ground received less of the volatile signals from the dual-infested plants and their response to the prey was less than the aboveground response.

Further Reading

Barbosa, P. & Castellanos, I. (eds.) (2005). *Ecology of Predator-Prey Interactions*. Oxford: Oxford University Press.

Brodeur, J. & Boivin, G. (eds.) (2006). *Trophic Guild Interactions in Biological Control*. Dordrecht, Netherlands: Springer.

Denno, R. F. & Lewis, D. (2009). Predator–prey interactions. In *The Princeton Guide to Ecology*, ed. S. A. Levin, pp. 202–212. Princeton, NJ: Princeton University Press.

Krebs, C. J. (2009). *Ecology: The Experimental Analysis of Distribution and Abundance*, 6th edn. San Francisco, CA: Benjamin Cummings.

Murdoch, W. (2009). Biological control: Theory and practice. In *The Princeton Guide to Ecology*, ed. S. A. Levin, pp. 683–688. Princeton, NJ: Princeton University Press.

Price, P. W., Denno, R. F., Eubanks, M. D., Finke, D. L., & Kaplan, I. (2011). *Insect Ecology: Behavior, Populations and Communities*. Cambridge: Cambridge University Press.
Schmitz, O. J. (2010). *Resolving Ecosystem Complexity*. Princeton, NJ: Princeton University Press.
Shapiro-Ilan, D. I., Bruck, D. J., & Lacey, L. A. (2012). Principles of epizootiology and microbial control. In *Insect Pathology*, ed. F. E. Vega & H. K. Kaya, 2nd edn., pp. 29–72. Amsterdam: Elsevier.

7 Predators

Use of invertebrate versus vertebrate predators has been strikingly different and these predators, themselves, have very different attributes. Therefore, these different types of predators will be discussed separately.

7.1 Vertebrate Predators

Vertebrate predators are better known to the general public than most invertebrate predators. However, the days for use of vertebrates for biological control are largely over; the prey eaten by vertebrate predators is too unpredictable. The behavior of vertebrates is usually more complex than that of most invertebrates. Especially of importance, vertebrates can learn in a new environment and switch to new types of prey. However, exactly because vertebrate predators are larger and more obvious, they were used for early biological control introductions. For example, since ancient times, domestic cats were housed at farms for controlling mice. As long ago as 1762, mynah birds, *Acridotheres tristis*, from India were introduced against red locusts, *Nomadacris septemfasciata*, on Mauritius, an island in the Indian Ocean. In 1872, the small Indian mongoose, *Herpestes javanicus*, was introduced from India to Trinidad to control rats in sugarcane and the activity of these voracious predators was said to prevent GB£45,000 of losses in sugar production, an enormous sum at that time. Unfortunately, both the mynah bird and the mongoose were introduced to many more locations and both went on to demonstrate the potential problems of introducing vertebrates for biological control. The mongoose was predominantly active during the days and rats were active at night. The mongoose became a pest after they quickly learned to kill chickens and the native ground-dwelling lizards and ground-nesting birds. The mynahs outcompeted native birds in Australia where they are known as "flying rats" and they also are major pests in fruit crops. In another disastrous introduction, the cane toad, *Rhinella marina*, was introduced from northern South America into the Caribbean and then, in 1935, it was introduced to Australia to control scarab larvae infesting sugarcane. As with the previous examples, the biology and behavior of this predator was not understood well enough before release and unforeseen side effects ensued without control of the pests (see Chapter 18).

Vertebrate predators are intelligent and also generally quite omnivorous. Therefore, vertebrate predators will switch the type of prey they eat fairly readily. As the unpredictable nature of vertebrate predators became apparent, their use for biological control largely ended. However, there remains one exceptional type of vertebrate predator that is still used today to some extent: small predaceous fish that feed on immature stages of mosquitoes (Box 7.1). Also, predaceous birds that feed on insect pests are sometimes protected as part of conservation biological control strategies.

7.2 Invertebrate Predators

The range of prey that will be attacked by predaceous invertebrates is often much more predictable than the range of prey attacked by vertebrate predators. These natural enemies have less ability to switch prey because they are less mobile or less able to control their mobility and are generally more restricted in the size of prey that can be caught and eaten, diet breadth, and habitat use. Some, like spiders, are generalists and feed on a variety of prey within a habitat.

Insect predators important for biological control have one of two major types of development. For both of these strategies, insects (and other invertebrates) shed their outer cuticles when developing (called molting) and each life stage between molts is called an instar. More primitive insects, the Hemimetabola, have immature stages called nymphs that are similar in appearance to adults, although adults are reproductively mature and have fully developed wings. This gradual type of development is seen with praying mantids and true bugs. Mites and spiders also have this type of gradual development. Predators are also found among the more evolutionarily advanced groups of insects with complete metamorphosis, having immature stages called larvae (singular = larva), that are very different from adults, with an intermediate pupal stage during which an extensive metamorphosis occurs. Predators with complete metamorphosis, the Holometabola, include groups such as wasps, ants, flies, and beetles. For the holometabolous strategy, the needs and habitats of immatures and adults can be very different; for example, for many species only larvae and not adults are predators. Because holometabolous immatures are often less mobile, adult females can have a huge impact on success of their offspring depending on their choice of locations for laying eggs.

Invertebrate predators are often not as adept at finding prey as many vertebrates. They must locate the habitat, then the prey, and thereafter attack, accept, and then eat the prey. They locate the general habitat, such as the host plant, where prey are usually found using chemical stimuli, including naturally produced plant volatiles. Once in the correct habitat, to find prey various invertebrate predators use vision, prey movement, and chemical stimuli requiring contact or not. A strategy that some predators use is locating their prey using volatiles emitted by plants after they are damaged by an herbivore (HIPVs, introduced in Chapter 5). Using this information can solve a problem for predators as volatiles from damaged plants can help predators to find hiding herbivores in a "sea" of plant material. Volatiles that are emitted only after damage are also a reliable indicator of herbivore presence that can be easier to detect than the herbivores themselves. For example, laboratory studies have shown that wild tobacco plants (*Nicotiana attenuata*) being attacked by leaf-feeding herbivores released additional volatiles. Exposing an important predator in this system, the big-eyed bug *Geocoris pallens*, to

Box 7.1 A Fishy Tale

The principal species of fish that have been exploited for biological control of arthropods are the mosquitofish, *Gambusia affinis*, native to southern North America, and *Gambusia holbrooki*, native to the southeastern United States. Both are commonly referred to as *Gambusia*. These are small (2.5–5 cm/1–2 in.) omnivorous species with high reproductive capacity that can live in shallow water and tolerate changes in temperature, salinity, and the presence of organic waste. *Gambusia* were first introduced from North Carolina to New Jersey in 1905. By 1975, *Gambusia* had been introduced for mosquito control in about fifty countries around the world, making them the most widely distributed biological control agent at the time.

Initially, small numbers of *Gambusia* were introduced (primarily as classical biological control agents) to locations with the goal that they would increase on their own over time. Later, when mosquito control efforts changed, mosquitofish were cultured, harvested, and stored over the winter so that inoculation biological control was possible with releases at specific times when mosquito control was needed, for example, after rice fields were flooded.

The resulting control of mosquitoes by these fish has been variable, although *Gambusia* are credited with some great successes. One success controlling mosquitoes that vectored malaria along the coast of the Black Sea was memorialized in Sochi, Russia, in 2010 with a brass statue of a *Gambusia* fish. Several additional fish species that feed on mosquito larvae, including *Poecilia* spp., have also been used in more than sixty countries. Control seems to work better when there is limited alternative food for the mosquitofish, so that they chiefly prey on mosquitoes. Unfortunately,

Figure Box 7.1.1 Statue of *Gambusia affinis* in Sochi, Russia. In the early 1900s, areas around Sochi were boggy with large mosquito populations and malaria (vectored by the mosquitoes) was a large problem. *Gambusia* were introduced and are credited with the fact that no cases of malaria were reported after the 1950s (photo by Cheyushova Marina).

Gambusia have exerted negative environmental impacts in some cases. *Gambusia* can directly impact populations of native fish through predation on fry (young) or they can exert an indirect effect by outcompeting native fish for resources, especially food. More than thirty species of native fish have been adversely affected after *Gambusia* introductions. In addition, the introduction of *Gambusia* has been linked to declines in aquatic amphibians and invertebrates. Because *Gambusia* feed on zooplankton, algal blooms can occur after they are introduced. The World Health Organization now does not recommend use of *Gambusia* for mosquito control because of the potential for environmental impacts. Thus, widespread use of *Gambusia* is not encouraged in natural habitats, although *Gambusia* is still used in private ponds. The lesson is that releases of an aggressive and versatile predator in natural habitats should not be done before a full evaluation of environmental risk has taken place, including in-depth studies to clarify the importance of the pest and whether the predator is significantly better at reducing the pest than the predators that are already present.

these volatiles in the laboratory resulted in higher predation levels during experiments. The researchers concluded that predators were using the HIPVs to help improve their ability to locate prey. Researchers have extended their studies of interactions between HIPVs and predators to agroecosystems. One study demonstrated that HIPVs from corn plants infested by mesophyll-feeding leafhoppers (*Zyginidia scutellaris*) or leaf-feeding caterpillars (*Spodoptera littoralis*) resulted in attraction by generalist predatory minute pirate bugs, *Orius* spp. Previous exposure to HIPVs helped these predators learn and subsequently find these prey more easily.

As a general rule, adults of invertebrate predators are often more mobile and have better vision than immatures, and adults lay eggs in locations where prey are present. Frequently, eggs are laid in areas with aggregations of prey so there will be plenty of food when eggs hatch. To assure this, adult female lady beetles not only detect whether prey are present but also can register the presence of eggs from other predators. If eggs of other predators are already present, a female will often lay fewer eggs of her own.

Invertebrate predators utilize a range of methods for capturing prey. In general, the body size of invertebrate predators is larger than that of their prey, especially for species that overwhelm their prey. Overwhelming prey could be considered the basic strategy of most invertebrate predators. However, some trickier predators do not have to be larger than their prey because they inject poison to kill prey. Social insects like ants successfully attack prey in groups, for example, army ants marching through tropical rainforests can subdue prey of large sizes because of the sheer numbers of ants simultaneously attacking in a coordinated way.

The habitat in which predators live impacts which prey they encounter and can capture, but invertebrate predators also employ several very different hunting methods for capturing prey. Many invertebrate predators use what has been called "active hunting." Mobile predators that have good vision, such as ground beetles and jumping spiders (Salticidae), chase after prey that they see. Others, that might have vision that is not as exact, use a combination of vision and chemical cues to find prey. For those with very poor vision, such as immature lady beetles, the principal method for detecting prey is tactile, so these predators roam incessantly. Because prey are often aggregated in distribution, as a strategy for finding prey lady beetle larvae first wander in the area of their last meal but, as hunger grows, they roam further and further from their last prey encounter in the hope of finding another group of prey (see below).

A second invertebrate hunting strategy called "sit and pursue" involves using a fixed location and then rushing out and pouncing on prey when the prey moves within chasing distance. Of course, the fixed location changes as prey distributions change. An example of this would be many wolf spiders that are often camouflage-colored for hiding but then rush out when prey are nearby and thus chase prey over shorter distances.

The third major way that invertebrate predators hunt prey is to "sit and wait," often remaining concealed during this time, and then attacking only when prey are present. This "ambush" strategy, well known from the praying mantid, is the best method for catching fast prey, although it requires a lot of patience and a fast response requiring limited movement, once prey are present. In an extreme, larvae of ant lions (Myrmeleontidae) sit and wait as they lie buried in the sandy soil at the bottoms of small pits with their mouths positioned so that any insects falling into the pit will readily be captured and eaten.

The flip side to the success of predators finding and catching prey is prey defense. Very mobile prey can simply evade capture, often by running or flying away. Many less-mobile prey have morphological features to deter predation, such as the hard covering of armored scales (Diaspididae) or abundant long hairs on tussock moth caterpillars (Erebidae; Lymantriinae). Herbivorous insects feeding on some plants can sequester noxious plant compounds. For example, oleander aphids, *Aphis nerii*, sequester toxic cardenolide steroids from host plants and are conspicuously colored yellow and black as a warning to predators. In a study where nine species of invertebrate predators were fed these aphids, three predator species did not survive, three had decreased growth, and only three had the physiological ability to eat this chemically defended prey and develop at a normal rate (Figure 7.1). Therefore, predators can be specialized for overcoming specific prey defenses although predators that are generalists may not be able to overcome specialized defenses. Predatory ants which are usually generalists often reject chemically defended caterpillars as prey. The types of defenses employed by pests can definitely have an impact on biological control and can determine which natural enemies will be successful. In an evaluation of classical biological control programs, caterpillars that were visually cryptic (blending in with their surroundings) and had smooth body surfaces had the highest levels of predation and were most successfully controlled by invertebrate predators.

Predators ingest prey in different ways. A general insectan model of eating involves use of mandibles for cutting and crushing food with a variety of additional mouthparts assisting in processing a meal (Figure 7.2). Alternately, some insects such as true bugs have tubular piercing-sucking mouthparts. For predators with piercing-sucking mouthparts, food must be liquid. So how do they eat prey? Saliva containing digestive enzymes is injected into the prey and the partially digested prey contents are then ingested. The saliva, in these cases, can also be paralytic or poisonous to arrest movement of prey

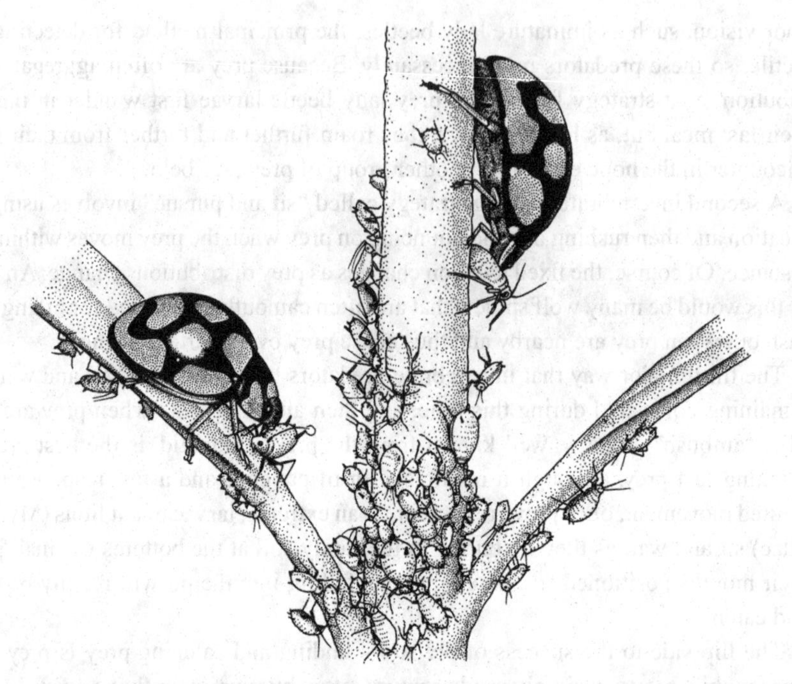

Figure 7.1 Chemically defended oleander aphids, *Aphis nerii*, cannot be eaten by some predators yet this lady beetle, *Cheilomenes lunata*, has no problems feeding on them (drawn by Karina H. McInnes; Gullan and Cranston, 2000).

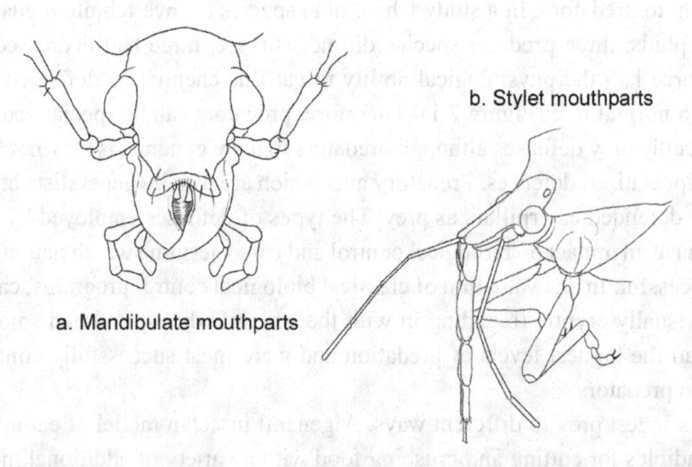

Figure 7.2 Mandibulate (chewing) versus stylet (piercing-sucking) mouthparts of insect predators. (a) Mandibulate mouthparts of the ground beetle (Carabidae), *Calosoma frigidum*. (b) Stylet mouthparts of the stink bug (Pentatomidae), *Podisus maculiventris*. Both predators feed on a variety of prey, including gypsy moth caterpillars (drawn by Alison Burke).

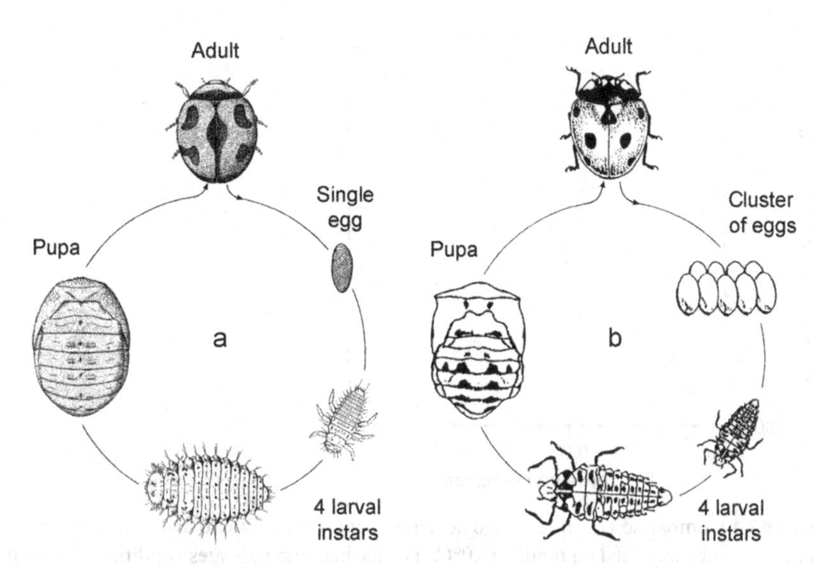

Figure 7.3 Life cycles of lady beetles demonstrating morphological variability throughout a life cycle. (a) The scale-feeding *Rodolia cardinalis*, the vedalia beetle (adults 3–4 mm long). (b) The aphid-feeding *Coccinella septempunctata*, the seven-spotted lady beetle (adults 7–8 mm long) (based on A. F. G. Dixon, 2000).

so that feeding can occur without interruption. Spiders and mites also inject digestive enzymes into prey and then ingest liquefied food.

Predation is a widespread life strategy among invertebrates. Below, we will describe some major groups of predators important in biological control. These will be presented as predators introduced for classical or augmentation biological control and then predators that play important roles in naturally occurring biological control as part of a community of natural enemies. The predators important in naturally occurring control have been the focus of habitat manipulations in some conservation biological control programs that work to increase populations of predators.

7.2.1 Arthropod Predators Used for Biological Control

Lady Beetles (Order Coleoptera: Family Coccinellidae)

Lady beetles, also called ladybugs or ladybird beetles, are some of the worlds' experts at eating small, soft-bodied prey such as aphids, whiteflies, mites, mealybugs, and scale insects (Figure 7.3). Adults of some species of aphid-feeding lady beetles can consume approximately 100 aphids per day. The well-known adult stages of lady beetles are shiny and convex, with short, clubbed antennae, and they are often depicted as an icon of biological control. The family name for lady beetles, Coccinellidae, means clothed in scarlet although many lady beetles are not red and, in fact, many are dull-colored, without markings and can be quite small (1 mm long). Patterns on the surfaces of adults help in identifying the species; once patterns on the wing covers (elytra) are formed, they do not change after an adult emerges from pupation and hardens.

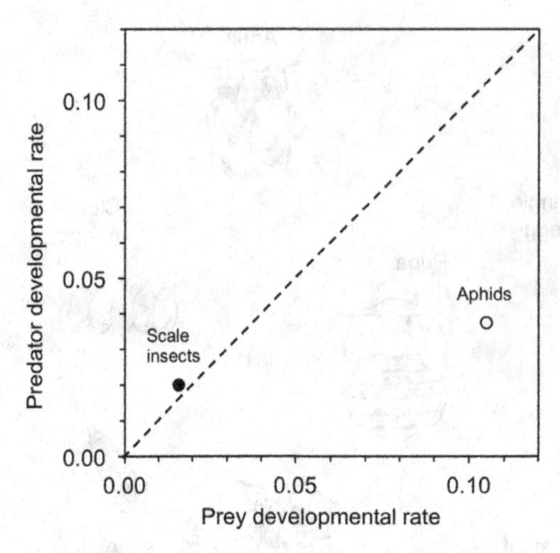

Figure 7.4 Comparison of the average developmental rates of aphid and scale prey and the lady beetle species that feed on them, at 20°C. The dashed line indicates conditions where the prey and predator develop at the same rate; predators developing faster than prey fall on the left of the line and predators developing more slowly than prey fall on the right (based on A. F. G. Dixon, 2000).

The flattened and more elongate immature stages of lady beetles resemble little dinosaurs or alligators more than the adults they will become. Larvae have reduced eyesight and, for many species, the larvae must touch their prey with their chemo-sensory mouthparts before they understand that what they have touched could be eaten. Aphid feeders often hunt by walking quickly, sometimes stopping to swing the front end of their body from side to side, to maximize chances of contacting prey with their mouthparts. Once a prey individual is found and has been eaten, then searching in that area becomes more concentrated; this strategy is well suited for specializing on prey that occur in aggregations, such as many aphids. Detailed studies have shown that if a larva has recently found and eaten a larger aphid, it will continue searching in an area longer than if it had found only a smaller aphid. To optimize use of aphid colonies, an adult female two-spotted lady beetle (*Adalia bipunctata*) searching for places to lay eggs will leave a colony if she detects the trail pheromone left by larvae of the same lady beetle species.

Young lady beetle larvae pierce their prey and suck the contents while older larvae and adults chew and eat the entire prey. Species of lady beetles often eat only certain types of prey, and this has been thought to increase their effectiveness in controlling pests (see Box 3.1). Many species of lady beetles are specialists on either aphids or scales, although a few species eat both. Both aphid- and scale-feeding lady beetles have been used for biological control. Classical biological control programs using scale feeders seem to have been more successful than programs using aphid feeders. This success is probably tied to the fact that scale feeders develop faster than their prey (Figure 7.4) and are more host specific. In contrast, aphids usually develop faster than the aphid-feeding

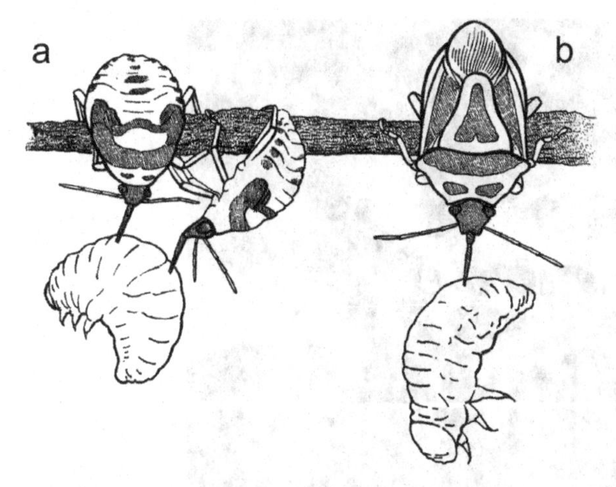

Figure 7.5 Nymphs (a) and an adult (b) of the stink bug *Perillus bioculatus* feeding on larvae of the Colorado potato beetle, *Leptinotarsa decemlineata*, impaled with their piercing-sucking mouthparts (drawn by F. Petre; Trouvelot, 1931).

lady beetles and the predators are often unable to keep up with their quickly increasing prey.

True Bugs (Order Hemiptera)

Although all insects are commonly called "bugs," to an entomologist, the Hemiptera are the only group containing true bugs. Many true bugs are plant feeding and others are blood sucking but there are also some important families of predatory bugs that feed on insects. Immature stages of Hemiptera resemble adults, being hemimetabolous, but do not have the fully formed wings. For adults, wings of predatory species are normally held in a flat position, on top of the abdomen. The most unique feature that differentiates bugs from other insects are their tubular mouthparts. Predaceous species extend their mouthparts forward and use them to spear their prey and inject enzymes to digest prey, sometimes also injecting poisons or compounds causing paralysis. Predatory bugs then suck out the body contents of the host (Figure 7.5). At rest, the mouthparts are held beneath the body and therefore are not readily visible.

Bugs are often general feeders, both immatures and adults eating eggs, immatures, and adults of a diversity of insects and mites. Predatory stink bugs (Pentatomidae) simply walk toward caterpillars with their mouthparts extended and pierce them. These hemipterans are especially well adapted to feed on prey covered with defensive hairs or spines because they can eat them while standing a short distance away. Most other terrestrial predatory insects have some modification so that legs are used for capturing prey and grasping them while eating, so they must be right next to prey. Interestingly, some hemipteran predators, such as the bug *Macrolophus* sold for augmentative control of whiteflies in Europe, can facultatively feed on plant materials, which is beneficial in instances where all prey in an area have been eaten already (see Section 7.3.4); their

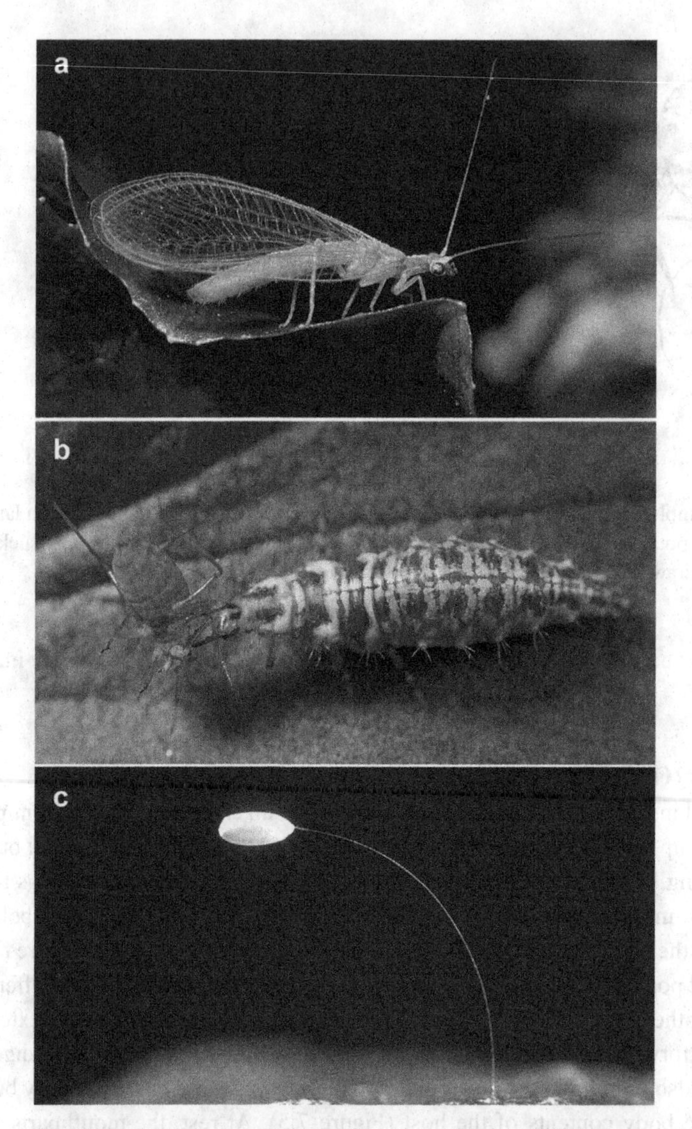

Figure 7.6 The lacewing *Chrysoperla carnea*. (a) Adult. (b) Voracious larva that often eats aphids. (c) Egg (photo by Jack Kelly Clark, courtesy of the University of California Statewide IPM Program).

damage to plants is minor and this ability to switch from predator to herbivore allows these bugs to persist in greenhouses so that they are present if prey populations increase again.

Lacewings (Order Neuroptera)

On spring and summer nights, among the moths at lights you can find green or brown insects with long delicate lacy wings folded tent-like over their abdomens (Figure 7.6a; see also the cover of this book). Adult lacewings lay their eggs and disperse at night.

These adults can be predaceous, but some feed on pollen and others do not feed at all. It is really the larvae of these holometabolous insects that are the important predatory stages of interest for pest control.

Lacewing larvae are 3–20 mm (1/8–4/5 in) long, with large pointed jaws for skewering their prey. They are perhaps best known for their appetite for aphids and are sometimes called aphid lions but they will also eat other small insects as well as mites. Lacewing larvae actively search for prey and, once they randomly bump into something, they can identify it as food only after contacting it with their mouthparts. The sickle-shaped mandibles of green and brown lacewing larvae are used to initially pierce prey and digestive salivary secretions are then injected (Figure 7.6b). Ultimately, only the predigested fluids are consumed by the larvae. Because lacewing larvae will eat each other, the adults usually lay eggs at the ends of small stalks so that the unhatched eggs dangle in the air (Figure 7.6c). This prevents the first larva that hatches from eating the nearby eggs of its unhatched brothers and sisters.

For biological control, green lacewings are released by shaking eggs onto foliage. They have been used in greenhouses and some row crops, but to date releases have not consistently provided control. Lacewings seem particularly vulnerable to predation by other predators (a process termed intraguild predation, see Section 7.3.2) and hence they may be better-suited for greenhouse releases where the presence of other predators can be managed. Naturally occurring populations of lacewings are considered important members of resident natural enemy communities.

Predatory Mites (Class Arachnida: Order Acarina)

Mites are arthropods, as are insects, but differ from insects in having eight legs, two body parts, and no antennae. With magnification, one can see that predatory mites are long-legged and often are pear-shaped and shiny (Figure 7.7). Mouthparts of predatory mites extend forward from their bodies while mouthparts of plant-feeding mites are directed downwards to the plants on which they feed. Predatory mites use their mouthparts to pierce their prey and inject digestive enzymes. The prey is therefore digested externally and the mite laps up the resulting liquefied mush. Eggs of predatory mites are often quite large relative to the mites and, with magnification, appear oblong and pearl-colored. Directly after hatching, immature mites (nymphs) have six legs but soon gain an extra pair of legs. Otherwise, immatures are very similar in appearance to the adults.

Most mites can just barely be seen with the naked eye. Yet, although small, pestiferous mites increase in numbers so readily they can create major problems. Predatory mites can be extremely effective natural enemies for control of plant-feeding mites. This has been clearly demonstrated by the ready occurrence of secondary pest outbreaks of mites in agriculture when insecticides kill predatory mites (see Chapter 1). After applications of pesticides, predatory mites will eventually increase in number again in the sprayed areas since either a few will survive the pesticide or they will recolonize the area through their ability to disperse over longer distances by riding on the wind.

Predatory mites are about the same size as the plant-feeding mites that they attack. Some of the best-known predatory mites are in the family Phytoseiidae but predators also occur in many other mite families. Some of these phytoseiids are generalist

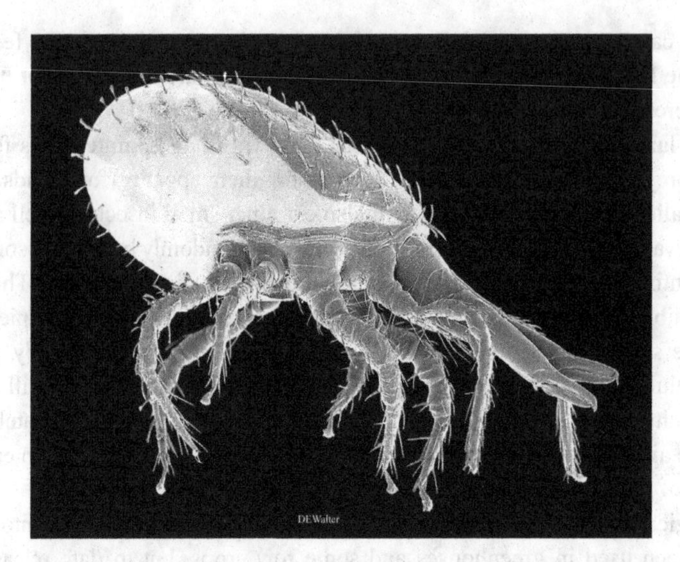

Figure 7.7 *Hypoaspis miles*, a soil-dwelling predatory mite used in greenhouses for control of western flower thrips (*Frankliniella occidentalis*) pupae and fungus gnat larvae (photo by David Evans Walter).

predators attacking plant-feeding mites as well as insect eggs and small immature insects. For example, the species *Phytoseiulus persimilis* is a predatory mite used worldwide to control spider mites feeding on plants. *Typhlodromus pyri* is even more of a generalist and may feed on pollen and fungi as well as pestiferous mites, thus persisting well in perennial systems such as orchards which can improve their efficacy as biological control agents (see Box 5.1). Populations of some predatory mites that readily eat pollen may be maintained in an area by making sure there are pollen sources for times when prey populations are low. Alternatively, some phytoseiids are more specific, preferring pestiferous spider mites and dispersing when prey populations decline. These last-mentioned species are known to use chemical cues from the prey mites themselves as well as host plant cues to locate prey.

Predatory Flies (Order Diptera)

Many adult flies are predators and are important members of naturally occurring food webs. However, for biological control, the larval stages of only a few types of flies have received most of the attention, that is, the flowerflies or hoverflies (Syrphidae) (Figure 7.8), aphid flies (Chamaemyiidae) and predaceous midges (Cecidomyiidae). While the adults of these groups are excellent fliers and feed on pollen, nectar, or do not feed the immatures are legless maggots. Larvae can move but certainly not fast or far so they really rely on adults to deposit eggs near hosts. Fly maggots have very reduced mouthparts but are still able to hold tightly onto small-bodied prey. Larvae of all of these flies are usually not very host specific, feeding on aphids, mites, scales, and other soft-bodied arthropods that are not very mobile and are smaller than the fly larvae.

Figure 7.8 Syrphid larva feeding on an aphid. The adult is a hoverfly, many of which are bee mimics (photo by Jack Kelly Clark, courtesy of the University of California Statewide IPM Program).

7.2.2 Invertebrate Predators Providing Naturally Occurring Biological Control

Praying Mantids (Order Mantodea)

Praying mantids can be quite large for insects, often 5–10 cm (2–4 in.) long and they are easier to notice than many insects for this reason. They are called praying mantids because their forelegs are held in an upraised position, similar to the posture assumed for praying by Christians. However, their forelegs are actually held in that position so they are ready to grab prey (Figure 7.9).

Praying mantids are "sit and wait" predators, often sitting on or next to flowers and eating bees and flies, sometimes including beneficials that visit the flowers. Therefore, they are effective at catching mobile and not sessile prey. However, mantids do not discriminate about what they eat. This extends to eating siblings as they hatch from egg masses. Adult males are even sometimes eaten by females after mating. Because praying mantids are such generalists, they are generally not considered very effective for controlling specific pests.

Ground Beetles (Order Coleoptera: Family Carabidae)

Ground beetles are the commonly found dark beetles that hide under stones or in dense plant material on the soil surface. Many are predators although some instead feed on seeds. Adults are medium- to large-sized and they are often dark with long legs. They are fast runners and rarely fly or sometimes cannot fly (Figure 7.10). Larvae of these holometabolous insects do not look anything like adults. Larvae are elongated and usually live within the soil. Both larvae and adults generally have prominent forward-directed mandibles and actively pursue their prey, especially if it is moving. Adults can cut their prey into pieces with their mandibles and swallow the pieces. Larvae usually use extra-oral digestion; enzymes are introduced into the prey and then the liquefied prey contents are ingested by the beetle.

Adults are usually active at night and can eat approximately their weight each day. Most species are generalists and eat a variety of different ground-dwelling prey. A few species climb vegetation to find prey, even climbing large tree trunks to eat caterpillars.

Figure 7.9 Praying mantids are some of the largest insect predators. They are ambush predators, remaining motionless while waiting for prey with front legs upraised (photo by Jack Kelly Clark, courtesy of the University of California Statewide IPM Program).

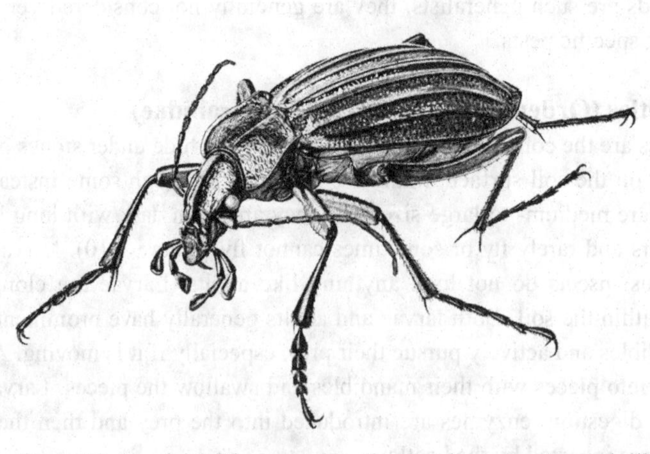

Figure 7.10 An adult of the ground beetle, *Carabus auronitens*, a brilliant green and golden species (18–26 mm long) common in the forests of central Europe (Essig, 1942).

Snail-feeders are morphologically adapted by having a small head that fits into the opening of a snail shell.

Carabids are important predators, providing natural control as part of complex food webs. Their populations can be strongly enhanced by conservation measures like maintaining beetle banks, hedges and flower strips as refuges and their numbers can be seriously reduced in monocultures where absence of shelters leaves no habitat for them.

Rove Beetles (Order Coleoptera: Family Staphylinidae)

Rove beetles occupy similar niches to ground beetles, namely some adults search for prey on top of the soil while for other species, adults and larvae find prey within the soil. Rove beetles are generally smaller than ground beetles and often have shortened hardened forewings (elytra), usually with wings used for flight folded beneath. The shortened forewings make it easier for these small beetles to move around in tight spaces but the fact that the abdomen is flexible also helps with orienting releases of chemicals used for defense or mimicry from abdominal glands. This is one of the largest families of beetles. They are especially important as predators of eggs and small fly larvae in the soil. Both rove and ground beetles are regularly monitored by pitfall traps that capture organisms walking on the soil surface.

Ants (Order Hymenoptera: Family Formicidae)

Ants are often extremely abundant and successful predators in many types of habitats. In some instances, they are considered the keystone predators in communities. This is true of endemic ants in native ecosystems but can also be seen with invasives. A dramatic example of an ant species that is a keystone predator is the red imported fire ant (*Solenopsis invicta*) native to South America, that was introduced to the southeastern United States, where this species has become abundant and to some extent has displaced many endemic predators.

Ants are related to bees and wasps, differing because they are usually wingless with the exception of those ants born to mate and disperse. Ants are of course social insects with two major female castes, the queens and workers, and there are usually morphological subdivisions within the workers.

Many ants are predatory or at least omnivorous but they are usually generalists in their choice of prey and in some forest systems they are regarded as important predators of pests. Therefore, ants in the genus *Formica* are protected in German forests to promote conservation biological control. There has been relatively little research on the use of ants for biological control. While some species are known to be beneficial, the majority are seen as nuisances or even detrimental to biological control. In particular, ants that tend aphids, soft scales, whiteflies, and mealybugs (all in the suborder Homoptera, order Hemiptera), and feed on the honeydew produced, in turn protect these homopterans from predators and parasitoids, so that ants are helping populations of these pests increase. Ants also move these honeydew-producing insects from plant to plant. Ants can also disrupt biological control of pests that are not ant-tended, such as mites and armored scales, if these occur on the same plants as honeydew-producing species that are being tended.

Spiders (Class Arachnida: Order Araneae)

Spiders are in the same class of arthropods as the mites, also having only two body parts, eight legs and no antennae. Spiders are ubiquitous, commonly found and all are predators. Spiders vary in their behavior, being suited for specific habitats and types of prey. Web builders make many types of webs but perhaps the best known are the orb weavers that create lovely spiral orb webs. Orb weavers have poor vision, but night or day they are sensitive to vibrations in their webs that potentially signal that struggling prey are caught in the web and cannot escape. Non-web-building spiders include the ground-dwelling wolf spiders that wander at night to find prey. During the day, crab spiders can often be found on or within flowers, waiting for unsuspecting flying insects to visit the flower for nectar or pollen. Crab spiders can change color and are often brightly colored so that they match the color of the flower in which they sit. The jumping spiders have their eight eyes arranged like headlights on their heads to find prey. They wander during the day in the vegetation and on the ground, and can jump impressive distances (50 times their own body length) to reach prey.

7.3 Predator Choices and Impacts

Predators range from specialists to omnivores and can have impacts as a result of eating prey or just as a result of being recognized as risky by prey. Below, we explore some of these aspects of predators in more depth relative to biological control.

7.3.1 Specialist versus Generalist Predators

Many predators feed on a broad range of prey and are then often called generalists. In contrast are the specialists, feeding on one species (monophagous) or only a limited variety (oligophagous) of prey. Actually, it can be difficult to categorize species using these subjective terms. Species that are actually generalists in prey choice can functionally be specialized in their prey use if they only utilize certain areas inhabited by few prey species or if their smaller size restricts them to only the few species of smaller prey within their habitat.

There has been much debate about whether specialist or generalist predators make better biological control agents. An advantage of generalists is that there is a better chance that they will persist in a system when the pest is not present. Therefore, when the pest increases or disperses into the habitat, the predators are already present. There are also some disadvantages to generalists, because they usually do not respond to prey populations in a density-dependent manner (Chapter 6) and might cause undesired effects if feeding on alternate prey (see Chapter 18). Specialist predators, on the other hand, have the benefit that direct impacts on nontarget organisms are negligible to nonexistent. However, specialist predators often do not persist as well in an environment once the prey are gone. This can be compensated for, to some extent, if the species readily disperses and would therefore reinvade soon, if the pest increases again. Also, some predators can survive when feeding part of the time on alternative food sources like plants, pollen, and fungal spores and this broad food use is called omnivory (see below).

7.3.2 Effects of Predators on Other Natural Enemies

It is important to consider whether predators are generalists or specialists for several practical reasons. Generalist predators, such as praying mantids and spiders, can feed on beneficials, including parasitoids and other predators. Therefore, their usefulness in controlling prey has to be considered relative to the extent to which they influence other natural enemies. In fact, interactions between natural enemies can take several forms: (1) there can be no interaction, (2) natural enemies can kill each other, (3) one natural enemy can interfere with foraging by another, or (4) a natural enemy could influence the behavior of pests making them more or less likely to be eaten by other natural enemies (see Section 7.3.3). Because predators tend more toward being generalists, they have been the focus of interest in such interactions. The principal concerns relative to biological control have been when one predator interferes with the ability of a second to capture prey and, more importantly, when one predator kills another predator, which has been called intraguild predation. Such negative interactions have been studied in systems with two or more predators known to prey on each other, by testing them singly and then in combinations to evaluate subsequent suppression of plant-feeding pests. These studies have clearly shown that adding predators to systems where other predators already occur does not consistently improve pest control although results seem to differ for different systems; manipulating the presence of different predators in one system has demonstrated that a higher level of intraguild predation in a system can be more important to biological control than greater predator diversity.

As an example of the effects of the same predator in different systems, red imported fire ants in cotton are voracious predators feeding on herbivores along with other natural enemies. However, while having high populations of fire ants in cotton maximized biological control of most pests, this did not extend to all pests. These fire ants tend cotton aphids, *Aphis gossypii*, protecting them from natural enemies and feeding on their honeydew, so aphid populations can increase to abundant densities. In contrast, in collards the red imported fire ants are not as voracious and although they eat some parasitized caterpillars, the effects of the fire ants and parasitoids are additive so that the resulting biological control is maximal when both fire ants and parasitoids are abundant.

7.3.3 Predators Also Affect Prey That They Don't Catch

Red-legged grasshoppers, *Melanoplus femurrubrum*, normally feed on a variety of different plants, but if hunting spiders, *Pisuarina mira*, are present, these grasshoppers will instead mostly feed in *Solidago* plants that are structurally complex; when they are feeding in *Solidago*, spiders have a more difficult time finding and capturing these grasshoppers. In this case, the predators are impacting the prey without killing them, which has been called a "nonconsumptive" effect. Nonconsumptive effects have also been referred to as "intimidation" or "insect fear." Although these term have perhaps principally been abandoned because they seem anthropomorphic, and they provide the idea that prey are changing their behavior to avoid predators.

Nonconsumptive effects can cause prey to change their habitats or diets in order to balance the risk of not eating with the risk of being eaten. Of course it has been much easier to document consumption of prey by predators rather than nonconsumption, which is more subtle to detect as it requires some kind of predator exclusion studies for confirmation. To show that nonconsumptive effects are occurring, scientists have to be able to manipulate the prey's detection of the predator without the prey also being eaten. One way this is done is by adding visual or acoustic cues characteristic of a predator. The presence of chemicals from the predator can also be used; for example, one study demonstrated that bird cherry-oat aphids (*Rhopalosiphum padi*) avoided leaves where their predators, the seven-spotted lady beetle (*Coccinella septempunctata*), had been walking. The prey could smell the predator's footprints for at least some period of time after the lady beetle was present.

In recent years, scientists have been finding that nonconsumptive effects can influence the dynamics of many types of ecosystems. We have been learning that avoidance of predators can have an impact that is as much or even more than the impact of direct consumption of prey by predators; in agricultural systems, these effects can influence the prey as well as the plants the prey would feed on (the plant being what we are trying to protect).

But how do these nonconsumptive effects, leading to predator avoidance or prevention of being eaten, influence biological control of pests? Defensive tactics used by prey in response to predators are varied with numerous of them resulting in decreased impacts by herbivores on plants, for example, reduced foraging effort by herbivores, which can then reduce the amount of energy the herbivores gain through feeding and which result in decreased reproduction. These effects can persist even after the predator is gone, so a transient predator moving through an area can have an impact. Alternatively, prey can disperse or emigrate to avoid predators. Adults can choose not to lay eggs where predators are present. Few studies have detected synergistic effects of multiple predators on pests; for synergism to occur, the total effects of two predator species should be significantly greater than adding together the effects of each species alone. However, two predators feeding on the same aphid prey have been shown to have a synergistic effect on predation, although only one of the predators feeds on the prey and the other has a nonconsumptive effect (Box 7.2).

7.3.4 The Importance of Zoophytophagy

Many predatory insects feed on plants as well as prey and this is called zoophytophagy. This varies by species so that plant feeding by predators can be only occasional or during specific parts of a holometabolous life cycle or it can instead be a regular occurrence. Many parasitoids or other insects such as ants feed on nectar from plants or honeydew, a liquid plant-derived waste product from hemipterans. Predatory mites are well known as feeding on pollen as well as prey. However, recent research has been focusing on insect predators that feed on plant tissues as well as prey. Whatever plant structures are eaten,

Box 7.2 Predators . . . Working Together?

Aphids and lady beetles have been popular among scientists studying predator–prey interactions. John Losey and Bob Denno conducted experiments with two aphid predators simultaneously to ask questions about their interactions. Pea aphids, *Acyrthosiphon pisum*, feeding on alfalfa plants are quite large for aphids, and are very active, readily dropping from plants if disturbed. In individual cages containing alfalfa plants and pea aphids calmly feeding, Losey introduced either an adult seven-spotted lady beetle, *Coccinella septempunctata*, a ground beetle, *Harpalus pensylvanicus*, or both predators. After 24 hours, he found that aphid populations had declined if the lady beetle alone was present but there was little effect if only the ground beetles were present. In contrast, if both predators were present, more aphids were eaten than if you added the effects of each predator together. What was happening? The lady beetles were disturbing the aphids that would then drop from plants. Without the ground beetles, the aphids that dropped to the ground simply walked over to the plant stem and climbed back up to the leaves. However, when ground beetles were present, aphids were an unexpected tasty meal that dropped from plants and hit the ground. The aphids were easy to catch since they aren't very well adapted to run away from fast-moving larger ground beetles on top of the soil. This study suggests that having both the ground beetles and the lady beetles yields a synergistic effect with more aphids killed than if the effects of each predator were added together.

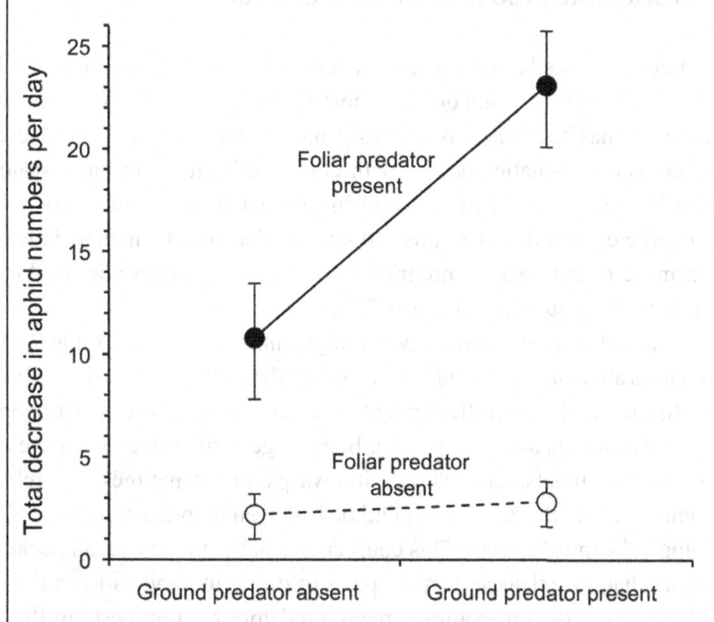

Figure Box 7.2.1 Comparison of the effect of predation by a foliar forager (the lady beetle *Coccinella septempunctata*), predation by a ground forager (the ground beetle *Harpalus pensylvanicus*), or both predators on consumption of pea aphids. More pea aphids were eaten with the combination of predators than the sum eaten when foliar and ground predators were tested separately (Losey & Denno, 1998).

this practice is thought to often provide essential nutrients for insect predators that they do not gain from feeding on prey alone.

Zoophytophagy can be viewed as very positive when it allows a predator to persist in an area where prey are few and if this practice leads to greater fitness because predators are receiving required nutrients. However, zoophytophagous predators have also required evaluation because if they are to be used for biological control, it is important that they do not switch and feed too much on plants instead of prey if that might potentially cause damage to crops. In addition, predators used for biological control that readily feed on plants might be less actively eating pests as prey.

Natural enemies feeding on herbivorous arthropod pests can be thought of as protectors, or bodyguards, for the plants that the pests are feeding on. Recent studies have questioned whether the zoophytophagous predators are protecting plants very well. It turns out that first zoophytophagous predators are eating the pests as prey, so this results in straightforward plant protection. However, there is also an indirect protective effect; experiments with the zoophytophagous mirid *Macrolophus pygmaeus* have shown that after these predators were feeding on tomato plants, the plants mounted defense responses so that two-spotted spider mites, *Tetranychus urticae*, subsequently exposed to plants had lower fitness. Thus, the plant feeding by the predator acted to protect the plant via activating the plant defenses to provide some resistance against the mites.

7.4 Use of Invertebrate Predators for Pest Control

Predators have been used quite extensively for classical biological control, with increasing emphasis through the years on more host-specific predators. A group that has been used extensively has been the lady beetles (Family Coccinellidae). This choice is in part owing to the early dramatic success with control of cottony cushion scale by a lady beetle (see Box 3.1) and, in part, owing to the fact that biological control programs have often targeted introduced aphids and scale insects on which many lady beetles specialize. In more recent years, a program against cassava green mite in Africa utilized predatory mites very successfully (Box 7.3).

Predators are also used extensively for augmentative releases (Table 7.1). However, many predators are generalists and this can dilute their effectiveness in controlling a specific pest. In the highly controlled greenhouse environment where invertebrate species that are present are usually not desired, being a generalist does not have to be detrimental. Other beneficials, such as parasitic wasps, are sometimes also released in the same greenhouses, so the releases should be balanced to avoid predation of these parasitoid biological control agents. This conflict is usually not a problem because many of the predators that are released prey on pests in different habitats from the pests being attacked by parasitoids, for example, many predators feed on pests in the soil, which would not impact parasitoids attacking foliage-feeding pests. Pest managers also have control over which species of natural enemies are released and when this occurs and can thus avoid intraguild predation.

Box 7.3 Mite against Mite

The starchy roots of cassava are a major staple food in much of central Africa. For many years after its introduction from South America, this plant had been relatively free of arthropod pests because it possesses high levels of poisonous cyanogenic glycosides and latex that deterred the native African phytophagous arthropods. Cassava green mite, *Mononychellus tanajoa*, was first found on cassava in east Africa in the early 1970s and this pest spread across the cassava-growing region causing up to 80 percent reduction in cassava root yield.

A classical biological control program was undertaken and the first challenge was obtaining a species name for the mite that occurred in Africa. It was critical to be able to recognize this same species when collecting in South America, where scientists assumed it had originated. After identification of the cassava green mite, exploration for predatory mites began. Between 1984 and 1988, more than 5.2 million predatory mites from Colombia, belonging to seven species, were imported into Africa and released but none of these species became established. Scientists hypothesized that problems were caused by low relative humidity in the cassava-growing area of Africa compared with Colombia, as well as lack of adequate alternate food sources for these predators when cassava green mite populations were low. However, in 1988, three species of predatory mites were collected in northeastern Brazil, an area drier than the previous collection areas, and shipped to Africa. Of these three species, clearly the most successful at providing control was *Typhlodromalus aripo*.

This was a big surprise because scientists had initially considered *T. aripo* the least likely to succeed of the three because it seemed less voracious and reproduced at a slower rate. However, this species turned out to establish, disperse, and persist better than the other two species owing, in part, to its specialized behavior. *Typhlodromalus aripo* resides in the growing tips of cassava plants during the day and forages on the leaves at night. The other two species live only on the leaves but mites like humid conditions and in the low relative humidities in the infested areas in Africa, this environmental factor played an important part. Scientists hypothesize that this behavior gave *T. aripo* an advantage because it inhabited protected locations during the driest times of the day, compared with the other two species that were not in protected microhabitats. In addition, *T. aripo* is able to persist at low prey densities because it is more of a generalist and will eat pollen and plant exudates as well as cassava green mites.

What about inundative or inoculative augmentation of predators in other than controlled environments? In central China, the seven-spotted lady beetle, *Coccinella septempunctata*, has been used extensively to control cotton aphids. This lady beetle is abundant in wheat in May so it is collected and then released in cotton for successful control. Releases of predatory mites against plant-feeding mites in orchards in California have also been very successful. In particular, use of strains of predatory

Table 7.1 Examples of predatory arthropods used for augmentative releases.

Predator group	Predator species	Target prey[1]	Predatory life stages	Use
Thysanoptera (thrips)	*Franklinothrips* spp.	Thysanoptera (thrips)	Nymph, adult	Greenhouses
Hemiptera (true bugs)	*Orius* spp.	Thysanoptera (thrips)	Nymph, adult	Solanaceous crops
	Anthocoris nemoralis	Hemiptera (aphids) Thysanoptera (thrips) Acari (mites)	Nymph, adult	Orchards
	Nesidiocoris tenuis	Lepidoptera (leafminers) Hemiptera (whiteflies)	Nymph, adult	Vegetables, greenhouses
Neuroptera (lacewings)	*Chrysopa rufilabris*	Hemiptera (aphids)	Larva	Many crops
Coleoptera (beetles)	*Adalia bipunctata*	Hemiptera (aphids)	Larva, adult	Vegetables, vineyards
	Atheta coriaria	Diptera (shore flies, fungus gnats) Thysanoptera (thrips)	Larva	Vegetables
Diptera (flies)	*Aphidoletes aphidimyza*	Hemiptera (aphids)	Larva	Greenhouses
	Ophyra aenescens	Diptera (stable flies)	Larva	Stables
Acari (mites)	*Phytoseiulus persimilis*	Acari (spider mites)	Nymph, adult	Greenhouses, plantscapes, crops
	Stratiolaelaps scimitus	Thysanoptera (thrips) Diptera (fungus gnats)	Nymph, adult	Greenhouses, nurseries, fields, poultry
	Neoseiulus californicus	Acari (spider mites)	Nymph, adult	Orchards

[1] Many of these predators will feed on numerous types of prey but they are listed here for a type of prey that they are often released specifically to control.

mites resistant to specific insecticides proved successful in crops where multiple pests needed to be controlled and insecticide applications were unavoidable. By using strains of predators selected for pesticide resistance, natural enemies were not killed when insecticides were applied.

It is often a little more difficult to document efficacy with larger invertebrate generalist predators than smaller more specialized invertebrate predators. One example would be the release of praying mantids in gardens. Although pleasant to maintain as residents, praying mantids are often not efficacious for pest control. Release of insectary-reared green lacewings to control leafhoppers in vineyards has also not been very effective, although lacewings are known to be voracious predators. Convergent lady beetles (*Hippodamia convergens*) provide an example of a predator that can control aphids in gardens under specific conditions (Box 7.4). The common wasp (*Vespula vulgaris*) is very efficient both as a generalist predator and as a pollinator. Owing to their stinging behavior, humans think of these voracious predators only as nuisance pests and they are generally never considered for purposeful releases.

Box 7.4 *Hippodamia convergens* **Takes Wing**

This species is named the convergent lady beetle because it has two converging white lines on its black thorax, forward of its black-spotted orange wing covers (elytra). These lady beetles prefer to feed on aphids, but aren't so choosy about where the aphids are living, ranging from gardens to trees to field crops. Convergent lady beetles are native to the western United States, where their specialized behavior preadapted them to be used for biological control. In summer, when aphid populations decline in the California Central Valley, these beetles fly to higher elevations in the foothills and mountains of the tall Sierra Nevada Mountains to the east. In the mountains, the beetles feed on pollen and nectar and then eventually aggregate in large numbers in canyons where they spend the winter. In early spring, when temperatures begin to warm and reach 18°C (65°F), adults mate and fly upwards to catch the winds that carry them to the floor of the Central Valley where they feed and reproduce.

Many years ago people found the large aggregations of adult beetles in the mountains and decided this seemed like a nice way to provide agents easily for biological control of aphids. The ease of collecting this species has led to the establishment of several companies that collect these lady beetles, store them under refrigeration and sell them for control of aphids. However, releasing lady beetles for aphid control has had mixed success. Most of these lady beetles will fly away when released and releases are not suitable for large farms; however, they have been used successfully in some landscape, garden, and greenhouse situations. In landscapes, lady beetles must be released on individual infested plants at dusk and provided with water during release by sprinkling plants. It is essential that treated plants have an adequate supply of aphids to support the lady beetles' nutritional needs. Lady beetles require large numbers of aphids in order to stay and the few lady beetles that remain on even heavily infested treated plants will clean up the aphid population in a few days and then fly elsewhere in search of more aphids. In release experiments, lady beetles did not lay eggs on released plants, probably because they did not get enough nutrition to stimulate egg production. Also, once lady beetles had eaten sufficiently and changed into flight mode after a release, they appeared to fly long distances and so did not self-disperse throughout a field, landscape, or garden. If aphid populations build up again in a landscape or garden, a re-release of lady beetles is necessary.

Conservation biological control regularly relies on naturally occurring predators whose populations can be conserved or enhanced. In this way, predators that would be far too difficult or expensive to mass-produce, even if techniques had been developed, can be used for control.

Further Reading

Brandmayr, P., Lövei, G. L., Zetto Brandmayr, T., Casale, A., & Vigna Taglianti, A. (eds.) (2000). *Natural History and Applied Ecology of Carabid Beetles*. Sofia, Bulgaria: Pensoft.

Carrillo, D., de Moraes, G. J., & Peña, J. E. (eds.) (2015). *Prospects for Biological Control of Harmful Organisms by Mites*. Dordrecht, Netherlands: Springer.

Dixon, A. F. G. (2000). *Insect Predator–Prey Dynamics: Ladybird Beetles and Biological Control*. Cambridge: Cambridge University Press.

Holland, J. M. (ed.) (2002). *The Agroecology of Carabid Beetles*. Andover, UK: Intercept.

Lundgren, J. G. (2009). *Relationships of Natural Enemies and Non-prey Foods*. Dordrecht, Netherlands: Springer.

Majerus, M. E. N. (2016). *A Natural History of Ladybird Beetles* (ed. H. E. Roy & P. M. J. Brown). Cambridge: Cambridge University Press.

McEwen, P., New, T. R., & Whittington, A. E. (eds.) (2001). *Lacewings in the Crop Environment*. Cambridge: Cambridge University Press.

Schmitz, O. J. (2010). *Resolving Ecosystem Complexity*. Princeton, NJ: Princeton University Press.

Wäckers, F. L., Van Rijn, P. C. J., & Bruin, J. (eds.) (2005). *Plant-Provided Food for Carnivorous Insects: A Protective Mutualism and its Applications*. Cambridge: Cambridge University Press.

Waldbauer, G. (2012). *How Not to be Eaten: The Insects Fight Back*. Berkeley: University of California Press.

8 Insect Parasitoids: Attack by Aliens

In the classic 1979 movie *Alien* starring Sigourney Weaver, a crew member traveling through space with Weaver becomes infested with an alien life-form. The alien develops within the crew member until approximately half the crew member's size and then emerges dramatically from the crew member's chest, killing him as it emerges. This screenplay could have been written by a parasitoid biologist. While the aliens are portrayed as bad and scary in the movie, in nature, parasitoids of insect hosts are part of complex food webs and their use in regulation of insect pest populations is a cornerstone of biological control.

Parasitoids are therefore a major type of natural enemy used to control invertebrates, usually insects. Parasitoid is a term derived from the more general term parasite. Parasites are organisms living in or on other organisms, from which they gain nourishment, and many parasites need their host organism to complete their life cycles. In the context of biological control of insects, the term parasitoid refers to insects that parasitize other insects or spiders and can be distinguished within the larger category of parasite because they eventually kill their host after completing development, often preventing host reproduction. Parasitoids use only a single host in contrast to predators that usually consume several hosts (prey). Most parasitoids have a holometabolous lifestyle that allows the different parasitoid life stages to specialize in different ways. Immature parasitoids are often soft-bodied, grub, or maggot-like in form and remain in close association with hosts to maximize their growth and development. The immatures feed on hosts either externally or internally and usually have no legs or eyes. The free-living adult parasitoids have eyes, antennae to detect chemical cues, legs, and wings. Adults are therefore usually the mobile stage, being better able to disperse, find a mate, and find healthy hosts for development of their progeny.

The sizes of adult parasitoids are often influenced by the size of their hosts. In all cases, the size of the host when it stops growing puts an upper limit on the sizes and numbers of parasitoids that can develop because the host is the sole food source for parasitoids during their development. Some adult parasitoids, feeding as immatures on large wood wasp larvae within the wood of tree trunks, can reach up to 10 cm in total length (including the long ovipositor). At the other extreme, the fairy flies (Mymaridae), which develop within insect eggs, are among the smallest multicellular eukaryotes, sometimes 0.2 mm long, actually smaller than some unicellular protists! Owing to their close physiological association with their hosts, many parasitoids are quite host specific, able only to develop in one stage of one or more host species.

8.1 Taxonomic Diversity in Parasitoids

Although the parasitoid life strategy seems rather specialized, it is a life history strategy that has been exploited by numerous groups of insects. Parasitoids are extremely common among wasps (Order Hymenoptera), less common among flies (Order Diptera), and, although found in the beetles (Order Coleoptera), moths and butterflies (Order Lepidoptera), and the lacewing order (Neuroptera), this lifestyle is rare in these last groups.

8.1.1 Parasitic Wasps (Order Hymenoptera)

It has been estimated that there are more than 65,000 species of Hymenoptera that develop as parasitoids and many of these species have not been described. The largest and most noticeable parasitoids generally belong to the Ichneumonoidea, the superfamily that includes the Ichneumonidae and Braconidae (Figure 8.1a). The superfamilies Chalcidoidea and Proctotrupoidea are other very diverse groups of wasp parasitoids, but are much less noticeable with many species only a few millimeters long (Figure 8.1b, c).

Once mated, females of parasitic Hymenoptera can control fertilization of their eggs. Since males develop from unfertilized eggs, mothers can therefore control the relative numbers of male and female offspring according to whether eggs are fertilized or not. Males in many species are smaller than females and adult females are known to lay male eggs in smaller hosts that could only support development of a smaller and probably less fecund female. However, more recent research has shown that bacteria residing within some parasitoids can have overriding influences on sex ratios. The bacteria are transmitted to the next generation maternally within or along with host eggs. Several different species of bacteria, for example, *Wolbachia*, have been found that either kill male eggs or change male eggs into females, thus promoting their own transmission to further generations within these hosts. These bacteria, which have been called reproductive parasites, thus can strongly influence biological control as sometimes when the bacteria are present, all parasitoids produced from a host would be the female sex, which finds and parasitizes new hosts.

Female parasitoids have an elongated tubular egg-laying structure, called an ovipositor. In some parasitoids, this structure extends far beyond the bodies of females and can be very conspicuous. In the bees and social wasps, all of which are related to hymenopteran parasitoids, the ovipositor has sometimes evolved for use as a sting for defense. Adult female parasitoid wasps instead use the ovipositor to inject eggs into hosts or lay eggs on top of them (Figure 8.2). Using an ovipositor, parasitoid wasps can be precise about depositing their offspring where they will have the best chance of survival. In parasitic Hymenoptera, the length of the ovipositor often reflects the type of host that is parasitized and where that host lives. For species laying eggs within hosts, the ovipositor is adapted to pierce the host cuticle and an egg then passes down the ovipositor and is deposited within the host. Other parasitic wasps lay eggs on top of

Figure 8.1 (a) The braconid parasitoid *Phanerotoma flavitestacea* (Ichneumonoidea) laying an egg in an egg of its host, the navel orangeworm, *Amyelois transitella*. The larva of this egg-larval parasitoid develops in the host larva and pupates after the host spins a cocoon to pupate (Caltagirone et al., 1964). (b) Tussock moth parasitoid *Spilochalcis* sp. (Chalcidoidea: Chalcididae). (c) The tiny *Anaphes iole* (Chalcidoidea: Mymaridae) (*c.* 0.5–0.6 mm long) laying an egg within an egg of a lygus bug (b and c, photos by Jack Kelly Clark, courtesy of the University of California Statewide IPM Program).

Figure 8.2 Braconid parasitoid ovipositing in the wheat aphid, *Schizaphis graminum*. The aphid is approximately 2.2 mm long (Webster, 1909).

Figure 8.3 Tachinid fly *Eucelatoria armigera* whose larvae develop within cotton bollworm larvae, *Helicoverpa zea* (van den Bosch & Hagen, 1966; © Regents, University of California).

host larvae, and often these hosts live within a concealed location such as a cocoon, a leaf mine, a plant gall, or even within the wood of a tree. In these instances, the ovipositor is used to drill through the material, often part of a plant, surrounding the host. It is a challenge for the parasitoid to determine that there is a host larva within a plant, exactly where the host is located, or even when the drilling ovipositor reaches the correct location to lay an egg. Wasp antennae are used to locate the general area, but the tips of the ovipositor and the ovipositor sheath contain different types of sensory cells that are thought to detect both mechanical and chemical stimuli, thus providing additional information for the adult female parasitoid as she probes before laying an egg.

8.1.2 Parasitic Flies (Diptera)

Flies are second only to the wasps in developing different strategies enabling life as a parasitoid (Figure 8.3). The family Tachinidae is perhaps the most diverse family of parasitoid flies, with over 8,000 known species. Adults of this group can easily be mistaken for large houseflies although their life cycle is certainly much more complex, requiring a living host for development.

Most flies do not have the advantages of a piercing ovipositor for injecting eggs into hosts and they may be less precise when depositing their young. Parasitic flies mostly attack exposed hosts and not hosts living in concealed locations within plants or plant galls. However, there is still diversity among the strategies used by parasitic flies for depositing young where they will successfully access hosts. Some species glue their

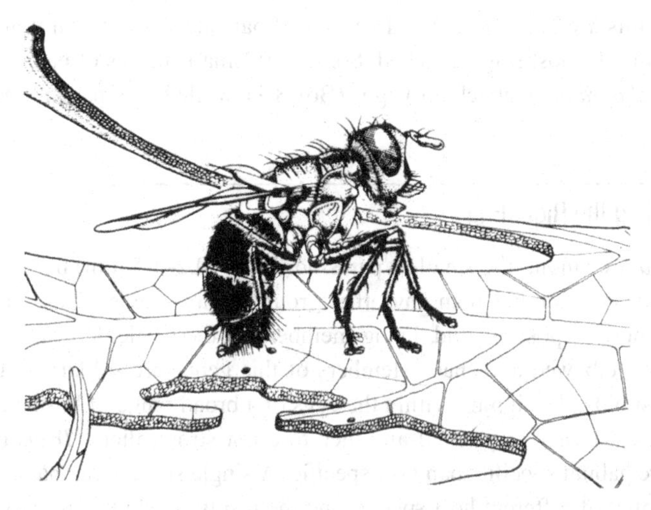

Figure 8.4 Female *Blepharipa pratensis* laying microtype eggs on a leaf, where they can be eaten by gypsy moth, *Lymantria dispar*, caterpillars (Godwin & Odell, 1981).

eggs externally on hosts and, after hatching, the larvae pierce the host cuticle and enter the host's body, where they grow and develop. Other parasitoid flies produce many tiny (0.2 mm) eggs, called microtype eggs, that are laid on foliage (Figure 8.4). Some eggs are then eaten by hosts where they hatch in the host gut and begin to develop within the host. In another strategy, the fly egg hatches within its mother, who then injects the larva into a host, using a piercing structure; this is referred to as larviposition.

8.1.3 Parasitic Beetles (Coleoptera)

Use of the parasitoid lifestyle among species of beetles is less common, but the very different method for finding hosts employed by some species is worth describing. Although the immature stages of most parasitoids are not mobile, in some parasitic beetle species, the very small first instars have legs and are very mobile (called triungulins). Some triungulins are adapted for attaching to adult bees so they are subsequently transported to bee nests where they locate and parasitize the bees' offspring. Subsequent instars are legless and thus remain living in or on their food supply. For such a species, eggs can therefore be laid away from hosts. The first instars do not require food, are mobile, and are attracted to moving objects. As might be expected, species with these mobile first instars lay many eggs as a result of the low chances of any individual eventually both finding and attaching to a host.

8.2 Diversity in Parasitoid Life Histories

Life history strategies among parasitoids are extremely diverse and can be quite intricate, often with finely tuned associations between parasitoids and hosts. Parasitoids most often develop in or on immature stages, such as eggs, larvae, or pupae, although occasionally adults are hosts. Individual species of parasitoids are usually highly specialized regarding the host stage attacked. Some very small species of parasitic wasps, such as tiny *Trichogramma*, attack host eggs (Box 8.1), while larger parasitoids develop

Box 8.1 Finding the Right Egg

Trichogramma are among the smallest parasitic wasps (0.2–1.5 mm) but have been the subjects of more studies than any other group of parasitoids. These egg parasitoids have short ovipositors and, being members of the Chalcidoidea, they have relatively short, elbowed antennae. Members of this genus are solitary or gregarious endoparasitoids, developing within the eggs of a broad range of hosts including many crop pests. Being very small and therefore not strong fliers, *Trichogramma* are often more habitat specific than host specific. A single species can parasitize the eggs of a number of different host species and the resulting adults will vary in size based on the sizes of the individual host eggs. When the female wasp emerges from a host egg in which she has developed, all of her own eggs are fully developed. In fact, as with all parasitic Hymenoptera, *Trichogramma* do not have to find mates to begin laying eggs. However, without fertilization, eggs develop but all will become males. Only fertilized eggs will become female.

Detailed studies have been conducted on recognition and acceptance of host eggs by adult female *Trichogramma*. *Trichogramma* look for small, rounded objects and will even attempt to lay eggs within glass beads of the correct size. After finding a host egg, a *Trichogramma* female examines the host surface, walking back and forth on it and drumming with her antennae for 10–40 seconds, with the length of examination based on the curvature of the egg surface. The female examines the egg for so long with good reason. She can detect marker chemicals deposited when other *Trichogramma* have been walking on the egg and she wants to avoid laying her eggs in previously parasitized host eggs, if possible. The external marker is water soluble so what if it has rained? When the female begins drilling with her ovipositor, with some experience she can detect whether parasitoid eggs have already been laid within the host egg or not. But host eggs can be difficult to find and females get frustrated. If a female does not find unparasitized host eggs within 10 minutes she will keep looking but after 90 nonproductive minutes, *Trichogramma* females give up and will lay eggs within a host egg regardless of previous parasitization. The number of eggs laid within a host egg is regulated by the size of the host egg, ranging from 1 to 40 *Trichogramma* eggs per host egg.

Trichogramma adults are winged but, being so small, they are not very capable of controlling where they go when dispersing. Host eggs are often aggregated so the *Trichogramma* just needs to find oviposition sites. Cleverly, the adult wasps can attach to mobile hosts and hitch a ride (called phoresy), only getting off once the host begins to lay eggs. This works well because the host takes the parasitoid to the new oviposition site and then the parasitoid is ensured of locating eggs that are freshly laid.

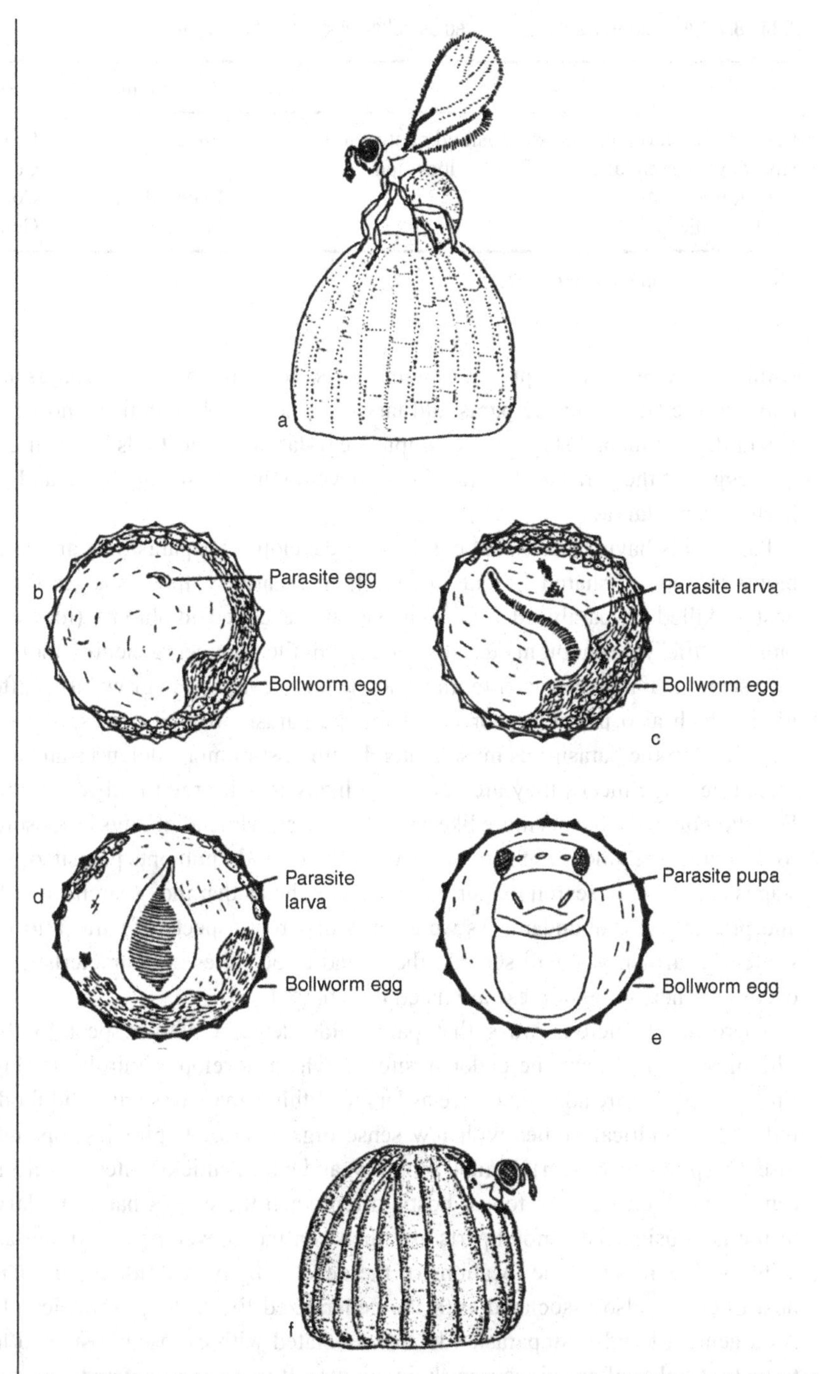

Figure Box 8.1.1 Life history of *Trichogramma*: (a) female ovipositing in a cotton bollworm (*Helicoverpa zea*) egg, (b) *Trichogramma* egg within bollworm egg, (c, d) parasitoid larva developing, (e) parasitoid pupates within the host egg shell, (f) adult wasp emerges from the egg (van den Bosch & Hagen, 1966; © Regents, University of California).

Table 8.1 Generalized life history strategies of koinobionts and idiobionts.

	Koinobiont	Idiobiont
Location for development of parasitoid relative to host	Endoparasitic	Ectoparasitic
Host development after parasitoid oviposition	Continues	Ceases
Location of host	Exposed	Concealed
Host specificity	Specialists	Generalists

Source: based on Quicke (1997).

within larvae or nymphs, pupae, or adult hosts. For some parasitoids, eggs are injected into one life stage, but the parasitoid larvae do not develop until the host has reached a later developmental stage. For example, "egg-larval" parasitoids lay their eggs within host eggs but the parasitoid larvae do not develop until host eggs hatch and parasitized hosts become larvae.

Parasitoids having hosts that continue to develop after parasitism are called koinobionts (koino = "shared," biont = "life"). Alternatively, those species for which the host is killed or paralyzed after oviposition are called idiobionts (idio = "single," biont = "life"). Koinobionts usually lay eggs in the younger, generally more abundant, host stages that may be easier to find, and the hosts continue to grow larger after oviposition, which also provides more food for the parasitoid as it grows. However, while they develop the parasitoids must contend with host immune defenses and a prolonged immature stage means they are also more likely to fall prey to other natural enemies. For idiobionts, the host is more like a piece of meat; while idiobionts are assured of their food source, the amount of food will not increase, and idiobionts parasitizing later host stages may have to search longer because many hosts that they search for will be older and probably less abundant. As we discuss different aspects of parasitoid life history strategies further, you will see that there tend to be suites of characteristics that often occur together; some of these are listed in Table 8.1.

There are numerous ways that parasitoids develop with respect to their hosts. The most common are the endoparasitoids, which develop within hosts (Figure 8.5). Endoparasitoids are adapted to live as larvae within a mass of semiliquid food and have reduced, cylindrical bodies with few sense organs, closed spiracles (openings of the insect respiratory system), limited mobility, and a thin cuticle. Alternatively, some parasitoids lay their eggs on top of hosts; often when these eggs hatch the larvae attach to the host using their mouthparts and then continue development, living as ectoparasitoids (Figure 8.6). The locations of the developing parasitoid larvae relative to the host often are also associated with the generalized life history strategies (Table 8.1). As a general trend, endoparasitoids are associated with exposed hosts and hosts continue to develop after oviposition (koinobionts). It is thought that endoparasitoids have limited host ranges because the immature stages must specialize to evade the immune responses of hosts. Ectoparasitoids are often associated with concealed hosts, such as caterpillars living within fruit or stems, or beetles within wood, where they kill or paralyze their hosts during or soon after oviposition (idiobionts). Thus, ectoparasitoids are

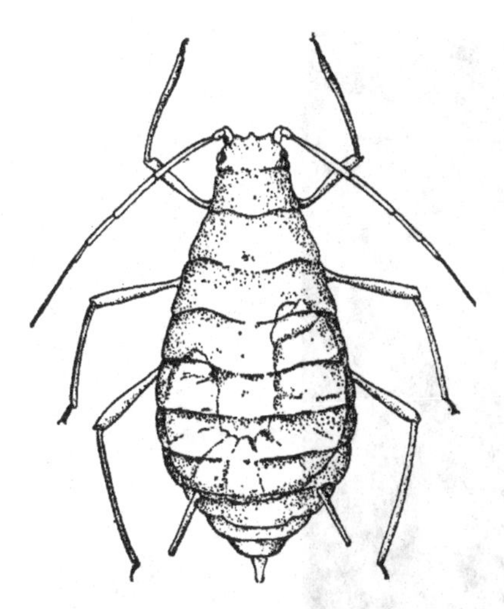

Figure 8.5 Larva of an aphidiid parasitoid (Ichneumonoidea) developing within the wheat aphid, *Schizaphis graminum*. The aphid is approximately 2.2 mm long (Webster, 1909).

more specialized by host location and when adult ectoparasitoids lay eggs they often search for a specific type of habitat and not a specific host. Because they live externally on hosts, evading the host immune system is not such a necessity. In fact, it is thought that ectoparasitoids are frequently more general in host range because they are not developing within the body of the host and their growth environment is thus not as specialized.

After consuming much or all of the host, endoparasitoids pupate within the host's body or exit the host to pupate. Those that leave the host may spin a silk cocoon in which they pupate on top of or next to the host's body (Figure 8.7). Parasitic wasps remaining within the host's body can use the hardened host cuticle for protection while

Figure 8.6 Larvae of the chalcidoid *Euplectrus* sp. (Family Eulophidae), a gregarious larval ectoparasitoid, on an armyworm larva (photo by James Carey).

Figure 8.7 Tomato hornworm caterpillar (*Manduca quinquemaculata*) covered with cocoons of the parasitic wasp *Cotesia congregata*. These parasitoids had developed as larvae within the caterpillar but exited to spin cocoons, within which they have pupated (photo by Donald C. Steinkraus).

they pupate. Parasitic flies often create a smooth hardened covering within which they pupate.

Parasitoids also are diverse in whether they share an individual host and how they do this. Solitary parasitoids lay one egg in a host and one parasitoid larva develops from it (Figure 8.8a). In gregarious parasitoid species, multiple eggs can be laid per host and many individuals can develop within or on one host. Most amazingly, in polyembryonic species, a single parasitoid egg laid within a host can divide to produce from 2 to more than 3,000 genetically identical individuals, with the number depending on the size of both the host and the parasitoid. We could hypothesize that there are trade-offs in development of parasitoids with a solitary versus polyembryonic type of life; a solitary parasitoid developing within a host larva would produce one adult wasp that would be quite large and better able to control its dispersal, while a polyembryonic species using the same size of host would produce many individuals but the resulting wasps are extremely small and would then have diminished capabilities for being able to control their dispersal.

In the case of multiple parasitism, individuals of different species oviposit in or on the same host individual. In most cases they compete and only one parasitoid species survives to emerge. One species may always be victorious or either species may win depending on which species laid its egg first. Superparasitism results when more than

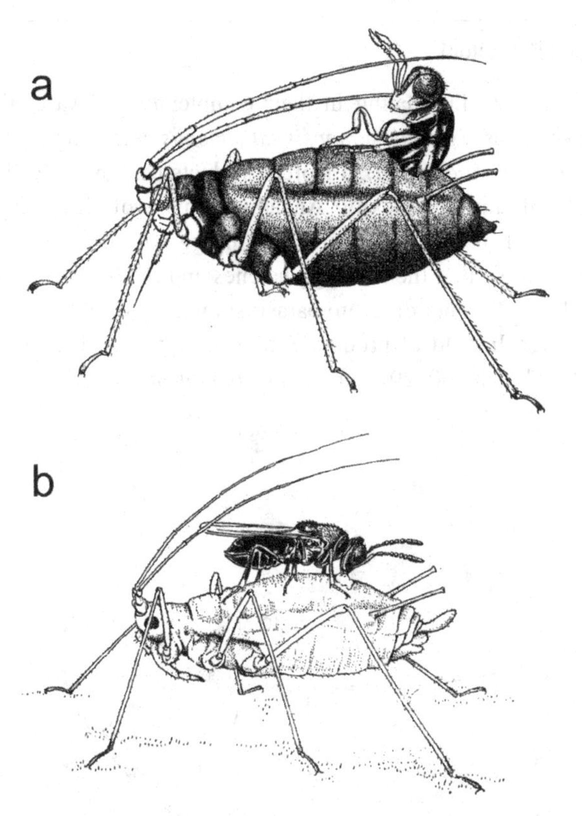

Figure 8.8 (a) The parasitoid *Aphelinus jucundus* (Chalcidoidea) emerging from a geranium aphid, *Acyrthosiphon malvae*. (b) The parasitoid *Aphidencyrtus aphidivorus* (Chalcidoidea) feeding at an ovipositional wound in its aphid host, *Acyrthosiphon malvae* (Griswold, 1929).

one egg or clutch of eggs of a particular parasitoid species are laid within or on one host. Superparasitism may result in all progeny being reduced in size or in one female's progeny being killed. Often the offspring of later-arriving females are disadvantaged relative to the earlier female's offspring. Researchers have been very interested in how superparasitism is avoided by adult female parasitoids. In many examples, this is accomplished by adult females using their ovipositors to deposit a chemical marker on the surface or inside a parasitized host. This marker acts as a deterrent to other parasitoids to avoid superparasitism. When superparasitism or multiple parasitism does occur, larvae of endoparasitoids may fight within hosts and can kill each other. A specialized instance of larval fighting occurs within some polyembryonic species with nonreproducing, specialized "defender-morph" larvae that act solely in defense of their genetically identical siblings and do not become adults (Box 8.2).

While most parasitoids attack insects feeding on plants or other noninsect resources in the environment (primary parasitoids), some parasitoids have taken this one step further and attack the parasitoids developing within or on previously parasitized hosts (these are called hyperparasites or hyperparasitoids). Some species are facultative hyperparasitoids and they can use either a parasitoid (in which case they are hyperparasitoids) or

Box 8.2 A Precocious Parasitoid

Parasitoid life cycles can be fascinating in their complexity and variability. The life cycle of the parasitic wasp, *Copidosomopsis tanytmenus*, attacking the Mediterranean flour moth (*Ephestia kuehniella*), a stored products pest, provides just such an excellent example of this complexity. An adult female of these tiny wasps (1.26 mm long to the end of the ovipositor) lays an egg within a host egg. This is an egg-larval parasitoid so that the host egg hatches and the host larva develops and, in fact, the host larva does not die from parasitism until it would normally spin a cocoon six weeks later. Instead of producing a cocoon, the host larva dies and becomes a mummy filled with 100–200 pupae of *C. tanytmenus*.

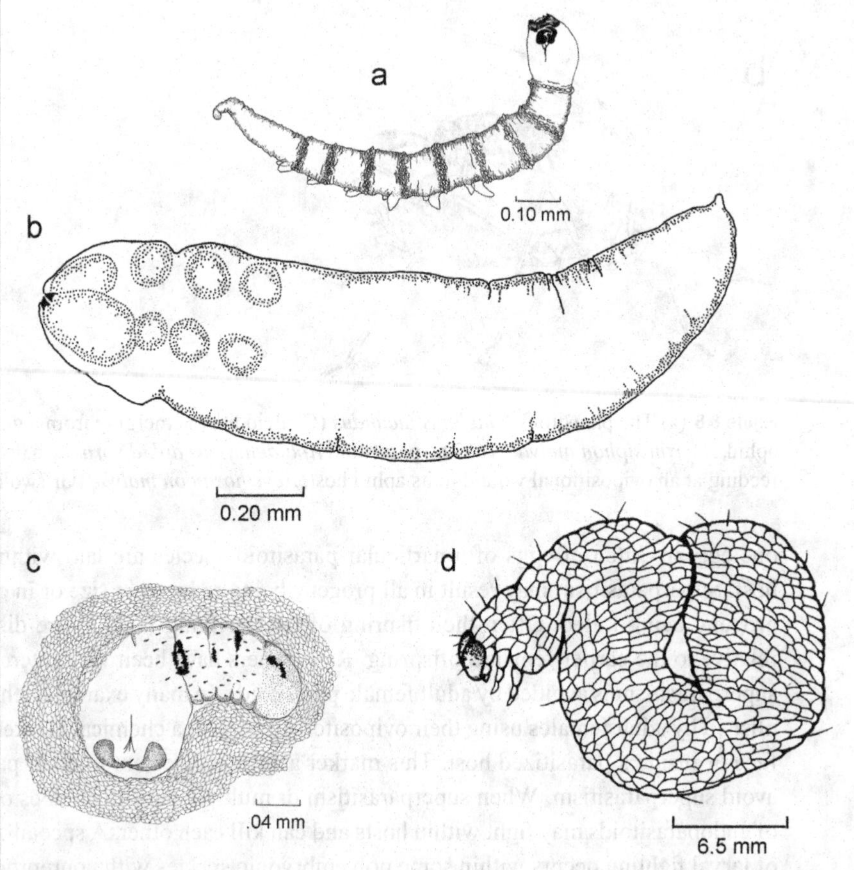

Figure Box 8.2.1 (a) A precocious larva of *Copidosomopsis tanytmenus* from a 14-day-old host Mediterranean flour moth caterpillar, *Ephestia kuehniella*. (b) One of the 164 mature normal *C. tanytmenus* larvae from a 40-day-old host caterpillar. (c) Capsule formed from caterpillar blood cells, containing dead, wounded larva of *Phanerotoma flavitestacea* after this competitor was attacked by the precocious larva of *C. tanytmenus*. (d) Cadaver of a larva of the silver-Y moth, *Autographa gamma*, filled with cocoons of the polyembryonic parasitoid *Copidosoma truncatellum* (a–c from Cruz, 1981; © Macmillan Publishers Ltd.; d from Silvestri, 1906).

After the parasitoid egg is laid within the host egg, each egg begins development as a mass of undifferentiated cells, a polygerm. Each cell will eventually grow into a wasp. The first one or two parasitoid larvae that develop are always precocious and are found in the host 10 days after the parasitoid oviposits and, with time, more precocious larvae can be found. The asexual precocious larvae are thin with large mandibles and as the polygerm increases in size these first emergers play the role of defenders. If another parasitoid lays an egg within the host, once that egg begins to develop, it is attacked by a precocious larva. The wounded invader is then recognized as non-self by the host and is encapsulated by the host's immune system. By 4.5 weeks after parasitoid oviposition, the individual cells of the polygerm differentiate into normal parasitoid larvae that are sac-like, with only very small mandibles. These normal larvae grow quickly and are adults by the sixth–seventh week. The initial defenders, the precocious larvae, do not develop further and are all dead by week 6. It has been suggested that the long developmental time of the normal larvae is possible through the protection provided by the precocious soldiers.

Thus, the precocious larvae prevent superparasitism as they and their siblings develop and they are altruistic, defending their sisters and brothers although they do not survive to reproduce.

The system where *Copidosoma floridanum* parasitizes the cabbage looper, *Trichoplusia ni*, is a bit different as these precocious larvae, which have large mandibles, provide protection via production of a physiologically suppressive factor that delays maturation of invading wasps by weakening and paralyzing them. The precocious soldiers in this system protect not only against invaders that are another species but also against parasitoids of the same species that are not close relatives.

the host of the parasitoid (and they are then primary parasitoids) for development. An extreme case of hyperparasitism has been found with some chalcidoid species in which males can only develop by feeding on females of their own species living within parasitized hosts. Hyperparasitoids can at times be extremely abundant and in some cases their activity has jeopardized the effectiveness of primary parasitoids being released for biological control.

Not only the immature stages of parasitoids require food. Many adult parasitoids live longer if they feed and food certainly also provides more energy for the extensive searching required to find low-density hosts. Foods for adult parasitoids are often nectar, honeydew, and pollen. A behavior called host feeding also occurs in species within 17 families of parasitic wasps. The adult female wasp usually creates a hole in the host body wall using her ovipositor and then turns around to eat the exposed host's blood and sometimes tissues too (see Figure 8.8b). This can occur after a parasitoid egg has been laid and, in these instances, feeding by the parasitoid does not kill the host so that the wasp progeny in that host can still successfully develop. More frequently, host feeding occurs when no egg is laid and then host feeding by the wasp usually kills the host. Host feeding is characteristic of parasitoid species that need food as adults to produce eggs that contain an abundance of yolk or species with adults that mature their eggs throughout their lives (synovigenic). Adult parasitoids that emerge from hosts as adults with their full complement of eggs (proovigenic) still require sugar for energy but do not require the lipid-rich nutrients that they would get from feeding on hosts. Idiobionts are often synovigenic while koinobionts are often proovigenic. Parasitoids that host feed may thus kill hosts when acting as predators as well as through parasitism. For inundative biological control, where rapid suppression of the host population is most important, parasitoid species that host feed may have an advantage over species that do not because they can potentially cause deaths of more hosts than the number in which they would lay eggs alone.

8.2.1 Life History Strategies in Parasitoid Communities

Many insect hosts have different parasitoids associated with different stages of their life cycles. Entomologists have been interested in the differences in life history strategies within parasitoid communities using the same host. For example, a group of parasitoids attacking larvae of Swaine jack pine sawflies, *Neodiprion swainei*, with larval stages similar to caterpillars feeding externally on foliage, in Quebec, Canada, were compared with the parasitoid community associated with oak galls and oak leaf miners in England. There seemed to be a similar pattern in both systems with "early succession" colonizers attacking larval hosts; these parasitoids had many eggs to lay (high fecundity) and their body sizes were larger but they were poor competitors. The smaller "late succession" colonizers used pupal hosts and had fewer eggs to lay but they were better competitors. From these relations, a "balanced mortality hypothesis" was proposed, suggesting that the fecundity level of the parasitoid species was balanced by the probability of survival of the progeny. Species using earlier host stages that would potentially be subject to mortality over a longer period of time had more eggs to lay. Species using later host stages, often in concealed locations that had less chance of being eaten or parasitized, had fewer eggs to lay. Later theorists connected the fact that most of the early succession colonizers were koinobionts and endoparasitoids while later succession colonizers were more frequently idiobionts and ectoparasitoids.

8.3 Locating and Accepting a Host

Successfully finding a host and parasitizing it are critical for reproduction by parasitoids. This can indeed be demanding for parasitoids with specialized requirements such as a limited number of acceptable host species, specific host life stages, or hosts having aggregated distributions so that they can be more difficult for parasitoids to find (although once an aggregation is found, a parasitoid has many hosts in one place to parasitize). The process of locating the correct hosts can be thought of in terms of a progression of several generalized steps: locating the correct habitat for the host, locating a host, and then evaluating the host to determine that it is the correct species, the right stage, and of sufficient quality for oviposition (Figure 8.9). As mentioned above, parasitoids can also check to make sure the host is not already parasitized.

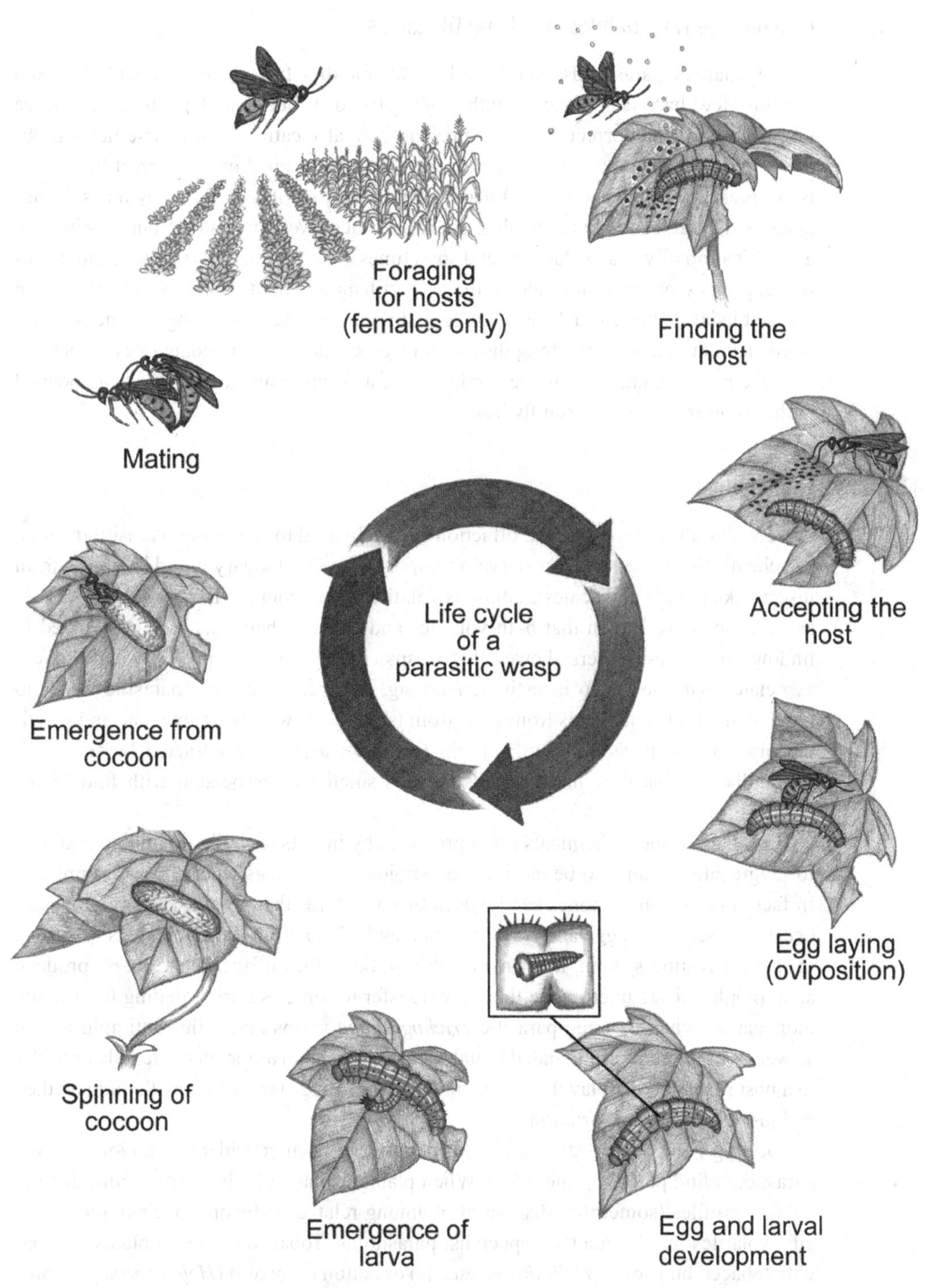

Figure 8.9 Example of the life cycle of a parasitic wasp. The female finds a host she will identify by touching her antennae to the caterpillar's feces. She injects eggs into the caterpillar, wasp larvae develop feeding within the caterpillar, pupate, and then emerge. From egg to adult death requires about 5 weeks (drawn by Patricia J. Wynne; Tumlinson et al., 1993).

8.3.1 Locating the Host Habitat over Long Distances

Even the largest parasitoids are relatively small and they face an environment that often contains few hard-to-find hosts with aggregated distributions. If parasitoids emerge where hosts of the correct stage are present, habitat location is of course not a problem. However, if the host population at that site has declined or the correct host stage is not present, the parasitoids will need to find a new habitat occupied by hosts. Long-distance habitat location is difficult to study and not so well understood, but may involve a combination of visual, olfactory, and sometimes auditory cues. The most reliable cues for long-range orientation to locate the correct habitat would be cues originating from the host itself, although such cues might be difficult to detect over long distances. More likely, habitat location over long distances involves the more abundant cues associated with the host habitat, such as the food plant of a herbivorous host or yeasts associated with rotting fruit hosting fruit fly hosts.

8.3.2 Finding Hosts

Once within the correct habitat, olfaction is widely used to find hosts. Parasitoids often use chemicals emitted by frass (larval feces) or honeydew (sugary liquid excretion from insects like aphids and scales), either as volatiles or on contact. Experiments with parasitic wasps have shown that both volatiles and contact chemicals can be involved in finding hosts. Researchers showed that wasps could learn that specific volatiles were associated with hosts. To investigate learning, researchers allowed parasitic wasps to touch nonvolatile chemicals from frass from larval hosts with their antennae and simultaneously let them smell vanilla. From then on wasps were attracted by the smell of vanilla because they had learned that this smell was associated with finding the correct host.

Host pheromones (chemicals often produced by insects to facilitate finding mates or for aggregation) can also be used by parasitoids to find areas where hosts are present. In fact, host sex pheromones can be used by parasitoids that do not attack adult stages because susceptible eggs and early instars may be located in the same areas as adults. In one interesting system, adult male cabbage butterflies of the genus *Pieris* produce an anti-aphrodisiac pheromone that they transfer to females during mating to promote monogamy. When tiny egg-parasitic *Trichogramma* wasps detect this anti-aphrodisiac as well as other odors from mated female *Pieris*, they hitch a ride on the female butterfly to a host plant and then lay their eggs into the *Pieris* eggs newly laid by the female they had just used for transportation.

Locating plants being attacked by hosts has long been considered a major way that parasitoids find plant-feeding hosts. When plants damaged by herbivores emit distinct volatile profiles (sometimes distinct even among related herbivore species), these specific volatiles may then attract specialist parasitoids. Tobacco or cotton plants infested with tobacco budworm (*Heliothis virescens*) or cotton bollworm (*Helicoverpa zea*) produce different volatiles. The parasitoid *Cardiochiles nigriceps* is quite specialized, predominantly attacking tobacco budworm instead of cotton bollworm. This parasitoid was attracted to plants on which tobacco budworm had fed much more than plants on which cotton bollworm had fed. To demonstrate that the attraction was not as a result of the presence of the caterpillars, both caterpillars and damaged leaves were removed and the preference remained for plants on which *H. virescens* had fed, triggered only by the plant volatiles that had been induced by caterpillar feeding.

Smells of host body parts such as moth wing scales can also aid parasitoids in locating hosts. Chemicals extracted from wing scales of European corn borers (*Ostrinia nubilalis*) are used by the egg parasitoid *Trichogramma nubilale* to remain in specific locations. Some insects, such as pine looper moth (*Bupalus piniarius*) larvae, leave behind a trail when they move from place to place. This chemical trail originating from the caterpillar cuticle is followed by large solitary parasitoids (*Poecilostictus cothurnatus*) to locate hosts. Silk is used by many arthropods for concealing and protecting their egg masses or pupae and is the source of volatile as well as contact chemicals and these chemicals are used by some parasitoids to locate hosts.

In addition to chemical signals, host vibrations, visual cues, and acoustical signals are used. Host vibrations are often used by parasitoids attacking concealed hosts. Parasitoids of beetle larvae feeding beneath tree bark often stand motionless on the bark surface to sense vibrations caused by movements of larval hosts within the wood. There is some evidence that some parasitoids may find hosts by looking for irregular outlines of leaves, caused by feeding damage from hosts. Parasitoids can also use auditory cues to find hosts. Tachinid flies (*Ormia ochracea*) parasitizing western trilling crickets (*Gryllus integer*) are attracted to male crickets singing to attract females. They lay living larvae on a singing male and the larvae quickly burrow inside to begin developing. Interestingly, satellite males that do not sing but try to steal females from singers are rewarded because they are rarely parasitized.

8.3.3 Accepting a Host

Generally, for parasitic wasps attacking motionless hosts, such as eggs, pupae, or scale insects, the hosts are first examined by the parasitoid tapping or drumming on the host with her antennae. These are very sensitive to chemical cues telling the parasitoid if this host is the correct species and whether or not it has already been parasitized (Box 8.1). Alternatively, the ovipositor can be used to probe within the host before an egg is laid.

Although examination using the antennae is common, parasitoids attacking more active or aggressive hosts may take less time for evaluation and various of these acceptance steps may not occur. In the extreme case of parasitoids attacking active or predaceous hosts, specialized behaviors have often evolved so that parasitoids can successfully oviposit but also survive to lay more eggs. Phorid flies in the genus *Pseudacteon* attacking ants are known to hover above potential hosts, much as a hawk hovers above its prey, and then quickly swoop down and oviposit very quickly into the ant's thorax using a hypodermic needle-style ovipositor before the ant can defend itself. The larva of this unusual parasitoid then migrates into the ant's head where it does not affect the ant's behavior as it develops, although the ant is killed once this parasitoid pupates.

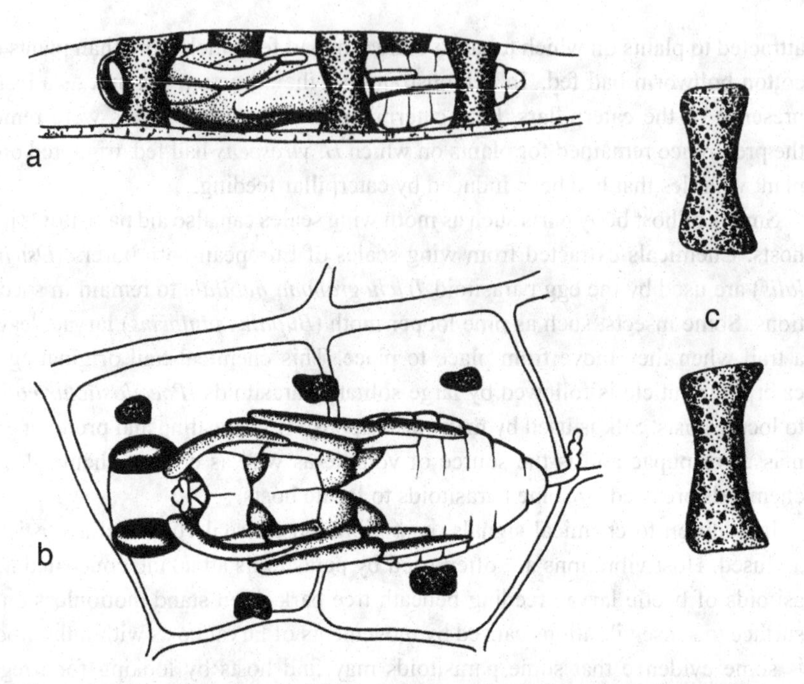

Figure 8.10 Supporting props of fecal matter made by the parasitoid *Chrysocharis gemma* for pupation, so that as a leaf dries out, the pupation chamber within the leaf does not collapse: (a) side view of pupa inside leaf; (b) view from above within leaf; (c) the fecal props (Viggiani, 1964; with kind permission of Kluwer Academic Publishers).

8.4 The Battle between Parasitoid and Host

Factors determining which host species can be used for development of a specific parasitoid can be varied and are often poorly understood. Host range for any parasitoid species is likely to be the outcome of the interplay between adaptations of the parasitoid to subdue, attack, and develop in particular hosts and adaptations by hosts to repel or resist the parasitoid. The diversity of reciprocal offense and defense that have been developed in this coevolutionary arms race is truly amazing.

8.4.1 Host Defense

Many different specialized features of insect species probably act as deterrents to parasitoids. Thick and hard host egg shells and cuticles or long hairs can act as physical barriers. Mobile hosts defend themselves by thrashing, kicking, shaking, dropping on silk threads, or simply falling when parasitoids attack. Hosts living in a concealed location such as a rolled leaf or leaf mine certainly gain some protection from these specialized habitats but some parasitoids, in turn, have evolved methods for overcoming these defenses. One small wasp parasitizing leaf miners has solved the problem of how safely to pupate within a flat leaf mine constructed between the two surfaces of a leaf. The larvae utilize their fecal material, which hardens as props to create a safe space within the leaf where the defenseless parasitoid pupa will not be squashed while metamorphosis occurs (Figure 8.10).

Potential hosts can be defended by other members of the community, sometimes quite effectively. Ants that tend aphids for the honeydew they produce may also protect the aphids from natural enemies, including parasitoids. Some hosts feed on plants with secondary plant compounds that are known to be toxic to many insects and they sequester these compounds. These so-called "nasty hosts" can then influence the developmental success of parasitoids; several species of parasitoids develop poorly within caterpillars with high levels of nicotine that they gained when feeding on tobacco plants. Although caterpillars can use these plant defenses to protect themselves, some parasitoids have evolved the ability to tolerate such compounds, thus utilizing a host with fewer competing parasitoids.

8.4.2 Parasitoid Attack

Parasitoids have developed specialized methods for successfully ovipositing in and on acceptable hosts. Ovipositors differ in length and morphology, so that parasitoids are able to lay eggs in or on their hosts, whether exposed or concealed deep within a habitat. Oviposition by some species can be very slow, requiring up to 30 minutes for the delicate ovipositor to drill with precision deep within tree trunks to lay eggs on host larvae feeding within (Box 8.3). At the other extreme, the phorid flies in the genus *Pseudacteon* can lay an egg in the thorax of a fire ant (*Solenopsis* spp.) in less than one second, presumably thus avoiding mass attack by the ant colony.

Box 8.3 Parasitoids Developing within Trees: How Do They Get There?

Some specialized challenges have been overcome by parasitoids attacking hosts that develop within tree trunks. In early June in forests in northeastern North America, you might be lucky enough to see a *Megarhyssa* adult flying. These are among the largest parasitoids (4–5 cm long with an ovipositor the same length, trailing behind females as they fly); these large wasps can look scary but in fact, their long ovipositors are only used for boring into trees and have nothing to do with stinging people! This group of parasitic wasps attack larvae of wood wasps, also called horntails, that develop within trees. Males of *Megarhyssa* emerge before females in spring and groups of males will aggregate around certain locations on the trunks of trees where they detect a female *Megarhyssa* that is almost ready to emerge, by chewing her way out from within the decomposing wood. With a closer look you'll find a small hole chewed in the tree trunk. A few males will have extended their abdomens into that hole where, inside the tree within a tight gallery, a female will be mating with one of these males. The female will eventually widen the hole in order to climb out and she will mate again. Then, she will fly away to find the correct locations to lay her eggs.

But how will a female *Megarhyssa* find these specialized wood-boring hosts in such hidden locations? The eggs and larvae of wood wasps are often at least several centimeters deep within tree trunks where they live with a symbiotic fungus (*Amylostereum*) injected by their mothers to help rot the wood. This fungus provides an olfactory cue to help adult females of some parasitoids fly to the cryptic locations of the immature wood wasps. Once in the right area, a female of the closely related parasitoid *Rhyssa persuasoria* uses the tips of her long antennae to help her find the exact location where a host larva is located within the tree, probably detecting vibrations caused by larval chewing, and she begins drilling with her ovipositor. She orients her body so that her ovipositor is perpendicular to the tree trunk, supported by the ovipositor sheaths and the bases of her hind legs. The tips of her ovipositor have cutting ridges and she cuts through the wood by working the two parts of the ovipositor back and forth against each other to saw into the wood. This is difficult and slow work that can take more than five hours; for adult female *Rhyssa persuasoria*, the full length of the ovipositor often must be inserted to reach a potential host. When a host is located, the female stings it to inject venom that paralyzes the host. The egg then passes down the egg canal in the middle of the ovipositor and is laid on top of the host where it will hatch and grow as an ectoparasitoid.

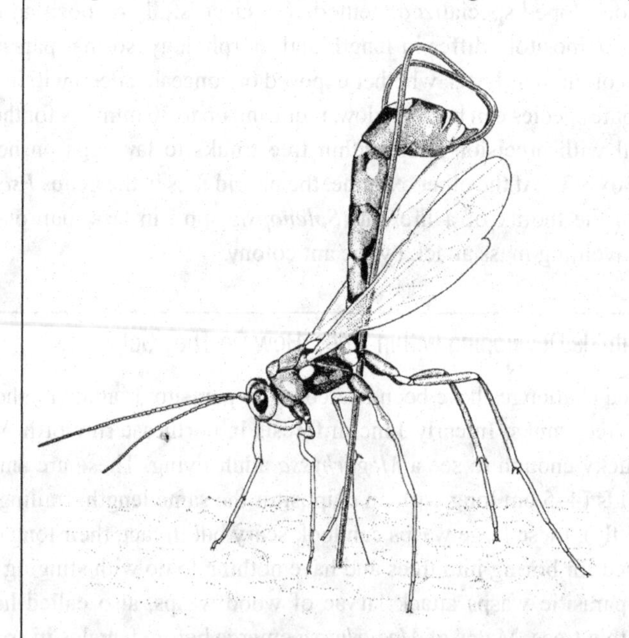

Figure Box 8.3.1 Female of the parasitic wasp *Megarhyssa nortoni* probing a pine log with her ovipositor in search of a larva of the wood wasp, *Sirex noctilio* (drawn by Karina H. McInnes; Gullan & Cranston, 2000).

There are numerous species of *Megarhyssa* and the closely related and slightly smaller *Rhyssa* that parasitize only wood wasps. A Eurasian wood wasp that can kill healthy pine trees, *Sirex noctilio*, was introduced to pine plantations in the southern hemisphere and both *Megarhyssa* and *Rhyssa* species have been introduced to numerous countries as parts of classical biological control campaigns. These big parasitoids, along with a smaller parasitoid (*Ibalia* spp.) that attacks earlier instars, parasitize the invasive wood wasps, but it was not until a parasitic nematode was also introduced that control was frequently achieved (Box 9.1).

Ectoparasitoids attacking concealed hosts often inject a permanent toxin when laying eggs that paralyzes the host to prevent host movement and molting. This makes sense because movements by the hosts in small spaces could damage externally attached parasitoid larvae that are quite incapable of defending themselves.

In contrast, endoparasitoids must contend with the host immune system because they are living within a host. Insect immune systems are less complex than mammalian immune systems, but insects can still effectively mount an assault on a parasitoid egg or immature parasitoid larva if their host blood cells recognize the intruder as non-self. If a parasitoid egg or larva is recognized as non-self, insect blood cells (hemocytes) can spread over the surface of the invader, effectively walling it off. The capsule of blood cells surrounding the invader often turns black in a process called melanization and the parasitoid within is killed, probably either by asphyxiation or as a result of the toxic compounds produced during the blackening process. Of course, endoparasitoids have developed methods for avoiding capsule formation (i.e., encapsulation). There seems to be variability among closely related parasitoids in whether they are encapsulated or not when within hosts. It has been hypothesized that some parasitoids escaping encapsulation have surfaces that are not recognized by the host as being foreign; in essence, the parasitoid can grow undetected within the host. Alternatively, some tricky species deposit their eggs within host tissues, such as the nervous tissue, the gut wall, and the fat body, so that the circulating blood cells that would be able to recognize them as non-self do not contact them.

8.4.3 Microbes in the Mix

To prevent encapsulation of parasitoid eggs and larvae within hosts, members of the larger ichneumonoid wasps have teamed up with viruses, named polydnaviruses, to help them survive within hosts. Adult females inject a polydnavirus into the host when ovipositing and the virus, along with additional materials injected by the female, acts to generally block the host immune response, including reducing the number of responsive blood cells and preventing melanization (Figure 8.11). The virus does not reproduce in the parasitized insect. It is incorporated in the parasitoid DNA and only reproduces within the reproductive tract of the adult female wasp. This strategy ensures persistence of the virus but also the distribution of this specialized virus to aid survival of wasp progeny, thus also ensuring survival of the virus.

In contrast, heritable bacterial symbionts in some hosts have been recognized as providing defense against parasitoids. The parasitoid *Aphidius ervi* cannot develop within pea aphids (*Acyrthosiphon pisum*) infected with the symbiotic bacteria *Hamiltonella* (these bacteria are themselves infected with a bacteriophage, that is vital to protection from the parasitoid). However, superparasitism (e.g., two eggs laid into an aphid instead of one) can overcome this defense so adult female parasitoids often choose to lay more than one egg in aphids hosting this symbiont. Thus, the aphids have found an associate to help them avoid parasitism while the parasitoid has figured out a way around this, suggesting the occurrence of a coevolutionary arms race between the aphid and parasitoid that is mediated by the symbiotic bacteria.

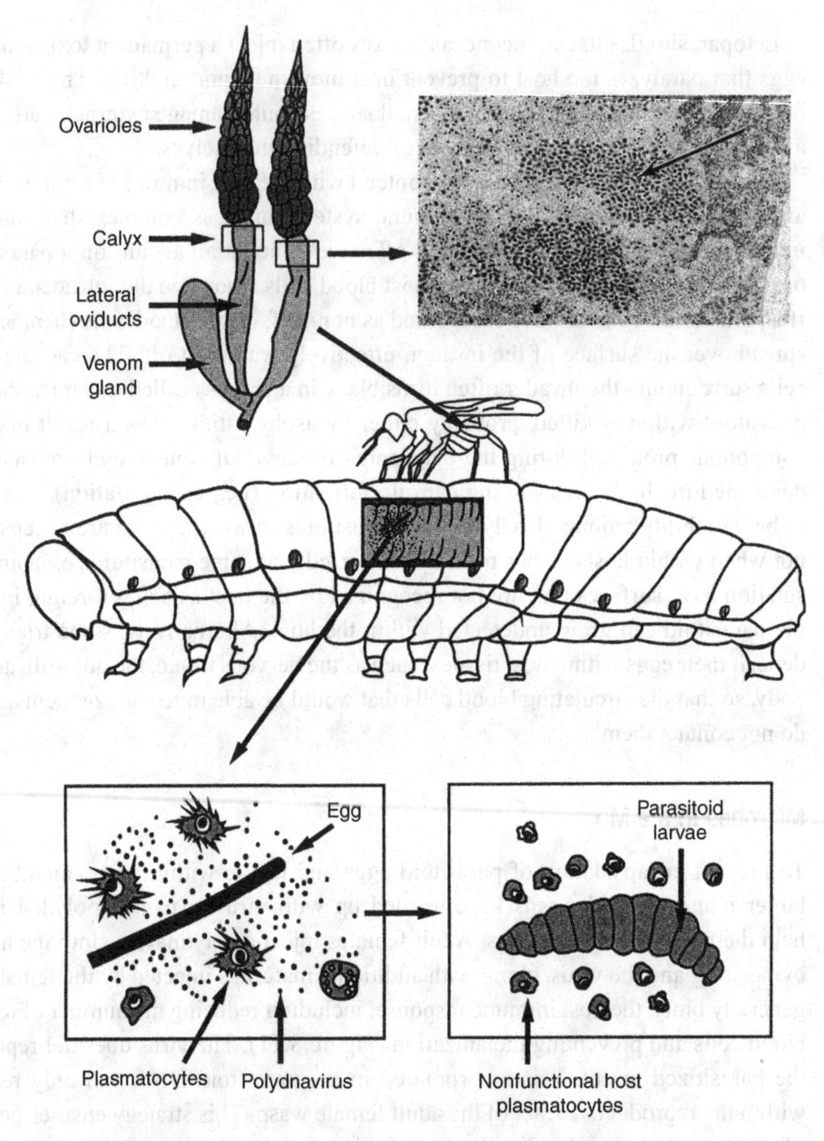

Figure 8.11 Polydnavirus–parasitoid–host relations. Polydnavirus is produced in the calyx region of the ovaries of an adult female parasitoid and is released within the host during oviposition to subsequently suppress the immune system of the parasitoid host (Boucias & Pendland, 1998).

8.5 Use of Parasitoids in Biological Control

Parasitoids have been central to the growth of biological control of arthropods. In particular, parasitoids that are prolific and actively search for and find hosts have been prized. Parasitoids that have been used for biological control are now those that are more host specific, so that their offspring target the pest and not nontarget species (see Chapter 18).

8.5.1 Classical Biological Control

Parasitoids have been used extensively for classical biological control (see Table 3.4). The high degree of host specificity characteristic of many parasitoid species makes these natural enemies first choices for classical biological control introductions. In an estimation made in 2016, a total of 1,512 species of parasitoids had been introduced for classical biological control. The majority of these were parasitic wasps, with fewer parasitic flies. Parasitoids have been used for good reason because there are many success stories from programs using parasitoids. Among the many successful classical biological control programs described in the literature, an example of an introduction of a tiny parasitic wasp for control of cassava mealybug (*Phenacoccus manihoti*) in Africa is described here (Box 8.4).

Box 8.4 Finding the Right Parasitoid to Control the Introduced Cassava Mealybug in Africa

Around the same time that the cassava green mite (see Box 7.3) was first found in Uganda, the phloem-feeding cassava mealybug, *Phenacoccus manihoti*, was first found in the Congo. As with the cassava green mite, this pest rapidly spread through central Africa and, by 1986, the cassava mealybug occurred in 25 countries, causing cassava yields to decline drastically. A classical biological control program was begun and the first step was to search for natural enemies. There are many species of *Phenacoccus* in central and northern South America but no one knew exactly where *P. manihoti* had come from. Scientists searched for *P. manihoti* to collect its natural enemies in Central America, northern Colombia, and Venezuela. They found a mealybug initially thought to be the new African pest. Several parasitoid species emerging from this mealybug were collected but they would not reproduce in *P. manihoti* in the insectary in the Congo. On closer examination, researchers found that there were both males and females of the mealybug that had been collected, while the African pest had only females, and these two mealybugs also differed in morphology. It was decided that the mealybug that had been collected was not the African pest. Foreign exploration continued in South America and *P. manihoti* was finally found in limited areas of Paraguay, Bolivia, and southern Brazil. Where it occurred, *P. manihoti* populations were hard to find. A highly host-specific parasitic wasp, *Apoanagyrus lopezi*, was found and first released in Africa in 1981. This parasitoid proved exceptionally effective at becoming established and controlling cassava mealybug. Attraction of these small wasps to the wax produced by cassava mealybugs helps them find hosts. This species parasitizes cassava mealybugs but also host feeds, which adds to mortality of hosts. Mass rearing methods were developed once it became evident that this wasp was particularly effective but would not spread so quickly on its own. However, distributing this wasp throughout the area with the pest was a problem; while the wasps spread on their own through agricultural areas, spread through rainforest zones was slower. To facilitate spread, wasps were mass-produced in insectaries, transported by air and then released on the ground. In some more remote areas, methods were developed to release wasps from aircraft flying over cassava-growing areas. Between 1981 and 1990, *A. lopezi* was released in over 100 areas and was documented as becoming established in 24 African countries. Total savings by these releases have been calculated to exceed US$10 billion and the release program is considered an exceptionally successful biological control program. Since 2008, the cassava mealybug has unfortunately also spread in Southeast Asia and *A. lopezi* is considered a very promising candidate for biological control there, building on the experience from Africa.

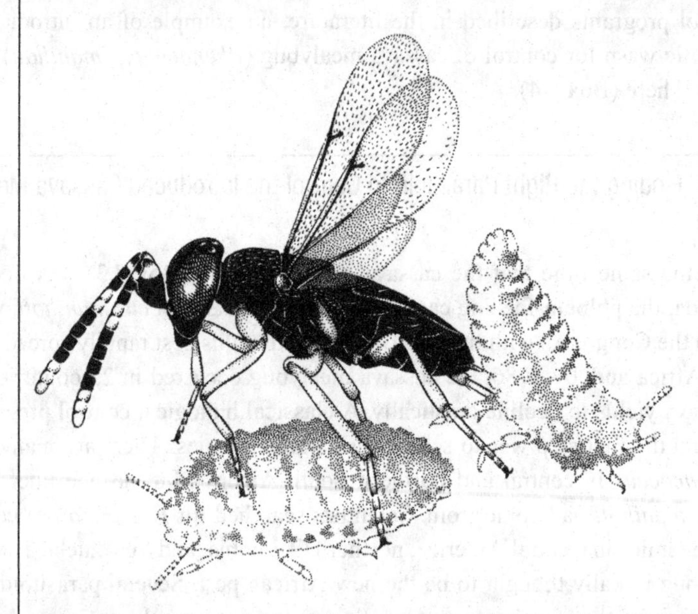

Figure Box 8.4.1 A female *Apoanagyrus lopezi* examines potential cassava mealybug hosts with her antennae and is shown inserting her ovipositor to lay an egg (Van Alphen & Jervis, 1996; with kind permission of Kluwer Academic Publishers).

With the great diversity in life history strategies among parasitoids, it has often been difficult to make decisions regarding which species to introduce for biological control, including how many species to introduce. Today, which parasitoid species can be released is also highly dependent on the host specificity of the parasitoid; any natural enemies with broader host ranges are not released (see Chapter 18). Relative to predicting parasitoid efficacy, recent models have demonstrated that making such a decision is, in fact, not simple and optimally requires information about factors such as fecundity, host feeding, and response to host density. Surprisingly, results from models disproved the dogma that host-feeding parasitoids are always superior; host feeders frequently have lower fecundity and they can require higher host densities before depressing pest populations.

8.5.2 Augmentative Releases

The other principal use of parasitoids for biological control has been their extensive development for augmentative releases, either inoculative or inundative. All parasitoids that have been developed for these purposes are parasitic wasps and predominantly include species with idiobiont strategies; since these species develop within hosts that are inactive, companies mass-producing biological control agents do not have to worry about feeding the hosts after they are parasitized and this facilitates mass production.

Augmentative use requires mass production of parasitoids, first producing hosts and then exposing them to parasitoids. In most cases, after parasitized hosts die, but while parasitoids are still inside the hosts, they are released for control. In field crops, *Trichogramma* species, tiny endoparasitoids attacking eggs of caterpillar pests, have been developed for control. Although *Trichogramma* can be difficult to identify there are a total of 19 known species and about 10 of these have been used for biological control. They have been used extensively around the world in quite a few crops, including beets, corn, soybeans, sugarcane, cereals, vegetables, and tree crops. However, the two major uses are against stem borers in graminaceous crops and against *Heliothis*/*Helicoverpa* in corn, cotton, and tomatoes. For many of the target species the larval stages that cause damage are hidden within the plant and thus are very difficult to control using chemical pesticides or other means. Using an egg parasitoid is wise because pests are then killed before they develop to a stage causing damage. However, it seems that inundative releases or seasonal inoculation of *Trichogramma* usually do not provide enough control on their own. Use of *Trichogramma* generally should be integrated with other methods such as (1) application of *Bacillus thuringiensis* to foliage, which would kill larvae (Chapter 10), (2) disrupting the ability of adult stages of the pests to mate through massive applications of pheromones (mating disruption), or (3) possibly well-timed use of pesticides that do not kill *Trichogramma*.

The chalcidoid *Encarsia formosa* is widely used by the greenhouse industry to combat the ubiquitous greenhouse whitefly, *Trialeurodes vaporariorum* (Box 8.5). As of 2015, a leading producer and distributor of biological control agents in Europe marketed 11 different species of parasitoids, among a total of 38 species of natural

Box 8.5 *Encarsia formosa* **against Greenhouse Whitefly**

The greenhouse whitefly (*Trialeurodes vaporariorum*) has a huge range of plants on which it feeds, but in greenhouses it is predominantly a major pest of vegetable crops. This pest sucks the phloem sap of plants and excretes sugar-laden honeydew that falls on the foliage below. The major injury from this pest is caused by sooty mold growing on the honeydew and subsequently reducing photosynthesis and respiration of the plants.

In greenhouses, there can be numerous pests on the same crop and spider mites are fairly regular culprits. Spider mites developed resistance to chemical insecticides in Europe beginning in 1949 so growers began using predatory mites for control. After greenhouse whiteflies emerged as a pest, growers had to use natural enemies to control greenhouse whitefly or spider mite control by predaceous mites was disrupted. The small parasitoid *Encarsia formosa* had first been investigated for whitefly control as early as 1927. Over time, it was found that this parasitoid can be very effective at controlling greenhouse whitefly. It uses the presence of immature whiteflies or honeydew to locate hosts. Adult females do not lay eggs in hosts that have previously been parasitized, thereby avoiding wasting eggs. As an added benefit, adult females can host feed on unparasitized hosts and thus kill hosts that are not used for progeny production.

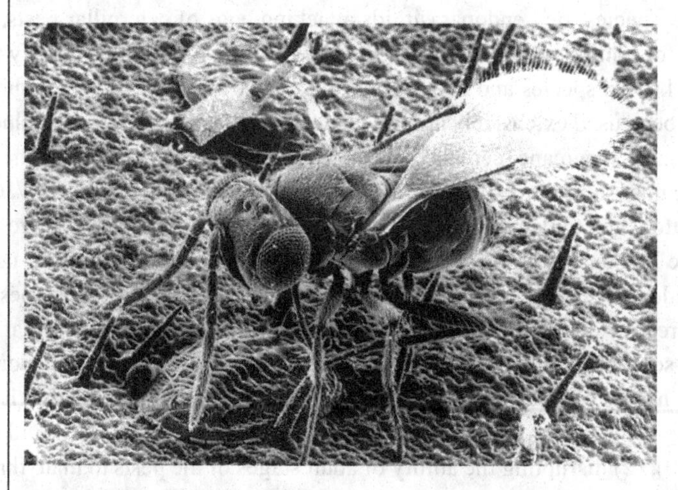

Figure Box 8.5.1 The whitefly parasitoid *Encarsia formosa* (photo by P. Sutherland; van Lenteren & Martin, 1999).

It was only in the 1970s that efficient and predictable control resulted after development of precise recommendations for use of *E. formosa* in greenhouses in the Netherlands and the United Kingdom. Use of this parasitoid increased so that in 1993 alone it was estimated that 4,800 ha of greenhouses were treated with *E. formosa*. Today, this parasitoid is universally accepted as a major option for control of whiteflies in greenhouses. *Encarsia formosa* is often inoculatively released and several specific methods for application have been suggested. For one, successive regular introductions of *E. formosa* begin soon after the crop has been planted (blind releases) or when the pest is first observed. As an alternative, plants with established populations of greenhouse whitefly parasitized by *E. formosa* are placed in the greenhouse throughout the year (banker plants). These are called banker plants because the plants host a population of parasitized whiteflies that will emerge to kill white flies over time, like having savings in a bank to withdraw over time.

Table 8.2 Examples of parasitic wasps used for augmentative release.

Parasitoid superfamily and family	Parasitoid species	Target host orders	Target hosts	Use
Ichneumonoidea				
Aphidiidae	*Aphidius colemani*	Hemiptera	Aphids	Greenhouse vegetables and ornamentals
	Aphidius matricariae	Hemiptera	Aphids	Greenhouse vegetables and ornamentals
Braconidae	*Cotesia flavipes*	Lepidoptera	Stem-boring caterpillars	Sugarcane
	Dacnusa sibirica	Diptera	Leaf-mining fly larvae	Vegetables
Chalcidoidea				
Aphelinidae	*Encarsia formosa*	Hemiptera	Whiteflies (*Trialeurodes* and *Bemisia*)	Many greenhouse crops
Trichogrammatidae	*Trichogramma* spp.	Lepidoptera	Lepidopteran larvae	Cereals, orchards, vegetables
Eulophidae	*Diglyphus isaea*	Diptera	Leaf-mining fly larvae	Vegetables
	Aprostocetus hagenowii	Blattodea	Cockroaches (*Periplaneta* spp.)	Indoor pests
Pteromalidae	*Muscidifurax* spp.	Diptera	House flies (*Musca domestica*)	Animal barns
Encyrtidae	*Metaphycus* spp.	Hemiptera	Soft scales (coccids)	Interior landscapes
	Leptomastix dactylopii	Hemiptera	Mealybugs	Conservatories, interior landscapes and greenhouses

enemies available, and these parasitoids were predominantly intended for insect control in indoor environments, such as greenhouses. Some parasitoid species commonly used in commercial biological control are listed in Table 8.2. As is evident, only parasitic Hymenoptera, and not Diptera or other taxonomic groups with a parasitoid life strategy, are produced for inoculative and inundative releases.

Further Reading

Cônsoli, F. L., Parra, J. R. P., & Zucchi, R. A. (eds.) (2010). *Egg Parasitoids in Agroecosystems with Emphasis on* Trichogramma. Dordrecht, Netherlands: Springer.

Hochberg, M. E. & Ives, A. R. (eds.) (2000). *Parasitoid Population Biology*. Princeton, NJ: Princeton University Press.

Quicke, D. L. J. (1997). *Parasitic Wasps*. London: Chapman & Hall.

Quicke, D. L. J. (2015). *The Braconid and Ichneumonid Parasitoid Wasps: Biology, Systematics, Evolution and Ecology*. Chichester, UK: Wiley Blackwell.

Wajnberg, E., Bernstein, C., & Van Alphen, J. (eds.) (2008). *Behavioral Ecology of Insect Parasitoids: From Theoretical Approaches to Field Applications*. Malden, MA: Blackwell Publishing.

Wajnberg, E. & Colazza, S. (eds.) (2013). *Chemical Ecology of Insect Parasitoids*. Oxford: Wiley Blackwell.

9 Parasitic Nematodes

The Phylum Nematoda is exceptionally diverse, including species adapted to just about every type of lifestyle imaginable. It is no surprise that many of these roundworms have adapted to lives as parasites of invertebrates. Nematodes, or roundworms, that attack arthropod pests range in size from those visible without magnification to microscopic species. They have reduced morphological features but one feature common to these species is that all are long and thin. Many can only enter hosts through body openings (mouth, anus, spiracles) or wounds, after which they often penetrate the body cavity (hemocoel). Others enter the host gut passively after nematode eggs are ingested with food. Some have a hardened "tooth" in their mouths that they use to penetrate actively through arthropod cuticle, although others without such a structure can also pass through cuticle.

Nematodes hatch from eggs and molt from one to another of four juvenile stages before molting to adults. For some species, adults are either male or female (amphimictic) while in others, adults are hermaphroditic, with each individual having reproductive organs of both sexes. As an alternative, in some species both types of sexuality can occur at different times in the life cycle. The nematode life cycle is often ordered such that only a specific stage, often a juvenile called an infective juvenile (or IJ), will leave a host to find a new host to infect. However, such departures often only happen when nutrients within a cadaver are exhausted.

All nematodes are basically aquatic, requiring at least a film of water in which to live and reproduce, although some insect parasitic nematodes can tolerate moderate desiccation for periods of time. Of course, while nematodes are living within arthropods or bodies of recently killed arthropods their surroundings are moist. During dispersal, nematodes are more at risk of desiccation and, for this reason, many nematodes occur in aquatic habitats or in the soil. Nematodes display some ability for finding or attacking hosts. After locating and infecting a host, some colonize the ovaries and/or eggs of adult female hosts and take advantage of the subsequent oviposition as a means for dispersal to locations where healthy immature hosts would occur. When the parasitized female hosts then lay eggs, infective juveniles are deposited instead of or along with host eggs (i.e., nemaposition). Other nematodes actively search for hosts using volatile cues to find them; for example, some species of *Steinernema* and *Heterorhabditis* are attracted by host fecal components, bacterial gradients, plant roots, or carbon dioxide. One study has shown that a species of *Heterorhabditis* is particularly attracted by caryophyllene produced when corn roots have been fed on by larvae of the western corn rootworm

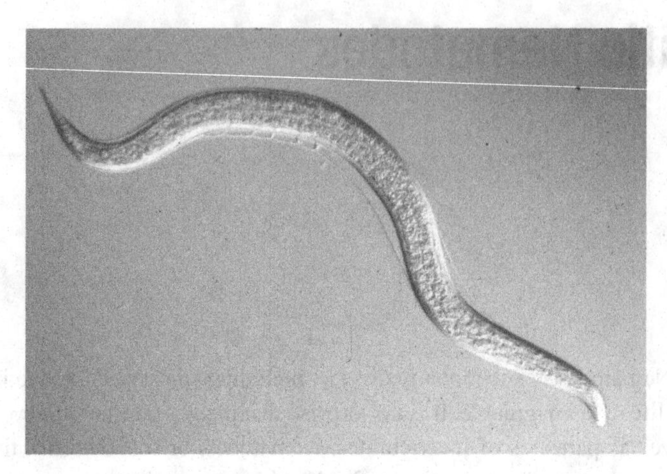

Figure 9.1 Infective juvenile nematode of *Steinernema carpocapsae* (photo by Patricia Timper; Hoffmann & Frodsham, 1993).

(*Diabrotica virgifera*); this chemical signal from the plant therefore attracts the nematodes who act as "bodyguards" for the plant by killing the root-feeding larvae.

Life history strategies of nematodes attacking invertebrates are diverse, ranging from species that live within hosts as parasites but do not cause mortality to species whose symbiotic bacteria help to kill hosts quickly.

9.1 Entomopathogenic Nematodes (EPNs)

Steinernematidae and Heterorhabditidae are two families of nematodes that have similar biologies and are used for biological control in similar ways; they are both referred to as entomopathogenic nematodes, or EPNs. Individuals of these nematode families are very small, less than 1–3 mm long (Figure 9.1). Although these two families are both within the order Rhabditida, they are not especially closely related. For species of *Steinernema*, both a male and a female nematode must enter a potential host for reproduction to take place while for species of *Heterorhabditis*, all infective juveniles (the stage entering new hosts) become hermaphrodites so only one individual is required to infect a new host for reproduction to ensue. For both genera, there can be numerous generations within a host before the dispersal stage, the infective juveniles, develop and leave. During this time, sometimes nematode eggs hatch within the mother and juveniles remain and grow within her, parasitizing her and only leaving once they themselves become adults.

A unique aspect of the biology of these nematodes is their symbiosis with very specific bacteria. The third stage infective juveniles carry symbiotic bacteria in their guts and, after invading a new host, regurgitate or defecate the bacteria, which are released into the new host. These bacteria, species of *Xenorhabdus* for Steinernematidae and *Photorhabdus* for Heterorhabditidae, are the main agents responsible for killing hosts very rapidly, within 2–3 days. Mortality of host insects is primarily caused by toxins

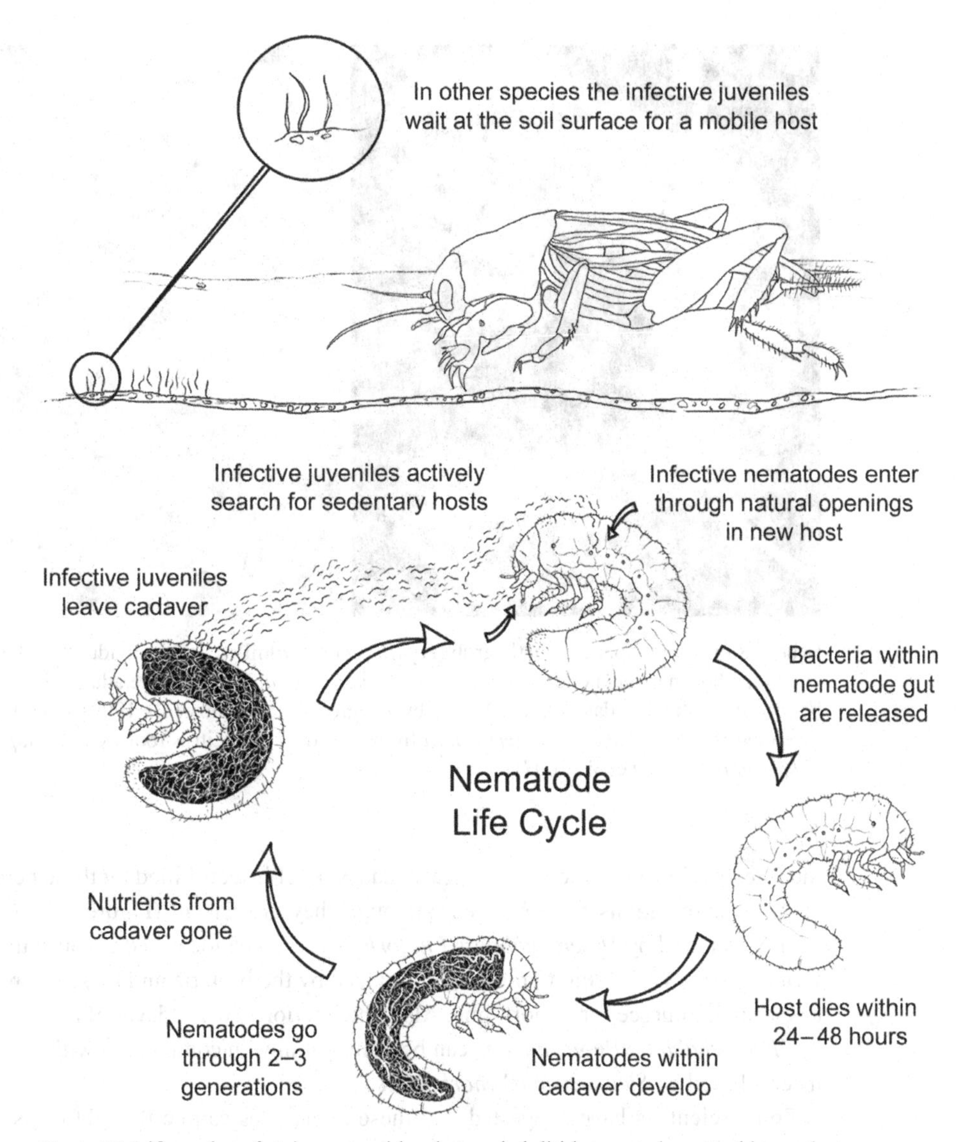

Figure 9.2 Life cycles of steinernematid or heterorhabditid nematodes, attacking sedentary or mobile hosts in or above the soil. Although infection through natural openings is shown, infection can also occur directly through cuticle (drawn by Alison Burke).

that kill the host (most are bacterial toxins but some are produced by nematodes) as well as by growth of the bacteria in the insect. Actually, nematodes and bacteria work together as teams because nematodes and bacteria both depress host immune responses before the host dies. The bacteria then continue to increase within the dead host, using the cadaver for nutrients and the nematodes principally feed on the bacteria. Nematode generations continue to develop within the same cadaver until nematode density is high and food is becoming less abundant, at which time infective juveniles exit to find a new host, taking some of the bacteria along with them in their guts (Figure 9.2). Until

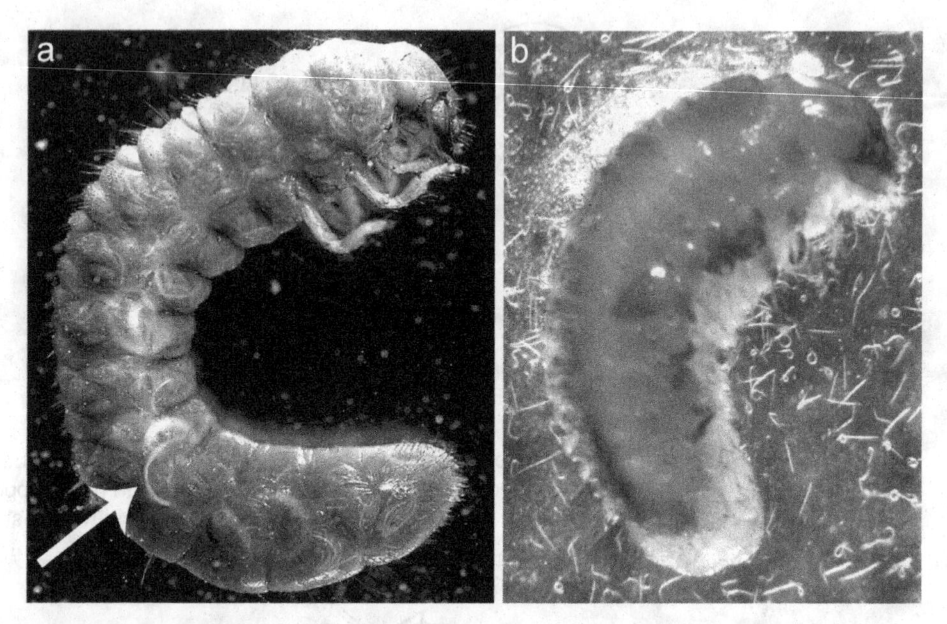

Figure 9.3 (a) Dead Japanese beetle grub (*Popillia japonica*) filled with individuals of the entomopathogenic nematode *Heterorhabditis bacteriophora*. Larger, white adult nematodes can be seen through the cadaver cuticle (photo by J. Ogrodnick; Vittum et al., 1999). (b) Infective juveniles of *Heterorhabditis bacteriophora* from a dead larva of citrus root weevil, *Diaprepes abbreviatus* (McCoy et al., 2007).

the infective juveniles are ready to leave, cadavers of insects killed by these nematode–bacterial associations remain intact, although they are flaccid (Figure 9.3). Cadavers of insects killed by *Heterorhabditis*/*Photorhabdus* can be identified because they often turn orange to red, owing to pigments produced by the bacteria and these cadavers also often can luminesce for a short time. Also, the interior of the cadaver of an insect killed by *Heterorhabditis*/*Photorhabdus* can be quite gummy, in comparison with cadavers of insects killed by *Steinernema*/*Xenorhabdus*.

Some scientists have suggested that these nematodes can be thought of as farmers bringing a crop to grow with them (i.e., the bacteria that will kill the host and which they will eat) when they move from insect host to insect host. The association between the nematodes and bacteria is mutualistic because both members benefit. Although the nematodes can often kill the host in the absence of bacteria, they do so more slowly. The nematodes cannot reproduce without feeding on the bacteria that supply them with required nutrients, such as sterols. With these bacteria, hosts are killed much more quickly and cadavers are kept free of other bacteria as a result of antibiotics produced by the symbiotic bacteria. The bacteria gain from the association because they cannot persist in the environment, disperse, locate a new host insect, and invade the hemocoel on their own, so the nematodes provide transport and protection for the bacteria.

Steinernema and *Heterorhabditis* species can have very different strategies for locating hosts. Most species of both genera actively search through the soil for hosts, usually targeting sedentary hosts, and these nematodes have been named cruisers. To locate and infect mobile hosts, some species of *Steinernema* display a radically different behavior. These ambushers go to the surface of the soil and stand vertically on their posterior ends on the top of soil particles, where some wave back and forth while others stand still, often for several hours (Figure 9.2). If a host walks by and contacts an ambusher nematode, the nematode attaches to it. If the host is nearby but not in contact the nematode can jump to reach it; it has been recorded that these nematodes can jump nine times as far as their body length and seven times as high as their body length. Of course there are species that use cruising part of the time and ambushing the rest.

Steinernematids and heterorhabditids survive for only a few hours on exposed surfaces. They are basically soil-dwelling and can be greatly affected by exposure to environmental extremes. As basics, they require moisture and oxygen. Dry soil can seriously impair mobility and survival, but nematodes within dry soil can often persist for 2–3 weeks. If nematodes are dried slowly under conditions of higher humidity but no free water, they survive and adapt and can enter a quiescent or dormant state that is more desiccation-tolerant. Soil structure can also greatly influence these nematodes, with enhanced movement and survival in lighter soils where there are larger pore spaces.

As an example of how our knowledge about these nematodes has been increasing, in 2002, 35 species had been described in the families Steinernematidae and Heterorhabditidae, but by 2012, 68 species had been described in Steinernematidae with 27 in Heterorhabditidae. In general each of these species is associated with a unique species of bacteria in the genera *Xenorhabdus* or *Photorhabdus*, respectively. However, because it is generally necessary to establish a living culture of a new species of EPN in order to identify the bacteria and because it is complicated to name new species of bacteria, the bacterial species associated with many of the described EPN have not yet been identified or described.

All EPNs are parasites of insects but they vary in their degrees of host specificity. Some have rather large host ranges while others seem to be more specific. Host specificity can be influenced at several levels. Some insects groom to remove nematodes before the nematodes penetrate. Soil-dwelling scarab larvae have fine sieve plates covering their respiratory openings (spiracles) that restrict nematode entry by that route. Thick and convoluted gut walls of some insect species are thought to deter nematode penetration.

It is common for *Steinernema* and *Heterorhabditis* to arouse a defense response by insect blood cells once they enter the body cavity of a potential host. Some nematodes can become encapsulated by blood cells and melanin, just as with parasitoid larvae, but other nematode species are not encapsulated, probably because they are not recognized as being foreign once they are within the host's body. Sometimes *Steinernema* or *Heterorhabditis* can overwhelm the encapsulation response; if enough nematodes enter the insect at one time, there are not sufficient blood cells to encapsulate all of them at the same time. It has recently been discovered that EPNs travel and search for new hosts in groups so that numerous nematodes would be able to penetrate a new host at the same time as another means for overwhelming the insect immune defenses. As a third alternative, the nematode or the bacteria can avoid host defenses by suppressing the immune

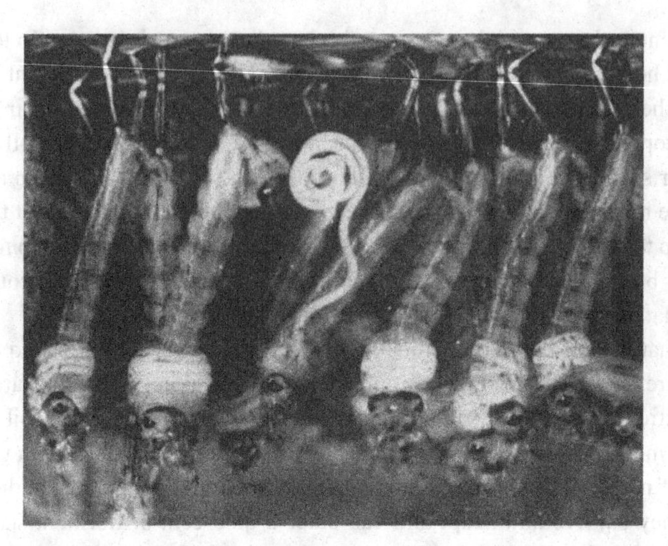

Figure 9.4 Juvenile mermithids (*Romanomermis culicivorax*) coiled within the thoraces of mosquito larvae (*Culex pipiens quinquefasciatus*) during laboratory mass rearing. One postparasite is just emerging (photo by J. J. Peterson, USDA ARS).

response of a potential host. In particular, some of the symbiotic bacteria are known to produce toxins that have this effect.

9.2 Mermithidae

A diversity of other groups of nematodes live more as long-term parasites within hosts and some are even ectoparasitic, attached to the surfaces of hosts. Among these longer-term parasites, the best known are the Mermithidae, obligate parasites that live in a variety of invertebrate hosts. Perhaps one of the reasons that these nematodes are fairly well known is that they are much larger and therefore more obvious than other nematodes. Mermithid adults are macroscopic, with adult females often 5–20 cm or more in length, although still very thin. They have been of great interest for biological control because hosts usually die once mermithids complete their development, leave the host, and enter the environment.

We know quite a bit about mermithid species attacking mosquitoes, biting midges, black flies, leafhoppers, and grasshoppers. For the mosquito-pathogenic *Romanomermis culicivorax*, juveniles live within mosquito larvae for only a few weeks (Figure 9.4) after which they emerge as postparasites, killing hosts as they emerge. They drop to the sediment at the bottom of aquatic habitats, develop into adult males or females, then mate and lay eggs that overwinter to produce preparasitic juveniles the next spring. The preparasites emerge from eggs and swim up to the surface of the water where they actively search for larval mosquitoes. The preparasites do not live for very long, usually less than a day, so they do not have much time to find a host. Preparasites that locate a host enter through the larval cuticle and develop within the body of the larval mosquito to continue the life cycle.

9.3 Use of Nematodes in Biological Control

Development of nematodes for biological control is an active area, concentrating on use of nematodes for control of soil-dwelling insects and slugs. In addition, nematodes have been used for control of pests in cryptic habitats, where the nematodes are protected from environmental extremes.

9.3.1 Classical Biological Control

Nematodes have been used for classical biological control of several pests, but perhaps the largest successful programs have been conducted using *Deladenus siricidicola* (Family Neotylenchidae). This nematode has been released in plantations of introduced pines in the southern hemisphere against the introduced wood wasp *Sirex noctilio* that has wood-boring larval stages (Box 9.1).

Another successful program has involved control of mole crickets (*Scapteriscus* spp.) that were introduced from South America to Florida where they are pests in lawns and pastures. *Steinernema scapterisci* from Uruguay and Argentina was introduced, became established and spread, resulting in up to 98 percent decreases in mole crickets over 3 years in some locations. However, in scattered lawn environments treated in various ways for multitudes of problems, *S. scapterisci* has not always persisted. For a more short-term approach, when this nematode is mass-produced and applied inundatively (2.5×10^9 nematodes/ha) it can provide the same level of control as standard insecticides.

9.3.2 Augmentative Releases

Emphasis with species of *Steinernema* and *Heterorhabditis* has been on their mass application as biopesticides, usually against soil-dwelling insect pests. This has been possible because these nematodes are easily grown in large quantities on relatively inexpensive media. During initial investigations, they were grown using solid media made of pork kidney or dog food but mass-production technology subsequently became more sophisticated. Mass production by larger industries has often involved production in liquid medium in fermentation tanks with capacities of up to 15,000 liters or more, yielding about 10^5 juveniles per milliliter. However, cottage industries producing nematodes still use solid media or insect hosts. Nematodes may be formulated by absorbing a highly concentrated nematode suspension onto a porous material (often something like vermiculite or diatomaceous earth) to provide ample oxygen so that nematodes survive well. In some preparations the nematodes are partially desiccated and mixed in powders or granules based on vermiculite or similar carriers. Methods for optimizing formulation,

Box 9.1 *Deladenus siricidicola* and *Sirex noctilio*

The native Eurasian wood wasp *Sirex noctilio* can kill healthy pine trees. A female *S. noctilio* drills with its ovipositor into the wood of a healthy tree and injects spores of a symbiotic fungus (*Amylostereum areolatum*) at the same time as she lays her eggs. The fungus is a tree pathogen and starts growing, eventually spreading throughout the tree. The female wood wasp also injects a toxin that shuts down tree defenses, and the toxin and fungus work together to kill the tree. The wood wasp egg is stimulated to hatch by the presence of the growing fungus and the larva subsequently develops by boring through the tree while feeding on the wood rotted by the fungus.

Pines are not native to Australia but they were planted there extensively as a source of wood. This worked well until *S. noctilio* was introduced to Australia in the early 1960s and, as this wood wasp spread, it left a wake of dead pines. Unfortunately, the Australian forestry industry had planted huge areas of species of pines that were highly susceptible to *S. noctilio*.

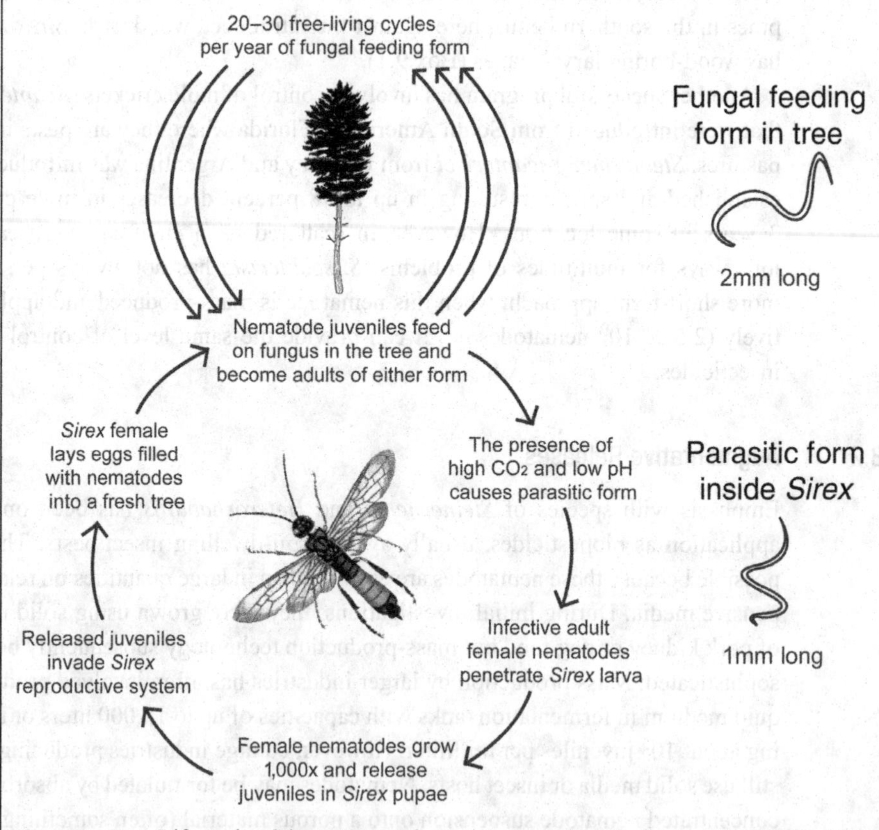

Figure Box 9.1.1 Life cycle of the entomopathogenic nematode *Deladenus siricidicola*, which has two different life forms, one of which can live as a parasite of the wood wasp *Sirex noctilio* and the second feeds on the wood rotting fungal symbiont of the host (drawn by Alison Burke, based on Bedding, 1993).

In Europe, researchers found that the nematode *Deladenus* (previously *Beddingia*) *siricidicola* sterilized these wood wasps so it was decided to introduce this nematode to Australia for classical biological control. To start its life cycle, juvenile nematodes produced within an adult female wood wasp invade the female's ovaries, with the result that each wood wasp egg can contain up to 200 juvenile nematodes and a viable wood wasp no longer hatches from that egg. Parasitized adult female wood wasps still have the urge to oviposit, so they travel to healthy trees where they deposit their nematode-filled eggs. Unparasitized wood wasps also lay eggs in these same trees and their healthy offspring become parasitized by the nematodes. Adult nematodes enter healthy wood wasp larvae by directly drilling through the body wall and they then grow and reproduce within, without killing the wood wasp larva.

It is difficult to imagine that the relatively few nematodes deposited when a female wood wasp nemaposits would be able to find the healthy larvae living within a tree. However, this nematode species can also develop solely by feeding on the wood wasp's symbiotic fungus. In fact, the nematode has two forms: one is parasitic and one feeds on this fungus. The nematodes that emerge from wood wasp eggs live and reproduce as fungal-feeding forms until stimulated by low pH and high carbon dioxide to become parasitic.

The fact that this nematode has two very different forms, that is, either parasitizing *S. noctilio* larvae or feeding on the symbiotic fungus, has been lucky for biological control efforts. The symbiotic fungus is fairly easy to grow in the laboratory and *D. siricidicola* is easily grown feeding on the fungus.

At the outset of the project to release *D. siricidicola* in Australia, a strain was chosen for mass production and release that caused total sterilization of adult female *S. noctilio*, did not interfere with female flight ability, and was compatible with parasitoids released against this host. After release, *D. siricidicola* was extremely effective at controlling *S. noctilio* but, because it was thought that long-distance spread of this nematode would be slow, the nematode was produced in the laboratory and redistributed through large biological control release programs. However, the efficacy of the nematode slowly declined. By the time this was noticed, after 20 years of mass production and distribution, nematode efficacy had dropped from 100 percent parasitism to 25 percent parasitism. Research demonstrated that the nematodes had become very efficient at growing while feeding on fungus but were less virulent against *S. noctilio*. This problem was rapidly corrected by re-isolating the nematode from wood wasps at the original release site and changing production methods. Today, virulence of nematodes is maintained because ample insect-virulent nematode cultures were frozen and each year a new virulent culture is thawed to use for mass production. Thus, before the virulence of a specific strain of nematode declines, a newly thawed, virulent culture is substituted for the culture that had been used over the past year. There is still a market for these nematodes because after they have decimated the host populations they do not persist well so they must be reintroduced if *S. noctilio* begin to increase again; therefore, in this case a classical biological control program has morphed into inoculative augmentation, when needed.

Table 9.1 Examples of commercialized species of nematodes and their target hosts.[1]

Pests	System	Nematode
Insecta		
Orthoptera		
Mole crickets	Turf	*Steinernema riobrave, S. scapterisci*
Lepidoptera		
Armyworms, cutworms, webworms	Turf	*S. carpocapsae, S. feltiae, S. riobrave*
Artichoke plume moth caterpillars	Artichoke	*S. carpocapsae*
Coleoptera		
Billbugs (weevils)	Turf	*Heterorhabditis bacteriophora, S. carpocapsae*
Cranberry girdler	Cranberries	*S. carpocapsae*
Root weevils	Berries	*H. bacteriophora, H. megidis*
	Citrus	*S. riobrave, H. indica*
	Cranberries	*H. bacteriophora, S. carpocapsae*
	Ornamentals	*H. bacteriophora, H. megidis*
Scarabs	Turf	*H. bacteriophora, H. megidis, S. glaseri*
Wood borers	Ornamentals, fruit trees	*S. carpocapsae, H. bacteriophora*
Diptera		
Fungus gnats	Mushrooms, ornamentals	*S. feltiae*
Gastropoda		
Slugs and snails	Variety of field crops	*Phasmarhabditis hermaphrodita*

Source: based on Grewal (2002); Shapiro-Ilan et al. (2017); and A. Koppenhofer (pers. comm.)
[1] Not all nematodes are available in all countries.

packaging, and shelf life were critical for development of these nematodes for control because these nematodes need to be living when applied. *Steinernema* and *Heterorhabditis* are generally more effective if applied to looser sandy or loamy soils that are moist and at moderate to warm soil temperatures. During application, using enough water to wash the nematodes off the soil surface and down into the soil is critical.

Work on use of Steinernematidae and Heterorhabditidae for control has only proceeded in earnest since the 1960s and 1970s. Many species have been commercialized to various degrees, with much of this commercial expansion since the 1990s (Table 9.1). Research has shown that species vary in efficacy against different pests, as a result of strategies for host location and inherent adaptation to host species (Box 9.2). Earlier attempts to use entomopathogenic nematodes against foliage-feeding insects were not successful. Pests against which nematodes are now used include insects like caterpillars, beetle grubs, flea larvae, and fly maggots associated with soil or in cryptic and moist habitats, for example, insects boring in stems. Use of nematodes often targets more specialized applications, such as pests in greenhouses, nurseries, and turf. Nematodes are simple to apply because they are compatible with conventional spray equipment and many pesticides. After application, in many instances the nematodes become established and will recycle in the pest population. Some species more widely used for control have

Box 9.2 The EPN/Pistachio Story

Managing insect pests often takes more than a single intervention; most successful integrated pest-management programs (see Chapter 19) employ an array of methods to reduce pest populations and more than a single life stage of the pest is targeted. The navel orangeworm, *Amyelois transitella*, is a pest in several different nut crops in California and the caterpillars are especially troublesome in pistachio production. Adults oviposit on developing nuts during the summer and sprays of insect growth regulators and pyrethroid insecticides are used to reduce their populations. At the end of the season, infested nuts fall to the ground and larvae overwinter in unharvested nuts (called "mummies"), protected from most sources of mortality – either natural or human-made. This is the life stage that is targeted using the entomopathogenic nematode *Steinernema carpocapsae*.

Entomopathogenic nematodes (EPNs) used in agriculture are very successful under the right conditions. They need a moist habitat that is protected from ultraviolet light, hosts need to be readily available, and the temperature should be between 15 and 30°C. These conditions are met for late winter/early spring applications of EPNs, but success depends on more than environmental conditions.

Four circumstances allow EPNs to be successful against navel orangeworm larvae:

1 *Susceptibility*: Larvae of the navel orangeworm are extremely susceptible to infection by *S. carpocapsae*.
2 *Competition with other products*: Because the overwintering late instars are found inside fallen nuts, any material used to control them must be able to get inside the nut to induce mortality. Neither conventional insecticides nor biopesticides other than EPN-based products are able to do this. Thus, EPNs are the only product that can achieve successful control of this part of the life cycle of navel orangeworm.
3 *Climate*: The climate of the Central Valley of California is cool and moist in late winter and early spring, the time when EPNs are applied, so these are ideal conditions. In the summer months, when the soil surface can exceed 50°C, there would be no chance of success with EPNs.
4 *Economics*: EPN products are more expensive than many other alternative controls. However, EPNs can be applied via irrigation systems, which offsets their cost because farmers do not have to pay for the fuel and labor associated with many other types of applications.

Therefore, under these conditions, EPNs can cause up to 70 percent mortality of overwintering navel orangeworm populations, demonstrating the importance of paying attention to the biology and ecology of both the pest and natural enemy in order to come up with a winning biological control program.

broad host ranges and their use must therefore be integrated with that of other soil-active natural enemies.

Compared with *Steinernema* and *Heterorhabditis*, few other species of nematodes have been considered for augmentative control of insects. However, *Phasmarhabditis hermaphrodita* has been developed for control of snails and slugs in the United Kingdom. This nematode lives as self-fertilizing hermaphrodites once it enters a new host and develops to the adult stage. As with *Steinernema* and *Heterorhabditis*, it enlists the help of bacteria; *P. hermaphrodita* carries bacteria to a new snail or slug and then feeds on the bacteria that multiply within the host. This nematode depends on bacteria to be able to kill hosts but it seems that the associations between bacteria and *P. hermaphrodita* are not as specific as in the cases of the EPNs. Although it can take from 7 to 21 days for this nematode species to kill a host, feeding by the host drops to only 10 percent of normal by four days after infection, thus preventing continued damage by the slug or snail before it dies.

The mermithid *Romanomermis culicivorax*, found attacking mosquito larvae in North America, was studied extensively and found to provide effective control in some areas (see Figure 9.4). Mass production was somewhat complex and expensive, but as a result of the high priority of mosquito control this mermithid was commercialized in the late 1970s as a product named Skeeter Doom. However, this product predominantly failed because of storage and transportation problems in addition to being outcompeted by a major new product on the market at that time, another natural enemy, the bacterium *Bacillus thuringiensis israelensis* (see Chapter 10). In recent years, another species of mermithid that kills mosquitoes, *Romanomermis iyengari*, which was originally found in India, has been investigated for use in Asia and Central America, as controlling mosquitoes always is a high priority.

Further Reading

Campbell, J. F. & Lewis, E. E. (2002). Entomopathogenic nematode host-search strategies. In *The Behavioural Ecology of Parasites*, ed. E. E. Lewis, J. F. Campbell, & M. V. K. Sukhdeo, pp. 13–38. Wallingford, UK: CABI Publishing.

Campos-Herrera, R. (ed.) (2015). *Nematode Pathogenesis of Insects and Other Pests: Ecology and Applied Technologies for Sustainable Plant and Crop Protection*. Dordrecht, Netherlands: Springer.

Frank, J. H. (2009). *Steinernema scapterisci* as a biological control agent of *Scapteriscus* mole crickets. In *Use of Microbes for Control and Eradication of Invasive Arthropods*, ed. A. E. Hajek, T. R. Glare, & M. O'Callaghan, pp. 115–132. Dordrecht, Netherlands: Springer.

Gaugler, R. (ed.) (2002). *Entomopathogenic Nematology*. Wallingford, UK: CABI Publishing

Grewal, P. S., Ehlers, R.-U., & Shapiro-Ilan, D. I. (2005). *Nematodes as Biocontrol Agents*. Wallingford, UK: CABI Publishing.

Lacey, L. A. & Georgis, R. (2012). Entomopathogenic nematodes for control of insect pests above and below ground with comments on commercial production. *Journal of Nematology*, **44**, 218–225.

Lewis, E. E. & Clarke, D. J. (2012). Nematode parasites and entomopathogens. In *Insect Pathology*, ed. F. E. Vega & H. K. Kaya, 2nd edn, pp. 395–443. Amsterdam: Elsevier.

Shapiro-Ilan, D. I., Hazir, S., & Glazer, I. (2017). Basic and applied research: Entomopathogenic nematodes. In *Microbial Agents for Control of Insect Pests: From Discovery to Commercial Development and Use*, ed. L. A. Lacey, pp. 91–105. Amsterdam: Academic Press.

Williams, D. W., Zylstra, K. E., & Mastro, V. C. (2012). Ecological considerations in using *Deladenus* (= *Beddingia*) *siricidicola* for the biological control of *Sirex noctilio* in North America. In *The Sirex Woodwasp and its Fungal Symbiont: Research and Management of a Worldwide Invasive Pest*, ed. B. Slippers, P. de Groot, & M. J. Wingfield, pp. 135–148. Dordrecht, Netherlands: Springer.

10 Bacterial Pathogens of Invertebrates

10.1 What is a Pathogen?

The following three chapters will cover some invertebrate and vertebrate pathogens important to biological control, so first we will review some of the general characteristics of pathogens. Many nematodes are microscopic and they are therefore often grouped with pathogenic microorganisms in the scientific literature although they are more complex eukaryotes. Several groups of microorganisms, including bacteria, viruses, and fungi (including microsporidia), that cause diseases in invertebrates and vertebrates are commonly called pathogens, indicating that these are microbial parasites, living at the expense of hosts and causing disease, a broad term meaning departure from a state of health or normality. Microorganisms utilize invertebrates for food as they do plants and other types of animals. Their relationships with hosts vary from obligate pathogens, which do not grow outside their hosts and cannot fulfill their life cycle in nature without their hosts, to facultative pathogens, which only act as pathogens when an opportunity presents itself. Many of the major microbial groups attacking invertebrates are the same as those that have adopted lifestyles as pathogens of vertebrates and plants. Virtually all species of pathogenic microorganisms infecting humans do not infect plants. Similarly, species of microorganisms causing disease in invertebrates usually specialize on invertebrates and do not infect or in other ways harm vertebrates. The pathogens vectored by insects, such as malaria and plant-pathogenic viruses, are special cases of more temporary associations of these microorganisms with insects. In fact, within the invertebrates, pathogens usually display specificity for certain groups of hosts and this is especially true of obligate pathogens that have close associations with hosts.

The hard exterior cuticle of many invertebrates provides a physical barrier to entry by microorganisms. So, either invertebrate pathogens must find ways to pass directly through the cuticle, or they must find other ways to infect (e.g., often through the gut when they are eaten). If a pathogen enters the body of an insect or mite, for example, these invertebrates then mount an immune response for protection. However, invertebrate immune systems are quite different from vertebrate immune systems and are not as complex and powerful. Numerous pathogens have developed the ability to overcome their invertebrate hosts and utilize the entire invertebrate body as a source of nutrients for microbial reproduction. Although some microorganisms can cause lingering, chronic infections, for use in pest control the focus has been on microbes causing rapid death. In addition to the bacteria (this chapter), pathogens that can kill hosts relatively quickly include some viruses (Chapter 11) and fungi (Chapter 12). These pathogens affect a diversity of invertebrates but it should be noted that all microbes except most fungi usually must be eaten in order to infect. Some viruses and fungi are especially known to cause dramatic epizootics in nature, so trying to utilize this potential has been one driving force toward development of pathogens for pest control. Among these pathogen groups, only viruses have been used for control of vertebrate pests and, while some excellent results have occurred, this approach has seldom been undertaken (Chapter 11).

10.2 General Biology of Insect-Pathogenic Bacteria

Bacteria are diverse unicellular organisms that have no internal membrane-bound organelles and this defines them as prokaryotes. They are very small and cannot be seen with the naked eye, ranging in size from less than one micron to several microns long. Bacteria are ubiquitous, being found in virtually any habitat where they use a great diversity of materials as nutrients and divide rapidly by fission. Some bacteria live in symbiotic associations with invertebrates, such as the saprophytic bacteria that live externally on animals and commensals that live within the guts of unwitting hosts, deriving protection and food from this association.

Some species of bacteria are more integrally involved with invertebrates, living as obligate or facultative symbionts within them. For example, bacteria in the genus *Buchnera* live within specialized host-produced cells, called mycetocytes, within aphids and are passed from mothers to offspring; virtually all aphids house symbiotic *Buchnera*. Phylogenetic studies using molecular techniques suggest that the association between *Buchnera* and aphids began 160–280 million years ago. This association is required by both the bacteria and the aphids; aphids without *Buchnera* grow poorly because they depend on these bacteria for essential amino acids and *Buchnera* cannot be grown outside of the aphids. Actually this association is so integrated that phylogenetic studies of *Buchnera* and their aphid hosts have proven useful to support hypotheses about evolution of aphid hosts.

Facultative bacterial symbionts in aphids can also play a part in biological control as aphids hosting the bacterium *Hamiltonella* can be more resistant to parasitoids and aphids hosting *Regiella* are more resistant to the fungal pathogen *Pandora neoaphidis*. Further studies being conducted suggest aphid protection by additional bacterial species. The bacterium *Wolbachia pipientis* is also a widespread inhabitant of many species of arthropods and nematodes, often acting as a reproductive parasite that adjusts sex ratios of hosts to favor its own transmission. While *W. pipientis* is parasitic in many hosts, in others the relationship has evolved to mutualism (see Chapter 20 for use of this bacterial species for control of mosquitoes vectoring dengue).

There are relatively few known species of bacteria specializing as invertebrate pathogens. However, numerous species of bacteria are opportunistic and can overcome insect defenses readily if they can gain entry to the body, as through wounds. Most bacteria that are pathogens of insects must be eaten by hosts. Once inside the gut, pathogenic bacteria usually invade the body cavity. However, most bacteria are not able to simply enter the hemocoel directly from the gut lumen, although some highly virulent pathogens have devised ways to breach the gut wall relatively rapidly. Illustrating the range in pathogenic strategies, a species from the genus *Paenibacillus* is an obligate parasite of Japanese beetle larvae and requires a long time, sometimes even more than a month, to kill hosts. In contrast, more virulent species in the genus *Bacillus* use toxins to damage the gut wall and kill hosts quickly, often within a few days. Other bacteria, for example, some species from the genus *Pseudomonas*, are opportunistic insect pathogens that seem mostly to occasionally infect insects that are in weakened, dense populations. There is thus a continuum of bacterial species infecting insects, from bacteria that are highly host specific or highly virulent to species of low virulence that can live and multiply outside the host and are pathogenic when opportunities occur.

When insects are infected with bacterial pathogens, their bodies can turn color from white to red, amber, black, or brown. Cadavers from recently killed insects can be flaccid and fragile but, as the body dries, it often shrivels and becomes hard and sometimes smelly. In fact, the bodies of insects killed in any way make excellent media for growth of saprophytic bacteria. Therefore, cadavers of any dead insects will soon be colonized by microorganisms, especially including bacteria, making diagnosis of the cause of death as a result of bacteria more difficult.

10.3 Use for Pest Control

The bacteria most widely used for biological control are spore formers (that are Gram positive), in the family Bacillaceae. There is diversity among diseases caused by spore-forming species, with mortality of hosts ranging from a matter of days to months for different bacterial species. Learning more about the cause of naturally occurring mortality of pests led to the discovery of a type of invertebrate/bacterial pathogen activity not reported previously. The nonspore-forming (Gram-negative) bacterium *Serratia entomophila* was found during investigations of deaths of pestiferous pasture scarabs in New Zealand; this bacterial species has a unique activity, using several toxins to clear the guts of infected hosts and eventually cross the gut wall.

Use of bacteria for pest control has focused on their application as sprayable suspensions. Only a few species of bacteria have been mass-produced and commercialized (Table 10.1) but one of these, *Bacillus thuringiensis*, is used for inundative release more than any other biological control product, including parasitoids and predators.

10.3.1 *Bacillus thuringiensis* (Bt)

Bt is a rod-shaped soil bacterium that can be found worldwide on plants, in insects, and in soil, surviving in the environment as resistant spores. Interestingly, Bt is only rarely found causing epizootics in insect populations, under natural conditions in unusually dense populations of susceptible species. Yet Bt has the power to kill many different kinds of insects and it has been developed extensively for pest control in a variety

Table 10.1 Examples of insect-pathogenic bacteria that have been used for biological control of insect pests.

Bacterial species	Target host group	Target hosts	Use
Bacillus thuringiensis israelensis	Diptera	Larvae of mosquitoes, black flies, fungus gnats	Streams and ponds, greenhouses
Bacillus thuringiensis kurstaki	Lepidoptera	Many species of caterpillars	Field crops, forests
Bacillus thuringiensis morrisoni	Coleoptera	Beetle larvae, especially leaf beetles	Tomato, potato, eucalyptus
Lysinibacillus sphaericus	Diptera	Mosquito larvae	Streams, lakes, contained water
Paenibacillus popilliae	Coleoptera	Japanese beetle larvae	Turf
Serratia entomophila	Coleoptera	New Zealand grass grubs	Pastures

of habitats from field crops to controlling insect vectors of human diseases, such as mosquitoes.

Bt is actually a complex of bacterial subspecies that can be differentiated based on serology or by molecular assays, but with the commonality that all of the subspecies produce a spore as well as a parasporal body within a sporangium (Figure 10.1). Parasporal bodies contain one or more proteinaceous protoxins in a crystalline structure and, therefore, these are frequently referred to as crystals. Crystals can account for 30 percent of the total protein content of the bacterial cell. Toxins in the crystals of Bt are

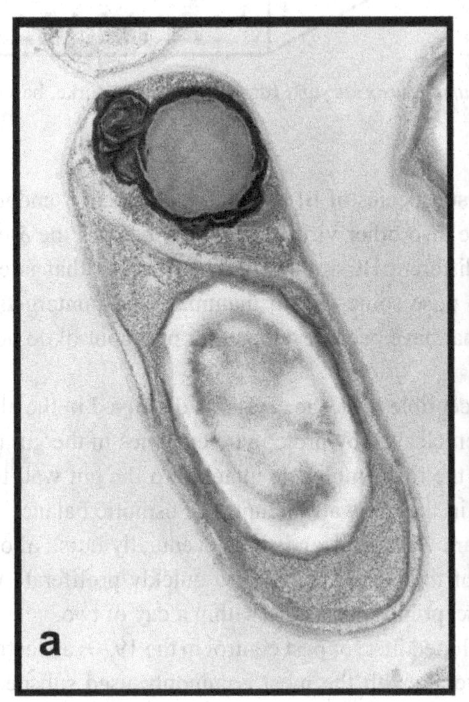

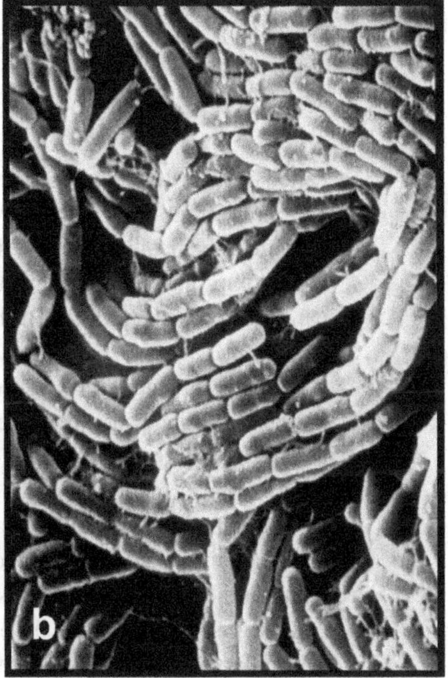

Figure 10.1 (a) Sporangium of *Bacillus thuringiensis*; note the crystal above and spore below (Feitelson et al., 1992). (b) Vegetative cells of *B. thuringiensis* (photo by Jean-Francois Charles, Institut Pasteur).

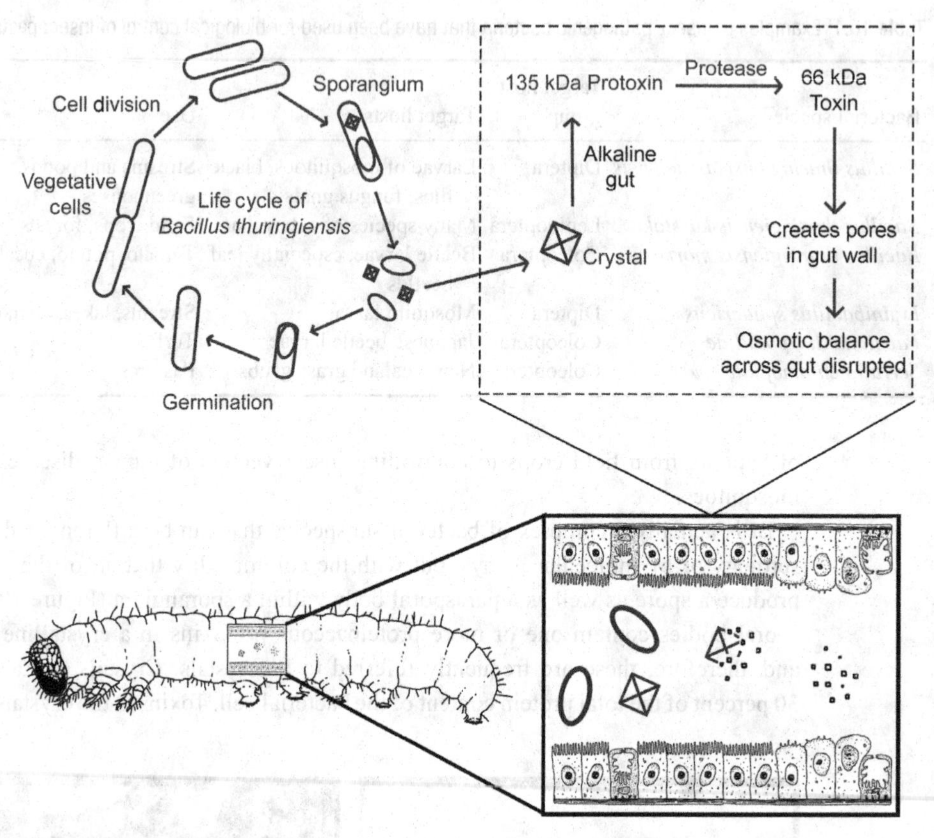

Figure 10.2 Life cycle of *Bacillus thuringiensis kurstaki* (drawn by Alison Burke, based on Tanada & Kaya, 1993).

called δ (delta)-endotoxins and subspecies of Bt have many different δ-endotoxins that act on different hosts. There are also other virulence factors besides the δ-endotoxins that are specifically made by different Bt strains. A type of toxin that is excreted, a β-exotoxin, has been shown to pose some risk to mammals and nontargets and thus care is taken that strains of Bt that have been developed for pest control do not produce this toxin.

When Bt is ingested by a susceptible host, the crystal is dissolved in the alkaline gut and the resulting protoxin is then cleaved by proteolytic enzymes in the gut to become activated (Figure 10.2). Part of the toxin molecule attaches to the gut wall to produce a pore. The formation of pores in the gut wall disrupts the osmotic balance across the midgut; gut cells subsequently are weakened, and some eventually burst, allowing bacterial cells to invade the body of the host. The bacteria quickly proliferate within the compromised host and most susceptible species die within a day or two.

The Bt isolates that were developed first for pest control in the 1920s are active against caterpillars (larvae of Lepidoptera), with the most commonly used subspecies being *B. t. kurstaki* (Btk) (Box 10.1). Lepidopteran species are differentially susceptible to the toxins, so finding individual strains of Bt that are active against different host species has been essential for targeting a diversity of specific pests. Dipteran-active strains were the

Box 10.1 An Unlikely Experiment

In 1950, Edward Steinhaus, Director of the Laboratory of Insect Pathology at the University of California, Berkeley was conducting studies with viruses for control of alfalfa caterpillar, *Colias eurytheme*. He was frustrated because the virus he was using was not killing the caterpillars fast enough to adequately prevent damage to the crop. One day he remembered that 8 years earlier, he had been sent a species of bacteria that was said to be a virulent pathogen. In fact, this strain of bacteria had been isolated from Mediterranean flour moth, *Ephestia kuehniella*, a stored products pest, in the province of Thuringia in Germany and had been named *Bacillus thuringiensis*. Laboratory bioassays in Europe had suggested that this bacterial species was a virulent pathogen against caterpillars and results from field trials against the pestiferous European corn borer, *Ostrinia nubilalis*, had been promising.

Figure Box 10.1.1 Edward A. Steinhaus, the father of modern insect pathology (photo by Elizabeth Davidson).

Steinhaus took the long-forgotten bacteria out of the refrigerator and sprayed it on a tray of alfalfa plants hosting several hundred caterpillars. When he came to work the next morning, most of the caterpillars had ceased feeding and were dying or already dead. Of course, Steinhaus was very excited but he knew that he needed to repeat this experiment before proceeding. He quickly began to grow more bacteria and that evening he sprayed the freshly grown bacteria on a new batch of caterpillars. The next day, to his surprise, there was no effect on the treated caterpillars. There was the possibility that the spray equipment he had used initially had been contaminated with chemical pesticides and had not been cleaned properly so the first batch of insects might have all died from pesticide poisoning. However, washing the spray equipment and spraying with the new culture again did not alter the results. Why did the first batch of caterpillars die when the second did not?

Steinhaus was not one to be deterred and his further work with this bacterial species showed that when cultures of these bacteria had matured, they formed spores and it was only at this point that they were insecticidal. With the spores are proteinaceous crystals (or parasporal bodies) and presence of these crystals had been documented by visual examination. However, the function of the crystal was not understood and it was not until 1955 that it was proved that the parasporal crystals were the cause of toxicity. Freshly growing cultures with plenty of nutrients lack spores and crystals and thus have no effect on caterpillars. The old cultures Steinhaus used initially certainly consisted primarily of spores and the crystalline toxins and we now know that this is why they were highly toxic. The young cultures he assayed second would have contained no spores and crystals, explaining their lack of effect on the caterpillars.

This was only the beginning of Steinhaus's work with *B. thuringiensis*. He went on to conduct basic and applied studies that demonstrated to both insect pathologists and industry the potential uses for this bacterium, which has grown to be the most frequently used type of biological control.

next to be found, in 1976, with the discovery of *B. t. israelensis* (Bti) infecting mosquitoes in Israel. The first Bt isolate active against beetles, *B. t. "tenebrionis"* (Btt; because of a name change, the proper name is *B. t. morrisoni*), was found in 1983 and thus began development of Bt for control of a few beetle pests. The search for new strains of Bt continues and now strains active against Hymenoptera, Hemiptera, Mallophaga, Nematoda, and Protista have been discovered, although not developed commercially. Owing to intensive prospecting for strains with novel activities, it has been estimated that at least 60,000 isolates of Bt are held in collections around the world. Today, *Bacillus thuringiensis* strains are known to produce a diversity of toxins, including no less than 732 crystal toxins (abbreviated as Cry), 38 cytotoxins (Cyt) and 125 vegetative inclusion proteins (VIP). At present, 85 subspecies (i.e., serovars) of Bt have been named but there are many strains having different activities available within each subspecies and new strains are found continually. Genes encoding some of these toxins have been sequenced, especially including toxins developed for pest control. The Bt insecticidal proteins are specific toxins, active in insect guts, and thus they are considered environmentally safe, especially with regard to vertebrates (see Box 18.1).

Products based on Bt cells are generally composed of bacterial cells that had produced spores and crystals and were then lysed. Thus, usually the parasporal bodies containing the δ-endotoxins have already been released from the sporangia but spores are present too. A variety of formulations of Bt are available, including emulsifiable concentrates, wettable powders, and granules for use in many different habitats against many different pests. As examples, Bt sprays have been extensively used to control a variety of lepidopteran pests in crops, orchards, and horticulture around the world. In northeastern North America, Bt is used extensively against defoliating caterpillars in forests, such as gypsy moth, *Lymantria dispar*, and spruce budworm, *Choristoneura fumiferana*. An important use of Bti has been its application for control of mosquitoes and blackflies affecting public health (Box 10.2), and sciarid flies in mushroom cultivation. Beetle-active strains of Bt have been used to control Colorado potato beetle, *Leptinotarsa decemlineata*, and leaf beetles in *Eucalyptus*.

Box 10.2 Icy-Pearl-Formulated Bacteria Prevent Mosquito Biting

Bacillus thuringiensis israelensis (Bti) and *Lysinibacillus sphaericus* are successful biological control agents for use against mosquitoes. The Upper Rhine Valley in Germany is a perfect place for mosquitoes to breed and since the valley is also densely populated by humans, mosquitoes are a problem. The frequency of mosquito bites can be high: more than 500 bites per minute have been counted. Where and when mosquitoes are potential vectors of human diseases and also in cases where mosquitoes are just annoying, biological control using these bacteria is a "green" option. In this case researchers decided to improve the formulation, the materials mixed with the bacteria, along with the application methods, so that when the mosquito-pathogenic bacteria were sprayed into water where mosquito larvae were growing, there were increased chances that they would be eaten by the mosquito larvae.

The formulation and application methods that were developed for this system are unique. The bacterium is suspended in water and frozen into pellets called "Icy-Pearls," which are then applied by helicopter to bodies of water. The ice pearls melt upon contact with water and infective bacterial spores are released below the surface – just at the place where the aquatic mosquito larvae take in nutrients. The larvae ingest the bacteria and become infected. This formulation and application strategy is environmentally friendly since it is created using frozen water and applied in such a way that only the mosquitoes are affected.

This strategy is particularly useful for controlling mosquitoes around human settlements. However, the biology and identity of each target mosquito must be taken into consideration. For species like *Aedes vexans*, which disperse more than 15 km, and which must be controlled using Bti, application must also be made in their breeding sites far from towns. *Culex pipiens* disperses only a few hundred meters so for this species, application of *L. sphaericus* can be restricted to a radius of 500 m from towns (see also 10.3.2).

The size of the area treated in the Upper Rhine Valley varies by year, with applications depending on whether it is a wet year favorable for mosquitoes or a dry year not favorable for mosquitoes. In 2011, a dry year, only 2,740 ha were treated, using 51 tons of ice granules. Then, from 2013 to 2015, 14,000–19,264 ha were treated, using from 200 to 311 tons. In addition, some areas are treated each year using more traditional application methods for Bti or *L. sphaericus*. A high level of regional coordination is needed for this program and this coordination lowers the overall cost, which amounted to less than US$1 per person per year in the treated areas, a reasonable cost to avoid mosquito bites.

The ability to grow Bt in large quantities, such as 50,000 liter batches in fermenters, makes this bacterium easy and relatively cheap to mass-produce. It is considered safe and thus has been widely accepted by users. It can be stored almost indefinitely and can be applied using the same equipment and techniques as synthetic commercial insecticides at a reasonable cost. As early as 1983, there were 410 registered formulations of Bt. In 1999 it was estimated that Bt was being applied as a bacterial insecticide to millions of hectares each year, leading to a yearly worldwide market of $100–200 million. However, in the last two decades, use of Bt sprays to control insect pests in major crops like corn and cotton has declined drastically because genes for Bt toxins have been inserted into plants and these plants are then protected. However, this switch from use of Bt sprays to growing crops genetically engineered using Bt has only occurred in major crops, in large part because of the high cost of developing and deploying a new transgenic crop (estimated in 2013 at $136 million) as well as continuing public controversy surrounding Bt-engineered crops.

Genetic Engineering Using Bt

As methods for manipulation of genes exploded in the 1980s, scientists learned that it was relatively simple to manipulate the genes encoding Bt toxins. These genes occur on plasmids within the bacterial cells and are therefore relatively easy to alter and move. Genes encoding Bt toxins have been manipulated by inserting them into new strains of Bt or by inserting them into other species of bacteria. However, expression of Bt toxin genes within plants has been the more long-lasting and influential development. The ultimate goals of this genetic engineering have been to increase stability in the activity of Bt, facilitate application (planting engineered seeds rather than spraying), and sometimes also to expand the host range. Some scientists and members of the public have questioned whether expression of Bt genes in plants should be called "biological control," which is perceived by many as the intentional spread/release of a living natural enemy of the target pest. People with this view suggest that Bt-engineered plants are instead examples of plants with resistance to pests and pathogens.

The first Bt-modified transgenic plants were developed in the mid-1980s and this technology quickly expanded in the United States, where transgenic cotton was first sold in 1996, closely followed by corn, and research and development continues with additional crops. The cost of developing Bt-engineered plants will probably limit the number of plant species developed in this way. To recoup their profits, companies selling seed of Bt-engineered plants require that farmers do not keep seed from transgenic crops that they grow. This new technology was rapidly adopted by US farmers because, although the seeds were costly, the resulting pest control was terrific, especially for combating pests living in concealed locations within plants that have been difficult to control using pesticides. By 2015, about 81 percent of all corn grown in the United States was Bt-corn and 84 percent of all cotton was Bt-cotton. Use in the United States is not alone as the area planted with Bt crops around the world was 1.1 million ha in 1996 and increased to 66 million ha in 2011. At present Bt-eggplant is being evaluated in Bangladesh as a replacement for the more than 100 sprays of chemical pesticides per year required to protect this crop.

Development of this new technology has not been without its critics. For example, farmers and the public in Europe have not favored the use of transgenic crops, maintaining that side effects of use of transgenic crops have not yet been researched adequately, and also that there is no reason to use transgenic crops if more "natural" solutions are available. So far, such plants are not allowed in the EU. It seems this opinion may change and use of this new technology is beginning to be considered in some European countries.

Development of Resistance

Researchers have worried that widespread use of Bt would lead to the development of resistance, just as resistance has developed to many synthetic chemical insecticides. Usually, development of resistance is not a concern for biological control agents. However, in some ways the activity of Bt is similar to synthetic chemical insecticides because its activity is often based on the activity of one or more specific chemicals, the toxins. It was hypothesized that with heavy use in small areas of a Bt strain with one toxin and with little genetic variation in target insects, resistance to a Bt toxin could develop. Resistance was first found in the Indian meal moth, *Plodia interpunctella*, caterpillars that were pests in stored grain. Subsequently, laboratory studies using many different host species repeatedly demonstrated that with repeated exposures to high doses, resistance could develop over some generations. Then, resistance was found in the field in populations of the diamondback moth (*Plutella xylostella*) in Asia and North America and populations of cabbage loopers, *Trichoplusia ni*, in greenhouses in North America. These pests had been exposed year round to Bt, often at high doses and on a regular basis. However, resistance has only been reported in some systems where Bt cells plus toxins are applied. Mosquito populations in many areas have received heavy exposure to Bti for years, for example, in the Rhine Valley of Germany (Box 10.2), but no resistance in field populations of mosquitoes has been documented.

Concern over development of resistance only began to reach a crescendo after Bt-engineered plants began being planted. Resistance to Bt-engineered plants develops in the same way as resistance to Bt when Bt cells and crystals are applied. It is generally considered that resistance develops as a result of the strong selective pressure because entire fields of Bt-engineered plants create exceptionally intense selection pressure which speeds the development of resistance. In 2005, field resistance had evolved in one pest species and by 2013, resistance to five of 13 major caterpillar pests had evolved around the world.

Development of resistance is especially a concern for growers wanting to use less chemical pesticides. Growers who market their produce as being free of chemical pesticides, such as organic growers, and who often rely on use of Bt for pest control also worry. If resistance to Bt develops, these growers would have fewer alternatives for pest control.

Management strategies that have been suggested to prevent or delay the development of resistance to Bt include switching between different Bt strains with the same target range, use of a mixture of several toxins within the same treatments, planting non-sprayed or non-Bt-expressing plants in certain areas so that not all insects are exposed to

Bt (i.e., providing refugia), using very high doses of Bt toxin, and planting mixtures of normal and transgenic seeds. In Asia, switching between *B.t. kurstaki* and *B.t. aizawai* strains, both of which are active against caterpillars, has proven useful. The strategy that has received the most support in the United States is a combination of high doses of Bt and providing refugia. This is how refugia work: resistance to Bt is genetically determined and is usually a recessive trait. Therefore, if insects that are resistant mate with insects that are not resistant, the offspring will not be resistant. By providing pests with crop plants not expressing Bt near the fields of transgenics, insects that are not resistant will continue to be present in the area and, should any insects from the Bt usage areas survive the exposure to high levels of toxin (i.e., a new resistant strain), they would mate with the refuge insects and their offspring would not be resistant. At present, research results suggest that the best way to delay the development of resistance is if resistance genes are in fact recessive, so that there is a low incidence of resistance to start, abundant refuges for nonresistant insects are provided, and Bt crops with more than one toxin are used before resistance develops.

10.3.2 Fighting Mosquito Wrigglers and Scarab Grubs

Three more species of bacteria have been developed for control of invertebrates. *Lysinibacillus sphaericus*, used against mosquito larvae (or wrigglers), actually has two different toxin proteins that are produced within each bacterial cell and both must be present for toxicity. *Lysinibacillus sphaericus* products are complementary to those products using Bti because *L. sphaericus* has different uses. This bacterium survives better in more polluted water than Bti and it targets mosquito species that Bti is not effective against, such as *Culex*. *Culex* species are thought to be the major vectors of the human pathogenic West Nile virus that was introduced to northeastern North America, and *L. sphaericus* has been used for control of vectors of this disease. One problem that has arisen with this species, however, has been that resistance to *L. sphaericus* in the southern house mosquito, *Culex quinquefasciatus*, has developed in several countries.

The spore-forming bacterium *Paenibacillus popilliae* was first found in 1933 in the northeastern United States, infecting larvae of the introduced scarab, Japanese beetle (*Popillia japonica*). These grubs live in the soil and feed on grass roots. Larvae must eat the bacteria and, when infected, their hemolymph (blood) and the end of their body is milky-colored instead of clear, so the resulting disease was named "milky disease" (Figure 10.3). After infection, this bacterium takes a long time to kill larvae unless the larvae are very young and ingest a large dose. *Paenibacillus popilliae* is an obligate pathogen and cannot be easily grown outside the host insects. This has presented problems for mass production of this bacterium. However, there are few ways to control such soil-dwelling pests and the lawns that Japanese beetle grubs damage are valuable. Therefore, in 1948 this bacterium became one of the first insect pathogens developed as a microbial control agent in the United States. After this pathogen was discovered, huge programs were undertaken to distribute *P. popilliae*, releasing 109 tons of spore powder to over 90,000 sites over a 14-year period. The subsequent decline in Japanese beetle populations was attributed in part to activity of this pathogen, which persists well

Figure 10.3 (a) Scarab beetle, *Rhopaea verreauxi*, grubs that are healthy (right) and infected with milky disease (left). Note milky appearance of the blood (hemolymph) seen through the abdomen and in droplet on leg (photo by R. Milner). (b) Sporangia of *P. popilliae*, often thought to look like shoe soles, with the crystal as the heel and the spore as the front of the sole (photo by M. Klein, USDA ARS).

in the soil, being found 25–30 years after original applications. Thus, an application of *P. popilliae* can help with controlling Japanese beetle larvae over numerous years. However, successful control of Japanese beetles owing to this species has been reported as variable, production of this obligate pathogen is difficult and it is expensive to produce, so availability has been limited.

The grass grub *Costelytra zealandica* is a major pest of grasslands in New Zealand. In this case, the grasses have been introduced but this beetle species is native. A native Gram-negative bacterial pathogen, *Serratia entomophila*, was found infecting these scarab grubs, which turn an amber color when infected (thus, the disease is called "amber disease"). After ingestion, toxins produced by these bacteria cause a scarab larva to stop feeding and void the gut. Digestive enzyme production is stopped and in this way the bacteria remain in the gut. Eventually, the gut cells weaken and the bacteria enter the insect's body and the grub dies. The pathogen will often build up on its own when there are high populations of grubs. However, when pastures are cultivated, both grass grubs and bacteria are killed leaving the new pastures vulnerable to reattack by grubs. In these circumstances, *S. entomophila* is applied (inoculated) in pastures so that it becomes reestablished, promotes early epizootics, and damage is prevented. Methods for mass production of *S. entomophila* in fermenters have been developed and this pathogen is registered for use in New Zealand.

In recent years, several new bacterial pathogens have been investigated for control of invertebrate pests. One of these is *Chromobacterium subtsugae*, a Gram-negative rod-shaped bacterium, initially isolated from soils, that produces several insecticidal toxins. When these bioactive compounds are eaten by numerous species of chewing and sucking insects, pests can be repelled, they will stop feeding, reproduction is reduced and death can occur within 2–7 days. A product including dead cells (with a few living), used fermentation media, and insecticidal compounds has been commercialized for control of numerous types of insect and mite pests.

Further Reading*

Charles, J.-F., Delécluse, A., & Nielsen-LeRoux, C. (eds.) (2000). *Entomopathogenic Bacteria: From Laboratory to Field Application*. Dordrecht, Netherlands: Kluwer Academic Publishers.

Fiuza, L. M., Polanczyk, R. A., & Crickmore, N. (eds.) (2017). Bacillus thuringiensis *and* Lysinibacillus sphaericus. Dordrecht, Netherlands: Springer.

Glare, T. R. & O'Callaghan, M. (2000). *Bacillus thuringiensis*: Biology, Ecology and Safety. Chichester, UK: John Wiley & Sons.

Jurat-Fuentes, J. L. & Jackson, T. A. (2012). Bacterial entomopathogens. In *Insect Pathology*, ed. F. E. Vega & H. K. Kaya, pp. 265–349. San Diego, CA: Academic Press.

Lacey, L. A. (ed.) (2017). *Microbial Control of Insect and Mite Pests: From Theory to Practice*. Amsterdam: Academic Press/Elsevier. (numerous chapters)

Lacey, L. A., Grzywacz, D., Shapiro-Ilan, D. I., Frutos, R., Brownbridge, M., & Goettel, M. S. (2015). Insect pathogens as biological control agents: Back to the future. *Journal of Invertebrate Pathology*, **132**, 1–41.

Vega, F. E. & Kaya, H. K. (eds.) (2012). *Insect Pathology*, 2nd edn. Amsterdam: Academic Press/Elsevier.

* Both general pathology and bacterial pathogens.

11 Viral Pathogens of Invertebrates and Vertebrates

11.1 General Biology of Viruses

Viruses differ from other groups of microorganisms since they are not themselves living cells and have no metabolism. Thus, viruses are genetic elements, containing either DNA or RNA, whose energy is derived from the host. Because viruses can only replicate themselves within a living cell and are dependent on hosts for reproduction, all viruses are obligate intracellular parasites. This close association within host cells fits with the fact that most viruses are very host specific.

Viruses are grouped based on their nucleic acid composition, their genome structure, and the morphology of their external coats. The basic structure of a virus is the viral DNA or RNA surrounded by a protein capsule and sometimes a membrane or envelope; this constitutes a virion. After viruses have replicated their DNA or RNA genome within host cells, it is then packaged into the virions that form the extracellular state that is infectious and is needed to reach new hosts. Virions are so small that the largest is barely visible with the light microscope. Some of the largest viruses infecting insects, the pox viruses, have virions up to 470 nanometers long. Thus, viral morphology must be investigated using the electron microscope and molecular biological techniques are a requirement for studying the activity of viruses.

Latin names for genus and species are not used for naming species of viruses. Viruses are classified by family and individual viruses are often named after the host or place in which they were first found, or sometimes they are named for the disease they cause, for example, influenza virus, smallpox virus. As is common in all sectors of pathology, a disease, the negative impact of a pathogen on a host, is often named before the causative pathogen is isolated and studied. Viruses have exploited a great diversity of hosts including vertebrates, invertebrates, plants, fungi, single-celled animals, and bacteria.

11.2 Diversity of Viruses Infecting Invertebrates

At least twenty viral families include pathogens of invertebrates. While some viruses attacking invertebrates occur in viral families that include viruses attacking vertebrates, the family Baculoviridae is known only from insects and other arthropods. Because of their pathogenicity, host specificity, and development for biological control, viruses in the Baculoviridae are among the best-studied invertebrate viruses. These viruses are

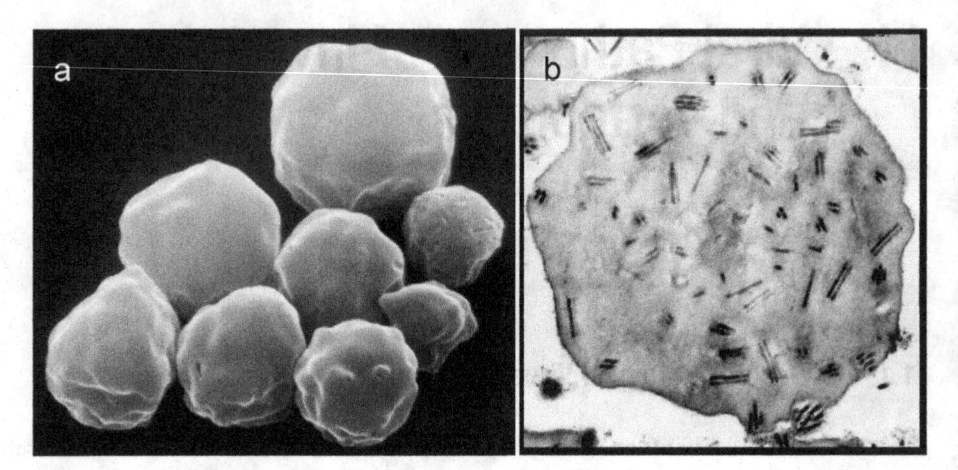

Figure 11.1 (a) Scanning electron micrograph of the occlusion bodies of a nucleopolyhedrovirus (photo by J. Podgwaite, USDA Forest Service). (b) Transmission electron micrograph of a cross section of an occlusion body from a nucleopolyhedrovirus. The small dark structures within the protein matrix of the occlusion body are the virions (photo by James Slavicek, USDA Forest Service).

known to infect a variety of insects, but especially caterpillars (butterfly and moth larvae of the order Lepidoptera), larvae of sawflies (Order Hymenoptera; relatives of wasps and bees having immature stages very similar to caterpillars), and mosquito larvae (Order Diptera). The high degree of host specificity of most baculoviruses makes them highly appropriate for numerous biological control purposes.

Among the viruses attacking insects, viruses in three families have special adaptations for survival in the environment. Invertebrate viruses in the families Baculoviridae, Poxviridae, and Reoviridae produce an occlusion body (OB), a structure that protects virions (Figure 11.1). The occlusion body is resistant to environmental insults and could, from that perspective, be considered functionally analogous to a bacterial spore. For the baculoviruses, cytoplasmic polyhedroviruses (within the family Reoviridae), and pox viruses, occlusion bodies are made of a protein matrix in which from one to many of the infectious virions are embedded. Occlusion bodies are produced within infected invertebrates and most are released into the environment after host death. Unprotected virions are fragile and are inactivated when exposed to sunlight. The proteinaceous occlusion body protects the virions in the environment before they infect another host, thus enhancing viral survival both within a season and for the longer term, between seasons or for many years. Occlusion bodies vary in size and shape for different groups. Within the Baculoviridae, the nucleopolyhedroviruses (NPVs) have many-sided occlusion bodies (*c.* 0.5–15 μm) that can contain many virions (Figure 11.1) while granuloviruses (GVs) have smaller, capsule-shaped occlusion bodies (*c.* 200 $\times$ 600 nm) that each contain one virion.

While most vertebrate viruses spread themselves from animal to animal by direct contact of a virus particle with a mucous membrane, viruses of arthropods generally must be eaten and they then infect through the gut wall. Because we know the most

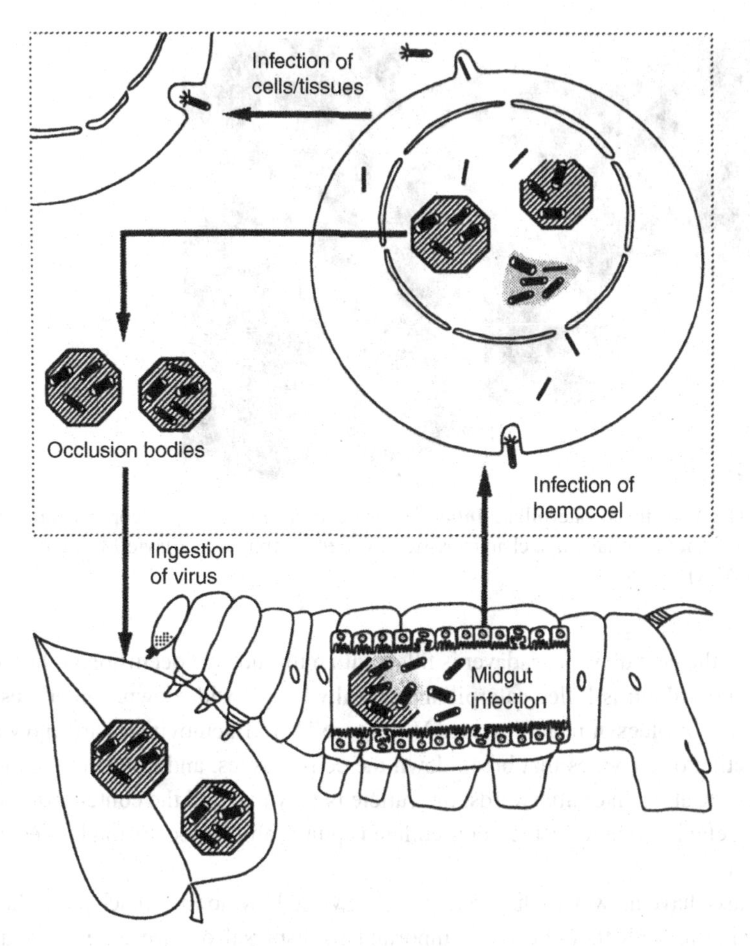

Figure 11.2 The cycle of a baculovirus infection. The occlusion bodies are ingested and the virus enters the midgut cells and replicates during primary infection. Nonoccluded forms of the virus are then released into the hemocoel and these spread to infect further cells within the host. In later stages of the infection, the occluded form of the virus is produced and released (Shuler et al., 1995).

about the Baculoviridae and this is the main group that is being exploited for biological control, species in this family will be used as examples of how viruses interact with arthropod hosts. When occlusion bodies are eaten by a caterpillar, alkaline conditions within the gut can cause the protein matrix of the occlusion body to dissolve, releasing the virions within the midgut lumen (Figure 11.2). The outer layer of a virion then binds to the surfaces of cells lining the midgut of the host, and the virion then enters the midgut cell. The virion moves to the nucleus where it loses the protein coat and then replicates. Progeny viruses are produced to spread infection throughout the host. While some viruses of insects can cause more chronic diseases, the baculoviruses are often virulent pathogens that kill hosts relatively quickly.

Some viruses are tissue specific and only attack certain tissues but many baculoviruses attacking lepidopteran larvae invade and appear to replicate in all tissues. By

Figure 11.3 Velvetbean caterpillar, *Anticarsia gemmatalis*, killed by nucleopolyhedrovirus and hanging from the foliage in a characteristic 'inverted V' (photo by Flavio Moscardi, EMBRAPA).

the time the host dies, the cadaver is filled with multitudes of occlusion bodies. Cadavers of dead baculovirus-killed caterpillars typically hang upside down, sometimes attached by the rear prolegs or in an 'inverted V' (Figure 11.3). Baculoviruses are known to cause production of enzymes that break down the cells, tissues, and inner cuticle of the host. Therefore, at death or afterwards, the cuticle is very thin and the contents of the cadaver are liquefied. Eventually, the outer cuticle ruptures and the occlusion bodies within are released.

Viruses have no way to disperse on their own so how do they reach a new host? Wind and rain are both thought to be important in dispersal of viruses. For baculoviruses infecting caterpillars and sawflies, the pathogen is transmitted quickly among hosts whose larvae feed gregariously. Occlusion bodies can be released when hosts defecate or when they die and their cuticles rupture. Occlusion bodies are then dispersed by being blown or washed from leaf to leaf, thus distributing the inoculum over a greater area. However, viruses also have some specialized methods for aiding dispersal. In Germany in the late 1800s, it was noted that before they died, caterpillars of the nun moth, *Lymantria monacha*, infected with a baculovirus climbed to the tops of spruce trees. Similar tree-topping behavior, called Wipfelkrankheit or summit disease, has also been observed for nun moth larvae on other coniferous tree species (Figure 11.4). This host behavior has clear advantages for the virus: when caterpillars die and the cadavers subsequently break open, virions are widely dispersed, being washed down onto lower foliage throughout the tree. Researchers have discovered a gene that is associated with the climbing behavior. While we still do not understand perfectly the interactions between host and virus that cause such behavior, climbing before death has been seen with many other species of baculovirus-infected caterpillars. To improve dispersal of occlusion bodies, other caterpillars infected with baculoviruses are known to disperse further. As a more exact method for dispersal, viruses can be picked up and

Figure 11.4 Nun moth caterpillars killed by a baculovirus, hanging from the top of a fir tree in Denmark (drawn by Claus Bering).

transferred as hitchhikers when a parasitoid oviposits into an infected host and then can be inoculated into a healthy host the next time the parasitoid oviposits. Another mode for baculovirus dispersal is by birds and small mammals that feed on infected insect larvae, fly some distance and when they defecate, occlusion bodies are deposited.

11.3 Use for Pest Control

Viruses have been used for long- as well as short-term insect pest control. With few exceptions, baculoviruses are the major virus group that has been developed (see Table 11.1), with some emphasis on baculoviruses that produce occlusion bodies containing many virions, the NPVs.

11.3.1 Classical Biological Control

Compared with parasitoids and predators, insect pathogens have not been used frequently for classical biological control (see Chapter 3). However, among the few instances where pathogens have been used, there have been some stunning successes

Table 11.1 Examples of insect-pathogenic viruses used for augmentation biological control.

Virus	Target host order	Target hosts	Uses
Baculoviridae: Nucleopolyhedrovirus			
AcMNPV	Lepidoptera	Alfalfa looper (*Autographa californica*) and cabbage looper (*Trichoplusia ni*) caterpillars	Vegetables
AngeMNPV	Lepidoptera	Velvetbean caterpillar (*Anticarsia gemmatalis*)	Soybeans
HezeSPV	Lepidoptera	Corn earworm (*Helicoverpa zea*) and tobacco budworm (*Heliothis virescens*) caterpillars	Field crops
LydiMNPV	Lepidoptera	Gypsy moth (*Lymantria dispar*) caterpillars	Deciduous trees
NeabNPV	Hymenoptera	Balsam fir sawfly (*Neodriprion abietis*) larvae	Balsam fir trees
SpliNPV	Lepidoptera	Egyptian cotton leafworm (*Spodoptera littoralis*) caterpillars	Vegetables and berries
Baculoviridae: Granulovirus			
AoGV	Lepidoptera	Summer fruit tortrix (*Adoxophyes orana*) caterpillars	Apples and other fruits
CpGV	Lepidoptera	Codling moth (*Cydia pomonella*) caterpillars	Apples and other pome fruit
PrGV	Lepidoptera	Small cabbage white butterfly (*Pieris rapae*) caterpillars	Vegetables

with introductions of baculoviruses, which are well known for causing epizootics in nature. The European spruce sawfly, *Gilpinia hercyniae*, was permanently controlled through introduction of an NPV (GiheNPV). A nonoccluded invertebrate virus (OrNV) was successfully used for classical biological control of the coconut palm rhinoceros beetle, *Oryctes rhinoceros*. These beetles are major pests of coconut and oil palms in the South Pacific and Southeast Asia. The adult beetle bores into the heart of the palm tree, and severe infestations can lead to death of palms. The larvae develop in the decaying palm or in other decaying vegetable matter such as compost. Originally found in Malaysia, a nonoccluded virus that principally develops in the gut cells of larvae and adults was found to be a potent biological control agent since infected adults do not live as long as uninfected adults and infected females have reduced fecundity. Adults spread the virus when they mate and when they defecate in feeding galleries or breeding sites. When eggs hatch and larvae eat the virus deposited by adults, they become infected. Because this virus is less stable in the environment, researchers found that it is best released by collecting adult beetles, infecting them, and then releasing them to disseminate the virus. This virus was released on many islands where the beetle had been accidentally introduced. As an example, in the Fijian Islands, palm frond damage declined for 24–30 months after virus introduction, with low damage for at least 24 more months. However, it was found that in areas where concentrations of breeding sites were present, in time beetle outbreaks could reoccur and the virus then needed to be inoculatively released again.

11.3.2 Augmentative Releases

The principal development of baculoviruses has been for use in inundative releases. While viruses can be applied with the same spray equipment as chemical pesticides, they do not kill immediately, as do many chemical pesticides, or even as quickly as Bt. However, baculoviruses are valued because most are very host specific. Insects infected with baculoviruses may take 5–9 days before dying from an infection. Therefore, viruses are appropriate for maintaining host populations at lower levels but generally not for rapidly controlling very large pest outbreaks requiring immediate control. In fact, in tea plantations in Japan, slower-killing viruses are preferred for controlling leaf-rolling caterpillars because these viruses have greater transmission than fast-killing viruses. Therefore, virus transmission occurs from generation to generation of the pests within a season, while keeping the pest populations low enough that damage is tolerable (Box 11.1).

Insect control through mass application of viruses for inundative augmentation has been developed quite extensively around the world (Table 11.1). At this time, the number of products available in Europe and North America has expanded but availability of baculoviral biopesticides has also been expanding in Asia, Australasia, and South America. Efforts have often focused on use of baculoviruses for control of species of foliar-feeding lepidopteran larvae that are known to rapidly develop resistance to chemical insecticides. Products are usually wettable powders, liquid concentrates, or granules and can therefore be applied using methods similar to those used for chemical insecticides. Applications of viruses can be calculated based on larval equivalents (LE, the average number of occlusion bodies from a single cadaver) per hectare or the number of occlusion bodies per hectare.

At present, large quantities of viruses are produced within their larval insect hosts. While many insect-pathogenic viruses can be grown in cell culture, thus far, this type of production is not being used for any product being marketed. Therefore, for mass production of insect-pathogenic viruses, a host colony must be maintained. In developed countries, virus production has been restricted only to those systems with hosts that can be mass-produced on artificial diets. Baculoviruses were previously considered as addressing narrow "niche" markets but today baculoviruses are moving toward use in mainstream commercial farming, especially to provide options for voids in control choices that opened when specific chemical pesticides could no longer be used. Although numerous baculoviruses have been commercialized in developed countries, there are concerns about how to produce the amounts of product necessary using host insects, especially if an entire industry depended on a baculovirus. For more extensive commercialization in industrialized countries, more efficient mass production must be developed as well as methods for enhancing survival of occlusion bodies after they have been sprayed. Nevertheless, in apple orchards in Europe, the virus CpGV is widely used to control the codling moth, *Cydia pomonella* (Box 11.2).

In developing countries, use of viruses for pest control has been quite successful, relative to the areas treated. Mass production of viruses is often a cottage industry or is done cooperatively by groups of farmers. The largest program for producing and applying viruses has been the program for use of a baculovirus (AngeNPV) for control of the velvetbean caterpillar, *Anticarsia gemmatalis*, on soybeans in Brazil. AngeNPV was used for this purpose on 2 million ha in 2004. Changes in farming practices that included use of broad-spectrum insecticides resulted in a decrease to use on 300,000 ha today.

Box 11.1 Slow-Killing Virus Protects Japanese Tea Fields

Insect-pathogenic viruses applied as biological control agents normally kill their hosts quickly and this is regarded as one of their assets. However, here is an example of an effective slow-killing virus product, in which a slower mode of action fits with the cropping system and the pest insects.

Tea, a perennial evergreen, is a major crop around the world. China is the biggest producer, but Japan is also important, with production of 364,500 tons of raw tea leaves in 2016. Chemical pesticides are heavily used, but some biological control is possible in Japan for the important small lepidopteran leafrollers *Homona magnanima* and *Adoxophyes honmai*, both from the family Tortricidae. The first tea harvest in May is the most important one, yielding the highest prices, but the tea is harvested several times during each year.

A mixture of two slow-killing granuloviruses (GVs), one isolated from *H. magnanima* and the other from a third tortricid species, *Adoxophyes orana*, have been used to control these leafrollers. A single application of the GV mixture 10 days after the peak of first adult flight of the pests ensures that the virus will proliferate in the two pest populations throughout their four or five discrete generations per year. The virus is transmitted successfully from one generation to the next and remains prevalent and active later in the season when densities of the insect populations have increased. Most infected larvae die inside the rolled leaves, which can thus function as a reservoir of viral occlusion bodies. Occlusion bodies are degraded by ultraviolet (UV) wavelengths from sunlight, but the rolled leaves protect the viruses from direct UV exposure until susceptible larvae in the next generation are present.

The control efficacy and transmission rate of viruses with different killing speeds were examined experimentally. *Adoxophyes honmai* larvae infected with a fast-killing NPV, originally isolated from *A. orana*, died sooner and caused less leaf damage than those larvae infected with a slow-killing NPV. On the other hand, the slower-killing NPV yielded greater transmission to the subsequent generations compared with the fast-killing NPV. Thus, the slow-killing viruses are transmitted from generation to generation of the pests within a season, and, while some damage occurs later in the season, this is tolerable because the later tea harvests are much less valuable. So, essentially, this system is an example of successful inoculative biological control program using insect viruses.

Box 11.2 Spraying Virus to Control Codling Moth around the World

Codling moth, *Cydia pomonella*, is a small grayish moth that has a huge impact. It is considered the most devastating pest of pome fruit, which particularly includes apples and pears, worldwide. After eggs hatch, the larvae bore into fruit as soon as possible so much of the larval life is spent protected within the fruit, making this a difficult pest to control. Unfortunately, this pest also has at least two generations each year. There is a long history of use of pesticides against codling moth with associated negative impacts on the environment, development of pesticide resistance in codling moth, and creation of secondary pests. Therefore, discovery of a baculovirus that kills codling moth larvae in the 1960s was exciting. The codling moth virus is a virulent granulovirus (CpGV) that kills codling moth larvae in just a few days, once it has been eaten.

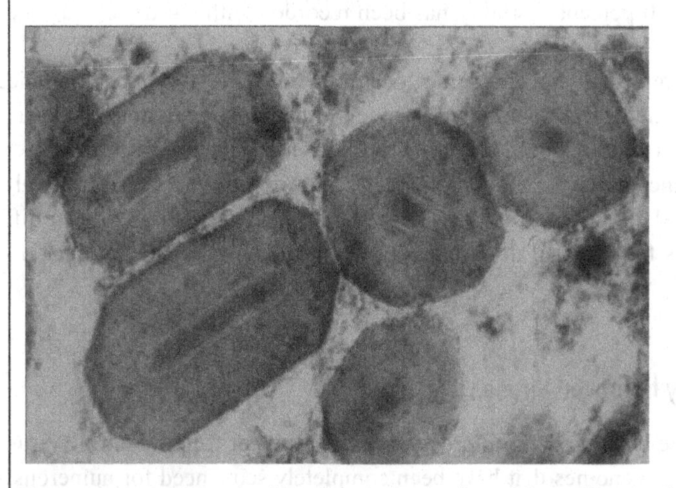

Figure Box 11.2.1 Capsule-shaped occlusion bodies of the codling moth granulovirus (CpGV) (in longitudinal and cross section) that each contain only one virion (each occlusion body is less than 1 micrometer long) (photo by Alois M. Huger, courtesy of Julius Kühn-Institut).

Methods for mass production of this virus were developed and successful field trials were undertaken in North America, Europe, South America, Oceania, and South Africa. The virus is applied during peak egg hatch to infect larvae before they enter the fruit. The larvae are not killed immediately but they die in a few days and the damage from entry of newly hatched larvae into the fruit can usually be acceptable to growers. The virus is pretty sensitive to ultraviolet rays in sunlight so it has a half-life of only a few days. Therefore, it has to be applied several times during the season and timing of applications is important.

CpGV was first produced in 1988 by a biological control company in Switzerland followed by products in Germany, France, the United States, and Canada. In 2008, it was estimated that CpGV was applied to about 100,000 ha annually in Europe and 8–12,000 ha annually in the United States and Canada. However, with prolonged intensive use, resistance to CpGV began cropping up after more than twenty years of use in some places in Europe. Under high pressure, some codling moth populations were no longer susceptible to the one specific strain of CpGV used in the product; this was solved by incorporating several new strains of the CpGV into a new product and it is also recommended that growers switch off with other environmentally safe control methods, when possible.

If a product is not mass-produced commercially, in some cases farmer training has enabled use of naturally occurring viruses for control. Farmers can collect cadavers of insects killed by the virus, store them in a refrigerator or freezer, then create a slurry and spray it on a crop at desired concentrations. For example, to control caterpillars of a large migratory hawk moth attacking cassava and rubber in Brazil, *Erinnyis ello*, farmers have used 20 ml of macerated cadavers of larvae killed by virus (ErelGV), diluted with 200 liters water on each hectare. The virus must be applied against younger caterpillars and 90 percent mortality has been recorded within 4 days, at a very low cost to farmers.

If used repeatedly in the same location for a long time, virus applications have resulted in the development of resistance in one host population. This was reported in Europe where codling moth had previously been controlled with CpGV (see Box 11.2). The resistance in the population only developed in response to one original virus strain. With several strains of this virus in a new product, each strain having different activity, chances for eventual development of resistance to the new product are considered negligible.

11.3.3 Genetically Improved Viruses

Efforts have been made to improve baculoviruses for use in control. Viruses have relatively small genomes that have been completely sequenced for numerous species and this provides a wealth of information for modifying viruses for agricultural applications. Emphasis has been on engineering viruses to increase speed of kill, thus decreasing insect damage. To date, field trials with engineered viruses have been limited and no transgenic baculoviruses have been commercialized.

11.4 Vertebrate Viral Pathogens

Many viruses infect vertebrate pests and a few have been exploited in biological control. These examples will be mentioned here with the warning that biological control of vertebrate pests with viruses today requires very thorough host specificity testing, risk analysis, and management to ensure safety toward humans and beneficial vertebrates associated with humans.

The most extensive biological control program against vertebrate pests has been the use of viruses for control of rabbits introduced to Australia. Rabbits are not native to Australia but, in 1859, European rabbits were purposefully introduced to create a more home-like environment for European settlers. There were no natural predators of rabbits in Australia so rabbits rapidly increased and spread, becoming the most important agricultural pest (Figure 11.5). Their feeding and burrowing destroyed pastures as well as causing erosion in the semidesert interior and their impact on vegetation was referred to as the "gray blanket." Numerous types of control, including the large-scale rabbit-proof fences, were unsuccessful and an alternative was needed.

Figure 11.5 European rabbits at an enclosed waterhole in South Australia in 1938. The abundant rabbit population destroyed pasture through plant consumption and burrowing (photo by M. W. Mules, 1938; courtesy of CSIRO).

11.4.1 Myxomatosis

In South America, a rabbit species closely related to the European rabbit (*Oryctolagus cuniculus*) was known to be infected by a pox virus called myxoma virus (MYXV), which caused small benign fibrous tumors that persisted for months but were not fatal. While this disease, called myxomatosis, is not virulent toward South American rabbits, the European rabbit species that had been introduced to Australia was extremely susceptible and few individuals survived more than 13 days after infection. This pathogen is specific to rabbits and has no effect on humans; to prove to the public that this virus was safe for release in Australia, researchers went as far as injecting themselves with the virus and publicizing the lack of any effects from the injections.

In May 1950, this South American virus was introduced to Australia and by December, hundreds of rabbits were found to be infected, dying, or dead many miles from release sites. During 1950–1951, the disease spread, causing 99 percent infection in some places, with an overall reduction in rabbit populations of 75–95 percent. This pathogen is predominantly vectored by mosquitoes and fleas that carry it from infected to healthy hosts; the relationship between vector and microbe is passive and the microbe does not reproduce within the vector. Directly after release, mosquitoes were the major vectors. To improve disease transmission, European rabbit fleas were introduced to Australia in 1968 and more xeric-adapted Spanish rabbit fleas were introduced in 1993.

By 1952, only two years after the initial release, scientists began to see a change in the interactions between rabbits and the myxoma virus. Since the virus was spread

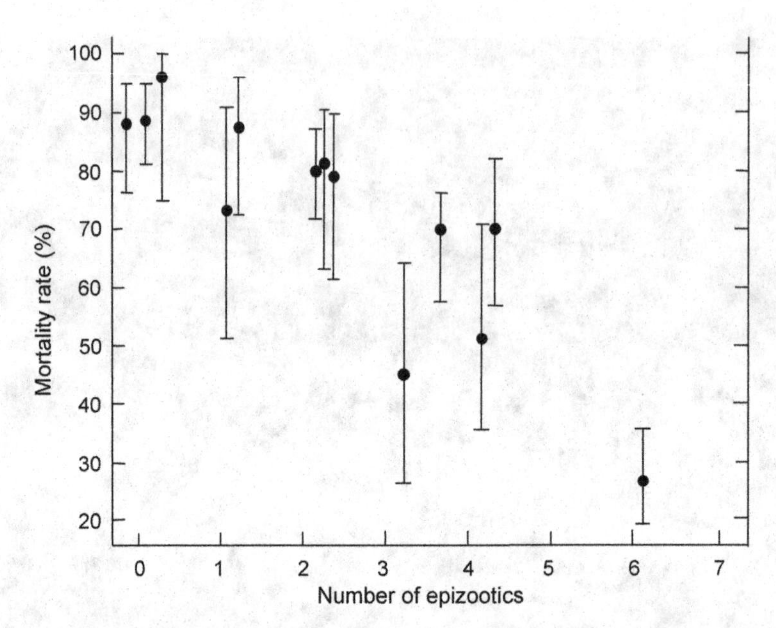

Figure 11.6 Mortality rates of Australian wild rabbit populations after exposure to varying numbers of myxomatosis epizootics. Mortality rate was quantified after challenge with the same virus strain, showing that when a population had been exposed to more epizootics, the rabbits were more resistant (Fenner & Myers, 1978).

when mosquitoes took a blood meal from a living infected host and then subsequently fed on a healthy host, if the disease killed hosts quickly, this allowed little time for mosquitoes to transmit the disease. Through natural selection, strains of virus evolved that killed fewer rabbits and rabbits that died lived longer before dying. These changes were beneficial to the virus, providing more time for it to be transmitted. Resistance to the pathogen was documented and it then increased in the rabbit populations fairly quickly (Figure 11.6). Thus, the virulence of the virus decreased while the resistance of the rabbits increased. This story provides a well-documented example of coevolution between host and pathogen, within a relatively short time period. The occurrence of coevolution raised the question of whether the myxoma virus was still effective in controlling the rabbits. To test this, rabbits in enclosures were immunized with either a weak strain of the virus or a virulent strain. Rabbit populations were then exposed to naturally occurring myxomatosis in the field. After two years, even with the genetic resistance from the immunizations, this disease was still important in suppressing rabbit populations despite the reduced effect of the virus in the field.

In summary, after the changes in the rabbit/myxoma virus relationship, it was estimated that there were still 300 million European rabbits in Australia. While rabbits were not causing problems as catastrophic as before the virus had been introduced, they once again had become major pests, causing an estimated AU$115 million loss to wool and meat industries as a result of pasture destruction. In addition, habitats for native plants and animals were decimated, all because of introductions of rabbits by settlers wanting to make Australia seem more like Europe.

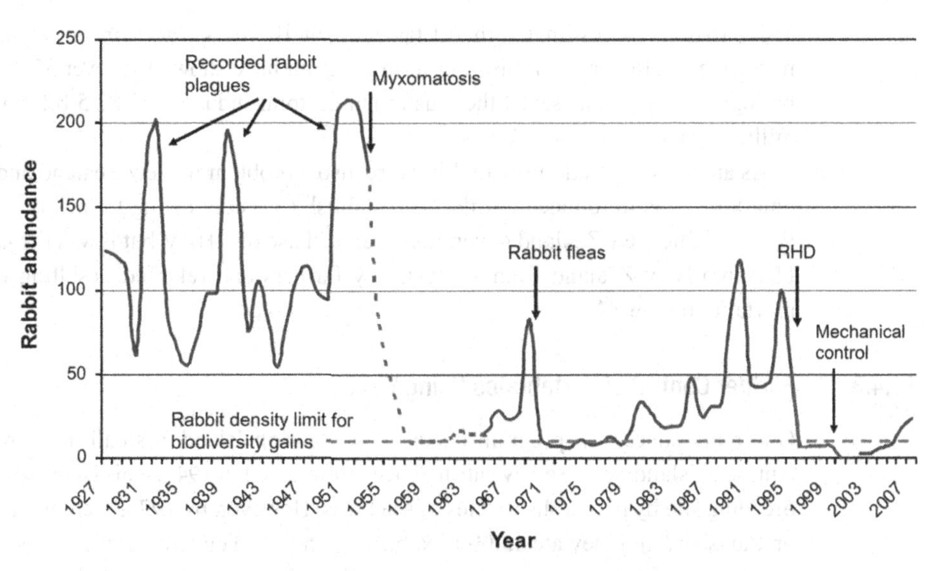

Figure 11.7 Long-term trends in rabbit densities in pasture lands in South Australia, showing changes in densities after releases of myxoma virus, fleas and RHDV (causing rabbit hemorrhagic disease; RHD), followed by destruction of rabbit warrens (mechanical control). Yet, as of 2010, the rabbit population was reported as increasing again to densities inhibiting native shrub revegetation (Saunders et al., 2010).

11.4.2 Rabbit Hemorrhagic Disease Virus (RHDV)

Owing to the host/pathogen coevolution between European rabbits and myxoma virus in Australia, control by myxomatosis was not complete. A new rabbit pathogen, a calicivirus, was discovered in angora rabbits exported from Germany to China in 1984. This calicivirus was very virulent, killing rabbits over 8 weeks old within 12–72 hours. Subsequent studies demonstrated that it was highly specific for only European rabbits. Australian scientists began investigations and in September 1995, this virus was accidentally established on the south coast of Australia when it escaped from quarantine on an offshore island, possibly carried to the mainland by some combination of flies and winds. This rabbit hemorrhagic disease virus (RHDV) immediately became established and began spreading and, in 1984 alone, this virus killed 140 million rabbits. The majority of the surviving rabbits were younger than 6 weeks old. This disease spread fastest in spring and autumn, at 10–18 km per day, probably aided by humans, and is now considered well established in Australia. RHDV has been most effective in semi-arid rangeland where the rabbit problems were the worst. After RHDV spread, in some areas there has been a noticeable regrowth of native vegetation, to a degree not seen within the memories of landholders. In 1996, RHDV was officially approved for biological control by the Australian government and the density of rabbits decreased to 5 percent of its previous density. Destruction of rabbit warrens was added to drive populations below a density that allows biodiversity to rebound (Figure 11.7). However, this problem is not over as some resistance has developed to RHDV and new biological control agents are being discussed, although the need for public involvement in decisions is understood.

This program has estimated that it takes about 10 years from concept to release of a new agent. Summaries of this long-term program have shown that over 57 years, rabbit biological control has saved the Australian pastoral industry AU$7.5 billion (AU$130 million/year).

As an interesting addition, rabbits were also a problem in New Zealand and the myxoma virus was introduced but did not establish (possibly owing to lack of appropriate vectors). The New Zealand government refused use of RHDV but it was illegally smuggled into New Zealand from Australia by farmers and released, resulting in a rabbit mortality rate of 84 percent.

11.4.3 Further Control of Vertebrates Using Viruses

One more successful program occurred on Marion Island, a small and isolated sub-Antarctic island, where a few cats had been introduced in 1949. This island was used as a breeding site by numerous species of sea birds. However, by 1975, there were 2,000 cats on the island and they ate 45,000 sea birds each year. Feline parvovirus was introduced in 1977 and five years later, there were only 600 cats, which were then eradicated by trapping and hunting.

In some parts of Europe, the native rabbits are regarded as pests. The myxoma virus was introduced privately in France, resulting in enormous effects in France and England, reducing the native rabbit density by 80 percent. Thirty years later RHDV naturally spread through the European rabbit populations but the impact was not as strong as that seen in Australia. In this case, the introduction to control a native species also impacted natural predators, including the Iberian lynx.

Further Reading

Cooke, B. D. (2014). *Australia's War Against Rabbits*. Melbourne: CSIRO Publishing.

Di Giallonardo, F. & Holmes, E. C. (2015). Exploring host–pathogen interactions through biological control. *PLOS Pathogens*, **11**(6), e1004865.

Eberle, K. E., Jehle, J. A., & Huber, J. (2012). Microbial control of crop pests using insect viruses. In *Integrated Pest Management*, ed. C. P. Abrol & U. Shankar, pp. 281–298. Wallingford, UK: CABI Publishing.

Fenner, F. & Fantini, B. (1999). *Biological Control of Vertebrate Pests: The History of Myxomatosis – An Experiment in Evolution*. Wallingford, UK: CABI Publishing.

Harrison, R. & Hoover, K. (2012). Baculoviruses and other occluded insect viruses. In *Insect Pathology*, ed. F. E. Vega & H. K. Kaya, pp. 73–131. San Diego, CA: Academic Press.

Hoddle, M. S. (1999). Biological control of vertebrate pests. In *Handbook of Biological Control: Principles and Applications of Biological Control*, ed. T. S. Bellows & T. W. Fisher, pp. 955–975. San Diego, CA: Academic Press.

Lacey, L. A. (ed.) (2017). *Microbial Control of Insect and Mite Pests: From Theory to Practice*. Amsterdam: Academic Press/Elsevier. (numerous chapters)

Lacey, L. A., Grzywacz, D., Shapiro-Ilan, D. I., Frutos, R., Brownbridge, M., & Goettel, M. S. (2015). Insect pathogens as biological control agents: Back to the future. *Journal of Invertebrate Pathology*, **132**, 1–41.

12 Fungal Pathogens of Invertebrates

The remaining group of invertebrate pathogens that has been exploited for biological control, the fungi, are eukaryotes. This group differs from viruses and bacteria as all fungi have nucleated cells. Although all fungal pathogens have this as a common feature, there is great diversity among the fungal pathogens of invertebrates important for biological control.

12.1 General Biology of Fungal Pathogens of Invertebrates

Fungi utilize a great diversity of resources as potential sources of nutrients but they need preformed organic matter for food. With this basic nutritional requirement, it seems logical that some fungi would have adopted lifestyles as pathogens. We will discuss fungi later as natural enemies attacking weeds (Chapter 15) and as both plant pathogens and antagonists of plant pathogens (Chapters 16 and 17). Although fungi are only very rarely lethal pathogens of vertebrates, they are important pathogens of many invertebrates. In recent years, using molecular-level information, a group called the Microsporidia is now known also to belong to the fungal kingdom. This group, which includes many pathogens of invertebrates, has very specialized biology and will be covered only at the end of this chapter; therefore, the beginning of this chapter will refer to those pathogens always known as fungi, with some discussion of some insect-pathogenic protists similar to fungi. In fact, there are probably over 750–1,000 species of fungi in more than a hundred genera that are pathogens of arthropods alone.

Among pathogens of arthropods, fungi are unique because most do not need to be eaten to infect since they can penetrate directly through the cuticles of hosts; therefore, fungi are major pathogens infecting insects with piercing-sucking mouthparts, such as aphids, whiteflies, and scale insects that generally would not ingest pathogens when eating. The microscopic reproductive units of fungi are called spores; spores are needed for infecting a new host, for persistence in the environment, and for dispersal. For infection, first, a fungal spore must land on the surface of a potential host. If it can attach to the cuticle and is not inhibited, the spore will begin to grow and differentiate. The fungus can then penetrate the host cuticle, using both mechanical pressure and enzymes to digest the cuticle as it penetrates. Once within the hemocoel, the fungus increases, initially as single cells within the insect blood. For some fungi, the fungus proliferates in the host blood and eventually invades the host's organs and tissues shortly before the

Figure 12.1 Cadaver of an adult Asian longhorned beetle, *Anoplophora glabripennis*, killed by *Beauveria brongniartii* (photo by Ann E. Hajek; Hajek et al., 2001).

host dies. Other fungi kill the insect more quickly, possibly through use of toxins that they produce, and only utilize the entire cadaver after the host is dead. As with baculovirus infections, it generally takes several days for a fungus-infected host to die, with the length of time depending on the fungal species and strain, the size and species of the host, and ambient conditions like temperature.

At some point in the process of infecting, utilizing, and killing hosts, fungi often switch over to growing as long tubes, or hyphae, a characteristic growth form for fungi. Generally, before an infected invertebrate host dies, there is little external evidence of fungal infection although the infected individual usually eats less. In some very interesting instances, as with some nucleopolyhedroviruses, host behavior can change and, before death, the host climbs so that the resulting cadaver will be at an elevated location. In some cases, the fungus even grows out of the host and attaches it to the substrate before the host dies. Elevated locations for cadavers certainly improve the potential for subsequent dispersal of fungal spores that are formed.

Just after death of the host, cadavers are often somewhat hard in consistency. Under humid conditions for most species, specialized fungal hyphae grow out through the host cuticle and spores are produced on these modified hyphae outside of the cadaver. The fungal growth on cadavers often appears fuzzy (Figure 12.1) and this growth can be variously colored depending on the fungal species, often from white to greens, pinks, reds, or oranges. Spores are actively shot off from cadavers by some fungi while for other fungal species, spores are dislodged by wind or rain or just when contacted. Sometimes spores are not produced on the cadaver surface and instead cadavers are filled with spores. As is typical of fungi, multitudes of spores are produced to increase the chances that a few might locate an appropriate food source, in this case an appropriate host, and successfully infect it. Fungal spores range from delicate short-lived structures that must remain wet to stay alive to thick-walled "resting" spores that can survive for many years in the soil.

Fungi differ from other pathogens because of their ability to infect by penetrating the cuticle but there is a downside to this strategy. Fungal spores are more exposed to the external environment than are pathogens that infect through the gut. Thus, these spores that require some moisture to germinate and infect can be exposed to potential desiccation. Fungal spores can also be exposed to ultraviolet radiation or high temperatures owing to their external, exposed locations before they infect and this is risky because fungi can be sensitive to both of these environmental factors. While these limitations make infection sound unlikely, fungal pathogens of invertebrates are actually quite common in nature because they utilize the moist and cooler conditions found overnight and in localized microhabitats. In particular, fungal spores often persist in the soil through periods when hosts are not present or environmental conditions are unfavorable and thus the soil can act like a reservoir, storing fungal inoculum until infection is again possible.

12.2 Diversity of Fungi and Fungal-Like Protists Infecting Invertebrate Pests

Several groups of fungi having very different characteristics have been exploited for pest control. The diverse characteristics of the different fungal groups strongly influence how they have been used.

12.2.1 Entomophthoralean Pathogens

The Entomophthorales are obligate pathogens that are often quite host specific and have multiple spore stages that are either short-lived or very long-lived. The long-lived resting spores provide a soil reservoir and short-lived asexual spores, or conidia, usually produced externally on cadavers, are actively ejected to become windborne (Figure 12.2). Using these specialized strategies, some Entomophthorales are able to respond to increases in host populations and cause dramatic epizootics. The most well-known species in the Entomophthorales infect caterpillars, aphids, flies, and beetles. Although most species of Entomophthorales can be grown outside of hosts (= *in vitro*) in the laboratory, mass production has been more difficult for this group because they often have complex nutritional requirements. Owing to the difficulties encountered with mass production of Entomophthorales, strategies requiring use of small amounts of inoculum, that is, classical biological control or inoculative release, are most suitable for this group.

12.2.2 Arthropod Pathogens in the Ascomycota

Fungi typically have complex life cycles with asexual and sexual stages but higher taxonomy has historically been based only on the sexual forms. However, many species of fungi infecting invertebrates are regularly found only as the asexual forms in nature and for these species, the sexual forms, when these have been found, often clearly belonged to a large fungal phylum called the Ascomycota. In the past, when only morphology could be used for identification, these fungi with unknown sexual forms were grouped

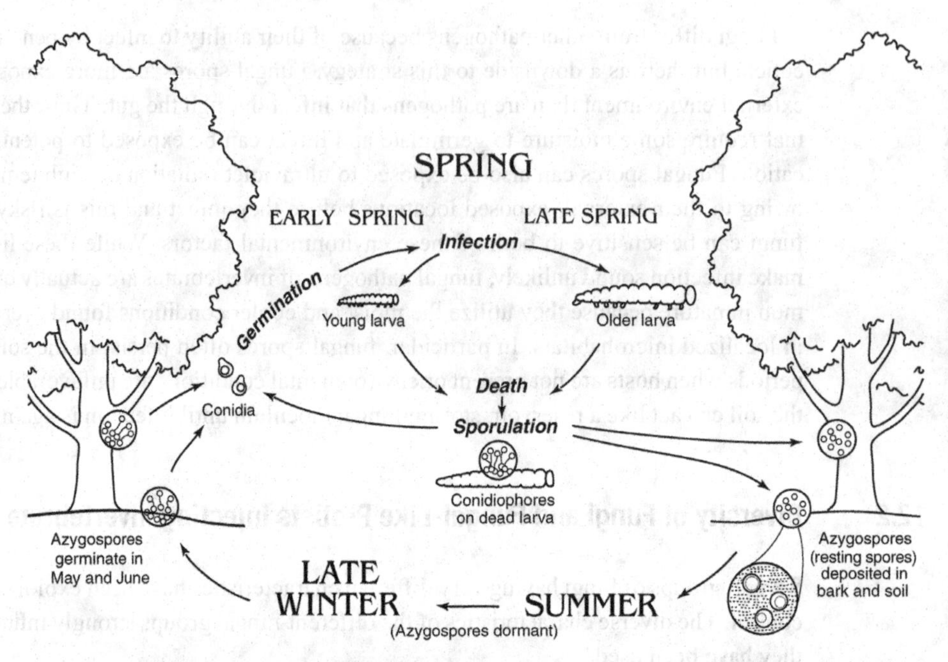

Figure 12.2 Life cycle of the entomophthoralean fungal pathogen *Entomophaga maimaiga* infecting gypsy moth, *Lymantria dispar* (drawn by Frances Fawcett).

in an informal phylum called the Deuteromycota, or the Fungi Imperfecti. When the sexual forms could be found and associated with the asexual forms, these fungi then often had two Latin names, frequently in two different genera, which of course led to confusion. This problem of identifying the sexual stage for a fungus, when only an asexual stage was present, has been solved with the advent of molecular methods. However, problems remain regarding what name to use for these fungi that rarely are found in the sexual forms, as according to the taxonomic rules, the name for the rare sexual form should be used, even when the species is well known by the name of a common asexual form. At present, mycologists are strongly encouraging use of just one Latin name for each of these fungal species (the "one fungus one name" principle) although in some cases it remains to be seen whether the name for the rare sexual form or the more well-known asexual form will be used. In addition, since molecular methods have been available, some individual species that have few morphological features to differentiate them have been split into numerous unique species, each with new names.

The most well-known and commonly found insect-pathogenic fungi all belong to the Ascomycota, with the majority in the order Hypocreales. The infective asexual spores (conidia) of many species in this group are generally somewhat more long-lasting than the short-lived infective conidia of the Entomophthorales. Pathogens in this group are often able to grow outside of hosts to some limited extent as saprophytes. In particular, these species are thus often considered facultative pathogens and, owing to their more flexible nutritional requirements, some species are more easily mass-produced. In recent years, fungi in this group have been found to occur within plants, as endophytes and to live around plant roots. When insect-pathogenic Hypocreales are living in these locations, they gain nutrients within plants or around roots and their presence can also provide some protection for plants from arthropods, so that these fungi can be thought of as "bodyguards" for the plants hosting them.

12.2.3 Primitive Fungi and Fungal Relatives

While many fungi attack terrestrial hosts, there are also species attacking aquatic hosts and these have especially been investigated for control of mosquito larvae. One group historically considered to be fungal, the water molds (Phylum Oomycota), has now been determined to be more closely related to the protists and algae than fungi. Spores of the oomycete pathogen infecting mosquitoes, *Lagenidium giganteum*, have flagellae and can swim to locate hosts. Another group of mosquito pathogens that also have flagellate spores that search for hosts is the genus *Coelomomyces*, which belongs to a primitive group of fungi. For some time, studies on *Coelomomyces* were stymied because this fungus could not be cultured in the laboratory. Then, Howard Whisler discovered that *Coelomomyces* from southern Alberta, Canada, also required an obligate alternate host, a small aquatic crustacean (copepod or ostracod); for a complete generation of this fungus, mosquitoes become infected followed by copepods becoming infected, and then back to mosquitoes, and so on, with different fungal forms associated with the different hosts. Therefore, the spores infecting mosquitoes cannot infect copepods and the spores infecting copepods cannot infect mosquitoes. While this life cycle is complex, it is functional in nature, where high levels of infection can occur in mosquito larvae.

12.3 Use of Fungal Pathogens for Pest Control

12.3.1 Classical Biological Control

As with all invertebrate pathogens, entomopathogenic fungi have not been used extensively for classical biological control. However, among the few attempts at classical biological control, the Entomophthorales have usually been used because these obligate pathogens can cause dramatic epizootics, responding to changes in host population density on their own. There have been some important successes in use of fungal pathogens for classical biological control. For example, the entomophthoralean *Entomophaga maimaiga* can cause dramatic epizootics important in controlling outbreak populations of gypsy moth (Box 12.1).

Box 12.1 A Fungus among Us

Owing to devastating outbreak populations, the invasive gypsy moth, *Lymantria dispar* (see Box 6.2), has been the target of biological control programs in North America since the early 1900s. In 1910 and 1911, a fungal pathogen specific to gypsy moth was introduced to North America from Japan. However, researchers did not think it became established and no one ever saw it. In 1989, populations of gypsy moth were on the rise but high levels of mortality were seen, with many cadavers of large gypsy moth larvae hanging on tree trunks. These cadavers were filled with the resting spores of a fungus in the Entomophthorales, *Entomophaga maimaiga*. This was the same fungal species that had been introduced from Japan in 1910–1911 but no one had ever found an infected gypsy moth larva in the United States in the intervening years. Based on some detective work using molecular-level studies of the fungus across its native distribution and evaluating weather, it is most likely that the fungus first seen in 1989 did come from Japan, but was introduced accidentally sometime after 1971.

Entomophaga maimaiga has two types of spores, short-lived conidia that are actively ejected from cadavers of younger larvae and long-lived resting spores that remain within cadavers of larger larvae. Cadavers of younger larvae are found in tree and shrub canopies, where the conidia actively ejected from cadavers can readily become airborne. Cadavers of larger larvae are found attached to tree trunks, eventually falling to the soil where they break open and resting spores are then washed into the soil at the bases of trees. Resting spores can persist in large numbers for years at the bases of trees, creating reservoirs of fungal inoculum ensuring that the fungus persists and, given adequate environmental conditions, is ready to infect gypsy moth caterpillars in subsequent years.

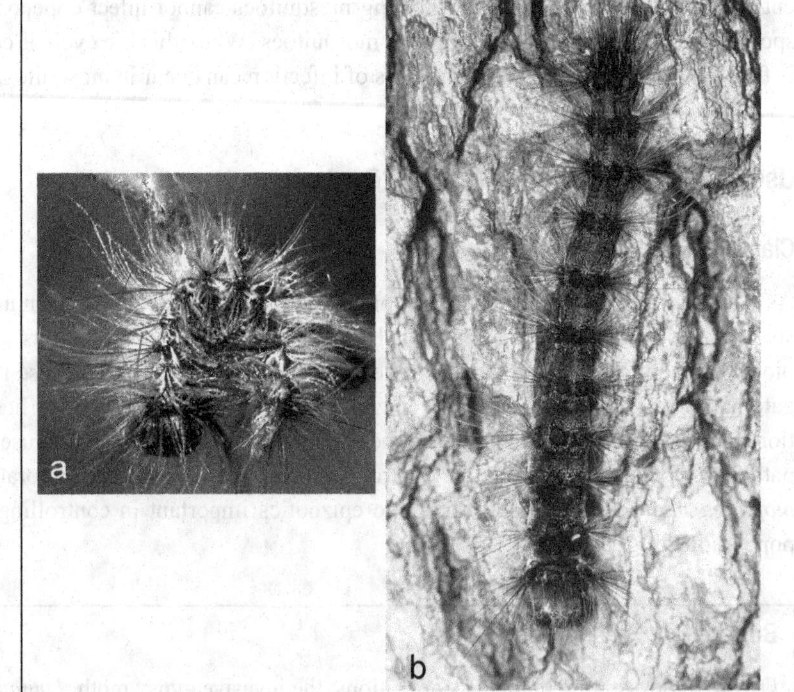

Figure Box 12.1.1 Gypsy moth larvae killed by *Entomophaga maimaiga*: (a) cadaver from which conidia were actively ejected; some are visible as they have adhered to the larval hairs; (b) cadaver of an older larva that is filled with resting spores, the overwintering stage of the fungus (photos by Donald Specker).

Since 1989, *E. maimaiga* has repeatedly caused epizootics in gypsy moth populations that were at both high and low densities. This fungus requires moisture for spores to germinate and it seems that *E. maimaiga* is more active during northeastern springs with at least average levels of rainfall. The activity of this fungus is reducing the frequency of gypsy moth outbreaks in parts of the northeastern United States. When gypsy moth outbreaks occur, it is possible that they will not be as severe and will not last as long because *E. maimaiga* is established in North America. However, time is needed to collect the long-term data to test this hypothesis.

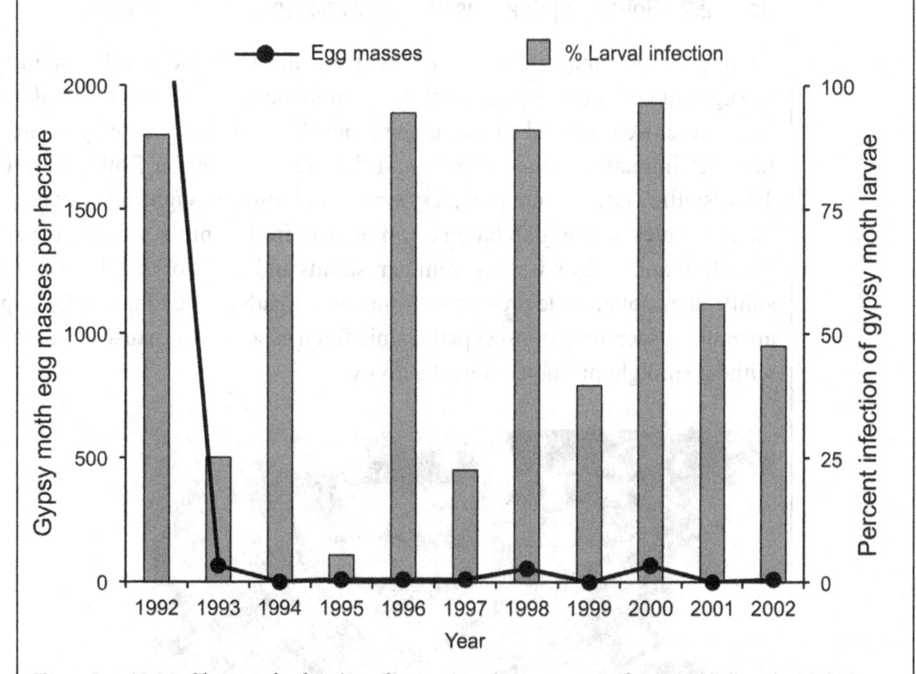

Figure Box 12.1.2 Changes in density of gypsy moth egg masses from 1992 (nearly 19,000 per hectare) to 2002 in a forest in central New York State, and the percent infection of larvae by *Entomophaga maimaiga*, beginning soon after the pathogen was first discovered in that region in 1990 (based on Hajek, 2004).

12.3.2 Augmentative Releases

Among the great diversity of species of fungi pathogenic to arthropods, only a handful of species, at least 12, have been developed for inundative release (Table 12.1). The vast majority of programs have utilized fungi in the genera *Beauveria* and *Metarhizium*. These fungi can often be grown on less expensive artificial media and have a relatively long shelf life, especially if stored in the cold.

While few species have been exploited, this does not mean that they have been used little. Since the 1960s, at least 171 products based on arthropod-pathogenic fungi have been developed, although by 2012, 26 of these were no longer in use. The regional distribution of the products demonstrates that arthropod-pathogenic fungi are used around the world. Use of *Metarhizium anisopliae* against sugarcane spittlebugs in Brazil has been ongoing since the 1970s (Box 12.2). Large programs have been developed for use of *Beauveria bassiana* against pine caterpillars, *Dendrolimus* spp., and corn borers in China, where this fungus is economically competitive with synthetic chemical pesticides and at a peak in the 1980s was applied to over 1 million ha per year. Especially in earlier

Box 12.2 Fighting Spittlebugs with *Metarhizium anisopliae* in Brazil

Among the most important pests of sugarcane in Brazil, the world's leading producer of sugarcane, is the sugarcane-infesting spittlebug *Mahanarva fimbriolata*. Spittlebugs are closely related to leafhoppers but they get their common name from the fact that immature stages secrete and then reside within a frothy mass of spittle. This "spittle" is a mixture of watery waste, air blown through abdominal openings to make bubbles, and a glandular secretion. The frothy spittle protects the spittlebugs from heat and cold as well as from parasitoids and predators but it turns out that the spittle does not provide protection from entomopathogenic fungi. In fact, spittlebugs are quite susceptible to insect pathogenic fungi especially because their moist habitat without sunlight promotes fungal activity.

Figure Box 12.2.1 Sugarcane root spittlebugs: (a) adult, (b) nymph with spittle removed from around it (photos by Alexandre de Sene Pinto).

Use of *Metarhizium anisopliae* against sugarcane spittlebugs in Brazil began in the 1970s and continues today. Although this fungus does not kill all of the spittlebugs, some spittlebugs can be present without significantly reducing sugarcane yield. Also, use of *M. anisopliae* is favored because this fungus is compatible with releases of hymenopteran and dipteran parasitoids against other serious pests on this crop, such as sugarcane borers. In fact, caterpillars of the sugarcane borer, *Diatraea saccharalis*, rival the spittlebugs as the most important pests of sugarcane and two parasitoids are very influential in their control. In 2014, the larval parasitoid *Cotesia flavipes* was reported as being released each year on 3.3 million ha of sugarcane and the egg parasitoid *Trichogramma galloi* was released on 500,000 ha. Parasitoids of sugarcane borers are very sensitive to chemical pesticides and use of *M. anisopliae* against spittlebugs on the same crop works well as these parasitoids are not harmed.

Since the 1970s *M. anisopliae* for spittlebug control has been mass-produced by private and public companies in Brazil as well as by sugarcane mills. Historically, sugarcane pests in Brazil were controlled by burning sugarcane leaves before harvest, which reduced transferring pests from one crop cycle to the next. However, as a result of air pollution problems, sugarcane burning is becoming less accepted and is now being banned in Sao Paulo State. Not burning the crop refuse could very well translate into increasing spittlebug problems, resulting in increasing use of *M. anisopliae*.

years, as *B. bassiana* was being developed for use in China, some creative application methods using firecrackers or explosives were tested as a means to disperse conidia in forests. However, eventually mechanical sprayers and dusters were found to be more effective.

Table 12.1 presents examples of some of the fungal pathogens that are mass-produced for control of a diversity of pests. In fact, fungal pathogens are used augmentatively against pests from all of the major insect orders, but especially the bugs and aphids, beetles, caterpillars, thrips, and grasshoppers. For these products, conidia are usually applied by spraying. Thus, survival of fungal spores is important and formulations to improve spore survival have been developed. Formulation of conidia in oil helps spores survive longer and ultra-low volume application facilitates application over large areas. Using oil and ultra-low volume application, *Metarhizium acridum* has been developed for control of locusts and grasshoppers in Africa where outbreaks of locusts cause extensive damage to vegetation at unpredictable intervals. One aspect of the aggregation behavior of locusts has been critical to development of this fungus for locust control. Young nymphs aggregate in "bands" before they develop into adults that can fly to create highly mobile, destructive swarms. These so-called "hopper bands" have been the main target for applications of *M. acridum*. The most efficient method for application of fungi in remote rangelands and other roadless areas has been ultra-low volume sprays from fixed-wing aircraft. The droplets applied in the concentrated sprays are very small and it is important that they not evaporate before reaching the target. To prevent evaporation, oils were investigated as formulations for fungal spores and paraffinic oils specifically

Table 12.1 Examples of ascomycete (Hypocreales) species used for augmentative release against arthropods and nematodes.

Fungus species	Target host group	Target hosts	Use
Beauveria bassiana	Many insect orders	Many arthropod species	Many crops indoor and outdoor
Beauveria brongniartii	Coleoptera	Scarab larvae, longhorned beetle larvae	Grasslands, orchards
Isaria fumosorosea	Hemiptera	Greenhouse whitefly, other hemipterans	Greenhouse
Lecanicillium longisporum	Hemiptera	Aphids	Greenhouses, outdoor vegetables
Lecanicillium muscarium	Hemiptera, Thysanoptera	Whiteflies, thrips	Greenhouses
Metarhizium acridum	Orthoptera	Locusts and grasshoppers	Pastures, grasslands
Metarhizium anisopliae	Many insect orders	Many insect species including termites, thrips	Many outdoor crops
Metarhizium brunneum	Coleoptera, Thysanoptera, Hemiptera, Parasitiformes	Scarab larvae, thrips, chinch bugs, ticks	Nurseries, greenhouses, grasslands
Purpureocillium lilacinum	Nematoda	Plant-parasitic nematodes, especially *Meloidogyne*	Horticultural and field crops

were found to be highly compatible with fungal spores. After an infected grasshopper has died from a fungal infection, the hyphae growing from the cadaver are white but the cadaver soon appears fuzzy and green, once the dark green spores are produced from the hyphae. Locusts are known to bask in the sun and can increase their body temperatures enough to kill fungal pathogens growing within them but *M. acridum* was chosen, at least in part, because of its great ability to withstand higher temperatures. The internal body temperatures of basking locusts can reach 100–104°F (38–40°C), but *M. acridum* can survive these temperatures and infected locusts thus cannot easily cure themselves of fungal infections by basking. The formulation of *M. acridum* that was developed can kill 70–90 percent of treated locusts in sparsely vegetated habitats, with no detectable impact on nontarget organisms. This product, named Green Muscle, has a shelf life of at least one year at 86°F (30°C). One drawback of this product is its slow speed of kill (10–20 days) although infected insects eat less before dying and are more prone to predation. Green Muscle is also more expensive than competitive synthetic chemical insecticides. However, Green Muscle provides the added benefit of so-called recycling in the field, meaning that once infected insects die, the fungus produces spores from the dead hosts and these spores can infect yet another group of insects (see Figure 4.1). In addition, some of the spores initially sprayed that do not contact hosts remain viable in the field as a pathogen reservoir.

A product based on the entomopathogenic fungus *Beauveria brongniartii* has been developed in Japan for control of citrus longhorned beetles, *Anoplophora chinensis*, whose larvae bore into the trunks of orchard trees. Adult citrus longhorned beetles normally walk on tree trunks and branches of trees. After emerging from within trees where they develop as immature stages, they must feed to become sexually mature. These beetles then mate and females lay their eggs under bark near the bases of trees. When the adult beetles are walking on trees, they walk across fiber bands containing cultures of *B. brongniartii* that encircle tree trunks. Beetles contacting the bands become inoculated with spores. The cuticles of these beetles are thick and hard but the spores send hyphae to penetrate through the thinner, bendable cuticle between the hard sclerotized plates of adult beetles, thus infecting the beetles. The bands containing fungal cultures hung on trees where beetles inoculate themselves, are covered with high densities of viable conidia for at least 30 days and cause significant population reductions without frequent band replacement.

12.4 Microsporidia

Species from a diversity of groups of single-celled animals, a loose assemblage sometimes referred to as the Kingdom Protista, infect insects. With the advent of molecular methods, relations of some of these groups have been clarified. The most common protistan pathogens of invertebrates and the main group investigated for use in pest control are the microsporidia. Although microsporidia have historically been considered within the Protista, based on molecular evidence they are now known to be fungi, although they are quite distantly related to any fungal species present today. Microsporidia are among the smallest eukaryotes, they have the smallest eukaryotic nuclear genomes, and they grow and reproduce within the cells of hosts. Therefore, as with viruses, microsporidia always live as pathogens.

Microsporidia have fascinating and intricate life histories. Intracellular stages can be multinucleate, going through very complicated life cycles alternating between producing spores (most less than 6 microns long) and proliferating vegetatively. Microsporidian species often grow only in specific host tissues such as the insect fat storage organ (fat body), the midgut wall, or the reproductive tissues. After proliferation in the host, microsporidia produce environmental spores that must be eaten by new hosts in order to infect. Much of the interior of the environmental spores is filled with a long, coiled tube, called a polar filament. Once a spore is within the lumen of the midgut of a potential insect host, the conditions within the digestive tract of the host act on the spore to cause the polar filament to be extruded. This tube acts as an extensible inoculating needle, piercing a cell in the midgut wall and the contents of the spore then pass through the tube to enter the host cell and begin a new infection (Figure 12.3).

As a group, microsporidia infect a diversity of hosts, from vertebrates to various single-celled organisms, but their most common hosts are arthropods. In recent years, microsporidia have been investigated extensively as a result of their involvement in the worldwide collapses in honey bee colonies. Most microsporidia cause chronic diseases

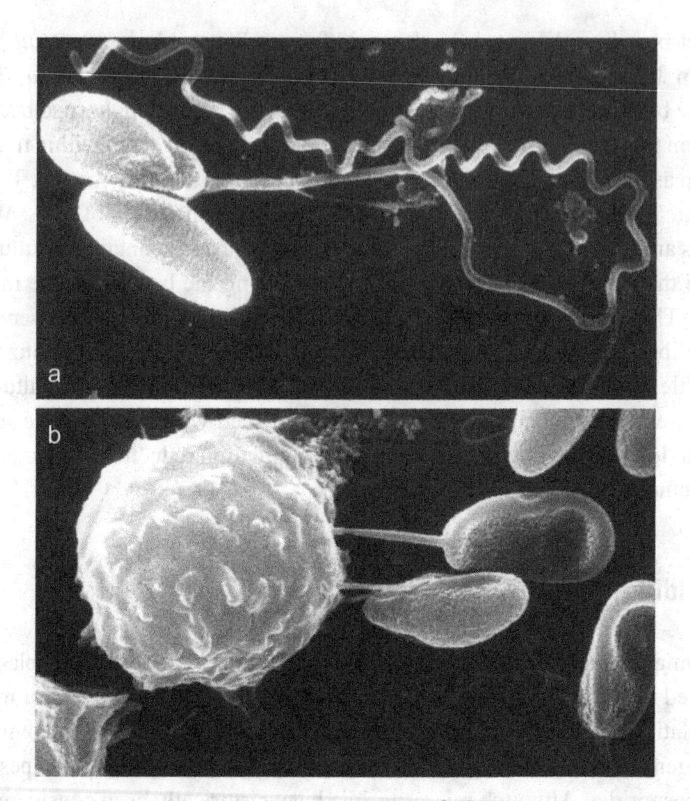

Figure 12.3 (a) Microsporidian spores, one of which has extruded its polar filament. (b) Two microsporidian spores that have extruded their polar filaments, which have impaled a target cell (photos by Andreas Linde).

that do not kill hosts and hosts often display few symptoms when infected. Infected arthropods often survive to adulthood but infected individuals are usually smaller than normal, they have shortened lives and their reproductive output is often reduced. To ensure that microsporidia persist in low-density or scattered host populations, these pathogens often are passed from adult hosts to their offspring within the host eggs, a process called vertical transmission.

Although microsporidia often do not kill hosts, they can have an overall debilitating effect on host populations. They have been exploited for control very seldom, in part because they cause chronic infections and not rapid host mortality, but also because they must be produced within hosts, which makes mass production more complicated and expensive. Some attempts at development for control have been made but the most long-standing attempt has been the use of the microsporidian *Nosema locustae*, developed for control of rangeland grasshoppers in the Great Plains of the United States. To produce this species for application, the environmental spores produced within infected grasshoppers are harvested from cadavers and mixed with wheat bran, which acts as a bait to attract grasshoppers which then become infected when eating the bait. When the bait is sprayed on a pasture, many grasshoppers became infected that same year.

In the field, after releases of the spores plus bran, *N. locustae* persists to the next year and the area with infected individuals expands. It is thought that spread and persistence of *N. locustae* after release has also been due in part to passage of the microsporidia through host eggs to the next generation. It is considered important to disease transmission that this microsporidian is also transmitted to healthy grasshoppers when they become infected after cannibalizing diseased individuals or eating cadavers of grasshoppers killed by *N. locustae*.

Further Reading

Butt, T. M., Jackson, C. W., & Magan, N. (eds.) (2001). *Fungi as Biocontrol Agents: Progress, Problems and Potential*. Wallingford, UK: CABI Publishers.

Cory, J. S. & Ericsson, J. D. (2010). Fungal entomopathogens in a tri-trophic context. *BioControl*, **55**, 75–88.

Ekesi, S. & Maniania, J. K. (eds.) (2007). *Use of Entomopathogenic Fungi in Biological Pest Management*. Trivandrum, India: Research Signpost.

Faria, M. R. & Wraight, S. P. (2007). Mycoinsecticides and mycoacaricides: A comprehensive list of worldwide coverage and international classification of formulation types. *Biological Control*, **43**, 237–256.

Hajek, A. E., Wraight, S. P., & Vandenberg, J. D. (2001). Control of arthropods using pathogenic fungi. In *Bio-Exploitation of Fungi*, ed. S. B. Pointing & K. D. Hyde, pp. 309–347. Hong Kong: Fungal Diversity Press.

Roy, H. E., Vega, F. E., Chandler, D., Goettel, M. S., Pell, J. K., & Wajnberg, E. (eds.) (2010). *The Ecology of Fungal Entomopathogens*. Dordrecht, Netherlands: Springer.

Scholte, E.-J., Knols, B. G. J., Samson, R. A., & Takken, W. (2004). Entomopathogenic fungi for mosquito control: A review. *Journal of Insect Science*, **4**, 1–24.

Solter, L. F., Becnel, J. J., & Oi, D. H. (2012). Microsporidian entomopathogens. In *Insect Pathology*, ed. F. E. Vega & H. K. Kaya, 2nd edn., pp. 221–264. Amsterdam: Elsevier.

Vega, F. E., Meyling, N. V., Luangsa-ard, J. J., & Blackwell, M. (2012). Fungal entomopathogens. In *Insect Pathology*, ed. F. E. Vega & H. K. Kaya, 2nd edn., pp. 171–220. Amsterdam: Elsevier.

Weiss, L. M. & Becnel, J. J. (2014). *Microsporidia: Pathogens of Opportunity*. Ames, IA: Wiley-Blackwell. (selected chapters)

Part IV

Biological Control of Weeds

13 Biology and Ecology of Herbivores Used for Biological Control of Weeds

13.1 Types of Agents

Plants are used by a great diversity of organisms that have different ways of exploiting them for food and shelter. The majority of herbivores that have been used for biological control of weeds are plant-feeding (also called phytophagous or herbivorous) invertebrates. These are often classified into guilds according to the plant resources they consume, such as roots, stems, leaves, flowers, seeds, and fruits. Secondarily, they can be considered by their style of feeding, such as sucking, chewing, leaf mining, stem mining, and root boring. One specialized type of feeding is gall formation. In this case, plant tissues are attacked by invertebrates early in development and the plant is tricked into creating a specialized enlarged structure called a gall, which often becomes home plus food for invertebrates living within. Affected plants can expend lots of energy making galls, at the expense of other vegetative growth. If a gall is in the reproductive tissues, this can limit or stop seed output.

Where and when an invertebrate attacks a plant can have profoundly different effects and of course this also varies by plant species. Insects attacking the growing tips of plants can have a greater effect than insects solely feeding on leaves, although some plants can compensate and create new growing tips. Sap-sucking or gall-forming insects often are thought to have lesser effects, although they cause a prolonged drain on plant nutrients (but see "Invertebrate natural enemies" in Section 14.3.1 Invertebrate Natural Enemies). Of course, fruit and flower feeders primarily affect plant reproduction so the next generation of plants would decrease in abundance and the plants that are being attacked would still be present.

Invertebrates feeding on plants rarely have an impact when feeding alone. Their effect is dependent on many individuals being present and active. Factors negatively affecting herbivore density such as predators, parasitoids, and weather all could negatively affect the impact of herbivorous species feeding on weeds. The ability to increase rapidly in numbers is not characteristic of all species of herbivores but is clearly a positive attribute for biological control agents when it results in more intensive effects on host plants.

Sometimes grazing by vertebrates such as sheep and goats has been used to control weeds, but this is usually nonspecific, with vertebrate herbivores feeding on a diversity of plant species and plant parts. Weeds in aquatic ecosystems can completely clog bodies of water and are difficult to control, but they often need to be controlled when they change ecosystems or disrupt human activities. Herbivorous fish for the control of aquatic weeds are the most common type of vertebrate used for the biological control of weeds.

Plant pathogens have also been used for controlling specific weed species. In a few cases, host-specific fungi have been used for classical biological control of weeds while, in some instances, products based on plant pathogens have been registered for inundative release (see Chapter 15).

Biological control programs have often considered introducing several different agents among the diversity of organisms available, often using several different feeding guilds based on the hypothesis that action of these would be complementary. Just as one individual herbivore usually does not kill an entire plant and many individuals of the same species are needed to have an impact, it has been hypothesized that within-species complementary activity can be extended to encompass different species of natural enemies working together to decimate weeds.

13.2 Weed Characteristics

Just as there is a huge diversity of species of phytophagous natural enemies, what we call weeds varies significantly, ranging from single-celled photosynthetic algae to trees. Many weeds are r-selected species, characterized by abundant reproduction but poor competitive abilities (see Chapters 6 and 16) and these species are excellent at colonizing disturbed habitats. A ubiquitous example of this plant life strategy would be the weeds readily colonizing overgrazed land – these weeds grow quickly when few to no other plants are present as competitors. However, while r-selected species have advantages in rapid colonization, this strategy also has weaknesses that can be exploited (see below).

What types of plants are most successfully controlled using natural enemies? Two researchers, J. J. Burdon and D. R. Marshall, decided to investigate whether plants that reproduce sexually or those that reproduce asexually were controlled more often using classical biological control. Sexually reproducing plants of course grow from seeds, while asexually reproducing plants do not require seeds but can grow directly from plant parts. As a result, there is more genetic diversity among populations of sexually reproducing plants than among asexually reproducing plants, many of which grow as genetically identical clones. Analyzing 81 separate control attempts, asexually reproducing plants were controlled more often than sexually reproducing plants. The authors hypothesized that this was due to the genetic similarity among individual plants within the same species, which allowed natural enemies to optimize their attacks more easily. In support of these findings, classical biological control programs against the asexually reproducing aquatic weed *Salvinia molesta* (see Chapter 13) have been extremely successful, while control of the sexually reproducing *Lantana camara* has been very difficult. *Lantana camara* has pretty flowers and is cloned by gardeners and grown horticulturally in the tropics. However, there is abundant genetic variability within the species *L. camara* and insects that are effective at damaging one clone can have little effect on others. However, in some areas such as on Ascension Island in the middle of the southern Atlantic Ocean, where we assume the genetic diversity of *L. camara* is limited, a small

Figure 13.1 *Teleonemia scrupulosa* commemorated on a postage stamp for Ascension Island in the middle of the Atlantic Ocean.

lace bug (Tingidae) named *Teleonemia scrupulosa* that was introduced for classical biological control of *L. camara* was considered effective enough to be commemorated on a national postage stamp (Figure 13.1).

Some weeds have proven especially difficult to control worldwide. Do these weeds have attributes in common to help predict the outcome of biological control programs? It seems that plants with better adaptations for withstanding stress can be more difficult to control. Especially among perennial weeds, resistance to control has been associated with a long growing period, large reserves (e.g., thick underground rhizomes or woody stems), good powers of regrowth after defoliation, good ability to replace fruit and seeds after injury, and large banks of dormant seeds. If weeds are of poor nutritional quality for herbivores, natural enemies are usually less successful for control. Annual weeds of arable crops are more difficult to control with classical biological controls because of the regular disruption in such cropping ecosystems. All of these characteristics can make it more difficult for phytophagous natural enemies to control weedy plants. However, none of these characteristics, alone or together, has entirely ruled out success of a biological control agent and thus it can be difficult to predict outcomes of classical biological control programs.

13.3 Types of Injury to Plants

While parasitoids and predators used for biological control of invertebrates often kill one or several hosts, individual biological control organisms attacking plants rarely kill them outright. In fact, plants usually can survive some degree of attack by herbivores or plant pathogens and still reproduce. Biological control agents are usually most effective in controlling weeds when they increase in numbers and act in concert. Although for weed control we are concerned with populations of weeds, the effects of natural enemies on weeds of course occur on individual plants. Plants are complex spatially and temporally and different structures that occur at different times can be attacked by different stages of natural enemies. There are several major ways that invertebrate natural enemies impact plants by feeding on different parts of a plant.

13.3.1 Reducing Flowers and Seeds

Attacking flowers and seed-producing structures can reduce the numbers of viable seeds set by plants, although of course this would not impact plants that are already established or plants that predominantly reproduce vegetatively. Therefore, this type of effect would have the greatest impact on annual plants that die each year. Likewise, when seeds are eaten or seed production is disrupted, spread by weed populations can be impacted.

For some plant species, destroying reproductive output has proven to be an effective strategy for controlling increase and spread. To control the leguminous tree *Sesbania punicea* in South Africa, a small weevil (*Trichapion lativentre*) from South America that voraciously attacks flower buds was introduced, reducing seed production by 98 percent. When this was followed by introduction of a seed-loving weevil (*Rhyssomatus marginatus*) that destroyed up to 88 percent of the remaining seeds, reproduction by this plant was almost entirely halted.

However, in many cases weed densities are not limited by the numbers of seeds; many weeds produce an overabundance of seeds and such species are therefore often limited by availability of new locations for seedlings to grow. Based on studies and models of the prolific seed-producer diffuse knapweed (*Centaurea diffusa*), J. Myers and C. Risley (2000) suggested that seed feeders are often not likely to be successful agents for reducing existing plant populations.

13.3.2 Direct Mortality of Plants

Biological control agents rarely directly kill plants, although this can occur if natural enemies increase in numbers and large populations then consume entire plants. Perhaps more commonly, large populations can eat all of the foliage on a plant, which is called defoliating the plant. Some plants, such as oak (*Quercus*) trees, can grow a second set of leaves after defoliation, but coniferous trees like pines (*Pinus*) die if completely defoliated. Even for plants that can refoliate, each time a new set of leaves is produced, the plant progressively uses up storage reserves and this weakens the plant.

Intensive site-specific feeding by herbivores can kill plants if vital functioning of the plant is impaired. For example, mortality of the plant could occur if the water transport system was destroyed by girdling or if the roots were killed. Yet, such directed and effective attacks that quickly kill host plants are not common strategies for the invertebrates that feed on plants. One could consider that it often behooves phytophagous invertebrates to keep their host plant alive so that they can develop while eating it and so that it is present for their offspring to eat. Therefore, for biological control of weeds, we are usually looking for the more unusual phytophagous invertebrates that for some reason or in some way strongly exploit their host plants, at least when the host plants are at high densities. Killing the plant is not necessarily required and just suppressing it can lead to the desired plants surviving and outcompeting the invasive.

13.3.3 Indirect Plant Mortality

Plants are generally thought to survive with some level of injury and they compensate by using storage reserves. Based on this fact, P. Harris (1986) proposed a "multiple stress hypothesis" for recommending how to impact weeds. He suggested that there is a damage threshold and above that threshold plants cannot compensate for damage and have an increased risk of death. The trick is to exceed that damage threshold for the weed species in question. This can be done indirectly by biological control agents attacking the specific tissues that are most sensitive, such as storage reserves. For example, if carbohydrates are stored in the leaves, then attacking the leaves would be most effective. Also, natural enemies that attack plants during time periods when they are more susceptible can be more effective. Over the winter and early in the year, when musk thistle (*Carduus nutans*) is growing as a rosette of leaves at ground level it can be killed by defoliation, but once musk thistles are producing fruiting stalks, defoliation is no longer effective for control.

Sometimes, for herbivores to "stress" a plant sufficiently to cause mortality, the plant must be growing optimally. One factor that limits growth and development of herbivores can be plant nutritional quality, especially the percentage of nitrogen. Although this seems counter-intuitive, there are several examples where fertilizer has been added to weeds to improve biological control (see Box 13.1). While some herbivores do well on

Box 13.1 The Aquatic Weed *Salvinia molesta*

Salvinia molesta is a free-floating aquatic fern originating in southeastern Brazil. It is a botanical curiosity and has been used extensively as an aquarium plant and thus it became distributed throughout the tropics over the past 50 years. Unexpectedly, if introduced to waterways, it often readily becomes established and can cause major problems. This plant forms colonies that grow very fast, doubling in size every 2.2 days, and it has grown to cover lakes and rivers with 1-m-thick mats. With a single layer of plants, waterways become impassable by rowed boats and with a thicker mat even diesel-powered boats cannot pass. It also impedes access to the water for domestic animals and humans. Control efforts were initiated in the 1940s after this weed became a serious problem in many countries.

Salvinia created the greatest problems in areas where the human inhabitants relied heavily on waterways for transportation. The floodplain of the Sepik River in Papua New Guinea is just such a place, where 80,000 people depend on transportation across the water for most goods and services. In addition, a staple food is made from the pith of sago palms and transport over water is essential for harvesting and distribution of palms. *Salvinia* became such a problem in this area that the lives and livelihoods of the inhabitants were threatened. At first, control was attempted with herbicides and removal and containment using a boom in the water, but these methods were not successful on a long-term basis.

In 1959, the *Salvinia* occurring on Lake Kariba on the Zambia/Zimbabwe border was identified as *Salvinia auriculata*. Therefore, the native home for this species, South America, was searched for natural enemies. A moth, a grasshopper, and a weevil were collected and introduced to various countries, but these agents often did not establish or established only very poorly and provided no control. By 1970, researchers realized that the cause of the problem was not *S. auriculata*, and the problematic weed was instead identified as *S. molesta*. In the native area for *S. molesta*, southeastern Brazil, the same three species of natural enemies were thought to be attacking *S. molesta*. However, further investigations showed that the weevil in southeastern Brazil was really a new species (*Cyrtobagous salviniae*) and an excellent biological control prospect as well. The 2-mm-long black adult *C. salviniae* feed selectively on growing points of *Salvinia*. *Salvinia* is very susceptible to this type of feeding because this plant relies on vegetative propagation. The tiny white weevil larvae tunnel through the rhizome, eating the vascular tissue, and the total effect of these weevils is devastating to *Salvinia*. These weevils are host specific, are very efficient at finding host plants, and can live at high densities; all of these are great characteristics for an effective phytophagous insect for weed control.

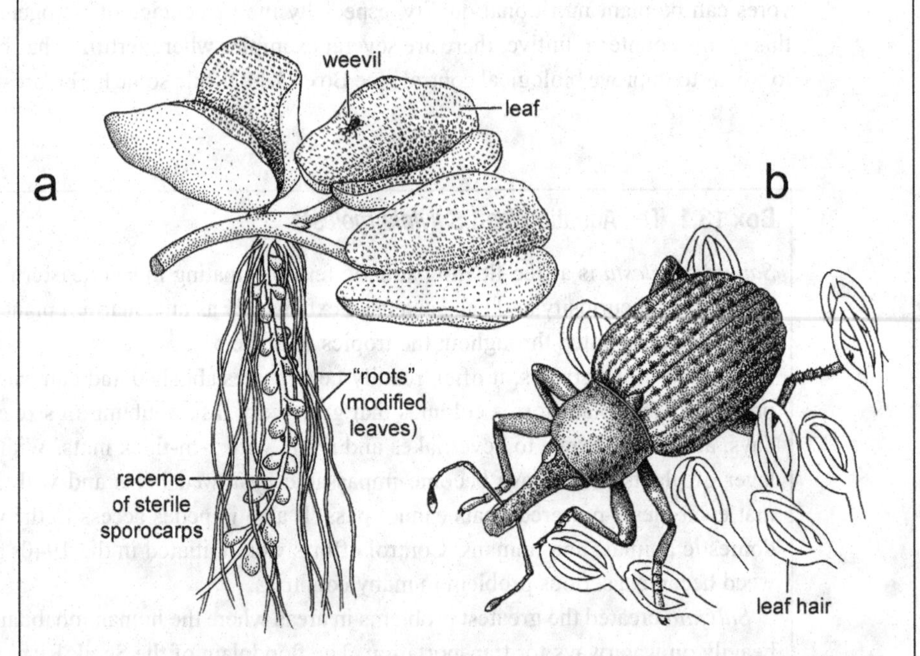

Figure Box 13.1.1 The aquatic weed *Salvinia molesta* and the small black weevil *Cyrtobagous salviniae* from southeastern Brazil, which was successfully released for classical biological control (by Karina H. McInnes; Gullan & Cranston, 2000).

This weevil was first introduced to Lake Moondarra in Australia. Within only 14 months, *Salvinia* was completely gone from the 200 infested hectares. With such stupendous results, the weevil was then introduced to many additional locations, including the Sepik River floodplain. Releases to control the Sepik infestation began in May 1983, and by August 1985 the areas covered by *Salvinia* had been reduced from 250 km^2 to 2 km^2; this weevil had destroyed 2 million tons of *Salvinia* in slightly over 2 years. During this program, researchers learned that if the plants were low in nitrogen, the beetles did not reproduce well. Therefore, urea was added to the water to enhance establishment and activity of the weevils. In the Sepik River floodplain, and in Australia, the control achieved was cost effective, environmentally sound, and permanent. The weevil has been introduced to additional areas where control was excellent. In one study in Sri Lanka, a cost: benefit analysis indicated returns on investments of 1:53 in terms of cash and 1:1,678 in terms of hours of labor. In recognition of the benefits derived from the program resulting in biological control of *Salvinia*, the team responsible for this introduction was awarded the UNESCO Science Prize in 1985.

Figure Box 13.1.2 (a) A creek in temperate Australia, near Sydney, infested with *Salvinia molesta*. (b) This same creek 3 years after the salvinia weevil, *Cyrtobagous salviniae*, had been released (photos by Mic H. Julien, CSIRO; Julien et al., 2002).

poorly growing plants, in other systems herbivores instead grow best on vigorous plants. In these examples, fertilizing weeds made all the difference in strong establishment and activity of phytophagous natural enemies that could provide very effective control when plants were growing optimally. However, when plants were growing without fertilizer, these herbivores were not very effective. In other instances, plants that were stressed by drought were less suitable for herbivore survival and otherwise effective phytophagous invertebrates provided little control when plants were drought-stressed. This is the case for the gall wasps *Trichilogaster acaciaelongifoliae*, introduced against invasive Sydney golden wattle, *Acacia longifolia*, trees in South Africa. These gall formers are much more effective in reducing acacia reproduction and vegetative growth in more humid coastal areas than in drier interior sites. These examples demonstrate that variability exists among those factors driving different weed/natural enemy systems and it is important to understand the relations between the herbivore, the weed, and plant quality to facilitate weed control by herbivores.

13.3.4 Interactions Leading to Increased Plant Stress

Multiple Agents

In many programs for biological control of weeds, numerous agents are released against one weed species, with the goal of increasing overall stress. Often the released agents attack the plant in different ways (seed feeders, leaf feeders, stem feeders, root feeders) or during different seasons, thereby increasing the overall plant stress. Examples have shown that this strategy can be very successful toward hastening the demise of unwanted plants (Box 13.2).

In some instances, a released herbivore can create an injury to a plant allowing entry by naturally occurring plant pathogens, leading to weed mortality. One famous example is the caterpillar of the cactus moth, *Cactoblastis cactorum*, attacking prickly pear cactus (*Opuntia* spp.) that had been introduced to Australia. The caterpillars feed gregariously within the fleshy cactus pads, but do not eat the entire pads, and then drop to the ground to pupate in the leaf litter. After the caterpillars create the wounds, native opportunistic microbes move into this nutrient-rich wound environment and go on to kill the cactus pad.

However, biological control practitioners need to understand interactions between multiple agents being released in order to avoid interactions that might lead to an overall decrease in their impact. Natural enemies can be separated in space and in time, but that does not mean that the activity of one does not impact the activity of another. In the case of musk thistle, *Carduus nutans*, a plant that reproduces from seeds, two weevils were studied that were separated in both space and time on the same plants. Larvae of the weevil *Trichosirocalus horridus* feed in the vegetative rosettes next to the ground from fall through spring and severely impact the ability of the plant to produce a stem and grow during summer, while larvae of the weevil *Rhinocyllus conicus* feed in the flower heads and reduce seed production. As can be imagined, the activity of *T. horridus* negatively impacted *R. conicus* in large part because there were fewer thistle heads for reproduction. In this case, although there was a negative impact between these

Box 13.2 Biological Control to Combat a Tree Threatening the Florida Everglades

The paperbark tea tree, *Melaleuca quinquenervia*, is usually just referred to by its generic name, *Melaleuca*, in the United States. This tree is native to northeastern Australia, New Guinea, and New Caledonia and was introduced to southern Florida in the late 1800s. There, it was propagated in natural areas by nurseries and was deliberately planted in marshes to prevent erosion. This tree grows well in moist wetland areas and that is the problem because it has been spreading and taking over one of the largest freshwater marshes in North America, the Florida Everglades. *Melaleuca* has been transforming the Everglades by growing in dense stands that accumulate soil and thus create raised islands, changing the habitat. The native sawgrasses that live throughout the Everglades are adapted to surviving grass fires. While *Melaleuca* is also adapted to surviving fires, the oils in the *Melaleuca* foliage and the corky bark make fires so hot that sawgrasses and other native plants and trees cannot survive. As a secondary blow, *Melaleuca* seeds are held in capsules that need fires to open. So, after a fire where *Melaleuca* has been burning, the normal grasses and plants native to the Everglades are all dead and the abundant *Melaleuca* seeds that have been released have lots of places to become established without competitors.

In the late 1980s deterioration of the Everglades as a consequence of this invasive tree started to be widely recognized and a plan was developed. This plan called for some use of traditional methods for weed control, that is, herbicide applications and mechanical harvesting, to remove trees from infested areas on public lands. However, these procedures resulted in drying of seed capsules and massive releases of *Melaleuca* seeds, which produced the opposite result because *Melaleuca* became even more pervasive. A biological control program was instituted with the specific goal of stopping regeneration of trees. This was a challenging system for biological control because the weed, in this case, was a tree. Most biological control of weeds programs have been focused on herbaceous annuals or perennials, or aquatic weeds.

Foreign exploration between 1989 and 1995 identified 400 species of insects feeding on *Melaleuca*. Researchers specifically evaluated general biology, host range, and efficacy in feeding on the *Melaleuca* tissues that would prevent regeneration, as well as matching the climate of the collection locations with the Florida Everglades. Eventually, only five species of insects were introduced and two did not establish. The weevil that was introduced, *Oxyops vitiosa*, feeds on young foliage and kills stem tips and this virtually eliminated flowering and seed production. Feeding by the phloem-feeding psyllid *Boreioglycaspis melaleucae* caused high seedling mortality and leaf drop from trees. The weevil and psyllid, together with a native fungal pathogen, killed sprouts being sent up from tree stumps. The midge *Lophodiplosis trifida* created so many galls on young growth that seedlings as well as sprouts from stumps were killed.

Figure Box 13.2.1 Seeds of an invasive weed tree and an effective seed-destroying beetle: (a) *Melaleuca quinquenervia*, showing the ripe seed capsules and abundant seed production (USDA/ARS Invasive Plant Research Laboratory). (b) The melaleuca weevil, *Oxyops vitiosa*, which can destroy as much as 95 percent of the seed crop (photo by Steve Ausmus, USDA/ARS; Van Driesche et al., 2016).

The result of the entire program has been that seed production was reduced by as much as 99 percent. Importantly, no significant feeding on nontarget plant species has occurred. The program still focuses on removing trees from infested areas; herbicides must still be used to kill mature trees but there is very limited regrowth afterward because the biological control agents move in, followed by the native plants. Since this program is dealing with a huge natural area and the invaders are trees, it is estimated that it will take decades to completely restore the infested areas of the Everglades to close to their natural state. However, there now appears to be a sustainable path toward reclaiming the Everglades as, overall, *Melaleuca* is experiencing negative regional population growth.

weevils, overall seed production was reduced by 59 percent when both weevils were present but only 45 percent when only *R. conicus* was present. Overall control was not impacted by the negative interactions between the two weevil species, but care should be taken so that negative interactions do not occur in other systems.

Abiotic Stress

Plants store carbohydrates to use as reserves when they need energy. As a part of the stress hypothesis, if plants are defoliated, they will use storage reserves for regrowth. If they subsequently are also stressed by drought or winter cold, this can result in mortality because reserves are gone. Cinnabar moth caterpillars (*Tyria jacobaeae*) reduce root reserves when eating weedy tansy ragwort (*Jacobaea vulgaris*) and this can lead to increased plant mortality when frost occurs.

Competing Plants

Biological control agents are also aided if highly competitive desirable plants occur along with weeds. All plants compete for light, water, nutrients, and space. If natural enemies stress weeds, the weeds will not be as effective at competing for these resources against strong competitors. When the weedy fiddleneck (*Amsinckia intermedia*) is attacked by a flower gall nematode (*Anguina amsinckiae*), wheat becomes the more successful competitor of these two plants and outcompetes this weed.

13.4 Regulation of Weed Density by Herbivores

To investigate the ecological theory behind how herbivores and pathogens control weeds, we turn to the study of plant ecology. However, in this case, we are specifically not investigating the normal flora composed of native plants in undisturbed natural ecosystems but instead we are interested in plants whose populations are "out of control," often alien species that have been introduced. The discussion of population regulation theory based on predators, parasitoids, or pathogens and their prey or hosts (Chapter 6) is directly applicable to weeds and herbivores or plant pathogens. The same issues about population regulation exist regarding the relative importance of density-dependent relations between natural enemies and pests, stability of the system, importance of metapopulations for recolonization, and importance of refuges in sheltering portions of populations. Below, we will explore some relevant issues more specific to the basis for regulation of weeds by phytophagous natural enemies.

13.4.1 How Do Herbivores Regulate Plant Populations?

Every plant species, including those most resistant to herbivory, is associated with organisms that have evolved to eat it. Our question then is why the terrestrial world is green if there are so many herbivores. Herbivores consume an average of only 10–20 percent of the annual net primary production in terrestrial ("green") ecosystems. In contrast, the aquatic world could be green but is not. It has been estimated that 80 percent of the available primary production under the water is removed by aquatic herbivores.

Understanding why terrestrial plants remain largely uneaten despite the diversity of herbivores and plant pathogens that occur is a fundamental question in plant ecology. For practical reasons, we are often interested in keeping areas green, which often means preventing herbivory in order to protect crop and horticultural plants or forests and natural areas. Relative to biological control of weeds, our interest is the opposite: can we provide whatever is necessary for herbivores and plant pathogens to be successful in controlling weedy vegetation?

There have been numerous theories proposed to explain why the terrestrial world is green. One idea is that predators, parasites, and pathogens attacking herbivores keep herbivores rare (this is a "top-down" theory; see Chapter 6). Another view is that herbivores are rare or they are basically ineffective at controlling plants because they exist "between the devil and the deep blue sea" (Lawton & McNeill, 1979). Under this idea, populations of terrestrial herbivores are controlled in part by their natural enemies (the devils) but also by the fact that plants are often not very good food (the deep blue sea). Many plants do not supply the abundant and diverse nutrients needed by herbivores and the supplies of nutrients for herbivores can vary drastically in time and space. Over evolutionary time, plants have also countered attacks by herbivores and pathogens through evolved defenses. Unlike most animal pests, plants cannot move to defend themselves, which is in part why they have relied on chemical and mechanical defenses. Plant defenses take many forms based on the type of herbivore exerting selective pressure. For example, thorns evolved to prevent feeding by vertebrate herbivores and many plants have indigestible or fibrous parts that also help to protect them. Some plants can simply tolerate loss of parts without this affecting their ability to produce offspring successfully. Many scientists have been fascinated by the chemicals in plants that are repellent or deterrent, or which inhibit digestion. These can be primary plant metabolites that have negative impacts when either too much or too little of compounds such as proteins, carbohydrates, and vitamins are present. Some plants also produce specialized secondary metabolites that can provide excellent protection. Examples of such secondary plant chemicals are alkaloids, cyanogens, and terpenoids. The furanocoumarins found in parsnip (*Pastinaca sativa*) or the cardiac glycosides found in milkweeds (*Asclepias* spp.) make these plant species unsuitable for many herbivores. However, a few herbivores have evolved the ability to feed on these chemically defended plants, despite the presence of the nasty secondary plant compounds, and these herbivores then encounter less competition for their source of food. An example would be the migratory monarch butterflies in North America (*Danaus plexippus*) with caterpillars feeding only on milkweeds, a genus of plant eaten by few other herbivores.

As with herbivore populations, plants remain at any given population density because of interactions between those factors killing them and those factors affecting their growth and reproduction. These are the "top-down" and "bottom-up" forces, respectively, that regulate plant populations. For example, with top-down regulation, the activities of herbivores could be key in determining densities of plants, while with bottom-up regulation, population densities could be affected most profoundly by the availability of space, nutrients, water, and light for the plant. This way of investigating regulation of plant populations can be useful for looking at conditions and interactions among herbivores and plant pathogens that at times effectively control weeds. In fact, there is agreement that in natural systems, plant populations are often controlled by some combination of bottom-up and top-down forces. But we've agreed that weeds (as opposed to native plants at some characteristic and stable density) are generally not being regulated well in the ecosystems in which they occur. Therefore, for biological control of weeds, we are trying to establish situations where top-down forces are successful in reducing and subsequently regulating weed populations.

For herbivores to be effective in controlling plant density, their populations must be able to respond to weed populations by being free from their own regulators. With weedy plants, there is often no lack of food for the herbivore and thus bottom-up forces are not limiting populations of the herbivore (although herbivores can be influenced by plant quality, so presence of a plant species does not always translate to optimal food for the herbivores). When exotic herbivores are released in new areas, the herbivore often escapes from the top-down forces with which it coevolved (predators and parasites) and that regulate its populations in its area of endemism. Thus, the herbivore can increase in response to abundant host plant populations and its effect on the weed it was introduced to control will largely be determined by the inherent ability of that herbivore to increase.

13.4.2 Weed Population Ecology

After studying insect and mite pests, there is the tendency to think of weeds based on individual plants. However, it is the plant biomass and not the individual plant that is important to reduce with weed control programs. Plants of the same species can vary dramatically in size and can be composed of many different parts that change with time. Many biological control agents attack only certain parts of plants, so thinking of plants as being composed of numerous modules (for example, leaves, roots, stems, flowers, seeds) can be helpful.

The population levels of plants can also be thought of as being determined by a balance between activation and inhibition. An example of this "activation–inhibition" hypothesis could be the balance between the occurrences of weed outbreaks caused by local disturbances that provide new, open habitats, such as humans building a road, with subsequent weed colonization in disturbed areas (activation) versus insect herbivory or plant competition reducing weed populations (inhibition). Of course, the goal with biological control is to shift this balance toward inhibition of weed populations.

How do we approach increasing weed inhibition, increasing the top-down forces, for weed control? The best-known general tactic is to work toward decreasing the growth rate of the target organisms, the weeds. As an example, we will use tansy ragwort, a poisonous pasture weed endemic to Europe that has been introduced to five continents and several islands. Tansy ragwort grows best if competition from other plants is reduced or absent and herbivores are absent. Tansy ragwort is therefore typical of the so-called r-selected, ruderal or "fugitive" species that are good colonizers and fast growers but cannot compete well with other plants. In northern California, the cinnabar moth, *Tyria jacobaeae*, was introduced from France in 1959 for control of tansy ragwort (Figure 13.2a). Caterpillars of this species fed on shoots and flowers during the summer but only

Figure 13.2 Two species released for control of tansy ragwort, *Jacobaea vulgaris*: (a) Larva of the cinnabar moth, *Tyria jacobaeae* (25 mm long). (b) The leaf beetle *Longitarsus jacobaeae* (adults 2–4 mm long) (photos by Noah Poritz).

partial control was achieved. In 1969, the leaf-feeding beetle *Longitarsus jacobaeae* was introduced from Italy to California, Oregon, and Washington in another attempt at control (Figure 13.2b). At sites in California and Oregon, this leaf beetle was extremely effective. Adults of *L. jacobaeae* feed on foliage and larvae feed on the roots. Field experiments were conducted with tansy ragwort caged with only the caterpillars, only the beetles, or both agents together and it was clear that instead of competing with each other, feeding by the beetles and caterpillars complemented each other. The root-feeding beetle larvae stressed tansy ragwort plants during winter and spring and, with sufficient summer defoliation by beetle adults plus cinnabar moth caterpillars, the ragwort plants were severely weakened. With ragwort under stress, the other plants in the area could outcompete ragwort and populations of this poisonous weed were significantly reduced. After both of these phytophagous insects became established, Oregon livestock losses because of tansy ragwort poisonings were reduced by 99 percent, saving producers more than US$5 million annually and herbicide applications were significantly reduced.

In keeping with Harris's stress theory, weed inhibition can be aided by increasing the overall stress impacting weeds so that the damage threshold (more damage than is acceptable) is exceeded. The action of plant-feeding individuals places a stress on plants but sometimes, unless other stresses also occur, plants do not die or growth and reproduction are not suppressed to acceptable levels. Such supplementary stresses can include climatic conditions, as found with the cinnabar moth attacking tansy ragwort in Canada. When this weed was growing in British Columbia, every summer the caterpillars of the cinnabar moth ate all of the leaves of the plants, but, after they finished feeding and pupated, plants still had enough storage reserves to rally and produce more leaves. In comparison, on well-drained sites in the colder Atlantic Canada area where cinnabar moth was released, plants were also defoliated. However, with the shorter growing season in this colder climate, the plants did not have enough time to regenerate leaves before colder weather set in. But they tried. The plants would be actively producing another set of leaves when cold weather began and in this condition they were more sensitive to cold and then died from the low winter temperatures. As a result, tansy ragwort virtually disappeared from cinnabar moth release sites in Atlantic Canada but was not as well controlled in British Columbia.

Owing to the inherent variability in biological systems, not all types of plant feeders are found attacking each plant species. However, in most cases, there are numerous species of natural enemies that feed on a weed species and the choice of which ones to pursue and introduce is not a trivial decision (see Chapter 18 regarding host specificity concerns; here we will emphasize efficacy). Biological and ecological information can help identify the "Achilles heel" in a weed life cycle in order to target the more sensitive growth stages or parts of the plant. However, it is important to release control agents that are not themselves attacked by local predators, parasitoids, and pathogens so that their populations can build and become as abundant as possible. For example, when the cinnabar moth was released in southern Australia to control tansy ragwort it did not become established because of parasites, diseases, and native predators, especially a species of predatory scorpionfly (*Harpobittacus nigriceps*).

A successful approach has been to study plant populations and understand which transitions among life stages are most sensitive to herbivory. Tansy ragwort is a long-lived plant and it seemed to Peter McEvoy that the transition from one-year-old plants to two-year-old plants was a sensitive stage, as these plants are undergoing the transition from being nonreproductive to being reproductive. This information added to the natural enemy species available pointed toward emphasizing introductions of the leaf beetle that stresses plant growth, especially in juvenile stages, instead of emphasizing the moth with caterpillars that feed on flowers. While the flower feeder can remove the potential production of many seeds, seeds are rarely a limiting part of the tansy ragwort life cycle so this natural enemy was considered to be of lesser importance. In addition, by understanding the ecology of this system, other limiting factors were identified that could then be used in tandem with releases of herbivores. Researchers suggested that tansy ragwort is a poor competitor but a good colonizer, so along with releasing herbivores, plant competition should be increased by identifying areas slated for control efforts and seeding them with nonweedy plants that are strong competitors. In addition, disturbance frequency and intensity in the weedy areas should be decreased so that new spaces for tansy ragwort to colonize were not created.

If multiple natural enemies are introduced, there is the potential for competition among herbivores both on the same plants and across plant populations. However, the most important factor in determining what types of natural enemies and how many species to introduce is what is necessary to increase overall damage to the weed. A study by J. Myers and P. Harris (1980) investigated two small, patterned-winged fruit flies (Tephritidae: *Urophora quadrifasciata* and *Urophora affinis*) attacking weedy knapweeds (*Centaurea diffusa* and *Centaurea maculosa*). Feeding by larvae of these flies causes galls to form in the seed heads, thereby greatly reducing seed production. When both species occurred together there was greater destruction of seeds than when either species occurred alone. While these flies coexisted and had a greater effect when combined, overall they did not control this weed. Models have suggested that better agents would be herbivores that feed on the basal rosettes of knapweeds, the persistent structure that must be killed so that plants do not grow back. However, practitioners in the biological control of weeds must always work within the bounds of the diversity of herbivores that have evolved to feed on that plant species and that can be manipulated. So, the ideal herbivore that a practitioner of biological control might have in mind does not always exist or perhaps is too difficult to rear and release. In addition, the host specificity of agents potentially being considered for release is a critical factor regarding whether a species can be used (see Chapter 18 for a more extensive discussion of this important aspect).

13.4.3 Weed Populations through Time

Weeds differ from insect pests because they occupy fixed locations for much longer than mobile insect or mite prey. In some ways, this situation is more similar to very sessile invertebrates such as scale insects or female whiteflies. The importance of one of the attributes considered critical for invertebrate predators and parasitoids, excellent searching ability, is less important with herbivores because weeds are relatively easier to locate; weeds are essentially sessile when compared with mobile pestiferous invertebrates. Interestingly, biological control programs against sessile invertebrates have been more successful, as has biological control of weeds (see Chapter 3). Of course, over longer periods of time, weeds do move when seeds are dispersed. The r-selected weed species often produce abundant seeds that can disperse and begin new populations, often in newly disturbed areas. Thus, there is generally a lag time before herbivores find newly

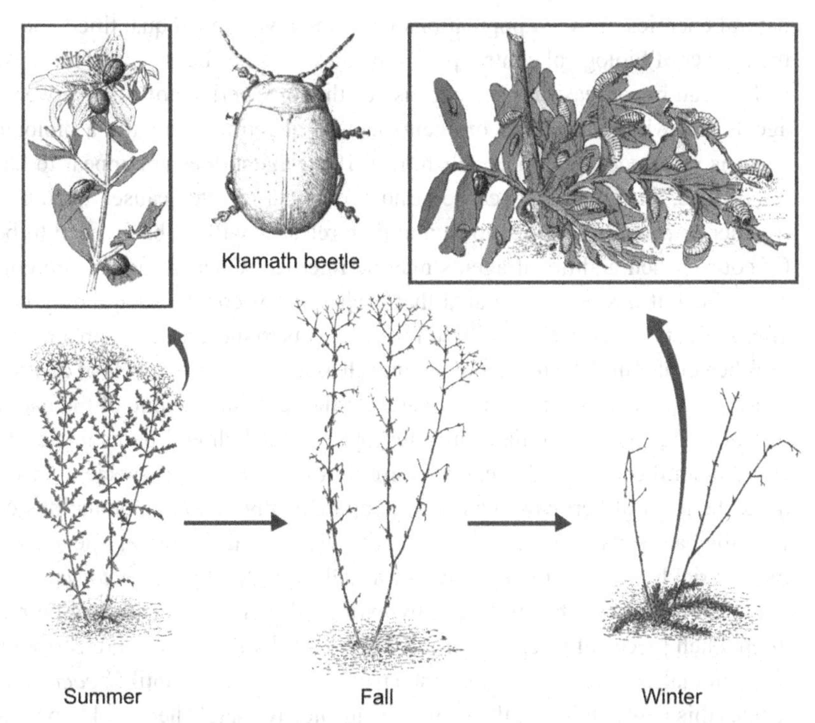

Figure 13.3 Leaf-feeding Klamathweed beetles, *Chrysolina quadrigemina*, which attack Klamath weed (common St. Johnswort), *Hypericum perforatum*, throughout the year (based on Holloway, 1957; drawings by John Langley Howard).

established populations of weeds that have created new metapopulations. An example of this is provided by a leaf-feeding beetle, *Chrysolina quadrigemina*, established for control of Klamath weed, *Hypericum perforatum* (Figure 13.3). After establishing and controlling this weed, populations of the beetle persisted by feeding on the few Klamath weed plants remaining that were growing in the shade, where the plants were suppressed but not killed. It is only when a new Klamath weed population became established in a sunny area that the beetle population once more increased to provide control of the weed growing in sunny areas.

13.5 Measuring Impact of Weed Biological Control

Evaluating the results of biological control projects is seldom straightforward but the challenges presented by weed control differ in some ways from those encountered with biological control of invertebrates and vertebrates. Usually, densities of plants in the native flora have not been measured prior to invasion of a weed. Although we often have no quantification of the plant density pre-invasion, it is still important to document the densities of weeds at the time of releases of biological control agents, as well as the effects of biological control programs after releases. Unfortunately, the effects of natural enemies on weed populations have not always been quantified and many times the impact of biological control programs on weeds has been judged more subjectively.

This can be, at least in part, because of the long periods of time between release of agents and when the effects of agents can be quantified. Initially a biological control agent is released to obtain establishment. If the agent does not appear to have become established, this could be because of numerous unpredicted causes, such as unseasonal weather or pesticide applications, and then releases will probably need to be repeated. Of course, then additional agents may be released. Once all agents are considered as established, it has been estimated that evaluation of control should only be conducted after at least 10 years to allow time for the agents to increase and spread.

When evaluating the impact of agents released, to quantify levels of control the weeds should be quantified and not the agents. At the least, at the time of the natural enemy's introduction, the area initially infested by the weed should be estimated with subsequent quantification of decreases in areas covered by weed species. Of course, over time, the level of herbivory can change dramatically as weed populations decrease, as with the caterpillars of *Cactoblastis cactorum* feeding on prickly pear cactus in Australia. Prickly pear cactus in an area can all be killed by *C. cactorum* so that the *C. cactorum* population becomes locally extinct. Prickly pears can then be reintroduced, often when pieces of prickly pear are carried by floods or seeds are carried by birds or other animals. Cactus populations then increase once again, until *C. cactorum* disperses, locates this new patch, and then controls this newly established prickly pear population. Thus, monitoring weed populations is never really finished, because weeds, once gone from an area, have great potential to reinvade, especially when disturbance creates new sites for weeds to colonize. In addition, once one invasive species has been controlled, it is possible that another invasive may take its place, a phenomenon that has been called the "invasive species treadmill." Therefore it is important to monitor the entire plant community, not just the weed species that was originally targeted, to determine whether the desired outcome has been achieved.

Further Reading

Coombs, E. M., Clark, J. K., Piper, G. L., & Cofrancesco, A. F., Jr. (eds.) (2004). *Biological Control of Invasive Plants in the United States*. Corvallis: Oregon State University Press.

Hinz, H. L., Schwarzländer, M., Gassmann, A., & Bourchier, R. S. (2014). Successes we may not have had: A retrospective analysis of selected weed biological control agents in the United States. *Invasive Plant Science and Management*, **7**, 565–579.

McEvoy, P. B., Grevstad, F. S., & Schooler, S. S. (2012). Insect invasions: Lessons from biological control of weeds. In *Insect Outbreaks Revisited*, ed. P. Barbosa, D. K. Letourneau, & A. A. Agrawal, pp. 395–428. Chichester, UK: Wiley-Blackwell.

Milbrath, L. R. & Nechols, J. R. (2014). Plant-mediated interactions: Considerations for agent selection in weed biological control programs. *Biological Control*, **72**, 80–90.

Myers, J. & Bazely, D. (2003). *Ecology and Control of Introduced Plants*. Cambridge: Cambridge University Press.

14 Phytophagous Invertebrates and Vertebrates

Among projects for the biological control of weeds, the vast majority would be classified as classical biological control and the majority of programs have used arthropods that eat plants.

14.1 Invertebrates

In nature there are many invertebrate species feeding on any plant species where it is native, and for biological control, choices are made regarding on which of these invertebrate species to concentrate. The invertebrates selected and released to control weeds are diverse, but insects have predominantly been used along with a few mites and nematodes (Table 14.1). In a few instances, snails, crayfish, and tadpole shrimp have also been tried against aquatic weeds. With such diversity in the invertebrates used, commonalities in their biologies are few. However, we can summarize by saying that different invertebrates can be found that utilize virtually all of the different parts of plants.

The majority of insects that are used for biological control bite off pieces of plants and chew them, as seen with weevils, leaf beetles, and caterpillars. Although the amount that each individual eats is quite small, populations of these species can increase to large numbers when weeds are abundant and control is then achieved by the synchrony of action of many individuals within the population. While some phytophagous insects used for biological control live externally on the foliage, larvae of others are often legless and live within protected locations in the plant tissues, for example, boring into stems or roots, and then dispersing as adults. Among those listed in Table 14.1, the groups most often used include beetles, caterpillars, different types of hemipterans (e.g., scale insects), and flies; the efficacy of organisms is important to whether they are used but the ability to ship and rear species is also critical (e.g., weevils can typically survive shipping for days with only a water source). Many of the most successful natural enemies are holometabolous, having immature stages capable of limited dispersal but metamorphosing into more mobile adults. In fact, holometabolous species comprise 87.7 percent of the successful invertebrate species released. For these groups, the immature stages can be extremely specialized in the areas they inhabit because they can later change into completely altered, highly mobile adults that will lay eggs in a new location. This is especially advantageous if, while these individuals were developing, they kill the host plant on which they fed and their offspring would then starve if the adult

Table 14.1 Diversity of invertebrates released for classical biological control of weeds.

Group	Total species	Number of species established (% of species)	Established releases[1]: Number of established releases	Established releases[1]: Number releases (%) providing control[2]
Insects (Insecta)				
Beetles (Coleoptera)	193	136 (70.5%)	463	266 (57.5%)
Caterpillars (Lepidoptera)	125	82 (65.6%)	192	81 (42.2%)
Bugs and scales (Hemiptera)	36	28 (77.8%)	129	87 (67.4%)
Flies (Diptera)	56	39 (69.6%)	106	31 (29.2%)
Thrips (Thysanoptera)	4	3 (75.0%)	11	4 (36.4%)
Sawflies, galling and seed-feeding wasps (Hymenoptera)	10	8 (80.0%)	13	6 (46.2%)
Grasshoppers (Orthoptera)	2	1 (50.0%)	4	0 (0.0%)
Mites (Acarina)	10	9 (90.0%)	22	6 (27.3%)
Nematodes (Nematoda)	2	1 (50.0%)	2	0 (0.0%)
Total	438	307 (70.1%)	942	481 (51.1%)

Source: data summarized by M. Schwarzländer, R. Winston and H. Hinz from Winston et al. (2014).

[1] Very frequently one species is released in numerous locations and results can differ by locations (i.e., 1,503 releases across all species; 62.7 percent of releases established).

[2] Established releases that resulted in variable, medium, or extensive control of the target weed.

could not disperse to find another appropriate plant. The basic biologies of some of the most commonly used groups are reviewed below.

14.1.1 Leaf Beetles (Order Coleoptera: Family Chrysomelidae)

Leaf beetle adults are generally less than 15 mm in length with bodies longer than wide and they can be rather robust. All beetles are holometabolous so that larval leaf beetles are very different in appearance from the adults. For some species the elongate larvae feed externally on the foliage and are able to hold on with their short legs. For other groups within this family, larval stages tunnel in stems and roots and these larvae usually have less morphological differentiation and reduced legs. Leaf beetles have chewing mouthparts as both adults and larvae, and both stages are phytophagous, although adults sometimes eat flowers as well as foliage. Therefore, although they are holometabolous and could specialize on very different types of food, adults and larval leaf beetles often feed on the same food. However, for some species, larvae and adults use different parts of the same plant. Having both adults and larvae feeding on the same plant could help to explain the effectiveness and environmental safety of these beetles for biological control. The flip side is that some species of this family of effective plant-feeding beetles can be serious pests of crops.

14.1.2 Weevils (Order Coleoptera: Family Curculionidae)

Adult weevils are often less than 15–20 mm in length, with the head narrowed into a protuberance called a rostrum that is characteristic of this family. Because of the appearance of the rostrum, these beetles are sometimes called snout beetles. For some species the rostrum is quite short and blunt (see Figure 18.2), while for others it is longer and narrower. The antennae are attached to the rostrum (see weevils in Boxes 13.1, 13.2, and 14.1). Weevils are usually dark or cryptically colored so that it is difficult to see them in nature. If disturbed, they frequently draw in their antennae and legs and fall to the ground remaining motionless and are then very difficult to see.

Box 14.1 Dealing with an Environmental Weed

Mile-a-minute weed, *Persicaria perfoliata*, is a vine from Asia that was first introduced in the east coast of the United States in the 1930s when its seeds hitchhiked along with holly (*Ilex*) seeds. It gained its name from its ability to grow 6 inches/day (15 cm/day; not really a mile-a-minute, but, still, the growth is quite fast for a plant!). This is sometimes referred to as an "environmental weed" which moves in, grows over, and often dominates natural areas and forest edges. This weed appeared to be so aggressive in part because of the lack of damage by any native herbivores. A biological control program was initiated in 1996 and a major goal for the program was to use highly specific agents for classical biological control. All stages of one weevil species from China, *Rhinoncomimus latipes*, attacked mile-a-minute vines, with larvae boring into stems and adults feeding on foliage. Adults of *R. latipes* barely fed on close relatives of mile-a-minute and when they were using these alternative plants for food, adults produced no eggs; larvae provided with close relatives of the weed as food did not survive. Therefore, this weevil was considered adequately host specific for release. However, one more step was necessary: would this weevil provide effective control? Caged trials suggested that the weevil significantly damaged plants and releases began in 2004, eight years after the program began.

Interestingly, mile-a-minute weed plants in the United States were shorter, with less biomass, compared with plants of this species in China, suggesting that these invasives could have changed during their decades in the United States and "evolved increased competitive ability," fitting this general hypothesis that had been proposed to explain success by some invasives. This attribute was considered as being positive because increased competitive ability is thought to often be accompanied by reduced investment in defenses. In agreement, studies showed that *R. latipes* caused more damage to the US mile-a-minute strain than to Chinese strains of this vine that were tested.

Biological control of weeds' programs require at least 10 years after the establishment of an agent before final results start to be possible so, as of 2017, results are still being evaluated and additional agents are being considered for release. However, it is clear that the weevil is spreading both on its own and with assistance and it is having an impact at some sites, with greater impacts in sunny sites and during droughts. A new development with this program occurred with recognition that as mile-a-minute plants declined, other invasive plants moved in. As a proactive solution, when weevils are released, native perennials and seeds of native plants are planted and a preemergent herbicide is applied so that as sites recover from being infested with mile-a-minute weeds, the treated areas will be filled with native plants and not invasives.

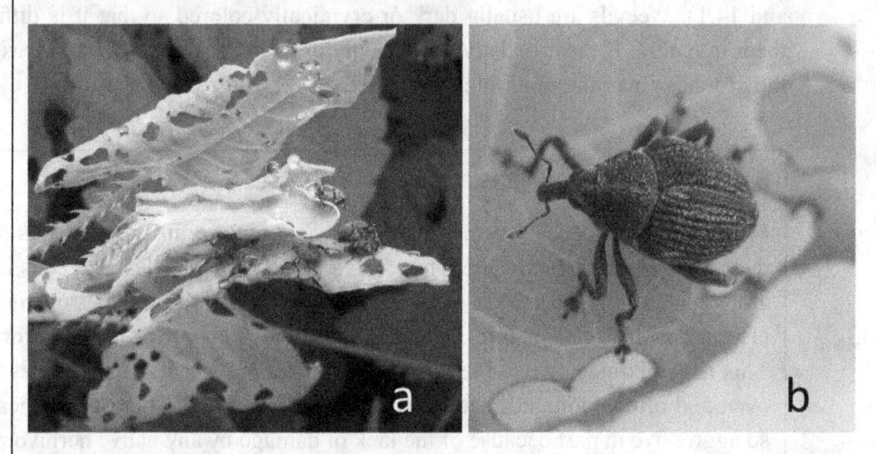

Figure Box 14.1.1 The weevil *Rhinoncomimus latipes* released for control of mile-a-minute weed along the Atlantic coast of the United States. (a) Damage caused by weevil feeding. (b) An adult *R. latipes* (about 1–2 mm long) (photos by Beatriz Moisset).

For plant-feeding weevils, both larvae and adults are phytophagous. The chewing mouthparts of the adult are somewhat hidden at the end of the rostrum and are often used to drill holes in fruits, nuts, and other plant parts. Larvae are C-shaped and whitish and almost always without legs, living in protected locations, chosen by the adult females when they lay eggs. Larvae often bore in roots and stems or fruiting structures.

14.1.3 Pyralid Caterpillars (Order Lepidoptera: Family Pyralidae)

Pyralids are small and rather delicate moths whose front wings are elongate or triangular. This is a large and ubiquitous family with species occupying both terrestrial and aquatic habitats. For many species, the larvae are smallish and usually live in shelters, either by webbing together leaves or shoots or by living within shoots, stems, or seed heads. For some species, larvae are gregarious, living in groups. Larvae have chewing mouthparts, although adults have drinking tubes, as do butterflies, and they therefore only imbibe water and nectar.

This family of moths includes among the most-used agents for biological control of weeds in part because it includes the cactus moth, *Cactoblastis cactorum*, which has been introduced successfully in many places. This is the moth that has been used extensively against prickly pear cactus (*Opuntia* spp.) in many different countries, with great success in Australia but with nontarget issues because of dispersal in North America where prickly pear cactus is native (see Section 18.2.4).

14.1.4 Scale Insects (Order Hemiptera: Suborder Sternorrhyncha: Superfamily Coccoidea)

Members of the hemipteran suborder Sternorrhyncha include scale insects, aphids, mealybugs, and whiteflies. These small insects all have a very specialized feeding strategy; they insert piercing-sucking mouthparts into plants to drink plant sap. Because the plant sap is low in essential nutrients, large quantities are imbibed and, after nutrients are removed within the hemipteran gut, the excess liquid is excreted. This excreted sugar solution, "honeydew," can frequently be found on plants near colonies of these insects. Ants that "tend" aphids, mealybugs, and scale insects feed on this copiously produced sugar solution and, in return, defend these herbivores.

Scale insects in the family Dactylopiidae have been used extensively for biological control of weeds. The life cycles of scale insects are exceedingly complex. Eggs are laid under the female covering or "shell" and these hatch to become mobile crawlers. The crawlers disperse and then settle down and begin feeding as nymphs, forming a shell to protect themselves. Although hemipterans are considered hemimetabolous, scale insects have an intermediate form of metamorphosis. Females, with few morphological features, gradually mature within their shells. Males have similar development to females through the settled nymph but after this males go through a pupa-like resting stage and then emerge as adults with wings. To many human observers, a colony of scale insects would look like only numerous small bumps on some part of a plant. As with the pyralids, the reason this group is among the most used is that species of *Dactylopius* have been used extensively for control of prickly pear cactus. Members of this scale genus are often called cochineal insects and all produce red pigment, historically leading to mass production of *Dactylopius coccus* as a source of carmine dye.

14.1.5 Less Frequently Used Groups

In addition, wood- and stem-boring beetles have occasionally been released as biological control agents as well as some types of true bugs, planthoppers, or other hemipterans that feed on seeds or plant fluids within leaves, petioles, and stems. Flies that have been used either live in mines in the small area between the top and bottom surfaces of leaves or form galls that disfigure plants while preventing abundant plant growth and reproduction. Some small wasps that have been used cause galls to form and wasp relatives, the caterpillar-like sawflies, have also been used to control weeds.

14.2 Successful Attributes of Invertebrate Herbivores

There are two characteristics particularly important when choosing insects to release for biological control of weeds: host specificity and the ability of the natural enemies to respond to weed populations in a density-dependent manner. Classical biological control has been the principal strategy for using invertebrates to control weeds; for this strategy to be successful it is often thought that natural enemies must increase in numbers when there are more weeds and decrease once weeds are rare (density dependence). It is considered optimal that after natural enemies decimate weed populations, they persist in the environment when weed populations are low. Therefore, if weed populations increase again, as, for example, after a disturbance, control agents are present and can respond by multiplying rapidly, resulting in subsequent suppression of the weed. Alternatively, if natural enemies do not persist but disperse readily or regularly, they can reinvade the area in case of rebound or reintroduction of the weed.

It is also optimal for phytophagous insects to be able to tolerate or even have a preference for creating and living in high densities as this characteristic is usually associated with more severe damage to host plants. For example, a leaf-mining buprestid beetle (*Taphrocerus schaefferi*) was considered a poor choice for control of yellow nutsedge (*Cyperus esculentus*) because larvae are cannibalistic within leaves, so population increase would of course be limited. In addition, larvae of this leaf miner were plagued by parasitoids that limited the ability of this species to increase in density and larvae principally caused damage when leaves were naturally senescing anyway.

Phytophagous species must be able to find new patches of weeds as the weed populations spread in order to be able to increase in response to increasing weed density. How do phytophagous insects find their host plants, especially when the plants are at low densities? Usually, this choice is made by the ovipositing adult female, which is important for holometabolous insects because immatures often have limited movement. Clearly, winged adult moths or flies can travel much further than their immature stages, caterpillars and maggots, respectively, so it makes sense that adults are the stage finding the host plants.

Dispersing phytophagous insects are known to utilize shapes and silhouettes over a long distance, such as trees on the horizon, for orientation. Research on shorter distance location of crucifers by phytophagous insects suggests that host plant selection by adults can be divided into three stages. First, host plant odors indicate to a dispersing insect that it is flying over appropriate host plants and landing is stimulated. Second, phytophagous insects use visual cues like color and shape to specifically land on plant surfaces and not on the soil. Third, using the chemical receptors on its legs and mouthparts, the adult samples the plant on which it has landed. The insect will lay eggs only after it has received sufficient stimuli; several landings on host plants can be required for reception of sufficient stimuli before eggs are laid.

Fascinating studies of chemical ecology have investigated the use of specialized plant chemicals by phytophagous natural enemies. It has been assumed that insect species with narrow host plant ranges, such as those that would be preferred for use in biological control, often rely on specific chemical cues for finding their host plants. Plant species produce a great diversity of different chemicals and many of these chemicals, sometimes called secondary plant compounds, are not required for growth and survival of the plant but are thought to have often evolved for plant protection. Individual phytophagous species responded to these compounds through time by evolving defenses to overcome the noxious secondary chemicals produced by certain plant species. In doing so, there is probably a cost because the phytophage must then detoxify or sequester these chemicals. However, adapting to a chemically defended species of plant has advantages because the phytophagous species then has fewer competitors utilizing the same food source. Another advantage for insects using these chemically defended plants is that the plant chemicals are species specific and can be used for accurate host plant location. Cruciferous plants, such as cabbage, mustard, and broccoli, all have glucosides that protect them against many insects but are used by specialized herbivores as phagostimulants (substances stimulating feeding) or as stimulants for oviposition.

Host plant choice is actually influenced by many factors with different priorities and perhaps no general rules are applicable for all species. Beside color and chemical clues, plant morphology and seasonal development can influence plant choice if the phytophagous invertebrates require only a specific stage in the plant life cycle such as the seeds or flowers. Environmental factors can influence whether a plant is chosen. Plants growing in sun or shade may be preferred, just as stressed, nonstressed, or vigorously growing plants may be preferred. Plants growing in dense populations can also attract more herbivores. While in an area packed with acceptable host plants, these phytophagous insects may also temporarily feed to some extent on less-preferred plants that unluckily happen to be in the same vicinity and this has been referred to as "spillover." Readiness of adults to oviposit can also influence plant choice; when a female holds many eggs that are mature and ready for oviposition, she may be less selective regarding which host plant to use for oviposition.

14.2.1 Host Specificity and Safety Testing

As with parasitoids and predators, some herbivores feed on only one or a few host plant species while others feed on numerous species. For biological control of weeds, understanding the host specificity of candidate organisms for release is of prime importance. Researchers must confirm that the phytophagous natural enemies being considered for release will not significantly impact the native flora, especially if rare, threatened, or endangered plant species are of concern. Of course, natural enemies are also always tested to make certain that they would not impact plants that provide food or fiber for humans. These issues will be discussed in more depth in Chapter 18, but, because host specificity is so important to biological control of weeds, here we will briefly discuss issues specific to phytophagous insects.

Methods based on natural enemy biology have been developed for testing the host specificity of herbivorous arthropods. Both acceptance of plants by adults for oviposition and suitability for development of immatures must be tested. Different stages of insects can react differently to a plant species. Adults are known to at times lay eggs on plants that are not optimal for development of immatures. Conversely, plants on which immatures can develop are not always accepted by adults for oviposition. For example, larvae of fruit flies (Tephritidae), longhorned beetles (Cerambycidae), and weevils (Curculionidae) show broader host acceptance than the adults, while for many moths and butterflies (Lepidoptera) the opposite is true and the egg-laying adult is more selective than the caterpillar. Determining host specificity using the developmental stage with the broadest host acceptance is therefore the safest approach for choosing biological control agents that will protect the environment. Methods used for testing different stages of a candidate biological control agent as well as for choosing which plant species to include in tests in addition to the target weed are discussed further in Chapter 18. Before any agent can be released there must be governmental approval, which varies by country and can be a slow process.

14.3 Strategies for Use of Phytophagous Invertebrates

14.3.1 Classical Biological Control

Some of the world's worst weeds have been successfully controlled using classical biological control, but generalizations about what makes for success are difficult because of the diversity of types of programs and the variety of ways in which results have been evaluated. There is no one correct way to evaluate classical biological control of weeds and many authors have evaluated programs from different angles or sometimes the oldest programs were not evaluated at all. Understandably, complex and often subjective results from classical biological control releases are difficult to group under the characteristic categories for success (see Chapter 3). The subjectivity in classification of results, for example, establishment, partial control, complete control, which come from many different sources and utilize different data for evaluation, leaves an unknown level of precision. In fact, reports of results are only as good as the data used for evaluation. Unfortunately, long-term collection of quantitative data after releasing natural enemies has been conducted too infrequently. Also, overall summarizations of success can be skewed by weeds for which control has been attempted with the same agents in country after country. Uses of the same agent repeatedly in different countries can either increase the numbers of successes (e.g., programs targeting prickly pear species, *Opuntia* spp., before 1950 in many countries) or can add to the numbers of unsuccessful programs (e.g., programs for control of the more recalcitrant *Lantana camara* in numerous countries between 1950 and 1970). For biological control of weeds, to analyze success properly, a number of years should be allowed after releases and before evaluation of establishment of a natural enemy. A period of time (up to 10 years after establishment) should be allowed for a natural enemy to have a chance to increase and demonstrate its contribution to control before a program is evaluated. As can be seen by the example of water hyacinth in Louisiana (Figure 14.1), two weevils and a pyralid were introduced over a span of 5 years. Although decreases in the area infested were precipitous in the first few years, this system required a total of 10 years before control seemed to have stabilized and evaluation could be realistic (see Figure 14.1).

Programs for the classical biological control of weeds can be evaluated in several ways. We will look at the success by weed species, use of different types of invertebrates, and use in different areas of the world.

Weed Species

In a 2014 summary of introductions, Winston et al. (2014) reported that approximately 224 weed species from 48 plant families had been the targets of classical biological

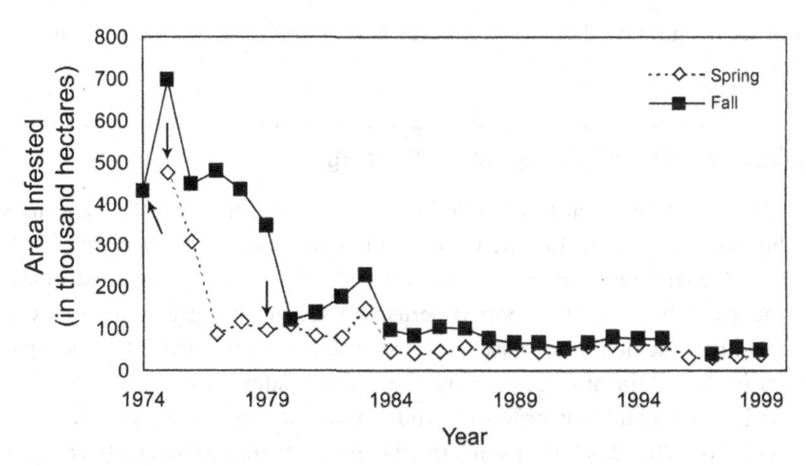

Figure 14.1 Reductions in area of Louisiana infested by the aquatic weed water hyacinth, *Eichhornia crassipes*, after introductions (see arrows) of two weevils (*Neochetina eichhorniae* in 1974 and *Neochetina bruchi* in 1975) and one pyralid (*Niphograpta albiguttalis* in 1979) (Center et al., 2002).

control projects using invertebrates. The majority of these projects had been aimed at perennial, terrestrial weeds. Weeds in the families Asteraceae and Cactaceae have frequently been targeted. There have also been some notable successes against aquatic weeds. However, there are numerous weeds that have not been targeted as often by biological control programs. A list naming 18 species as the world's worst weeds reported that 10 of these were grasses. Historically, weedy grasses have rarely been targets for classical biological control programs although this trend could be changing. With their simple architecture, grasses have fewer herbivores per grass species and weedy grasses also often have close relatives among the grasses grown as agricultural crops so attention must be paid to natural enemies that will not attack valued nontarget species.

Annual weeds of arable agriculture are also not often targeted for classical biological control. These weeds generally have a short lifespan and occur in environments unstable for establishment and persistence of natural enemies. In addition, biological control of weeds requires a period of time for establishment and increase of natural enemies before control is evident and normal cropping practices with annual crops do not allow for such a slow response.

Invertebrate Natural Enemies

Based on the 2014 summary, Coleoptera (beetles) is the order of insects most frequently utilized, with 57.5 percent of established releases providing some degree of control (Table 14.1). In comparison, the next most frequently used orders, moths and butterflies (Lepidoptera) and bugs, aphids, and scales (Hemiptera) produced successful control in 42.2 percent and 67.4 percent of established releases, respectively. Although this figure for Hemiptera appears to be high, it is based on 36 species released, while for beetles 193 species have been used. Among the beetles, the most successful families have been the leaf-feeding beetles and the weevils (for examples see Boxes 14.2 and 14.1, respectively).

Box 14.2 Flea Beetles Against Leafy Spurge

Many invasive species in North America first seen in the 1800s were most probably introduced from Europe via the ballast soil used to balance ships and this was the case with leafy spurge, *Euphorbia esula*. Leafy spurge was first found in Massachusetts but it was transported, perhaps mixed in with agricultural seeds, across the country. It settled in to cause the most problems in the plains that occupy the center of the continent in Canada and the United States. Leafy spurge is a deep-rooted herbaceous plant that grows 60–100 cm tall. It can densely populate large areas, with from 200–2,000 stems/m^2. In 1997 it was estimated that leafy spurge occupied 5 million acres (2 million ha). At this level of coverage, leafy spurge displaces native plants, changes native communities, and also has an impact on agriculture because it contains latex, making it lethal food for livestock.

The deep roots of this plant and its ability to reproduce both vegetatively and by seed help explain its great success as an invader. In addition, herbicide application is not cost effective over the large low-value areas occupied by leafy spurge. Also, using herbicides, which have to be repeatedly applied, does not provide a sustainable solution.

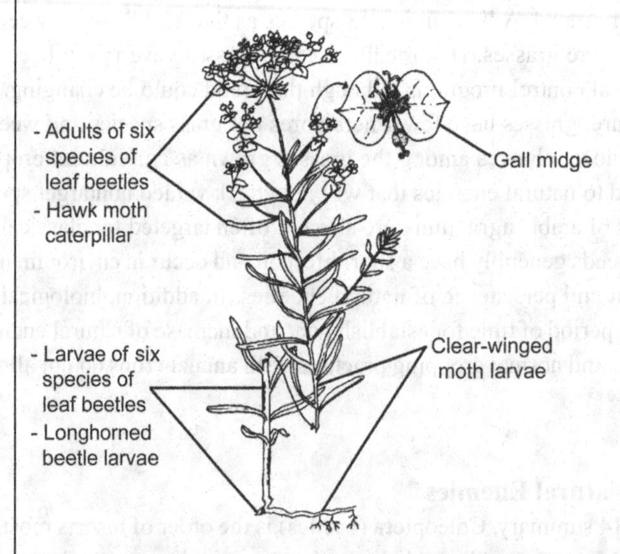

Figure Box 14.2.1 The diversity of natural enemies introduced against leafy spurge, *Euphorbia esula*, in North America. Agents released include leaf and flower feeders, stem and root miners, and a gall former (drawn by Rodney G. Lym).

Numerous natural enemies from Europe and Asia have been introduced against leafy spurge, beginning in 1965. By far the most effective have been several species of *Aphthona* flea beetles (a type of leaf beetle; Chrysomelidae). Adults of these beetles feed on foliage and have enlarged hind legs so that they spring away, reminding us of activity of insects from a different order, the fleas, when they are disturbed. However, the larval stage is the more important stage for controlling leafy spurge as *Aphthona* larvae tunnel through roots and, in doing so, structurally weaken plants as well as opening wounds for entry by plant pathogens in the soil. One of these species of flea beetle, *A. nigriscutis*, is more effective in areas with drier, sandy soil while another, *A. lacertosa*, is more effective in areas with moister soil. Once the beetles have been released in an area, beetle populations persist as successive generations attack the plants. Leafy spurge density can be greatly reduced although the plants may eventually reappear if segments of living roots remain in the soil. While leafy spurge will never be completely eradicated, the flea beetles have been able to reduce leafy spurge stem densities by up to 80–90 percent over large areas.

Leafy spurge is an aggressive colonizer of disturbed habitats. For example, the leafy spurge in an area might be under control, but after a flood, when much of the native vegetation has been killed, leafy spurge can quickly dominate again. The flea beetles being used for biological control don't disperse well on their own so it can take them a while to get to new patches of host plants and increase in numbers to control them. Another option that is available is to purchase *Aphthona* from private companies to release in order to speed up reestablishment of the beetles in disturbed areas.

In one summation, the single most successful individual natural enemy was the scale insect *Dactylopius ceylonicus* which, along with some of its close relatives, has been repeatedly used to control prickly pear cactus in numerous parts of the world. These scale insects increase in numbers, covering plants and sucking the phloem sap. Their feeding often leaves openings for disease organisms to invade the fleshy cactus pads, leading to rot and death of the infested parts of the plant.

Using these data, the biological attributes of phytophagous arthropods that are needed for success are not crystal clear. However, biological factors associated with establishment of agents seem to be a high rate of increase, long-lived adults, numerous generations per year, and low feeding rate associated with small sizes of individuals. Interestingly, in many cases, targeted weeds are uncommon in their native lands but, when found, phytophagous insect species associated with the weeds are also commonly found. As can be seen, success of a natural enemy cannot always be predicted according to the type of agent or the type of damage. This was certainly the case with a chalcid gall wasp (*Trichilogaster acaciaelongifoliae*) from Australia attacking an invasive leguminous tree in South Africa, *Acacia longifolia*, or golden wattle. Adult gall wasps were introduced, although they seemed sluggish and there seemed little chance for success based on historically poor success in weed biological control using wasps and the fact that gall formers were not often known for significantly aiding control. Surprisingly, within a few years, the acacia branches were weighed down with galls and the infestation of this gall wasp had stopped seed production.

Practically all types of phytophagous insects have been considered as biological control agents. While host specificity is the first concern, the most host-specific insects are not always used, in part because these species might be more challenging for handling, transport, and rearing. Bernd Blossey and Tamaru Hunt-Joshi compared the establishment and success of root feeders versus foliage feeders used for biological control of weeds. Perhaps because the activity of root feeders goes undetected without specific evaluation and because they are more difficult to rear and study, root feeders have been neglected for many years. There were only four releases of root feeders in biological control programs between 1902 and 1960 and, while this increased to 46 from 1960 through 1999, releases of aboveground feeders were almost 10 times as numerous. In the 100 years from about 1900 to 2000, approximately the same percentages of root feeders became established after release as aboveground herbivores. However, a large difference was found in the percentages of established species contributing to control, with 53.7 percent for root feeders and 33.6 percent for aboveground feeders. Perhaps this summarization will encourage more biological control programs to spend the extra effort necessary to work with cryptic (hidden) root feeders and reexamine emphasis on other types of invertebrates that are more difficult to work with. The question remains as to why root feeders seem to be more successful. One of the major reasons cited for lack of establishment and success of all phytophagous herbivores released for classical biological control is predation and parasitism on natural enemies that have been released. Certainly, within their underground homes, root feeders would be more protected from aboveground predators and parasitoids so perhaps this could at least in part explain why their survival is better. However, we know that living in the soil certainly does not always translate into complete protection from natural enemies but would instead result in exposure to soil-dwelling natural enemies, including insect pathogens.

By far the greatest successes in biological control of weeds using arthropods have occurred when using introduced agents against introduced and not native weeds. Only four species of native weeds have been successfully controlled using intentionally introduced exotic herbivores and all four of these weeds belong to the prickly pear cactus genus *Opuntia*.

Habitats

Early biological control of weeds programs were often targeting pasture situations, where often when the weed was suppressed, the desired outcome would occur: grasses for livestock consumption would reinvade the area. In recent years, programs have often targeted natural areas (also referred to as wildlands) where communities of plants are more complex. The goal is to reestablish a native community but often when an invasive is removed, other invasive weeds are quick to invade the area. In these circumstances, such eventualities must be understood and plans made accordingly, in order for native plants to replace the invasives (see Box 14.1).

International Cooperation

Biological control of weeds has been used around the world, with releases in 130 different countries (see Table 3.1). In fact, many plant species that are weeds in one country have also been introduced to numerous other countries where they have become troublesome. Cooperation among countries has in the past often been excellent (but see Section 20.3 for recent issues), with the result that once a natural enemy has proven successful in one country, it has been released for weed control in numerous additional countries. However, the door swings both ways and we often find that a weed that is difficult to control in one country can be similarly difficult to control elsewhere. Many programs have released agents against *Lantana camara* in different countries but with little success, except in some isolated islands. In these cases, the international cooperation leading to releasing similarly unsuccessful agents again and again in different countries has decreased the overall success rates of biological control of weeds.

14.3.2 Commercial Availability and Release

Classical biological control releases are conducted by governmental or academic organizations. The goal is to release each natural enemy so that it will spread and increase on its own. However, resources for rearing and releasing natural enemies are limited. Some natural enemies do not disperse quickly, or at least not quickly enough for land managers wanting to get rid of weeds. Therefore, in the United States small companies have been developed to provide the same phytophagous insects that have previously been released through classical biological control programs. As an example, one US company in the northwest markets 16 natural enemies for control of 8 weed species: beetles eating leaves or mining in roots and stems (12), a few caterpillars that defoliate plants (2), one fly creating galls in stems, and one fly creating galls in shoots.

In some cases, natural enemies may already be present in an area of concern but growers want to boost populations to improve or hasten control; we would then call this inoculative augmentation. Such augmentation would certainly be necessary if insecticides had been applied to the land, killing phytophagous natural enemies that had previously been present. There is also the possibility that although the weed in question had previously been eliminated, with disturbance at a location, that same weed species could recolonize and once more increase in abundance. In such a case, land managers might want to re-release natural enemies to speed control (Box 14.2).

14.3.3 Conservation/Integrating Controls

Conservation biological control strategies have not been used widely for weed control, yet interest in this type of control is growing. Conservation biological control of weeds using invertebrates has taken two major forms. Most programs are based on directly enhancing the efficacy of introduced natural enemies that are released. Of course, knowledge of the factors limiting populations of natural enemies is critical before being able to facilitate an increase in the herbivore population and for this reason conservation tactics are tailored to specific systems. One well-known method for improving the efficacy of some natural enemies is manipulating plant quality. Nitrogen fertilization of weeds seems counter-intuitive but this has improved control of both prickly pear (*Opuntia* spp.) and *Salvinia* (Box 13.1) by the herbivorous insects released against them.

The use of pesticides may interfere with the first priority in biological control introductions, establishment of control agents. In some cases, insecticides being used to control insect pests can disrupt weed control by herbivorous insects. In South Africa, a weevil, *Trichapion lativentre*, is able to control the leguminous weed *Sesbania punicea* but is very sensitive to insecticides applied in citrus orchards to control insect pests. Drift from citrus orchards can cause declines in weevil populations during the summer and the weed then can reach high densities up to several meters from orchards. Clearly, controlling insecticide applications would conserve populations of this weevil. On the other hand, insecticides have been used in South Africa to help conserve a natural enemy in a different system. The scale insect *Dactylopius opuntiae* was not providing control of prickly pear in specific areas where two species of lady beetles that feed on this scale were abundant. Insecticides were applied at low rates, which reduced predator populations but did not kill the scales, thus enabling control of the prickly pear by the scales.

The following example demonstrates an alternative approach when pesticides killing natural enemies seemed unavoidable. Researchers in Florida found that herbicide applications against water hyacinth, *Eichhornia crassipes*, were directly detrimental to biological control. The three established biological control agents all had immature sessile stages that feed internally within plants and thus were often killed when frequent herbicide applications killed plants. Therefore, a new species has been investigated with all stages feeding externally on plants and having mechanisms for dispersal following plant death or reductions in host quality. This small planthopper (*Megamelus scutellaris*) was released in 2010 and evaluations are still under way.

14.4 Vertebrates

Larger vertebrates, like goats and sheep, can be used for weed control but this is never the sole use of these animals and there are certainly drawbacks to this approach. These animals do not specifically feed on only one plant species and their activity is generally useless for controlling weeds unpalatable to livestock. Regardless, sheep and goats have been used to control tansy ragwort and blackberry (*Rubus* spp.), and geese have been used to control weeds in cotton. In contrast, fish are the one major group of vertebrates that has been used quite extensively for aquatic weed control.

Herbivorous fish also do not feed on only one plant species but usually graze on a diversity of aquatic plants. Eleven species of fish have been used to control aquatic weeds and algae, although by far the most commonly used species is the grass carp (*Ctenopharyngodon idella*). This species is native to the Amur River in Asia and prefers moving rather than still water. Grass carp, also called white amurs, can grow to up to 40 pounds (18 kg) and they live for over 10 years. They are omnivorous when very young but, after the 8–12-inch (20–30 cm) stage that is released for weed control, they are strict vegetarians. They eat a wide variety of plants but are somewhat selective, preferring softer, submerged plants, and they are not effective against floating plants. They prefer not to feed in areas with regular human activity, such as boat docks or swimming areas.

Grass carp have been released for weed control in at least 36 countries, including the United States.

In the United States, there is concern regarding whether this species will negatively affect native species of fish. Although no direct negative influence has been shown, grass carp can alter the aquatic habitat, potentially making it unacceptable for native fish species. Therefore, grass carp are released only in confined bodies of water, especially those not connecting with other bodies of water during floods. Release of grass carp always requires permits and in the United States most grass carp that are released have been sterilized so that they will not reproduce. To this end, grass carp are raised in fish hatcheries and their eggs are subjected to temperature or pressure shock so that an extra chromosome is present, making the eggs triploid instead of diploid. Triploid grass carp are sterile; thus, although the individuals released can live a long time, they will not be able to reproduce. As an extra precaution, sterilized grass carp are tested to insure that the sterilization treatment was effective and that only triploid fish are released.

Use of grass carp to control aquatic vegetation can be very effective but it can require several years before an effect is seen. The correct number of fish must be released so overgrazing does not occur. It is important that grass carp are released where the weeds that must be controlled are the species preferred by grass carp, as grass carp will not control algae or floating, surface plants. Thus, grass carp can offer an inexpensive long-term solution to difficult problems with aquatic weeds although they are certainly not the answer for all situations.

Further Reading

Blossey, B. & Hunt-Joshi, T. R. (2003). Belowground herbivory by insects: Influence on plants and aboveground herbivores. *Annual Review of Entomology*, **48**, 521–547.

Landcare Research (New Zealand) (2013). *The Biological Control of Weeds Book*. Retrieved from www.landcareresearch.co.nz/publications/books/biocontrol-of-weeds-book.

Myers, J. & Bazely, D. (2003). *Ecology and Control of Introduced Plants*. Cambridge: Cambridge University Press.

Schwarzländer, M., Hinz, H. L., Winston, R. L., & Day, M. D. (2018). Biological control of weeds: An analysis of all catalogued introductions, rates of establishment and estimates of success, worldwide. *BioControl* (in press).

15 Plant Pathogens for Controlling Weeds

Just as microorganisms can decimate populations of insect pests, some microbe species are very effective at growing and reproducing using plants as nutrient sources. Plant pathology is the study of microbes causing plant disease, with emphasis on how to control these microbes and manage their populations. However, some plant pathologists use their training to investigate the use of living microorganisms to control weedy plants. These plant pathologists are working to increase populations of plant-pathogenic microbes, quite the opposite from the activities of many plant pathologists.

15.1 Plant Pathogens and Target Weeds for Biological Control

Most major categories of microscopic organisms (viruses, bacteria, fungi, protists, and nematodes) have members that are plant pathogens. However, the microorganisms most commonly selected for control of weeds are the fungi, with less emphasis on viruses and bacteria. As we will discuss, use of microbes for biological control of weeds has focused both on classical and augmentative approaches and groups of fungi having very different attributes have been utilized for these very different biological control approaches.

As you have learned, herbivorous invertebrates (Chapters 13 and 14) use plants as a source of food and a habitat for reproduction. They are successful for biological control because they reduce plant biomass and density, but when these species are host specific, their populations suffer if they eat all of their preferred plants. One basic life strategy for plant pathogens, biotrophy, has some similarities with many invertebrates. Biotrophs are obligate plant parasites that must have living plants for growth and reproduction. They therefore usually do not kill their hosts outright but instead cause a constant energy drain on their hosts; they depend on the plant for their food and are in trouble if they kill all of their hosts all of the time. Therefore, biotrophic plant pathogens are usually used for classical biological control as they often have life strategies that allow them to persist in the environment and respond to increases in host populations. The second major life strategy is the necrotroph which instead kills the host or host tissues to derive nourishment. The majority of plant pathogens used for augmentative biological control of weeds are necrotrophs although short-term persistence with even these species works so that secondary cycling can occur (see below and Chapter 4).

Once a species of pathogen has been recognized as having good potential for development for weed control, researchers have learned that understanding the genetic variability in both the target weed and the pathogen species must be the next step. A weed species might have little genetic variability in its populations if, when it was introduced as an invasive species, few individuals were introduced or survived to reproduce. Alternatively, a native weed might have abundant genetic variability. The problem is that different strains of a weed (often called biotypes) can have varying levels of susceptibility or resistance to the same species or strain of a pathogen. Alternately, different strains of a pathogen (called pathotypes) can have very different levels of virulence against a weed biotype. Therefore, plant pathologists have found that initial research efforts need to be concentrated on the genetic diversity in both host and pathogen before proceeding. For the weed, this is determined by identifying the biotype or biotypes already present and their relative distributions and abundance so that potential pathogens can be tested against specific weed biotypes. For example, two different genotypes of Russian thistle, *Salsola tragus*, were introduced to California. This variability was recognized after biological control using pathogens began being investigated. A fungal pathogen that could be used via a classical biological control approach was *Colletotrichum salsolae*, which had been collected in Hungary. This Hungarian pathotype provided good control against one biotype of the thistle, with infections causing 60 percent reduction in plant weight, while only a 9 percent reduction occurred when the second biotype was challenged. Interestingly, this information led to a reevaluation of the taxonomy of this host genus, resulting in realizing that three distinct species of *Salsola* occurred in California, with this result based on both morphological and molecular methods.

For pathogens, one specific pathogen strain is used for a control program. However, numerous pathotypes that are biologically different, often from different geographic locations, usually exist. Today, a diversity of strains of a pathogen are collected for testing in order to identify the most appropriate pathotype (most aggressive while being adequately host specific) to develop for control of the targeted weed. Because different weed biotypes within a species can vary in resistance to a pathotype, it seems that weed species that are more genetically homogeneous in the area of concern are better targets for biological control.

15.2 Augmentation Biological Control

The main goal of an augmentative release strategy is to attack an entire weed population with an application of a pathogen. Similar application equipment can often be used for applying microbes as the equipment used for applying chemical pesticides. Pathogens for weed control are generally applied annually. Because of the similarities of these biopesticides with chemical herbicides, products are often called bioherbicides but as use of this strategy against weeds has often employed fungi, some products are called mycoherbicides (the prefix "myco" refers to fungi). Of course, these products differ significantly from synthetic chemical pesticides. In particular, only pathogens that are moderately host specific are developed and used to target a single species of weed or several closely related weed species, while chemical herbicides often affect many different species of plants. Also, bioherbicides must be handled to make sure that the organisms remain viable while such specialized handling is not necessary with chemical herbicides. The goal for applications of bioherbicides is to match infective stages of the pathogen with susceptible growth stages of hosts under optimal environmental conditions. For annual or perennial weeds, applications are most effective when made soon after plants start growing in spring when the weeds are still small and this timing is called postemergent application.

Plant pathologists have found that when planning their work, concentrating their efforts on weeds with specific characteristics can lead to more successful programs. Some weeds are able to overcome pathogen pressure if they are not killed by application of a bioherbicide, often by rapid growth or active regeneration after new damage by the pathogen has declined; these partially resistant weeds wait while disease is the worst and then can outgrow the damage. Troubles as a result of this recovery capacity can be augmented in plant species with several different types of reproduction. For example, a weed that produces seeds as well as growing from a tuber or a broken piece of the plant stem lying on the soil can be more difficult to control than a weed reproducing only from seeds. Therefore, based on several decades of experience, if possible, plant pathologists develop bioherbicides against weeds with characteristics associated with more successes (no or limited capacity to overcome disease and fewer means for reproduction) as well as choosing aggressive but specific pathotypes.

In biological control of weeds by plant pathogens, terminology for strategies takes a different twist. An "inundative" release means releasing lots of inoculum in order to start disease epidemics in the target weeds. However, plant pathologists working on these pathogens have found that particularly for pathogens attacking foliage, additional attributes are characteristic of successful systems. As mentioned, the pathogen must be highly aggressive; different species and strains of pathogens vary in their impacts on host plants (weeds in this case) and highly effective bioherbicides are made with pathogens that are "natural born killers," a term coined by Raghavan Charudattan (2010) to refer to pathogens that kill or cause near-lethal disease severity. Second, it seems that although pathogens are inundatively applied, the initial inoculum is generally not enough to cause symptoms over all targeted plants. Secondary cycling (the pathogen grows in the plant and then reproduces and disperses to infect another area on the plant or another plant) that is preferably quite rapid is necessary for the pathogen to increase and spread to more areas on individual weeds and to more weeds in the area. In fact, these plant pathogens are microbes with limited dispersal ability that are attacking stationary plants and secondary cycling provides a means for increase and spread by pathogens after release. However, it should be mentioned that in some cases pathogens get additional help with dispersal, as with the fungus *Colletotrichum gloeosporioides* f. sp. *aeschynomene* that is dispersed by rain, green treefrogs, and grasshoppers in the field. So, in successful biological control of weeds using plant pathogens as bioherbicides, augmentation or even use of the term inundative release usually refers to release of an inundative amount of inoculum followed by secondary cycling.

Most of the pathogens developed to date as bioherbicides have been fungi. Generally these are species native to the areas where they will be used so issues regarding nontarget effects do not occur or, if they do, they can be managed without any harm

Table 15.1 Examples of plant pathogens used successfully for augmentative biological control, registered for use in the respective countries.

Plant pathogen (product name)	Weedy species	Target sites	Country
Bacteria			
Lactobacillus spp.	Numerous clover species	Turf, lawns	Canada
Fungi[1]			
Chondrostereum purpureum	Stumps of broad-leaved trees	Tree plantations	Canada
Colletotrichum gloeosporioides f. sp. *hakeae*	Silky hakea (*Hakea sericea*)	Conservation areas	South Africa
Cylindrobasidium laeve	Black and golden wattles (*Acacia mearnsii* and *A. pycnantha*)	Natural areas, along water courses, and catchment areas	South Africa
Virus			
Tobacco mild green mosaic tobamovirus, strain U2	Tropical soda apple (*Solanum viarum*)	Pastures	USA

Source: based on Charudattan (2015).

[1] The genera *Chondrostereum* and *Cylindrobasidium* belong to the Basidiomycota; *Colletotrichum* (sexual stage: *Glomerella*) belongs to the Ascomycota.

occurring and therefore regulatory issues around registration are more focused on efficacy. These necrotrophic fungal pathogens are generally easy to mass-produce on inexpensive media. No matter where and when these pathogens are applied, it must be remembered that these are living organisms. Fungal spores, the active ingredients in many products, are generally sensitive to ultraviolet radiation and require free moisture (such as dew) to cause infection. Some products therefore have been improved to retain moisture so that the fungal inoculum remains alive and active.

The first microbial herbicides began being used in China in the 1960s, and in the 1980s, DeVine, using *Phytophthora palmivora*, was developed against stranglervine, *Morrenia odorata*, in Florida citrus orchards and Collego, using *Colletotrichum gloeosporioides*, was used against northern jointvetch, *Aeschynomene virginica*, in rice and soybeans in the southeastern United States. There was continuing interest in increasing the number of bioherbicides on the market. By 2007, over 200 plant pathogens, that included foliar pathogens, soil-borne fungal and bacterial pathogens, and deleterious rhizobacteria (bacteria colonizing roots), were reported as having been evaluated as potential bioherbicides. By 2015, 19 products had been developed based on 16 microbial species registered for use. Of the 19 products, six are still on the market in the United States and Canada and in South Africa, including two products that are made available when needed. Four of these are fungal, one is bacterial, and the other is a first product containing a plant virus (Table 15.1). These products are intended for very specific purposes. For example, *Chondrostereum purpureum*, which has been registered in Canada, is based on a wood-inhabiting fungus that is applied to cut stumps to prevent the stumps from resprouting (Box 15.1).

Box 15.1 Control of Unwanted Deciduous Trees

The fungus *Chondrostereum purpureum* has been developed as a mycoherbicide (named Chontrol) for augmentative use in forestry and in rights-of-way vegetation management, for example, around power lines, pipe lines, and roadways, to control growth of deciduous "weed trees and bushes," like red alders and aspens. It is also used in locations where chemical herbicide use is restricted, such as urban areas and community watersheds. The strain of this fungus that was developed as a product was isolated in British Columbia in 1989. Researchers spent time making sure that this product was effective and had good shelf life, and this mycoherbicide was approved for governmental registration in 2004. This was the first mycoherbicide registered for use in Canada.

Figure Box 15.1.1 The basidiomycete fungus *Chondrostereum purpureum*, contained in the product Chontrol, producing fruiting bodies on an infected branch. A suspension of this saprophytic fungus is applied to stumps of weedy trees to prevent resprouting. This fungus was initially used to control American black cherry (*Prunus serotina*) in Europe under the name BioChon and is now registered as Chontrol for use in Canada (photo by Clive Shirley).

The biological attributes of the fungal strain being used run counter to what would often be expected of a fungal pathogen for killing unwanted plants. Instead of being highly virulent, the isolate of this fungus being used is weakly pathogenic and primarily saprophytic. This works well in this situation as the fungus is not a virulent pathogen that will attack other plants. This product is applied as a paste of spores directly to preexisting wounds, often recently cut stumps. It takes advantage of the large open area to infect the cambium layer (the growing part of the tree), after which it colonizes the entire stump, decaying it. Eventually the fungus grows outside of the wood (see figure), where it produces short-lived spores that are airborne. When growing within the cut stump, *C. purpureum* kills what remains of the living tree, thus preventing the stump from producing sprouts to regrow.

The Canadian government does not consider that this fungus is environmentally risky because *C. purpureum* is ubiquitous in temperate climates in the northern and southern hemispheres and healthy trees are resistant to attack. As another bonus, trials have shown that Chontrol is just as effective as chemical herbicides in controlling regrowth of targeted deciduous trees.

Many of the pathogens that have been investigated were not found to be effective for control (in some cases the pathogen was not aggressive enough or secondary cycling was not effective) and some have been too difficult to mass-produce. It is a challenge for companies to manufacture a product with a long shelf life, especially with fungal pathogens of weeds. In contrast, the oomycete (a protist with biology similar to fungi) *P. palmivora* used in DeVine was a little too effective at persisting. End users did not always purchase this product on a yearly basis because this fungus has a stage that persists in undisturbed field sites long after application. Thus, the company that produced this fungus was not able to sell this product to the extent intended, because users did not buy it on a regular basis and the product is therefore no longer produced.

A new type of pathology has been investigated recently with a virus that attacks tropical soda apple, *Solanum viarum*, an invasive weed in pastures and rangelands in the southeastern United States, where it outcompetes grasses needed for livestock. This virus, tobacco mild green mosaic tobamovirus, usually causes mild symptoms in host plants but tropical soda apple plants over-react to infections with a systemic necrosis response that kills the plant. This virus has recently been registered as SolviNix. It must be applied to actively growing plants to be effective. Some type of wounding is necessary for the virus to infect and this is usually accomplished through high pressure spraying of plants that are about 30 cm tall. The very recent development of this new approach (a virus eliciting a hypersensitive systemic necrosis response) encourages researchers in this young field to keep their eyes open for new opportunities.

15.3 Classical Biological Control

Classical biological control introductions of exotic herbivorous arthropods have been successful for controlling myriad weed species. This same approach has also been used with fungal pathogens that attack weeds (Table 15.2). A major requirement for such programs when herbivorous arthropods are chosen as biological control agents is to make certain the agent is very host specific and this is also an overriding issue for pathogens to be introduced against weeds. Although there are numerous groups of fungi that are host specific, the group used most extensively for classical biological control has

Table 15.2 Successful classical biological control introductions of fungal pathogens for weed control.

Fungal pathogen	Target weed	Country of introduction
Basidiomycota		
Rusts		
Maravalia cryptostegiae	*Cryptostegia grandiflora* (rubber-vine weed)	Australia
Puccinia carduorum	Nodding (musk) thistle[1] (*Carduus nutans*)	Continental USA
Puccinia chondrillina	Rush skeletonweed (*Chondrilla juncea*)	Australia and continental USA
Phragmidium violaceum	Blackberry (*Rubus* spp.)	Chile
Uromycladium tepperianum	Port Jackson willow (*Acacia saligna*)	South Africa
Smuts		
Entyloma ageratinae	Hamakua pamakani (*Ageratina riparia*)	USA (Hawaii), New Zealand, and South Africa
Ascomycota		
Septoria passiflorae	Banana passionflower (*Passiflora tarminiana*)	USA (Hawaii)

Source: based on Charudattan (2005) and Kok (2001)

[1] Success was in combination with three species of herbivorous beetle that were also part of the classical biological control program against nodding thistle (Kok, 2001).

been the rust fungi (Figure 15.1). Rusts can be very damaging to their hosts and are also usually highly host specific.

Rust fungi are actually quite closely related to mushrooms based on fruiting structures and spore forms (both belonging to the Basidiomycota), but to the human eye they do not look like mushrooms at all. Much of their growth occurs within host plants. Rusts can have complicated life cycles with up to five different types of fruiting structures and five different types of spores that occur in a specific sequence. Although some rusts require two different species of hosts in their life cycle, rusts considered for biological control have only one host species; they are autoecious. Rust fungi are obligate pathogens in nature and only some have been grown in the laboratory on special media.

Rusts infect plants by entering through stomates, the microscopic openings in the plant epidermis used for gas exchange, as well as through intact leaf surfaces by forming infection structures called appressoria that help fungi punch through the leaf surface. The rust grows within the plant and infected plants subsequently develop rust-colored spots on leaves and stems, where the rust spores are produced. These areas, called pustules, can become black toward the end of the season when more persistent spores are produced. Other rusts growing inside hosts cause the host plants to form galls. In severe cases, infected plants become weakened and stunted and can even be killed. Rust spores are principally spread on the wind, an aspect that is very helpful in spreading these pathogens after they are introduced. The capacity of rusts to spread readily is especially important because rust inoculum is not easily mass-produced.

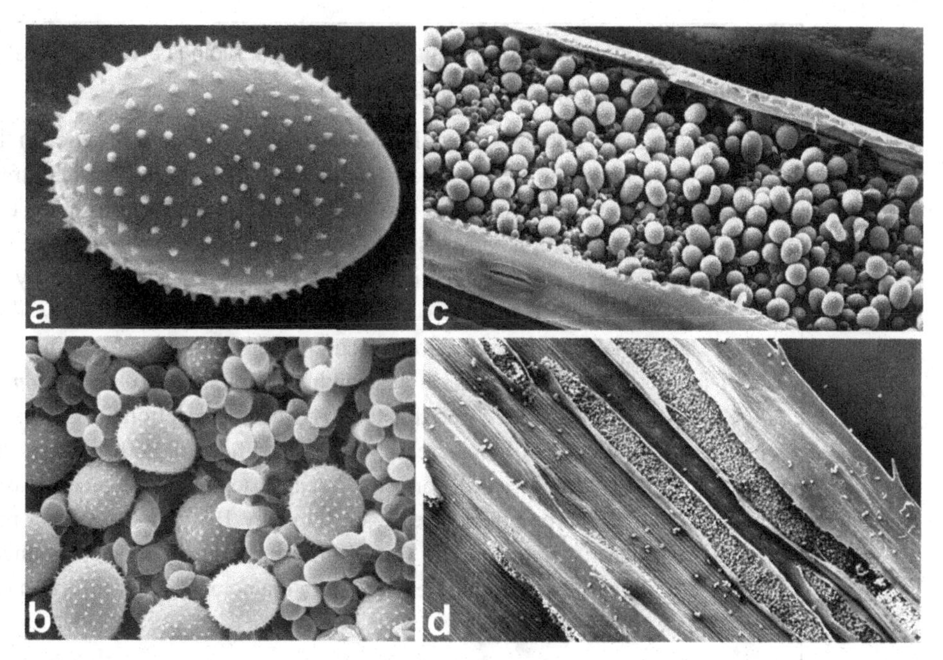

Figure 15.1 The rust fungus *Puccinia canaliculata* infecting the weed yellow nutsedge, *Cyperus esculentus*. (a) Urediniospore that will be windborne (approximately 21 × 16 micrometers). (b and c) Urediniospores being produced from a leaf of yellow nutsedge. (d) Severe infection with urediniospores being produced in eruptions on the leaf surface, parallel with the main axis of the leaf blade (Wetzstein & Phatak, 1987).

How are rusts for classical biological control found? Because the symptoms of rusts are often quite evident, they are relatively well known and described in the literature. This makes it easier during foreign exploration because scientists often know whether a rust species is associated with a certain weed. Microbes are sensitive to environmental conditions and they generally cannot seek shelter to protect themselves if conditions become harmful, as when free moisture is not available or temperatures are either too high or too low. Therefore, for classical biological programs, it is important to understand the environmental needs of the rust being considered for classical biological control and compare these needs to the environment in the release area. To release rusts, spores in water are sprayed onto plants or spores are dusted onto plants. Alternatively, harvested infected host plant parts or potted infected plants are placed in the field. After the rust becomes established in the weed population, it will spread on its own.

A plant pathogen used for classical biological control of weeds that became highly successful, the rust fungus *Uromycladium tepperianum*, was released in South Africa in 1987 (Box 15.2). Other pathogens are being developed for classical biological control of weeds, and this has resulted in several additional successes (Table 15.2). Most of these successes involve obligate pathogens that, without human intervention, increase on their own within the host population. Almost all of the successful programs have used rust and smut fungi.

Box 15.2 Introducing a Rust Fungus as a Classical Biological Control Agent to Control Introduced Acacias in South Africa

Acacia saligna, the Port Jackson willow, was introduced to South Africa in the nineteenth century for use in tanning and to stabilize sandy soils. These small trees grew well in the Cape Provinces, in fact too well, creating large, dense stands that were problematic in agricultural areas, as well as outcompeting the native flora in areas of sensitive native vegetation. The native biota in the invaded areas primarily includes low-growing plants with few trees and these invaders readily overgrew the native plants. These small acacia trees were difficult to control chemically or mechanically. A rust fungus that caused galls on acacias in Australasia, *Uromycladium tepperianum*, was introduced in 1987 and this fungus became exceptionally successful. The numbers of galls per tree often increase, with a maximum of 2,000 galls per tree, and the rust infections ultimately kill some trees. The acacia populations decreased quite quickly as the rust spread from release sites and, by 1997, an 80 percent reduction in *A. saligna* density was reported. Over time, researchers realized that they found galls on 100 percent of trees with ≥ 5 cm stem diameter but less infection occurred when trees were smaller.

Figure Box 15.2.1 (a) A *Uromycladium tepperianum* gall on Port Jackson willow, *Acacia saligna*. (b) *Acacia saligna* with heavily galled branches (photos by Raghavan Charudattan).

Researchers followed populations over 20 years and the infection prevalence in Port Jackson willows decreased over time but, in large part, that was because the numbers of these willows decreased. One wrinkle in this story is that *A. saligna* trees produce abundant seeds that create a seed bank in the soil and the seeds are stimulated to germinate after fires. After the rust was well established, *A. saligna* would decline until after a fire had occurred, when massive batches of new seedlings would start growing. Thankfully, the rust responds to the increased density of hosts and as the seedlings become larger, the rust will attack these too.

Since the first release of a plant pathogen for classical biological control of a weed occurred as recently as 1971, use of pathogens is a relatively new approach for classical biological control of weeds, considering that the first release of an arthropod to control a weed occurred in 1795. Interestingly, this later development of plant pathogens for classical biological control of weeds is similar to the lag in use of pathogens for controlling arthropods, when compared with the earlier and greater efforts in use of arthropod natural enemies for controlling arthropods.

Further Reading

Bailey, K. J. (2014). The bioherbicide approach to weed control using plant pathogens. In *Integrated Pest Management: Current Concepts and Ecological Perspectives*, ed. D. P. Abrol, pp. 245–266. San Diego, CA: Academic Press-Elsevier.

Charudattan, R. (2005). Ecological, practical and political inputs into selection of weed targets: What makes a good biological control agent? *Biological Control*, **35**, 183–196.

Charudattan, R. (2015). Weed control with microbial bioherbicides. In *Weed Science for Sustainable Agriculture, Environment and Biodiversity*, ed. A. N. Rao & N. T. Yaduraju, vol. 1, pp. 79–96. Hyderabad, India: Proceedings of the Plenary and Lead Papers of the 25th Asian-Pacific Weed Science Society Conference.

Yandoc-Ables, C. B., Rosskopf, E. N., & Charudattan, R. (2007). Plant pathogens at work: Progress and possibilities for weed biocontrol classical versus bioherbicidal approaches. *Plant Health Progress*, online. DOI: 10.1094/PHP-2007-0822-01-RV.

Part V

Biological Control of Plant Pathogens and Plant Parasitic Nematodes

16 Biology and Ecology of Microorganisms for Control of Plant Diseases

16.1 Types of Plant Pathogens and Their Antagonists

The term disease, in plant pathology, is defined as the malfunctioning of plant cells and tissues, resulting from continuous irritation and leading to development of symptoms such as reduced growth or mortality. Plant diseases can be caused by several types of microorganism, but the most frequently encountered and diverse are fungi, bacteria, and viruses. However, oomycetes (earlier regarded as fungi), phytoplasmas, and viroids can also cause diseases in plants. Based on common treatments of the subject, we will also consider plant parasitic nematodes in the chapters on biological control of plant pathogens.

Plant pathogens vary in their relations with host plants with interactions ranging from parasites that require living hosts to saprophytes that feed on dead organic matter. Thus, some plant pathogens are biotrophs that feed on living cells and can only grow and multiply in nature when living on or in living organisms. These include viruses, powdery and downy mildews, rust fungi, and some bacteria, among others. Other plant pathogens are necrotrophs that kill plant tissues before ingesting their contents. Some plant pathogens, including many fungi and bacteria, can live on dead as well as living plant material and these are therefore facultative pathogens, while others can only live as pathogens and are therefore obligate.

Both fungal and bacterial plant pathogens exist in the environment inside or outside of plants or in or on dead plant material. Many bacterial and fungal plant pathogens have active epiphytic (occurring on the plant surface without causing infection) and saprophytic phases in their life cycles. Some species have special persistent stages that may persist for many years. Dispersal of bacteria and fungi is by wind and rain and sometimes they are vectored by higher organisms. In contrast, plant-pathogenic viruses develop within plant cells and often depend on other types of organisms, frequently insects, for dispersal or they are transferred from plant to plant when plant roots grow together. Plant parasitic nematodes are soil-dwelling, generally attacking plant roots, and they can disperse on plant parts or can be vectored by insects.

Plant pathogens attack plant parts above the ground (leaves, stems, flowers, and fruits), at the soil surface (lower parts of stems), as well as in the soil (roots) and are said to have infected a plant only after microorganisms have entered the host plant's tissues. There are also many plant-damaging nematodes that are free-living in the soil or within stems or roots. Some nematode species live within specialized structures on roots that offer protection and common species are appropriately named cyst nematodes or root-knot nematodes. Many studies on biological control of plant pathogens have focused on diseases caused by pathogens in the soil. Soil-borne plant pathogens are especially well known for causing serious problems with germinating seeds and young, tender plants. For all plant parts, both aerial and within the soil, wounds provide an easy site for entry of pathogens into plant tissues. Thus, tissues that are damaged during growth, manipulation, or harvest are particularly susceptible to infection. Several major groups of plant pathogens, including viruses and phytoplasmas, are almost entirely dependent on wounds or vectors to gain entry to hosts. Another group of plant pathogens dependent on wounds are those fungi and bacteria causing post-harvest diseases of fruit. After plant products are harvested, plant pathogens can cause serious damage during storage. Of particular concern are those saprophytic bacteria and fungi that are experts at rapidly colonizing organic matter, such as stored crops, to use as a source of nutrients.

Plant pathologists commonly call those microorganisms that suppress plant disease "biological control agents" (which we will refer to as BCAs, as is common usage), or "antagonists," rather than "natural enemies." The term antagonist is appropriate for this field because these organisms frequently do not act to kill the plant pathogens themselves but instead have an impact due to a variety of mechanisms. Many BCAs belong to the same major taxonomic groups as many plant pathogens, the fungi and bacteria, and occupy the same niches as the pathogens. Yet, they do not cause plant disease but instead can suppress disease in several different ways. Disease suppression in many cases is caused by interactions between BCAs and plant-pathogenic microorganisms. Some of these BCAs have developed life strategies of either feeding on or outcompeting the primary pathogens. One specific strategy of BCAs is parasitism, frequently referred to as hyperparasitism because these antagonists are parasites of plant parasites. A second specific strategy used by BCAs is competition. Antagonists competing with plant pathogens are often specialized in their ability to exploit defined niches. Life histories of antagonists that impede the development of other microorganisms can be amazing in their complexity and diversity. Suppression of plant diseases can also be caused by induced resistance responses in the plant that are initiated by microorganisms that either can or cannot cause diseases. As one example, this effect can be caused by viruses. Plants inoculated with mild forms of viruses can develop resistance against further viral disease.

16.2 Ecology of Macroorganisms Versus Microorganisms

We have discussed the ecologies of natural enemies and pests that are macroorganisms, as well as interactions between macroorganismal pests (e.g., animals and weeds) and microscopic natural enemies. To discuss the basis for biological control of plant pathogens, we now switch to interactions between pests and their antagonists when both are microorganisms.

Studies of the interactions of plants and animals with their environment have dominated the general field of ecology. However, as early as 1934, George Gause published results from population dynamics studies using two species of ciliate protistans, during which he found cyclic oscillations characteristic of predator–prey interactions. The use of microorganisms to investigate general ecological theories has once again become an active area of research in more recent years.

The types of studies that have been and can readily be conducted with macroorganisms and microorganisms differ because of scale. Studies of the ecology of macroorganisms are often conducted as investigations of associations using correlations to derive mechanistic explanations for observed patterns. This correlative approach is because of the complexity of larger organisms and their interactions with their surroundings, and, perhaps most importantly, the difficulty in conducting realistic experiments with them. In contrast, when studying the ecology of microorganisms, experimentation is used more extensively and frequently fairly realistic model systems constructed in the laboratory (in microcosms) or greenhouses (in mesocosms) are used to investigate interactions.

Ecological studies of microorganisms have other challenges compared with working with macroorganisms, which to some extent can be related to the scale at which we must work. Individual lady beetles or parasitoids can be observed as they attack individual pests. It is often straightforward to count parasitoids and their host caterpillars and know exactly how many individuals are present at certain times to evaluate changes in density, and in addition perform observational studies to record their interactions. Studying interactions among microorganisms differs because of the size of individual microorganisms compared with macroorganisms. In fact, plant-pathogenic microorganisms are seldom considered as individuals and are usually considered as groups; for example, the individual bacterial cells infecting a plant are considered together as one population or colony. Interactions among colonies of microorganisms are not studied on the basis of individual cell by individual cell but rather population versus population. In some cases, for a plant pathogen to damage a plant or for an antagonist to deter a plant pathogen, an effect can be initiated by only one cell present in an auspicious location where it can readily reproduce. Alternatively, for some diseases it seems that many cells (comparable to a high dose) are necessary to initiate an infection successfully. However, plant disease does not develop because of one or a few microbial cells being present, but only after those first cells infect. Many pathogen cells are generally required before disease symptoms on the host plants are visible to the human eye. Likewise, effects caused by antagonists are as a result of the populations of cells. This group-based approach for infection by pathogens or protection by antagonists is supported by the fact that groups of microorganisms growing in the same microhabitat are often thought to be genetically almost identical, or even clonal. For example, the fungal cells at one lesion on a plant are often genetically identical.

16.3 Ecology of Plant Pathogens and Their Antagonists

Antagonists are extremely diverse in their activity, making it difficult to summarize their relations with their respective environments. A "place to thrive" is critical for survival and growth of all pathogens and antagonists. In fact, all organisms occur in a niche, a concept including whatever that organism needs to live. Niches are theoretical constructs determined by the physiological properties of the organism, its environment, and the resources needed by that organism. The ideal conditions for an organism comprise its fundamental niche, while the conditions actually present are called the realized niche.

For microorganisms needing a home, the vast majority of an aerial plant surface can be a hostile environment, dry, and often with few nutrients and with UV exposure. The roots of a plant provide different challenges because of the diverse community of competing organisms in the soil. Thus, at any location on a plant, microhabitats where moisture and nutrients are present, for example, as a result of leakage from a wound or natural opening of the plant like a stomate, would present a valuable habitat for many microorganisms.

16.3.1 Life History Strategies of Biological Control Agents

Biological control of plant pathogens effected through competition is based on limiting the realized niches of pathogens, due to the activity of other microorganisms. Among both plant pathogens and antagonists, some have high reproductive capacity and can quickly occupy a new resource (r-selected species). These organisms are characteristic of disturbed sites where they can arrive and occupy a site quickly before other microorganisms arrive. At the other end of this spectrum are organisms that are characteristic of stable situations (K-selected species). These species are good at persisting and include both highly successful competitors and microorganisms that are stress tolerant. For example, while the opportunistic foliar fungal pathogen *Botrytis cinerea* (causing grey mold disease) colonizes new locations quickly, it is a poor competitor when compared with some other fungal species, for example, from the genus *Penicillium*, which produce secondary metabolites inhibiting potential competitors over a longer period of time.

Plant pathogens occurring along the continuum from rapid colonizers (r-selected species) to microorganisms successful at persisting over time include species that attack young or weakened plants. For example, species from the genera *Botrytis*, *Pythium*, and *Rhizoctonia* attack young and weakened plants, while they often are not good at attacking healthy, well-established plants. Also, these species are often not effective at competing with other microorganisms. In contrast, stress-tolerant species of plant pathogens grow slowly in marginal habitats where it is difficult to live. For example, rust fungi in part grow externally in the dry, nutrient-poor environment of leaves where little competition occurs. Good competitors such as the K-selected wood-rotting fungi (e.g., *Armillaria mellea*) are often not quick at colonizing new sites but, once present, grow slowly and persist, while defending against secondary invaders.

The r-selected species of microorganisms have proven to be successful agents for biological control in numerous instances. Although they are generally poor competitors, they are adapted to disturbed situations that they colonize quickly. These species are also the easiest to mass-produce in the laboratory because growth requirements are often not restrictive. Owing to their ability to colonize new habitats readily, antagonists that are r-selected are often used protectively, with application occurring before infection by a plant pathogen might occur. After infection by a plant pathogen, biological control agents that are more competitive and those that are more persistent (K-selected species) are often more appropriate for control. In practice though, most biological control agents are not used after a disease is well established but are more effective when used earlier, either before infection occurs or early in the course of an infection. Using BCAs against plant pathogens therefore often relies on the ability of the biological control agent to reproduce and occupy a site. For this reason, the strategy resembles what is called "inoculation biological control" for biological control of insects and weeds (see Chapter 4), because the offspring of the natural enemy that is released have an effect but permanent establishment is not assumed (as would be indicative of classical biological control).

The concept of the r- versus K-selected species is often linked with concepts of succession, with r-selected species being considered as early colonists and K-selected species as the climax species. Three species of antagonists were found attacking the cereal cyst nematode (*Heterodera avenae*) in suppressive soils. The obligate bacterial pathogen *Pasteuria penetrans* infected second-stage juveniles, the fungus *Nematophthora gynophila* infected developing females, and the facultative fungal pathogen *Pochonia chlamydosporia* infected eggs. These antagonists were thus affecting different stages in the nematode life cycle and have been thought of as specializing at different stages in succession, as the nematodes developed. Knowledge of this specialization would clearly be essential when considering use of these antagonists for control purposes.

16.3.2 Abiotic and Biotic Conditions

The ecology of a plant pathogen or plant-pathogenic nematode must be understood to develop a successful biological control strategy. The conditions that are optimal for growth of a pathogen attacking a plant should be similar to those that are optimal for its antagonists and the best antagonists can function well under the complete range of conditions favoring a pathogen. Otherwise, if conditions occur that only favor the pathogen, the antagonist would lose advantages it might have and disease would develop.

Antagonists of plant pathogens are strongly influenced by both abiotic and biotic conditions, as with previous microbes that have been discussed (Chapters 9, 10, 11, 12, and 15). For all antagonists applied to the soil or plant surfaces, appropriate temperature and moisture levels are critical for survival and activity. Additional abiotic conditions that are important for optimal activity of antagonists in the soil include the soil type, soil pH, and nutritional status of the soil. Also, presence of pollutants in the soil, such as heavy metals, can threaten activity of antagonists.

Antagonists can also be impacted by biotic factors such as the community of microorganisms residing in the area of application. These can impact the survival and activity of antagonists, for example, secondary metabolites from some common soil fungi can reduce the activity of *Pseudomonas fluorescens* against root-knot nematodes on tomatoes. On the other hand, presence of some other soil fungi increased production of antimicrobials by *P. fluorescens*. Biotic communities together can act to control plant-pathogenic microorganisms and will be discussed in more detail as disease suppressive soils (see Section 17.4.1). In addition, the species of plant that is being attacked by the pathogen or nematode can strongly influence antagonists. In soil systems, different plants can leak nutrients to different extents which can alter the composition and inhabitants of the rhizosphere, the soil around a living root.

16.4 Studying Plant Pathogens and Biological Control Agents

Molecular biology is a critically important tool for biologists in recent years but especially for microbial ecologists. For example, molecular techniques provide marker genes (also called molecular markers) that allow microbiologists to follow the populations of a specific microbial strain, for example, in the soil in a growth chamber, or on plant surfaces in natural environments (e.g., leaves or roots). Using molecular markers, plant pathologists have developed a far better understanding of the population dynamics of specific plant pathogens and interactions of plant pathogens with naturally occurring or introduced antagonists in soil, within plants, or on plant surfaces, where biological control occurs. Some molecular markers, such as the gene encoding the green fluorescent protein, "gfp" (obtained from a jellyfish), allow microbiologists to monitor individual microbial cells on plant surfaces using fluorescence microscopy. These markers are becoming increasingly important in evaluating the ecology of microorganisms in natural habitats. It should be mentioned that to study ecology, releases of genetically modified organisms require specific approval, which in some countries can be difficult to obtain.

Using molecular biology, plant pathologists have also been able to develop mutants of antagonists that differ from the parental strain in only very specific ways. Such mutants have become extremely important tools for identifying the mechanisms used by antagonists to suppress plant disease. For example, some antagonists produce antibiotics that are toxic to their target pathogens, inhibiting pathogen growth on culture media in a laboratory. Because microorganisms need sufficient nutrition to produce antibiotics, however, there were good reasons to question whether antibiotics were produced by antagonists in natural habitats in the quantities needed to suppress target pathogens. Using molecular biology, plant pathologists created mutants of antagonists that are deficient only in antibiotic production. If an antibiotic-producing antagonist suppresses plant disease, while an antibiotic-deficient mutant does not suppress disease, then plant pathologists can conclude that antibiotic production is an important determinant of biological control in this pathogen/antagonist system.

Many microorganisms cannot be cultured and this fact has severely hampered scientists' abilities to study communities of microorganisms (also called microbiomes) in different natural environments. More recently, molecular methods have been available to do just this. Now, the diversity and numbers of microorganisms present in the soil or on leaves of healthy versus infected plants can be evaluated to help study these communities, with a goal of understanding better the microbial interactions leading to plant disease resistance.

We should, however, mention that molecular methods should be conducted along with methods of study that have been available for much longer: studying morphology, growth characteristics, and interactions with hosts and, of course, conducting experiments. Diverse methods are needed to characterize a microorganism and its activity fully.

16.5 Modes of Antagonism among Microorganisms

Antagonists interact with plant pathogens in numerous ways. In some instances interactions are directly caused by one microorganism suppressing the activity of another. These interactions are based on either (1) antagonists producing antimicrobials to kill competitors (= antibiosis), (2) antagonists directly destroying pathogens via parasitization, or (3) antagonists excluding a pathogen through competition for space and/or nutrients. Interactions can also be indirect, as when the presence of one microbe induces resistance in the host plant toward a different strain of that same species or a different plant pathogen altogether. Indirect effects are also possible when manipulation of the environment alters the microhabitat so that the species composition and abundance of the community of microorganisms changes, with the end result of decreased densities of plant pathogens (see Chapter 17). We will discuss each of the direct and indirect mechanisms of interaction used by biological control agents separately, although it is thought that in nature several mechanisms are often used by the same species.

16.5.1 Antibiosis

Antibiosis is one of the major mechanisms underlying biological control of plant pathogens. The term antibiosis has its roots in the term antibiotics, which refers to organic substances produced by microorganisms that, even at low concentrations, are deleterious to the growth and metabolic activity of other microorganisms. Antibiosis therefore refers to the inhibition of one microorganism by an antibiotic produced by another microorganism. The result of antibiosis is often death of microbial cells by breakdown of the cell cytoplasm.

One advantage of using antibiosis for control is that antibiotics produced by an antagonist can diffuse in films of water or through moist soil so that actual physical contact between the antagonist and the pathogen does not have to occur. This is perhaps best seen when cultures of an antagonist grow on an agar plate near a susceptible microorganism and an unoccupied zone of inhibition surrounds the colony of the antagonist: this zone is caused by diffusion of an antibiotic produced by the antagonist into the surrounding agar medium (Figure 16.1). There is compelling evidence that certain antagonists produce antimicrobials when they are colonizing the rhizosphere, seed surfaces, or plant wounds, and that the concentrations of antibiotics produced are adequate to suppress targeted plant pathogens. One example of antibiosis being harnessed very effectively to control a pathogen is biological control of the disease called crown gall which is caused by the bacterial pathogen *Agrobacterium tumefaciens* (Box 16.1). Another example is the fungus *Penicillium oxalicum* which produces secondary

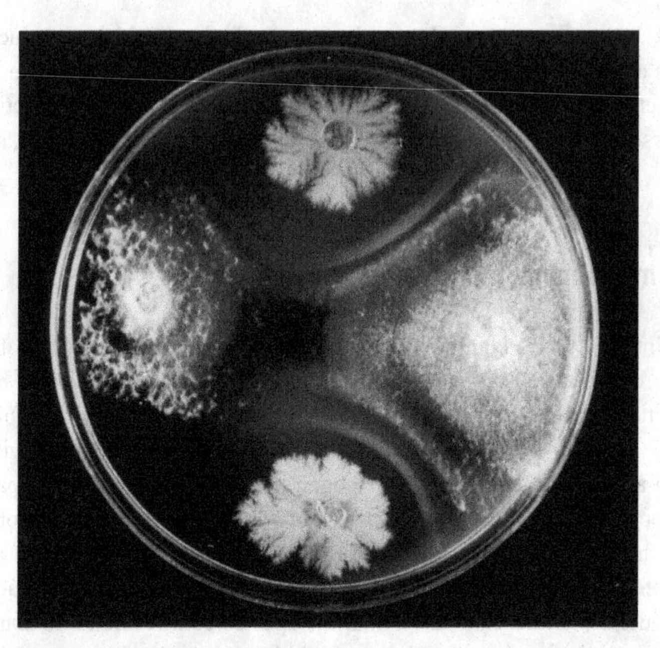

Figure 16.1 Example demonstrating antibiosis between pathogen strains in a petri dish. Different strains of a pathogen have been inoculated at right and left and a potential antagonist has been inoculated at top and bottom. Note the wide inhibition zones caused by the diffusion of antibiotics from the antagonist (Campbell, 1989).

metabolites that inhibit the plant pathogen *Sclerotinia sclerotiorum* (causing white mold disease). Microorganisms that produce metabolites that interfere with human or veterinary medicine are not usually considered as BCAs.

Box 16.1 Biological Control of Crown Gall

When rootstocks or wood of nursery stock are cut or grafted, an excellent location is provided for the bacterium *Agrobacterium tumefaciens* to invade. Once this bacterial species has entered a wound, it causes uncontrolled growth in that area of the plant stem, resulting in production of a gall, often at the crown of the plant where the stem enters the soil. Valuable perennial plants such as peach, plum, almond, and other fruit trees are susceptible, as well as vines and other herbaceous plants, making a total of 140 species of plants that can be affected by this gall-forming species of bacteria.

The virulence of *A. tumefaciens* is because of a plasmid, an extrachromosomal piece of DNA that can replicate independently of the bacterial chromosome and can be transferred between organisms. Strains of *A. tumefaciens* can gain or lose virulence based on presence or absence of the tumor-inducing, or Ti-plasmid. When the bacterial cells enter a potential host plant, part of the plasmid is transferred to the host plant genome and is expressed to stimulate the uncontrolled plant growth creating the gall.

Figure Box 16.1.1 Crown gall occurring at the junction of an apple stem with the ground (courtesy of Cornell University Department of Plant Pathology).

The bacterial antagonist *Agrobacterium radiobacter* strain K-84 is closely related to *A. tumefaciens*, but does not have the Ti-plasmid and thus does not cause crown galls. *Agrobacterium radiobacter* K-84 can colonize surfaces of wounds on roots or stems of nursery stock and, if it is present at a wound or a site where *A. tumefaciens* could potentially infect, it will prevent the pathogen from becoming established and thus no galling occurs. *Agrobacterium radiobacter* K-84 produces an antibiotic, agrocin 84, that is toxic to many strains of *A. tumefaciens* and this antibiotic plays an important role in the antagonism.

Agrobacterium radiobacter K-84 was recognized as a valuable control agent and was first developed commercially for control of crown gall in 1973. Cuttings, transplants, or root-pruned seedlings are dipped into a water-based suspension of *A. radiobacter* and then planted. *Agrobacterium radiobacter* then colonizes any wounds and *A. tumefaciens*, which is common in the soil, cannot establish infections, although *A. radiobacter* must be present at the wound first. Antibiosis can be very specific; the strain of *A. radiobacter* that has been commercially available since 1973, K-84, is not effective against strains of *A. tumefaciens* that attack grapes, pome fruit (e.g., apples), and some ornamentals, although it is very effective against *A. tumefaciens* attacking stone fruit (e.g., peaches and plums). Although the strain K-84 has been used very successfully for biological control, there have been reports of breakdown in its effectiveness. This has been attributed to transfer of some genetic material responsible for producing the antibiotic agrocin 84 from K-84 to the pathogen, thus making the pathogen immune to agrocin 84. However, it has been possible to genetically splice out the small region of the DNA in K-84 that allowed the transfer of the gene encoding agrocin. The new bacterial strain with this deletion, strain K1026, is now commercially available and is effective on plants such as roses and raspberries, in addition to peaches and plums.

Antibiosis has many advantages, so why not always use antagonists for biological control of plant pathogens that use antibiosis as a strategy? First, antibiosis seems to be more effective when nutrients are abundant or excessive in the antagonist's microhabitat. Antagonists produce antibiotics only when nutrients are available and not all microbial habitats on plant surfaces or in the soil have adequate nutrients for antimicrobial production. Antimicrobials also are not normally produced by microorganisms in the soil and generally do not persist well. Therefore, microhabitats where antibiotics are present are limited in area and in time. In addition, antibiotics differ in the number of microorganisms that they damage and some antibiotics have narrow specificities. Lastly, many antagonists do not produce antibiotics but still effectively suppress disease through other mechanisms. An additional aspect of antibiotics produced by BCAs is that they can protect the BCAs themselves from predation by protists or nematodes.

16.5.2 Parasitism

Some microorganisms directly attack other microorganisms and use them as sources of nutrients. Such parasitism requires direct contact between microorganisms; it is often referred to as hyperparasitism, or mycoparasitism when interactions involve a fungus. Mycoparasitism can be visually demonstrated by fungal parasites that penetrate through the tubular body form (hyphae) of plant-pathogenic fungi for access to the hyphal contents (Figure 16.2).

The BCA *Gliocladium virens*, which is closely related to *Trichoderma*, produces antibiotics as well as acting as a mycoparasite. When *G. virens* was altered to study its mode of action, a mutant that no longer formed coils around host hyphae still provided effective disease control, suggesting that the readily visible mycoparasitism is not required and is only one part of the antagonism caused by *G. virens*. At the other end of the spectrum is *Ampelomyces quisqualis*, an antagonist growing only as a hyperparasite of powdery mildews that grow on plant surfaces.

Some of the major antagonists affecting plant parasitic nematodes are fungal pathogens that attack nematodes. Some of these fungi produce spores that adhere to the nematode cuticle before infecting the nematode. Several fungal species effective against root-knot nematodes are endoparasites, growing within nematodes before killing them. In particular, *Pochonia chlamydosporia* shows promise because this endoparasite produces thick-walled resting spores for persistence in the soil, a valuable feature for an antagonist. Most fascinating are those fungi whose hyphae grow into rings when

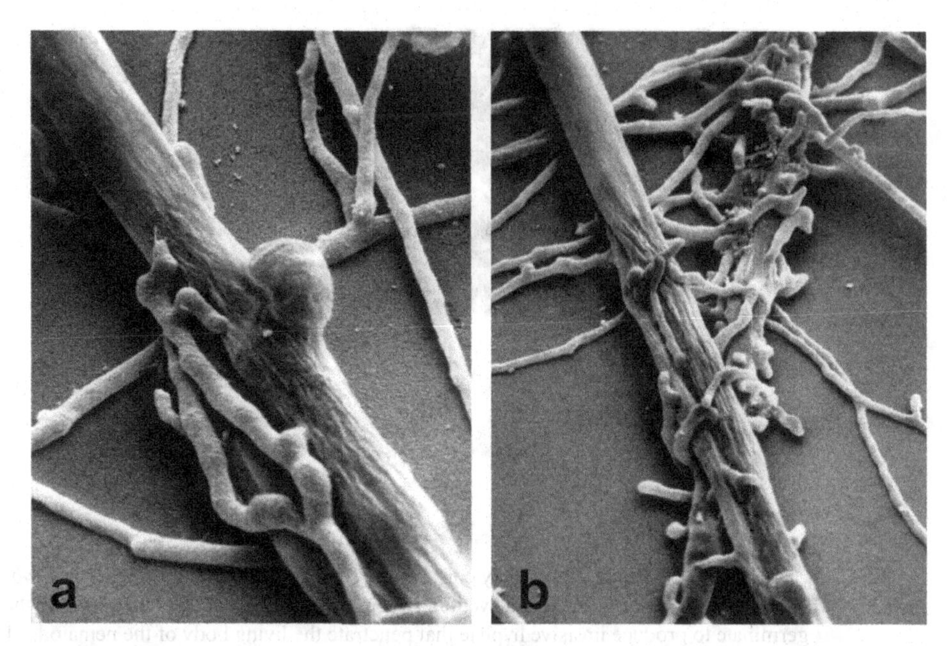

Figure 16.2 Scanning electron micrographs illustrating mycoparasitism. (a) The oomycete *Pythium acanthicum* parasitizing hyphae of the basidiomycete *Corticium*. (b) *Pythium acanthicum* parasitizing *Phycomyces blakeleeanus* hyphae that appear collapsed. Parasitic hyphae create holes in the host hypha to penetrate (Hoch & Fuller, 1977).

nematodes are present, such as *Arthrobotrys dactyloides*. When a nematode swims through a ring, the ring constricts around the nematode, trapping it (Figure 16.3). The body of the nematode is subsequently invaded by the fungus and the nematode dies and is consumed by the fungus. These beneficial fungi naturally occur in soils and efforts to increase their efficacy in controlling nematodes are focused on making the soil environment more amenable to these nematode-attacking fungi (see Chapter 17).

16.5.3 Resource Competition

It has now been estimated that the rhizospheres of plants grown in agricultural soils can harbor tens of thousands of different bacterial species; and that is without counting the other microorganisms that are present in that habitat. Bacteria present on leaf surfaces are also more diverse and abundant than previously considered.

Species living in these habitats have the same resources available to them and thus competition for these resources (also called exploitative competition in Chapter 6) occurs. When two microorganisms need the same limiting resource, once one microorganism has gained access, that resource is not totally available to the other. Microorganisms generally are thought to compete for nutrients, in particular carbon and nitrogen. Microorganisms are also frequently quite sensitive to environmental conditions so space that provides protection as well as conditions for growth can be limiting. Water is required by virtually all growing stages of microorganisms and competition can

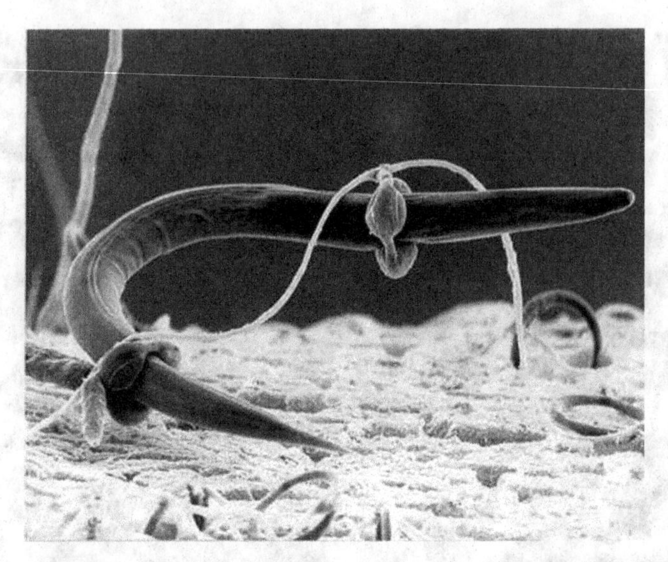

Figure 16.3 A nematode captured by constricting rings of the predatory fungus *Arthrobotrys dactyloides*. Ring cells swell and eventually constrict the body of the nematode and then germinate to produce invasive hyphae that penetrate the living body of the nematode. The fungus subsequently grows through the host's body, digesting the contents (Barron, 1979).

occur for microhabitats where water and nutrients are available. In particular, surfaces of leaves can be very heterogeneous as habitats for microorganisms. On leaf surfaces, microbes that are exposed to drying and UV experience more stressful conditions, while microbes invading the interiors of leaves gain protection there. Regardless of the exact location of the microbe on or in a leaf, we are learning that microbes inhabit microhabitats for which we are only beginning to be able to measure conditions.

Roots provide a different situation because drying and exposure to UV are not as critical as on plant surfaces. Also exudates that can be used by microorganisms can be produced along root surfaces. Microorganisms live along the surfaces of roots but still only a limited amount of space and nutrients are available in this defined habitat. The area around a root is critical for soil-borne pathogens as well as for their antagonists and the microbial dynamics around roots for limited space and nutrients can be intense. As roots grow, some microorganisms can spread with them and retain their activity. This is especially important because many pathogens infect at the root tip, which is not a static entity. Antagonists with this ability to spread as plants grow are often applied to seeds or to plants at the soil surface or to upper soil levels. Those microorganisms whose populations persist over time, increasing to large populations in the rhizosphere and spreading to lower portions of the root, such as the root tip, are called rhizosphere competent.

For control purposes, competitive BCAs are applied before pathogens arrive to exclude them from sites where they would normally colonize. This strategy has been used to protect seeds of crop plants, flowers that will become apples and pears, and fruits after harvest. This site exclusion strategy is also used in forestry to protect planned wound sites, such as when cuttings are made, during replanting, or when trees are cut and stumps need to be protected. In these cases, nonpathogenic microorganisms that are good competitors are applied directly after these types of wounding caused during management. These BCAs colonize this excellent growth environment and, in doing so, restrict colonization by pathogenic microorganisms that would colonize but, in addition, cause disease. The first biological control agent commercialized to control plant pathogens uses this strategy to control spread of root pathogens in forest stands (Box 16.2).

Box 16.2 Fungus Stumps Fungus

After forests are harvested, tree stumps remain. What we cannot see is that under the ground, as trees had grown, their roots spread out and had grown together with roots of neighboring trees via root grafts, before the trees were harvested. After a pine is cut, spores of the pathogenic fungus *Heterobasidion annosum* are soon deposited on the freshly cut stump by the wind and these spores then establish an infection. This virulent pathogen does not stop in that stump, but grows into the root system and continues growing through root grafts and thus spreads to the root systems of nearby healthy trees. Disease caused by this fungus, called root and butt rot of conifers, is a problem in tree stands where only some of the trees are cut and others remain. The goal of this silvicultural practice called thinning is to promote growth of the trees that remain, so exposing them to pathogens as a result of thinning is certainly not the desired outcome.

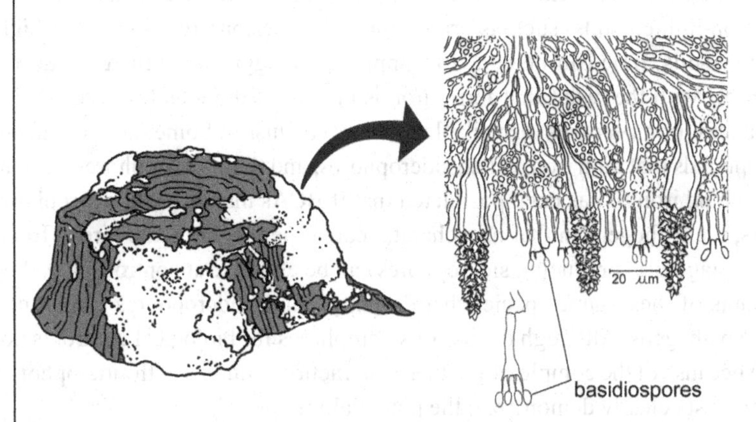

Figure Box 16.2.1 *Phlebiopsis gigantea* growing on a tree stump. In some areas of the fungal growth, infective basidiospores are produced (stump drawn by Alison Burke; close-up of fungus from Eriksson et al., 1981).

The fungus *Phlebiopsis gigantea* (previously known as belonging to the genera *Phanerochaete*, *Peniophora*, and *Phlebia*) is a great competitor and colonizer of wounds but is not pathogenic to living trees. If this fungus is applied to tree stumps just after cutting them, it will spread through the stumps and roots and will become well established, occupying this resource before *H. annosum* arrives. Even if *H. annosum* is already in the stump but is not well established, *Phlebiopsis gigantea* can exclude or replace it. *Phlebiopsis gigantea* antagonizes *H. annosum* through a phenomenon known as hyphal interference; when hyphae of *Phlebiopsis gigantea* contact hyphae of *H. annosum*, the protoplasm within the *H. annosum* hyphal cells becomes disorganized and the cell membranes appear leaky.

Phlebiopsis gigantea is common in the environment but its natural levels are too low and sporadic to provide control of *H. annosum* naturally. *Phlebiopsis gigantea* began to be mass-produced for use against *H. annosum*, but developing a method for applying this fungus to cut stumps was a challenge. In time, several different methods for application were developed. Fungal spores can be applied as a powder or water suspension to a cut stump. A clever and faster alternative was mixing the spores in the lubricating oil of the chainsaw used for cutting the tree and the fungus was then deposited in the wound as the tree was cut. *Phlebiopsis gigantea* for control of *H. annosum* was first commercialized in 1962, becoming the first product for biological control of a plant pathogen. Described in the literature in 1963, this system also provided the first operational use of biological control for control of a plant pathogen. This fungus is available today in England, Sweden, Norway, Switzerland, and Finland. It was available until 1995 in the United States, at which time the US Environmental Protection Agency discontinued its availability until registration requirements were addressed and satisfied.

One form of nutrient competition involving access to iron has been investigated for iron-limited soils, such as arable soils on limestone rocks or other high pH soils. In clay soils, lime can be added to improve the aggregate structure but this causes a high pH with the result that ferric iron is precipitated out as ferric hydroxide. In these instances, iron becomes unavailable to microorganisms. Some microorganisms produce compounds that bind iron, called siderophores, and these siderophores vary in their ability to bind iron. It has been postulated that if BCAs that produce siderophores colonize roots, the pathogens in that microhabitat could then be limited by iron. To demonstrate this, antagonists producing siderophores can be effective at controlling pathogens while mutants of these same species that do not produce siderophores cannot control these same pathogens. Although the use of siderophores for biological control is not yet available because of the complexity of these interactions, studies with siderophore-producing antagonists clearly demonstrate the potential for control.

16.5.4 Indirect Effects that Promote Disease Resistance

In some cases, antagonists help suppress diseases indirectly by inducing resistance responses in the plant rather than by directly suppressing populations of a target plant pathogen. Although plants do not naturally produce antibodies, they have their own methods for building protection against microbial invaders.

If certain plant species are inoculated with specific pathogens, this can lead to temporary or sometimes almost permanent immunization, called systemic acquired resistance (SAR), in the plant. One type of induced resistance that has been used to manage diseases caused by viruses, for which few other types of control are effective, is called cross protection. There can be many strains of any one plant virus and these strains can vary dramatically in the severity of disease that they cause. The minor damage caused by mild strains of some viral pathogens can in some cases be tolerated by growers, and it has been found that those mild strains can be used as BCAs. When a plant is inoculated with a mild strain of some viral pathogens, the plant becomes protected from infection by other strains of the same virus, leading to the name mild-strain cross protection. Cross protection has been tested for use against numerous diseases caused by plant-pathogenic viruses, but the biggest success story has been its use against citrus tristeza. Citrus tristeza is a very serious disease affecting sensitive varieties of citrus and is caused by a virus principally vectored by aphids. It is native to Africa but was introduced to South America some time in the 1920s and became widespread. Mild strains of the virus were found and these became the basis for the cross-protection program. Cross protection is somewhat like human vaccination: a bud from a tree infected with the mild strain of the virus is grafted onto a healthy tree and the mild virus spreads throughout the tree to protect it against infection by virulent strains of the virus. Researchers found that citrus tristeza cross protection has not broken down through time. The first cross-protected citrus plants were distributed in Brazil in 1968. By 1980, eight million Pera sweet orange trees (*Citrus sinensis*) in Brazil were cross protected and, in 2008, that cross protection was reported as being effective for at least 40 years. Use of cross protection against citrus tristeza continues around the world today. Cross protection is not used on more types of plants because mild strains of appropriate viruses are not always available or effective. This strategy also might not be appropriate for field crops where resistance would have to be induced in many individual plants that often will not be present the next year. In perennial crops, resistance owing to cross protection can fade after a few years and, when it is active, cross protection sometimes can be unevenly distributed throughout larger plants such as trees. Infection of strains of plant-pathogenic fungi with mycoviruses that can spread throughout plants, limiting fungal fitness and decreasing disease symptoms, provides a somewhat different systemic protection, called hypovirulence (see Box 17.1).

Plants have means for defending themselves against pathogens, some of which are constitutive (always present, such as cell walls) versus inducible (produced in reaction to damage or stress). Relative to biological control, some nonpathogenic microorganisms are known to elicit induced systemic resistance (ISR) when plants are exposed to them prior to infection. This biological control effect does not involve directly killing the plant pathogen. ISR-promoting microorganisms occur in the soil and can also promote plant growth via mechanisms like production of plant hormones and facilitating availability of nutrients (i.e., by plant growth-promoting rhizobacteria [PGPR] and plant growth-promoting fungi [PGPF]). Resistance-inducing mechanisms are not created immediately by the plants, but with exposure to pathogens, the plant's immune system is primed to create a faster and stronger immune response to subsequent pathogen attack. Exposure to ISR-inducing microorganisms in the root zone can lead to protection against foliar or systemic pathogens as well as root pathogens. Once activated, the induced systemic response is maintained for variable lengths of time and can be effective against multiple pathogen species, even if populations of the inducing agent decline over time.

Actually part of the activity of some biological control agents for suppressing plant pathogens includes ISR. Some PGPR strains, including some that are commercially available, such as *Bacillus subtilis*, have demonstrated biological control activity against soil-borne pathogens. The ISR being induced seems to differ based on the microorganism inciting the defense and since it also can be effective against a diversity of types of pathogens the effectiveness of ISR that is enabled differs for different inducer/pathogen/plant host systems. ISR seems to hold great potential, especially in cases such as insect-transmitted plant pathogens that are often difficult or impossible to control with pesticides.

Many plants maintain symbiotic relationships with certain kinds of fungi associated with their roots. These relations are generally thought to be mutualistic, with both of the partners benefiting from the relationship. Fungal root associates, or mycorrhizae, either grow extracellularly (ectomycorrhizae) or, more commonly, intracellularly (endomycorrhizae). Mycorrhizae obtain organic nutrients from the plant and they help the plant by enhancing nutrient uptake and water transport. Mycorrhizae gain a sheltered place to live with abundant nutrients and the plant increases in growth and yield. This relates in some cases to biological control because plants with mycorrhizae have also been found to be protected against certain soil-borne pathogens, perhaps in part due to being stronger and healthier and thus better able to resist pathogens themselves. For example, pine seedlings with mycorrhizae were protected from the pathogenic oomycete *Phytophthora cinnamomi* and cotton with mycorrhizae was protected against the fungus *Verticillium dahliae* that causes wilt, as well as from root-knot nematodes. Some mycorrhizae have been available commercially for application to promote plant health. However, these fungi are often difficult to work with, mass-produce, and apply and, while they are used to promote plant health, they are not yet specifically used for protection against or control of plant pathogens.

16.5.5 Employing Multiple Strategies

While some BCAs are thought to predominantly use only one type of the strategies described above, some use diverse strategies. For example, *Trichoderma asperellum* and *Trichoderma atroviride* are fungal BCAs that are especially used to control numerous species of plant-pathogenic fungi and oomycetes that kill seedlings. These BCAs have amazingly diverse activities, including strengthening plant defenses and solubilizing nutrients (so that the plants will be stronger) and directly reducing plant-pathogenic fungi in the soil via parasitism and competition for nutrients and space in the soil (see also Box 17.2). Plant growth-promoting rhizobacteria also have numerous effects on plants. While they directly promote growth, they also can use many of the antagonistic strategies described above.

Further Reading

Barron, G. L. (1977). *The Nematode-Destroying Fungi*. Guelph, Canada: Canadian Biological Publications.

Choudhary, D. I. & Varma, A. (2016). *Microbial-Mediated Induced Systemic Resistance in Plants*. Singapore: Springer.

Li, J., Zou, C., Xu, J., Ji, X., Niu, X., Yang, J., Huang, X., & Zhang, K. Q. (2015). Molecular mechanisms of nematode–nematophagous microbe interactions: Basis for biological control of plant-parasitic nematodes. *Annual Review of Phytopathology*, **53**, 67–95.

Mavrodi, D. V., Mavrodi, O. V., De La Fuente, L., Landa, B. B., Thomashow, L. S., & Weller, D. M. (2013). Management of plant pathogens and pests using microbial biological control agents. In *Plant Pathology Concepts and Laboratory Exercises*, ed. B. H. Ownley & R. N. Trigiano, 3rd edn., pp. 425–440. Boca Raton, FL: CRC Press.

Mazzola, M. & Freilich, S. (2017). Prospects for biological soil borne disease control: Application of indigenous versus synthetic microbiomes. *Phytopathology*, **107**, 256–263.

Moosavi, M. R. & Zare, R. (2015). Factors affecting commercial success of biocontrol agents of phytonematodes. In *Biocontrol Agents of Phytonematodes*, ed. T. Askary & P. R. P. Martinelli, pp. 423–445. Wallingford, UK: CABI Publishing.

Narayanasamy, P. (2013). *Biological Management of Diseases of Crops. Volume 1: Characteristics of Biological Control Agents*. Dordrecht, Netherlands: Springer.

Pal, K. K. & Gardener, B. M. (2006). Biological control of plant pathogens. *The Plant Health Instructor*. DOI: 10.1094/PH1-A-2006-1117-02.

Whipps, J. M. & McQuilken, M. P. (2009). Biological control agents in plant disease control. In *Disease Control in Crops: Biological and Environmentally Friendly Approaches*, ed. D. Walters, pp. 27–61. Oxford: Wiley-Blackwell.

17 Microbial Antagonists Combating Plant Pathogens and Plant Parasitic Nematodes

In the field of plant pathology, the focus for biological control is on suppressing the development of plant diseases, much more than on thinking of biological control as controlling specific individual organisms that cause disease. It follows that emphasis is placed on use of biological control agents to make certain that plants are not injured, while entomologists working in biological control of plant-feeding pests often place more emphasis on impacts of natural enemies on densities of specific pests, which then leads to reduced damage.

17.1 Finding Antagonists

When antagonists are present in the correct microhabitat and are active, it is thought that plant pathogens will not occur at levels harming plants. However, if a plant pathogen is occurring at higher levels, antagonists might still be present yet inactive because conditions for growth are not adequate. So, how are antagonists of plant pathogens found? The search for effective antagonists of plant pathogens is an endeavor shared by university and private industry researchers alike. Typically, antagonists are isolated from the same habitats where their target pathogens live – from plant tissues, on plant surfaces, or from the soil. Sometimes in a field of diseased plants, just a few plants are very obviously healthy. These healthy plants might have escaped disease because of the presence of naturally occurring antagonists that suppress the pathogen affecting neighboring diseased plants and microorganisms isolated from these healthy plants hold the promise of being antagonists. Antagonists occurring in the soil have been identified in much the same way as those isolated from healthy plants. When diseases that might be expected never occur in specific geographic regions, antagonists of plant pathogens have been found in the soils. In these instances, soils have been called suppressive and suppressive soils have been one focus of research in biological control of plant disease (see below).

In some cases, the same species of microorganism as the pathogen has been isolated from a healthy plant, but it is a weak strain causing little to no negative impact on host plants. In fact, many microorganisms have strains within the same species that vary in the relative ability to cause disease, from highly aggressive strains to strains living within hosts but not having detectable negative impacts. This is the case with a fungal pathogen infecting chestnut trees that can occur as either highly virulent strains or strains with reduced virulence (Box 17.1).

Box 17.1 Virus Weakens Fungus Attacking Chestnut Trees

The American chestnut, *Castanea dentata*, used to be a common and valuable tree species in many forests of the eastern United States. In the early 1900s a fungal pathogen, *Cryphonectria parasitica*, was first found in New York City, attacking American chestnut trees. *Cryphonectria parasitica* swept across North America and after it had spread, the once-common American chestnut trees were rare; the mighty chestnuts had been killed by chestnut blight, the disease caused by this fungus. Spores of this virulent fungus infect susceptible trees through preexisting wounds in the bark. The fungus grows in the cambium, the growing part of the tree which is just below the bark, and girdles branches or the main trunk, killing the aboveground parts of the tree. New sprouts may grow from the root system, but they eventually become blighted and die.

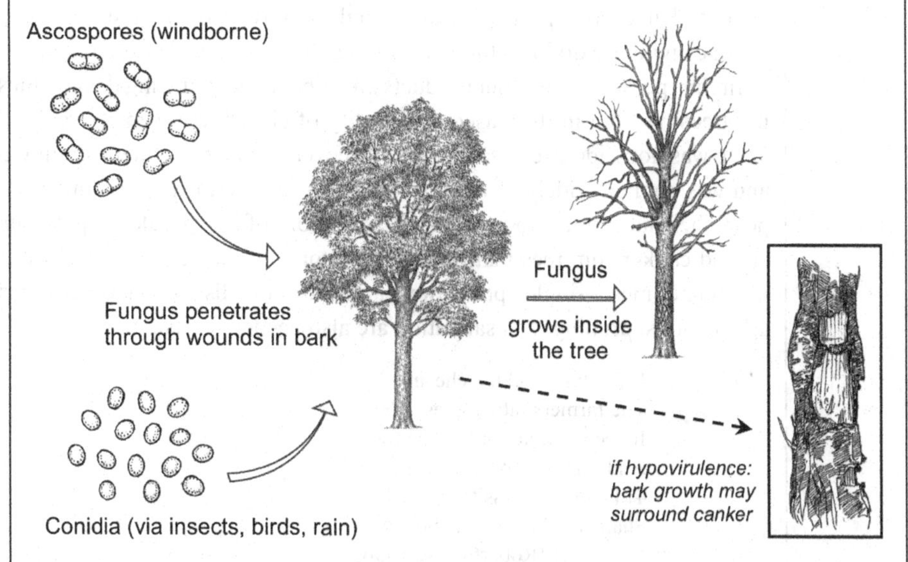

Figure Box 17.1.1 Chestnut blight. Sexual spores (ascospores), transported on the wind and asexual spores (conidia) transported by insects, birds, and rain, land on the bark of a chestnut tree and penetrate via preexisting wounds and the fungus grows in and under the bark, killing the cambium. Cankers, delimited dead areas on the trunk, eventually girdle and kill infected trees by blocking the exchange of water and nutrients between leaves and roots (drawn by Tom Prentiss; Newhouse, 1990). Inset at lower right: With viral hypovirulence, the fungus can be weakened and the tree may be able to survive by surrounding the canker with bark growth; however, new cankers can develop and tree health can be noticeably reduced (canker drawn by Frederick L. Paillet).

In the late 1930s, for the first time the native chestnut trees in Europe began to show signs of chestnut blight and they died rapidly. However, about 20 years later, it was noticed that some sucker sprouts from infected trees were surviving well in Italy. Researchers learned that although the same fungus occurred in these sprouts, it was not as virulent. Detailed studies went on to demonstrate that this fungal strain with reduced virulence (i.e., a hypovirulent strain) was morphologically distinct, having white instead of orange coloration, reduced spore production, and slower growth. Furthermore, in the laboratory when the hypovirulent strain grew in the same culture dishes as the virulent strain, the virulent strain would become hypovirulent if the hyphae of the two strains grew together. Research demonstrated that double-stranded (ds) RNA typical of fungal viruses resided within the hypovirulent strain and this dsRNA could be transferred when cells of virulent and hypovirulent strains grew together, thus transferring the trait for reduced virulence. Wonderfully, this transfer also occurred within infected trees. If a tree was infected with the virulent strain and then inoculated with the hypovirulent strain, although the fungus remained present, it no longer harmed the tree.

Further research demonstrated that this system is more complex than originally thought; several viruses are known to affect the virulence of this fungus, viral RNA genomes are present but not viral particles themselves. Some viruses that cause hypovirulence in *C. parasitica* are found in North America but are not present in Europe and these differ in their ability to reduce virulence of the fungus.

In Europe, where chestnut products are popular as food, fungal epidemics in chestnut trees resulted in decreased availability of chestnuts for many years. Hypovirulence was not widespread so, after its discovery, the hypovirulent strain was isolated and was applied widely throughout southern Europe to suppress the disease. European chestnut farmers today can purchase tubes of hypovirulence paste and apply it around cankers on trees that display symptoms of the virulent strain of *C. parasitica*. Sometimes a single application can control the disease but other times multiple applications, pruning, and sanitation are also needed to save trees.

Will blight end the chestnut?
The farmers rather guess not
It keeps smoldering at the roots
And sending up new shoots
Till another parasite
Shall come to end the blight.
(Robert Frost, 1936)

Once a collection of potential antagonists has been isolated, they are subjected to various tests to determine which ones show potential as biological control agents. Potential agents are first tested for antagonism against the target pathogen in petri dishes, followed by greenhouse experiments where researchers spray them onto plants to determine if they can reduce disease symptom development. Sometimes they are also tested for nutrient utilization to determine if they might be able to directly compete with pathogens for specific resources. In other cases, potential antagonists are grown under varying conditions to determine if they produce antifungal compounds and these compounds are then identified and studied further by chemists. Antagonists showing promise from initial trials are often evaluated side-by-side in field trials where they are evaluated based on their ability to reduce disease symptoms in real-world environments.

Finding an antagonist can be the easier step compared with fully understanding the mechanisms it uses to prevent disease. Some mechanisms are also easier to identify than others, such as antibiosis or parasitism, because these can be visualized with experiments in petri dishes or under a microscope. Plant pathologists have learned that this obvious activity is not always the only mechanism used by an antagonist against a plant pathogen (see Section 16.5.5).

17.2 Types of Antagonists

Antagonists are taxonomically diverse, as are the pathogens they are used against. Fungi and bacteria are principally used for biological control against fungal and bacterial plant pathogens. Fungi can be used to control either fungi or bacteria and bacteria can be used to control either fungi or bacteria. No one mechanism of action is used only by bacteria or only by fungi. Aside from some of the species of microorganisms that are obligate parasites and are thus more specialized, many antagonists of plant pathogens occur naturally in the environment and can be ubiquitous and common members of the microbial community because their occurrence is often not dependent on the presence of a pathogen. Viruses can act as antagonists too but their mode of action is completely different from fungi and bacteria.

17.2.1 Fungal Antagonists

The first major discovery of antagonism of microorganisms toward plant pathogens was the finding by Alexander Fleming in 1928 that *Penicillium notatum*, an ascomycete fungus, inhibited the growth of some bacteria. This discovery led to the development of penicillin, a very successful drug against human diseases caused by bacteria. For that discovery, Fleming was awarded the Nobel Prize in Medicine in 1945.

Many of the fungal pathogens used as antagonists belong to the phylum Ascomycota, or the sac fungi. This group includes many species that can look like molds. The sexual spores of the Ascomycota are usually produced in groups of eight within small sacs, or asci. A complete fungal life cycle usually includes both sexual and asexual stages, but the formal taxonomy of fungi is historically based on the sexual stages. As explained in Chapter 12, in the past many of these fungal species had two names (one for the sexual stage and one for asexual) but mycologists are deciding which only one name will remain in use, under the "one fungus – one name" rule.

The most common fungal genus used as a biological control agent is probably the green-colored mold *Trichoderma*. *Trichoderma* antagonists are particularly effective because they are fast growing and aggressive, and because they are versatile because of their employment of multiple modes of action (see Box 17.2, p. 316). Aside from *Trichoderma*, another group of fungi commonly studied as antagonists include the nematode-trapping fungi, discussed in Sections 16.5.2 and 17.5. Finally, some fungi used as antagonists actually belong to the same genera as fungal pathogens, but they do not cause plant diseases. An example of this is a nonpathogenic strain of *Aspergillus flavus*, which occupies the same niche as pathogenic strains of *A. flavus* and can therefore compete with this devastating cereal pathogen.

Although many of the fungal species used for biological control belong to the Ascomycota, antagonists from other groups of fungi are also important. A fungus growing as single cells, the yeast *Candida oleophila*, is used to control pathogens attacking fruits and vegetables after they have been harvested. The fungus applied to stumps of conifers, *Phlebiopsis gigantea* (see Box 16.2), belongs to the group of fungi that includes mushrooms, the Basidiomycota.

Those natural enemy species providing control through competition are usually not specifically associated with just one species of plant pathogen; their association is based on using a microhabitat that is also preferred by the plant pathogen. Many antagonists thus occur commonly in nature. Different species are important for controlling pathogens attacking virtually all of the different locations on plants. Perhaps the greatest use for antagonistic fungi has been in suppression of soil-borne plant pathogens. However, fungal antagonists have also been developed for control of pathogens attacking tree wounds, foliage, and harvested produce.

17.2.2 Bacterial Antagonists

Many of the bacterial species utilized as antagonists are active against soil-borne plant pathogens. Members of the family Pseudomonadaceae (e.g., *Pseudomonas fluorescens*) and order Actinomycetes (e.g., *Streptomyces griseoviridis*) often use combinations of competition and antibiosis to prevail in the complex soil environment. *Pseudomonas* species are also used for their ability to simulate systemic plant defenses that can reach aerial plant tissues and protect against foliar diseases as well as soil-borne diseases.

Many of the bacterial antagonists of plant pathogens belong to the family Bacillaceae, the same family as the insect pathogen *Bacillus thuringiensis*, although not closely related to exactly that species. The Bacillaceae, which are antagonists to plant pathogens, are spore-forming bacteria and several are known to produce antibiotics. These spore-forming species have the ability to persist in the environment and this characteristic prolongs their presence after application. Members of the Bacillaceae are applied against foliar as well as soil-borne pathogens.

17.2.3 Oomycete Antagonists

Some antagonists are water molds, a group called the oomycetes that were previously thought to be fungi since they had many characteristics similar to fungi. However, these species are protists, now classified within the eukaryotic kingdom Chromista, class Oomycota. For example, the oomycete *Pythium oligandrum* is an antagonist that suppresses diseases caused by phytopathogenic fungi and oomycetes through competitive interactions.

17.3 Strategies for Using Antagonists to Control Plant Pathogens

Methods for use of these different types of microorganisms for control differ by the type of plant and pathogen. For biological control of plant pathogens, the major ways that generalist antagonists are released is through augmentation of natural enemy populations, and in specific systems antagonists are conserved or enhanced. However, augmentation, rather than conservation, is the most common approach for biological control of plant diseases, although naturally suppressive soils can be seen as a type of conservation biological control.

17.3.1 Augmentation: Inundative Versus Inoculative Releases

The principal way that antagonists of plant pathogens are used is by releasing them either where the plant pathogen already occurs or, prophylactically, on the plant surfaces, where they can prevent future infection by the pathogen. While antagonists used to impact plant pathogens already present often have been called "biopesticides," those applied prophylactically are sometimes called "bioprotectants" instead. Some plant pathologists have called all of these mass applications of antagonists "introductions" but this general term could be used in reference to several strategies, including classical biological control, inoculative or inundative releases. So what do we call releases of antagonists against plant pathogens? Using our definitions for release strategies, an inundative release would occur when control is achieved exclusively by the organisms themselves that have been released (see Chapter 4). Thus, for an inundative release of insect natural enemies to succeed, large numbers of natural enemies are usually released. In contrast, most releases of microbial antagonists fit our definition of inoculative releases. Microorganisms increase so quickly that it is not the released organisms that are effective against the host but rather the generations of microorganisms produced after release that increase to colonize the complete habitat, sometimes producing antibiotics as they increase. We have defined inoculative releases as releases of natural enemies that need to multiply to be effective and that will thus control the pest for a more extended period than an inundative release. Antagonists that are applied in large amounts, either as protectants to prevent infection by a pathogen or as biopesticides to control preexisting pathogens, must grow to colonize the sites where they will be active; in this way both types of applications (applied preventively or after disease is present) fit our definition of inoculative releases. While the exact length of a so-called "extended period" of control can be subjective, many antagonists can protect against or inhibit plant pathogens for longer than an immediate effect. Therefore, in the vast majority of cases, releases of antagonists to control plant pathogens are inoculative releases; antagonists that are released inoculate the area to be protected, much as aliquots of microorganisms are used to inoculate media, and these microorganisms then increase and spread throughout the area necessary for colonization.

Antagonists of plant pathogens first began to be developed for augmentative release in 1962, when *P. gigantea* became available for suppression of root and butt rot of conifers (see Box 16.2). The number of biological control agents for control of plant pathogens available as commercial products has proliferated since then. Early products were certainly those species and strains of species that were easy to grow in culture and that seemed effective during laboratory studies. However, applications in the field did not consistently provide control. Scientists realized that a better understanding of activity in the field, especially within the soil, was necessary to produce effective biological control agents. For example, the microorganisms released needed to be able to survive in the highly competitive root zone and laboratory studies did not always predict activity in the field. Therefore, detailed studies of persistence in this area were necessary, ultimately yielding improved products.

Numerous commercial microbial products are produced worldwide for inoculative releases to control plant pathogens. While new products are created, older ones are discontinued but there is a general trend of increase. In 2014, 66 products containing antagonists (including 37 based on fungi and 25 based on bacteria) were reported as being on the market. Species of the fungus *Trichoderma* and the bacteria *Bacillus* predominated. A sample of some of the commonly used antagonists is presented in Table 17.1. Products affect different types of plant pathogens on different plant parts using different mechanisms of action, although the antagonists for control of diseases caused by seed- or root-infecting pathogens predominate.

Antagonists that affect only one or few plant pathogen species are generally those that act principally as obligate parasites while antagonists that employ numerous modes of antagonism, sometimes including parasitism, often have broader host ranges. Antagonists that suppress a number of diseases are more likely to be commercialized than those antagonists that suppress only one disease because of the expanded market size offered by a broader spectrum of activity. Thus, many commercial biological control agents suppress more than one disease, often due to their production of broad-spectrum antibiotics or their superior abilities to outcompete numerous pathogens for limited resources on plant surfaces. Perhaps of more importance to their development for control, those antagonists with broader ranges of activity (sort of "generalists") are usually easier to mass-produce so these are the species that have been simplest to develop into commercial products. In addition, species producing long-lived survival structures, such as the chlamydospores of *Trichoderma* spp., or the endospores of *Bacillus* spp., are preferred for commercialization because these species can persist in a commercial formulation throughout production and storage, thereby ensuring that a viable product is made available to the consumer. In summary, companies compare many isolates of different species with respect to relevant biological characteristics to find and select "winner isolates" that provide effective control, are environmentally safe, and are cost effective to mass-produce.

Registration requirements, often originally developed for chemical pesticides, can pose a serious burden before commercialization of antagonists. Some microorganisms that previously were thought of primarily as antagonists have been found to also effectively promote plant growth or plant health (Box 17.2), and when sold for these last two purposes, toxicity testing required to register a microbial biological control

Table 17.1 Examples of some important microbial antagonists of plant pathogens commercially available in 2014.

Antagonist	Plant part	Disease and pathogen	Mode of action[1]
Bacteria			
Agrobacterium radiobacter **K1026**	Woody stems and roots	*Agrobacterium tumefaciens* (causes crown gall)	Competition, antibiosis, resistance promotion
Bacillus amyloliquefaciens plantarum[2]	Seedlings	Damping off	Induces systemic resistance, antibiosis
Bacillus subtilis	Foliage	*Botrytis*, mildews, scab, and other fungi	Competition, induces systemic resistance
Pseudomonas fluorescens	Flowers and plant surfaces	*Botrytis, Erwinia amylovora* (cause of fire blight)	Competition, frost damage prevention
Pseudomonas syringae	Harvested fruit and vegetables	Fungal pathogens of stored products	Mechanism unknown, probably competition
Streptomyces griseoviridis	Seeds, roots	Seed- and soil-borne fungi	Competition, parasitism, antibiosis
Fungi			
Ampelomyces quisqualis	Foliage	Powdery mildew	Hyperparasite
Candida oleophila	Harvested fruit	Fungi attacking stored fruit	Competition
Coniothyrium minitans	Roots	*Sclerotinia*	Mycoparasite of long-lived sclerotia in soil
Trichoderma harzianum	Roots, seeds, foliage	Damping off, soil and foliar fungal pathogens	Mycoparasitism, competition, induced resistance, increased plant growth
Oomycota			
Pythium oligandrum	Roots	Soil-borne fungal pathogens	Competition, mycoparasitism, antibiosis, increased plant resistance

Source: based on Gwynn (2014).

[1] Mode of action not always completely understood.

[2] Subspecies previously included in *B. subtilis* (Borriss et al., 2011).

agent might be avoided, making it easier for them to reach the market. In China, for example, between 1985 and 1993, "yield increasing bacteria," largely based on *Bacillus* species, were applied on over 50 different crops covering a cumulative area of 40 million ha.

Box 17.2 *Trichoderma* as a Friend of Higher Plants and Enemy of Plant Pathogens

Trichoderma species had long been known to grow as mycoparasites, often coiling around hyphae of a plant-pathogenic fungus and producing enzymes to penetrate into cells of fungal hosts which then die. *Trichoderma* then enters the dead cell to feed on it as a necrotroph, living as a saprophyte. By 1997, the species *T. harzianum* was found also to use antibiosis and to compete with plant pathogens for nutrients and space. By 2002, research results suggested that this mycoparasite also has direct effects on the plants themselves and this initiated gaining exciting new information on the activities of *Trichoderma*. Scientists learned that in addition to living as an antagonist, *Trichoderma* can also live as a beneficial, asymptomatic plant symbiont. It colonizes leaves, roots, and stems and produces signals that activate the plant genome to increase plant growth and promote disease resistance. The disease resistance can be induced by plant pathogens via systemic acquired resistance (SAR), a whole plant resistance response that can reduce the severity of disease. The presence of nonpathogenic rhizobacteria can also induce or enhance resistance to disease. However, resistance can also be in response to abiotic conditions as, for example, plants inoculated with *Trichoderma* are more tolerant of drought, high temperature, or salinity. So, *Trichoderma* can be very beneficial to plants, in addition to acting as an antagonist.

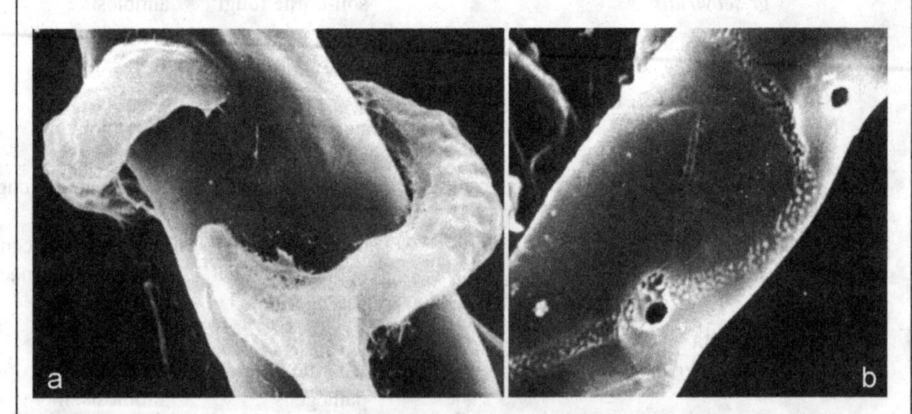

Figure Box 17.2.1 (a) Hypha of *Trichoderma hamatum* coiling around and penetrating a hypha of the plant pathogen *Rhizoctonia solani*. Partial degradation of the host cell wall can be seen. (b) Hypha of the plant pathogen *Sclerotium rolfsii* from which a coiled hypha of *Trichoderma harzianum* was removed, showing a digested area with holes where the antagonist penetrated (Elad et al., 1983).

Trichoderma harzianum was the first antagonist developed as a seed treatment for controlling fungi causing damping off of seedlings. Uses diversified with improvements to the fungus as a biopesticide and as understanding of this fungus increased. In 2015, 56 *Trichoderma*-based agricultural products were listed from around the world for use on a diversity of crops, often as foliar sprays or for application to soil. Some products are applied to wounds on trees. These products are variously sold for promoting plant health or for protecting against plant pathogens or both. For some products, several species of *Trichoderma* are combined to expand efficacy. As with many other biological control agents used for augmentation, products based on *Trichoderma* work best if applied before plant pathogens are well established and are best used as part of integrated pest management programs.

As with other types of biological control, antagonists are effective when plant pathogen populations are low and their use is not intended for controlling pathogens during disease epidemics. Of course, many of these antagonists are used to protect plants (as bioprotectants) so that pathogens do not even become established, a type of strategy not used for microorganisms in biological control of arthropods (Chapters 9, 10, 11, and 12) or weeds (Chapter 15). The antagonists developed for inoculative release will be discussed below, based on the areas of the plant where they are effective.

Seeds and Roots

Common diseases targeted using biological control agents are soil-borne diseases. Historically, soil-borne pathogens have often provided the greater challenges for control compared with foliar pathogens as there are limited options available for control. Both fungal and bacterial antagonists are sold for control of soil-borne pathogens and they usually compete with plant pathogens as well as using other mechanisms of action. While this all sounds simple, the soil is a hostile and very diverse environment for biological control agents and initial attempts to use biological control agents in this arena met with erratic performance. Costs of products were high because this was new technology and therefore markets were not developed. Improvements are still being made as optimal strains are chosen and more is learned about modes of action. Plant pathologists, without good alternative controls, were undaunted and proceeded to improve biological control methods. They found that pathogens affecting the greenhouse and nursery markets were often a better target than pathogens affecting field crops, perhaps because the horticultural industry can control environmental conditions and can often sustain the costs, a situation similar to biological control of insect and mite pests which has been very successful in controlled greenhouse environments worldwide.

A pathogen that is well established in a plant can be very difficult to control using any method. A clever change in designing BCAs was to apply natural enemies before disease organisms had infected, for example, by application to seeds before planting. With this type of application, the seed surface is colonized by the antagonist and when the plant begins to grow, it is protected against pathogens that might kill sensitive seedlings, a general disease condition called damping off. Damping off, caused by fungi such as *Rhizoctonia* and oomycetes such as *Pythium* or *Phytophthora* spp., is a very serious disease in horticulture, agriculture, and forestry and can lead to almost total mortality in a planting. Many of the antagonists that control pathogens causing damping off are excellent competitors and rapidly colonize new locations after being introduced. The goal of such releases has been called "preemptive exclusion" of pathogens. This same idea has been used to protect cuttings and bulbs when they are transplanted by treating them with liquid suspensions of antagonists directly before planting. Alternatively, the BCAs can be mixed directly into the soil or seed furrow before planting.

Similarly, control of root diseases is most successful if BCAs are applied so that they are established in the rhizosphere before roots are colonized by a pathogen; roots that support established populations of a pathogen before an antagonist is applied are much more difficult to protect. Roots do not have a well-defined surface layer of connected cells, as do leaves and stems, and so they are initially not as well defended. Root tips are especially susceptible to infection by many soil-borne fungal pathogens; the susceptible root tips are constantly growing and moving through the soil where they encounter new pathogens. One strategy is to treat the entire rooting medium with an antagonist before planting. When applied before planting, rhizosphere-competent antagonists can colonize the root surfaces and spread along the root as it grows to offer protection. Some fluorescent pseudomonad bacteria that are antagonists are "rhizosphere competent" and will spread with growing roots, although, if applied to seeds or the soil surface, these same bacteria require percolating water to move below 3 cm depth in the soil. In this case, the method of application is critical for effective use of these antagonists. Many fungal antagonists are not rhizosphere competent although some *Trichoderma* spp. are exceptions to this generality. Some strains of the fungal species *T. harzianum* will spread to colonize growing roots where they employ a variety of methods for attacking or outcompeting plant pathogens. Several products based on *Trichoderma* species are available worldwide for suppression of numerous plant diseases (Box 17.2).

Stems and Crowns

Plant stems vary in size and complexity from stems of annuals to tree trunks. As with transplanted cuttings, an open wound in a stem or crown provides an excellent location for entry and establishment of a pathogen. The antagonist products that have been most successful for the longest time are those applied at wound sites on trunks or stems to protect against pathogen invasion, as with the fungus *P. gigantea* colonizing pine and spruce stumps (see Box 16.2) and the bacterium *A. radiobacter* protecting against crown gall (see Box 16.1).

Post-Harvest Fruit

Both fungal and bacterial antagonists have also been developed to protect fruit against decay after harvest and during storage. Yeasts are good at colonizing plant surfaces under adverse environmental conditions and several yeast-based products have been developed (Figure 17.1). Post-harvest pathogens are usually weak pathogens that need wounds on fruit to gain access. For suppression of post-harvest disease, fruit are often coated with antagonists before harvest to protect wounds that typically accompany harvest and post-harvest handling. Alternatively, application can occur after harvest by dipping fruit into solutions containing the antagonist.

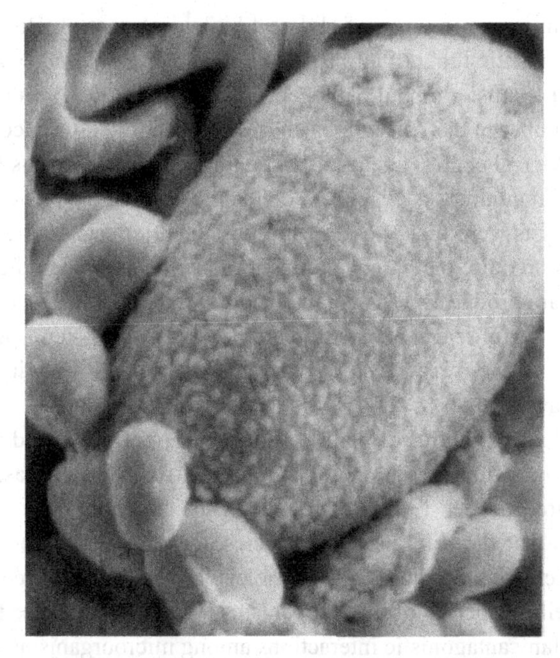

Figure 17.1 Antagonistic yeast cells surrounding a spore of *Botrytis cinerea*, an important fungal pathogen causing post-harvest decay of fruits and vegetables (photo by Y. Elad and E. Fischer; Albajes et al., 1999).

Foliage

The most commonly used BCAs for control of foliar plant pathogens are used to target the fungi, especially *Botrytis* and powdery mildews. Perhaps least well developed for biological control are the natural enemies combating foliar pathogens, possibly because this is the most common location where fungicides are applied so there is the least incentive to provide alternate controls. Some plant pathogens (e.g., *Botrytis*) that attack foliage are necrotrophic, with the ability to increase in dead plant tissues. In contrast, the commonly occurring powdery and downy mildews attacking foliage require living hosts (i.e., biotrophs); while they seldom kill hosts, they can cause substantial declines in growth and yield. Fungal antagonists can be applied to prevent damage by each of these types of pathogen. At least two of these antagonists, *Ampelomyces quisqualis* and *Pseudomyza flocculosa*, require high humidity after application for spore germination.

Bacterial diseases affecting aerial plant parts, such as fire blight (caused by *Erwinia amylovora*), can be extremely destructive. There is a very real lack of chemicals to treat these diseases and development of biological control agents against foliar bacterial pathogens is critically needed.

17.3.2 Classical Biological Control

In stark contrast with biological control of invertebrates or weeds, classical biological control has been applied against plant pathogens in only a few instances. Biological control of chestnut blight is one instance of classical biological control (Box 17.1). In a second instance, two parasitic fungi (*Dicyma pulvinata* and *Cylindrosporium concentricum*) were introduced against a fungal pathogen, *Phyllachora huberi*, causing black crust on the foliage of rubber in the Amazon basin of Brazil and long-term control was documented. These biological control agents are obligate pathogens living as hyperparasites and their increase in numbers is dependent on the presence of the plant pathogen. After the fungal antagonists were first released, the subsequent effects on the pathogen were cyclical. The BCAs require the pathogen for reproduction and when they kill the pathogen, many spores are produced that then proceed to protect the next set of leaves. However, on this second set of leaves, because there is little disease, few spores were produced. The third set of leaves was therefore not well protected, resulting in pathogen development in these leaves.

Perhaps classical biological control of plant pathogens has not been used more frequently because, in many instances, the major requirements of typical classical biological control agents are not met. Most importantly, a major requirement for classical biological control is that the natural enemy species does not already occur in the area where it would be released. Studies based on molecular techniques have suggested that generalist, soil-dwelling species of microorganisms are widely dispersed, perhaps occurring worldwide. Many antagonistic interactions among microorganisms are based on competition for limiting resources and not only on host-specific interactions, so their occurrence is not tied to presence of a host. Classical biological control using obligate microbial pathogens with narrow host ranges is appropriate for controlling some weeds and insects with localized distributions. However, it seems that generalist antagonists of plant pathogens on crops grown around the world are probably already almost worldwide in distribution. Therefore, there is no need to introduce them to new areas because they can be assumed to already occur there.

17.4 Conservation/Environmental Manipulation

Some plant pathologists have argued that because augmentative releases are based on only one strain of one species of antagonist, this strategy fails to take advantage of the great diversity of antagonists in the environment. Conservation strategies, also commonly considered as cultural practices in plant pathology, take advantage of the broad diversity present among naturally occurring antagonists in the soil or on plant surfaces.

The environment is full of antagonists that could potentially inhibit activity of plant pathogens, but to take advantage of their activity conditions must be adjusted to favor persistence and increase of antagonists. Crop management strategies to conserve and enhance antagonists have actually been used for a long time. An extremely important disease management strategy worldwide, crop rotation, can take advantage of the activity of antagonistic microorganisms in the soil. Quite simply, when a crop is rotated every second or third year, the resident community of antagonists has a chance to lower the inoculum levels of root- and foliage-infecting microorganisms while that crop is absent. Early in the twentieth century, researchers found that by adding compost, barnyard manure, and green manure to the soil, many soil-borne diseases could be suppressed. Since then, plant pathologists have learned that the addition of these amendments to soil enhances microbial activity, which is considered largely responsible for the observed disease suppression. Soil amendments can be especially useful in systems where crop rotation is not practical.

17.4.1 Disease Suppressive Soils

Plant pathologists have long known that when specific crops are grown in certain areas, disease problems that regularly occur elsewhere are not seen. In particular, this effect has been recognized for soil-borne diseases. Suppression of the plant pathogen is thought to often be caused by rhizosphere microbiomes. Communities of microorganisms plus plant associates are now known to work together as "superorganisms." The microbiota in the soil provide specific functions while plants exude some of their photosynthetically fixed carbon which is used by the microbes. Studies are showing that the soil microbiome around plant roots includes tens of thousands of microbial species, some of which help protect plant roots before or when a pathogen attacks. Soils displaying resulting phenomena of disease inhibition are said to be suppressive soils. In contrast, soils where a disease occurs are called conducive (nonsuppressive) soils.

Agricultural soils suppressive to plant pathogens are known from around the world. There are two major types of suppressive soils: general and specific. General suppression can occur as a result of the total microbial biomass in the soil and this effect cannot be transferred to other soils. Specific suppression is a result of the effects of specific groups of microorganisms. If this latter type of suppressive soil is added to sterile soil, the previously sterile soil will become suppressive and this activity is then said to be transferable. Generally, sterilization of suppressive soils often destroys this activity, demonstrating the biotic origin of the suppression. For specific suppression to develop in a soil, a specific pathogen needs to be present for some time, thus allowing the antagonists in the soil to increase.

The oldest known example of a suppressive soil is from the Chateaurenard region in southern France. Melon plants are frequently susceptible to the soil-dwelling fungal pathogen *Fusarium oxysporum* f. sp. *melonis* but, although the pathogen is present in this area, it does not cause the disease problems commonly found elsewhere. Saprophytic growth of the pathogen is suppressed along with germination of the persistent soil-borne stage, the chlamydospores. While some other varieties of *Fusarium* are suppressed in this area, other species of soil-dwelling pathogens are not affected. This suppressiveness can be described as a biological property of the soil. Many years of research were necessary to pinpoint the agents in the soil responsible for suppression. A nonpathogenic strain of *F. oxysporum* and fluorescent pseudomonads (i.e., species of *Pseudomonas*) are thought to limit the pathogen through competition for nutrients and induced systemic resistance in plants. In fact, if organic matter is added to this suppressive soil so that nutrients are plentiful, suppression of the pathogen ends and *F. oxysporum* f. sp. *melonis* once more becomes a problem. Interestingly, the nonpathogenic strain of *F. oxysporum* seems to remain quite stable in its activity, which is consistent with the long-standing occurrence of these suppressive soils.

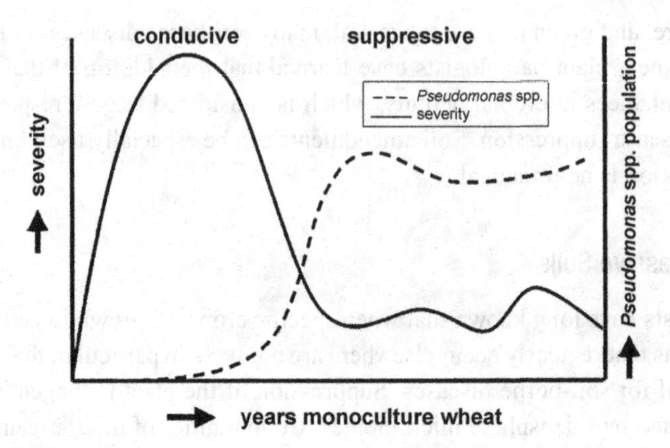

Figure 17.2 Hypothetical model of the influence of a soil-dwelling fluorescent species of the bacterium *Pseudomonas* on take-all of wheat caused by the fungus *Gaeumannomyces graminis* var. *tritici*. With monoculture of a take-all susceptible host over several years, after one or more disease outbreaks, populations of these bacteria increase to densities over a threshold and soils that were conducive to growth of this pathogen become suppressive and take-all is controlled (Weller et al., 2002).

Suppressive soils have been described for many soil-borne pathogens. Another excellent example of a suppressive soil involves a disease named "take-all," caused by the fungal pathogen *Gaeumannomyces graminis* var. *tritici* that attacks the roots of wheat and occurs worldwide. Take-all can be managed by crop rotation, but this approach is not always practical economically. However, this is one example where monoculture without crop rotation has been shown to be beneficial for pest control. If wheat is grown continuously year after year, the severity of take-all increases for 4 to 6 years, but then begins to decline, a phenomenon called "take-all decline." Take-all decline is attributed to suppressive soils that are built up during continuous wheat monoculture. Although numerous antagonists may contribute to take-all decline, one group of antagonists stands out; take-all decline occurs when certain antibiotic-producing strains of the fluorescent bacteria *Pseudomonas* spp. increase in soils (Figure 17.2). In Washington state, where take-all suppressive soils are well characterized, there is compelling evidence that strains of *Pseudomonas* spp. that produce the antifungal metabolite 2,4-diacetylphloroglucinol, which is inhibitory to *G. graminis* var. *tritici*, are largely responsible for take-all decline.

Utilizing naturally occurring suppressive soils by planting a crop in an appropriate soil type is one way to take advantage of suppressive soils. Another idea is to manipulate the system and transfer the microorganisms that confer suppression. A simple alternative that is suitable in some cases is to add organic amendments to the soil. Although the benefits seem to be system specific, if green organic material is added to some soils, the activity of resident antagonists is increased and pathogens can be suppressed, often by antibiosis. Such an effect of boosting the activity of resident antagonistic microorganisms has also been shown when adding chitin from shellfish production or compost to soils.

17.5 Biological Control of Plant Parasitic Nematodes

The Phylum Nematoda includes many species occupying a diversity of habitats. While many nematode species are free-living, some of the most important plant parasitic nematodes are quite sedentary, remaining closely associated with plant roots. Many plant parasitic nematodes have developed defenses for protection in the complex soil environment, such as burrowing within plant tissues or producing their own protective structures on plant roots. Thus, plant parasitic nematodes are difficult to control and growers have often depended on soil fumigation. However, many products used for soil fumigation can no longer be used as they are considered hazardous to the environment. Therefore, availability of antagonists that breach nematode defenses is extremely helpful.

Numerous specialist and generalist microorganisms are known to attack plant parasitic nematodes within the soil and some soils are generally suppressive to certain plant parasitic nematodes. Our knowledge about the ecology of these antagonists and the conditions necessary for them to provide effective control is increasing, but studying these systems around plant roots presents difficulties. Major groups of antagonists include (1) obligate nematode pathogens such as the bacterium *Pasteuria penetrans*, the fungus *Hirsutella rhossiliensis*, and most of the nematode-trapping fungi (see Figure 16.3); (2) facultative nematode pathogens such as *Purpureocillium lilacinus* and fungi in the genus *Pochonia*; (3) bacteria that inhabit the root zone, called rhizobacteria, that often produce toxins or modify root exudates used by nematodes; and (4) competing microorganisms that colonize roots, such as mycorrhizal fungi.

Let's consider the potential of these first two groups for commercial production. First, obligate pathogens do not grow outside of their host naturally and they are often difficult to culture. Obligate pathogens also often have narrower host ranges, making the activity of a commercial product very specific. While there are environmental advantages to specificity, less of the product would be sold compared with an antagonist with broader specificity. For facultative pathogens, one additional asset is that they can increase in the soil on their own and perhaps less material would have to be applied. Facultative pathogens are usually easier to mass-produce and are less specific.

The soil environment and host plant can have a strong influence on nematodes as well as on the community of antagonists in the rhizosphere. As an example, in soils where the cereal cyst nematode (*Heterodera avenae*) is not abundant, nematode populations appear to decline over a period of a few weeks after the new generation of females and eggs are exposed in the root zone before moving into cysts. It was hypothesized that during this time of exposure, females and eggs might be attacked by fungal pathogens. Modeling the dynamics of various fungi in the soil, the facultative pathogen *Pochonia chlamydosporia* was found to increase in density when nematodes had increased and vice versa. Nematode-suppressive soils have been identified numerous times but further study has not always attributed the antagonistic activity to the same natural enemy. These findings suggested that *P. chlamydosporia* could be reacting to nematodes in a density-dependent manner and thus regulating pest densities. However, density dependence in other fungi that are parasites, such as the nematode-trapping fungi

(*Arthrobotrys* and others), has been difficult to demonstrate and densities of organisms potentially competing with nematodes are not always associated with nematode densities.

There are currently few commercial products for biological control of plant parasitic nematodes and their use to date is limited. The field is advancing through the identification of key antagonists and study of their ecology. This is critically needed because the soil fumigants previously widely used for nematode control have been banned in many countries. As examples of some species that are sold for control of plant-pathogenic nematodes, the facultative fungal pathogen *Purpureocillium lilacinus* is an egg parasite of root-knot as well as cyst nematodes and also attacks fungal root pathogens. The rhizosphere-dwelling bacterium *Burkholderia cepacia* is an aggressive colonizer of the root zone and outcompetes plant-pathogenic fungi as well as plant parasitic nematodes. As an alternative approach, researchers have discussed methods for improving soil to conserve and enhance antagonists of nematodes and strategies are actively being investigated.

Further Reading

Askary, T. & Martinelli, P. R. P. (eds.) (2015). *Biocontrol Agents of Phytonematodes*. Wallingford, UK: CABI Publishing.

Gajendran, G., Ramakrishnan, S., Subramanian, S., Jonathan, E. I., Sivakumar, M., & Kumar, S. (eds.) (2009). *Biological Control of Plant Parasitic Nematodes*. Udaipur, India: Agrotech Publishing Academy.

Harman, G. E. (2000). Myths and dogmas of biocontrol: Changes in perceptions derived from research on *Trichoderma harzianum* T-22. *Plant Disease*, **84**, 377–393.

Kerry, B. R. (2000). Rhizosphere interactions and the exploitation of microbial agents for the biological control of plant-parasitic nematodes. *Annual Review of Phytopathology*, **38**, 423–441.

Mérillon, J. M. & Ramawat, K. G. (eds.) (2012). *Plant Defence: Biological Control*. Dordrecht, Netherlands: Springer.

Narayanasamy, P. (2013). *Biological Management of Diseases of Crops. Volume 2: Integration of Biological Control Strategies with Crop Disease Management Systems*. Dordrecht, Netherlands: Springer.

Stirling, G. R. (2014). *Biological Control of Plant-Parasitic Nematodes: Soil Ecosystem Management in Sustainable Agriculture*, 2nd edn. Wallingford, UK: CABI Publishing.

Weller, D. M., Raaijmakers, J. M., McSpadden Gardener, B. B., & Thomashow, L. S. (2002). Microbial populations responsible for specific soil suppressiveness to plant pathogens. *Annual Review of Phytopathology*, **40**, 309–348.

Wojciech, J. J. & Korsten, L. (2002). Biological control of postharvest diseases of fruit. *Annual Review of Phytopathology*, **40**, 411–441.

Zaidi, N. W. & Singh, U. S. (2013). *Trichoderma* in plant health management. In Trichoderma: *Biology and Applications*, ed. P. K. Mukherjee, B. A. Horwitz, U. S. Singh, M. Mukherjee, & M. Schmoll, pp. 230–241. Wallingford, UK: CABI Publishing.

Part VI

Biological Control: Concerns, Changes, and Challenges

18 Making Biological Control Safe

18.1 What are Nontarget Impacts?

Most natural enemies have hosts that they preferentially attack but many are also able to use a secondary set of hosts when the most common, preferred hosts are unavailable. The use of a secondary host or prey species is sometimes referred to as a spillover event. To determine what are primary versus secondary options, the specificity of natural enemies being considered for biological control of pests must be tested. Initially the natural enemies are evaluated to ensure that they do not target humans or other mammals as first, second, or even distant alternatives. Very few biological control programs are aimed at controlling vertebrates (but see Chapter 11 for controlling invasive rabbits and cats); therefore, natural enemies that would affect any vertebrates are only rarely considered for biological control.

The acute and chronic toxicities toward vertebrates associated with some synthetic chemical pesticides are unheard of for biological control agents. Herbivores attacking weeds and parasitoids and predators attacking arthropod pests, which represent the vast majority of biological control agents used nowadays, do not represent any significant danger for humans and other vertebrates (but see in Box 18.2 for the case of the *Harmonia* lady beetle causing annoyance through aggregations in buildings). However, a few of the species of microorganisms that have been considered for biological control of plant pathogens or arthropods can act as opportunistic pathogens of humans, especially affecting immune-compromised individuals. Some of these fungi and bacteria also produce secondary metabolites that can act as toxins or carcinogens. Because of these variations in biology among microbial strains, researchers must ensure that the microbial strains used for biological control are not pathogenic or otherwise harmful to humans (Box 18.1). Perhaps of greater concern regarding effects of biological control agents on vertebrates are immune responses by workers regularly exposed to massive amounts of biological control agents during mass production or during field applications. Precautions have been developed to make sure workers are not hypersensitive and to prevent high levels of exposure to any agent that could provoke such a response. No matter what type of agent is being used for control, label directions for application should always be followed.

Today, the vast majority of organisms of concern regarding unintended effects of biological control programs are invertebrates and plants. An important group of nontarget species that are considered for testing if at all appropriate to the natural enemy are

Box 18.1 Safety of *Bacillus thuringiensis*

The most widely used biopesticide, *Bacillus thuringiensis* (Bt) (Chapter 10), merits discussion regarding safety, a subject that has been investigated extensively. Since this pathogen is a common soil inhabitant worldwide, the issues are whether higher doses than naturally occur will have deleterious nontarget effects. Of course, results from any studies must be viewed with regard to whether the dose being used for studies reflects doses that could be encountered in nature. As everyone knows, while nutrients at low concentrations are needed by our bodies, in very high doses the same compounds can be toxic. In addition, would the nontarget organisms of concern living in the environment ever be exposed to the high doses of Bt used in laboratory tests?

Bt has been shown by testing and many years of experience to be neither toxic nor harmful in other ways to mammals, including humans. Proven cases of Bt causing clinical symptoms in humans are very rare and seem related to instances where the bacterium could enter and grow as a saprotroph in, for example, a burn wound.

Of greatest concern is the effect of Bt on nontarget invertebrates. Individual strains of Bt exhibit quite different characteristics and there are multitudes of strains. Surprisingly, Bt almost exclusively affects phytophagous and never carnivorous hosts, and therefore has little direct impact on predators. Overall, Bt has rarely been found to directly harm any beneficial invertebrates. Bt can indirectly impact beneficials by robbing them of their prey or hosts but, for integrated control with the goal of lowering pest populations, Bt is useful in the majority of circumstances. As for honey bees and earthworms, the Bt strains that are used for pest control have shown no adverse effects.

Further studies have addressed the diversity of species that might be impacted by use of Bt in noncrop areas hosting native species, such as aquatic habitats that would be sprayed for mosquito control or forests that could be sprayed to control outbreak populations of caterpillars. No study has ever detected nontarget species at risk of eradication after Bt applications. Extensive studies have been conducted evaluating the effects of Bt active against gypsy moth (*Lymantria dispar*) on the diverse native caterpillar fauna living in the same habitat as gypsy moths in the eastern United States. In West Virginia, Linda Butler and associates (1995) found that the populations of native caterpillars present when Bt sprays were applied declined the year of application. Dave Wagner and associates (1996) took this one step further and followed populations of caterpillars in forests during a Bt treatment year as well as the following two years. During the treatment year, 19 of the 20 common caterpillar species on foliage decreased slightly but most species recovered the year after. In fact, for gypsy moth, it has been suggested that not controlling outbreaks in some way would have at least as great an impact on the native fauna as spraying Bt, as a result of alterations in the habitat when outbreak populations of gypsy moth larvae eat most of the leaves in the forest in spring.

Another part of risk evaluation is determining persistence. Bt is a spore-producing bacterium and Bt spores do an excellent job of persisting in protected locations such as the soil and can be re-isolated at least several years after release. The Bt toxin released from Bt cells survives only a relatively short time. Bt spores persisting in an area do not seem to pose a problem because this pathogen is found worldwide and rarely causes epizootics on its own.

In summary, studies have proven that Bt has reason to be considered an environmentally friendly means for pest control. In "Keeping an eye on *B. thuringiensis*," Bernard Dixon (1994) points out that even when he considered safety of Bt with a critical eye, any organism so widely disseminated with virtually no adverse incidents over several decades is likely to be relatively safe.

beneficial organisms, such as crop plants, horticultural plants, honey bees, silkworms (*Bombyx mori*), and other biological control agents, like parasitoids and predators. All of these organisms are valued because they are directly used by humans or they provide ecosystem services. Close watch has always been paid prior to use of biological control agents to prevent negative effects of natural enemies on beneficial nontargets. The potential for effects on beneficial organisms by a natural enemy considered for release is determined prior to release. In recent years, the spectrum of species being evaluated as nontargets has expanded. With growing interests in protecting the biodiversity of native ecosystems, effects of biological control agents on the native flora and fauna, especially threatened or endangered and rare species, are of concern.

Concerns about safety have focused on both classical biological control as well as more recently, releases for augmentation biological control, largely because the results of both of these types of releases are considered irreversible in the event of establishment of an exotic biological control agent. In particular, nontarget effects to economic plants have always been scrutinized closely during programs for the biological control of weeds to make certain that the natural enemies being investigated would not attack economically important plants, such as crops. Historically, effects from inundative or inoculative releases were considered temporary, occurring during the period of time when natural enemies that are released are abundant, but then declining afterward. However, there has been discussion regarding exotic natural enemies released inundatively that can then persist and spread, while impacting native biodiversity (Box 18.2). In the past there has also been an assumption that natural enemies used in greenhouses were warm-adapted and would not survive if they escaped in colder climates, although we are finding that this is not always correct. Nontarget effects of conservation biological control have not historically been evaluated because natural enemies are not released, although it is now recognized that manipulations can change the densities of native species. Because many microorganisms are considered ubiquitous (see Chapter 16) and conserving biodiversity usually does not include microorganisms, biological control programs utilizing microorganisms to control plant diseases often have not focused on nontarget effects once it is determined that the macroflora and fauna are not impacted.

For classical biological control, introducing a specialized exotic natural enemy to control a pest from the area of origin of the natural enemy (an 'old association') is considered much safer than introducing an exotic natural enemy to control a pest that the

Box 18.2 Lack of Harmony with *Harmonia*

The lady beetle *Harmonia axyridis* is native to Asia but is now well known in both Europe and North America and can also be found in South America and South Africa. Relatively few insect species have a common name but the fact that this species has numerous common names, including "multicolored Asian lady beetle," "harlequin ladybird," and "Halloween lady beetle," is an indication of how familiar it has become; in the scientific community it is often just referred to as *Harmonia*. Introductions of *H. axyridis* for control of aphids began in North America in 1916 but this species was only considered established in 1988 around New Orleans and, in 1991, in the northwestern United States, after which *H. axyridis* spread across the North American continent. In Europe, *H. axyridis* was introduced in France in 1961 and the Ukraine in 1964 and began being sold as a natural enemy for indoor purposes by biological control companies from 1995 in France, Belgium, and the Netherlands. Some of these beetles escaped and since then this species has spread across much of Europe, even moving north into Scandinavia.

Among lady beetles, this species is considered fairly large, with adults being 6–9 mm long. In fact, adults can be a bit difficult to recognize as there is extensive variability in their appearance since they can have from 0–22 spots and the coloration of their elytra (wing covers) varies from red or orange with black spots to black with red spots. The immature and adult stages are generalist predators and they are a bit unusual among lady beetles because they will colonize many types of habitats, including feeding on aphid colonies on leaves in trees.

Eventually, after this species became established in Europe and North America, several different kinds of problems began to occur. First, this species is able to outcompete and even eat other lady beetles, especially the eggs, larvae, and pupae, so in general the abundance of native lady beetles has often declined after *Harmonia* became established. Studies are often specifically conducted by sampling lady beetles in trees, where the declines in native species have resulted in "habitat displacement," as the native species are no longer in habitats that *Harmonia* prefers. *Harmonia axyridis*, however, isn't that particular about what it eats and has been seen eating butterfly eggs and larvae along with herbivorous beetles introduced for weed biological control. However, perhaps *H. axyridis* is best known by the public because of the fact that adults move to protected mass-aggregation sites where they spend the winter. Unfortunately, the beetles often choose to aggregate in human houses and people generally do not like sharing their living spaces with insects. These home invaders can also produce a defensive secretion that can stain surfaces within the house and allergic responses to their presence have been reported at times. A third problem caused by *H. axyridis* also directly impacts people by damaging crops. In fall, the adult beetles sometimes feed on wine grapes and if bodies of adults contaminate grape clusters at harvest, the flavor of the resulting wine is tainted.

Figure Box 18.2.1 An overwintering indoor aggregation of *Harmonia axyridis* in the corner of a windowsill (photo by Stephen Cresswell).

In summary, when *H. axyridis* was collected in Asia with the purpose of introducing it to the west for biological control of aphids and scales, researchers knew that it was a generalist and they also knew that overwintering aggregations of these beetles invaded houses. It was incorrectly assumed that these beetles would mostly stay indoors in greenhouses and would not survive outside in colder climates and therefore did not pose a threat to the environment. Researchers did not know that *H. axyridis* would impact wine production and this lady beetle also surpassed expectations about the nuisance it has caused to humans. Therefore, this species has created an excellent example that both direct and indirect effects of a generalist predator can be unexpected and unacceptable.

natural enemy did not evolve with (a 'new association'). For a 'new association', also called neoclassical biological control, the exotic natural enemy must have a somewhat broader host range so that it will accept the native pest as a host, although it has not previously encountered that pest species. The extension of this argument is that if an exotic natural enemy will accept a native pest with which it did not evolve, what other, nontarget species might it use as host or prey? For this reason, the neoclassical approach is usually only undertaken when other approaches have failed or are not available.

18.1.1 Host Specificity

Central to concerns about nontarget effects is understanding the host specificity of a natural enemy. Of course, this term refers to the range of plant species attacked by an herbivore or plant pathogen as well as the range of host or prey animals attacked by predators, parasitoids, and pathogens. The breadth of hosts utilized by a natural enemy can be categorized along a continuum ranging from monophagous to oligophagous to polyphagous. Control programs generally strive to use natural enemies that are monophagous, those species utilizing only one host or prey species (or sometimes this term can be expanded to several host or prey species, but from the same genus). However, efficient, strictly monophagous natural enemies that can be used in biological control programs are seldom available. Sometimes this is because the strictly monophagous natural enemies simply do not exist in the community of natural enemies associated with a pest. Highly specific natural enemies can pose challenges for biological control use because not all species considered for biological control are easy to culture successfully and sometimes a monophagous species that appears promising is just too difficult to work with; this concern is mainly relevant for augmentation biological control as for classical biological control the amount that natural enemies are produced before release is less. More generally, species that usually exploit only one host or prey species occasionally switch, or "drift," and use different host/prey species. This behavior of using a limited number of host/prey species is called oligophagy. The other end of the spectrum is polyphagy, eating many types of food, which in this sense could be many prey species, like a generalist predator. This type of feeding strategy is now never considered for introductions of exotic species.

18.1.2 History of the "1980s–2000s Nontarget Controversy"

During the earlier years of biological control, broad host specificity (polyphagy) in natural enemies was usually not considered an impediment to their use. Pests at outbreak densities can cause extensive damage to crops that provide food and fiber and the effects of natural enemies on nontarget invertebrates were often not even considered detrimental during attempts to control devastating pests. Instead, emphasis was on whether the natural enemy was controlling the pest. When less was known about biodiversity and our environment, and humans were struggling to battle pests to provide food and shelter for themselves, a broad host range was even considered beneficial because the natural enemy population could persist after target pests were under control and would then be able to respond more quickly if and when another pest outbreak occurred.

During the twentieth century, food production became more dependable and the environmental movement initiated by Rachel Carson (see Box 1.1) developed a stronger voice toward valuing and protecting the environment. In truth, the potential for nontarget effects on the native invertebrate fauna were noted as early as the 1890s in Hawaii. By the mid-1980s, the first influential paper from an environmental viewpoint (Howarth, 1983) publicized the fact that biological control agents were attacking native organisms that were not pests and stated that this should not occur. Everyone agreed that introductions of generalist vertebrate predators like the mongoose and cane toad (see Section 18.2.2) had been huge mistakes that should not have occurred and that vertebrates should mostly not be used for classical and augmentation biological control in the future (except in specific circumstances; see Box 7.1 and Section 14.4). However, there was much discussion in the scientific literature about other biological control agents attacking nonpestiferous invertebrates and plants, and these unintended hosts/prey were referred to as nontargets.

Through the 1990s and 2000s, arguments against biological control by some conservation-minded scientists became quite heated. A few exceptionally bad examples, where nontarget effects became abundantly clear, received attention and some of these are described below (or in more detail in Heimpel & Mills, 2017). However, as discussed in the chapters on classical and augmentation biological control, these practices have been used to control pests in multitudes of systems and with many different natural enemies. Extensive study confirmed that host switching was definitely occurring to some extent but, for the majority of instances, the natural enemy population was not causing significant population-level changes in nontarget species (see below for more detail). After much discussion, it seems that the major concerns about the environmental safety of biological control can be narrowed down to three main issues: (1) whether native species are now extinct because of the activity of introduced natural enemies, (2) whether there are population-level reductions in nontargets because of biological control agents, and (3) whether there are indirect effects impacting nontargets, caused by introduced natural enemies.

Extinction

There are no irrefutable examples where biological control agents have been solely responsible for driving nontargets to extinction. Perhaps the closest case is the introduction of the predatory snail *Euglandina rosea* to Moorea, an island near Tahiti in the Society Islands, for control of the giant African snail, *Lissachatina fulica*. Instead of controlling the giant African snail, the predator instead ate snails that were much smaller, in the genus *Partula*, which had been the focus of studies of evolutionary biology. Today, there are no longer any *Partula* snails on Moorea. (Interestingly, those *Partula* were not native to the island of Moorea but had been introduced.) It is clear that *Euglandina* played a part in the extinction of *Partula* from Moorea, although it was not the only culprit in *Partula*'s demise; habitat destruction as well as human collecting are also thought to play a part. Also, in this instance, there are numerous islands in the Society Islands and seven of the *Partula* species that had been extirpated (i.e., extinct only in specific areas but not everywhere) from Moorea were still present on other islands, although two are now thought to be extinct globally.

In another often-mentioned example of extinction caused by classical biological control, a closer look has not proved that the nontargets were, in fact, extinct. The parasitic fly *Bessa remota* that was released against the coconut moth, *Levuana iridescens*, in Fiji in 1925 (Box 18.3) was thought to have attacked a related native moth (*Heteropan dolens*), causing its apparent extirpation (i.e., local extinction). However, in this case, it is questionable whether the coconut moth is native to Fiji and whether this parasitoid even attacks *H. dolens*, the native moth.

Population Reduction

As for whether classical biological control decimates populations of nontargets, a closer look at the data has shown that there is little proof, at least for insect introductions. One recent review reported that out of over 2,000 exotic insect natural enemies released for classical biological control against insects, only eight have caused

Box 18.3 Controlling the Coconut Moth in Fiji

The purple-winged coconut moth, *Levuana iridescens*, was a devastating outbreaking pest of coconut palms in the island of Viti Levu in the Republic of Fiji in the early 1920s and began spreading to other Fijian islands. Because the ethnic Fijian culture depended on coconut palms, a biological control program was undertaken. Initially, researchers thought that the coconut moth was probably native to Viti Levu, so this is where searches for natural enemies were conducted. Actually, since that time, thinking has changed and researchers would now assume that if a pest was out of control in an area (as this moth was on Viti Levu), that was probably not where the pest originated because in its area of endemism the natural enemies would keep moth population densities low or at least lower than severe outbreaks.

Because no natural enemies were found attacking coconut moths on Viti Levu, the frustrated researchers next hypothesized that the coconut moth was actually an invasive species, and it must be rare wherever it was native because focused searches were not finding it elsewhere. The search for natural enemies was thus expanded to natural enemies attacking closely related caterpillars in Indonesia and Melanesia. In 1925, during an outbreak of the related moth species, *Palmartona catoxantha*, near Kuala Lumpur, Malaysia, both wasp and fly parasitoids were found in abundance. Seventeen large cages were constructed and filled with 85 young palm trees hosting 20,000 parasitized and unparasitized caterpillars. These cages traveled by rail to Singapore and then spent 25 more days on a ship headed for Fiji. By the time the cages were opened in the quarantine on Viti Levu, no wasps had survived but a total of 315 individuals of the parasitic fly *Bessa remota* were still alive. These flies were easy to rear on coconut moth caterpillars and by 1926, 15,000 flies had been released in coconut palm-growing areas of Fiji. The flies established and then dispersed in the Fijian archipelago on their own and did a great job of controlling the coconut moth. By 1929, a final report from the project leaders stated that no new outbreak of the coconut moth had occurred since 1926.

There have been criticisms of the impact of this program on Fijian biodiversity but these were based in part on some mistaken impressions. It now seems most likely that the coconut moth was indeed an invasive species on Viti Levu and in fact is probably still there but at low levels, especially because competitive invasive species have been introduced that use the same resources for larval development. So, this pest is probably not extirpated from Fiji. However, it was also claimed that the parasitic fly *B. remota* caused the extinction of a related native moth, *Heteropan dolens*, which had not been collected after *B. remota* was established. However, there are no records of *B. remota* attacking this native species. Also, this native moth is thought to possibly be extirpated from Viti Levu but it can be found on other Fijian islands, so it is not actually extinct (which would mean that it did not occur anywhere). This is also a case of lack of data; the programs supporting entomologists in the South Pacific waned and then ended and there is a lack of recent data from this difficult-to-access system on islands in the middle of the Pacific Ocean. Finally, at issue is also a question of priorities: without this classical biological control program, either the coconut palms would continue to be decimated, with drastic effects on the ethnic Fijian culture, or continued insecticide spraying to control coconut moths would have impacted a broader range of the invertebrate fauna.

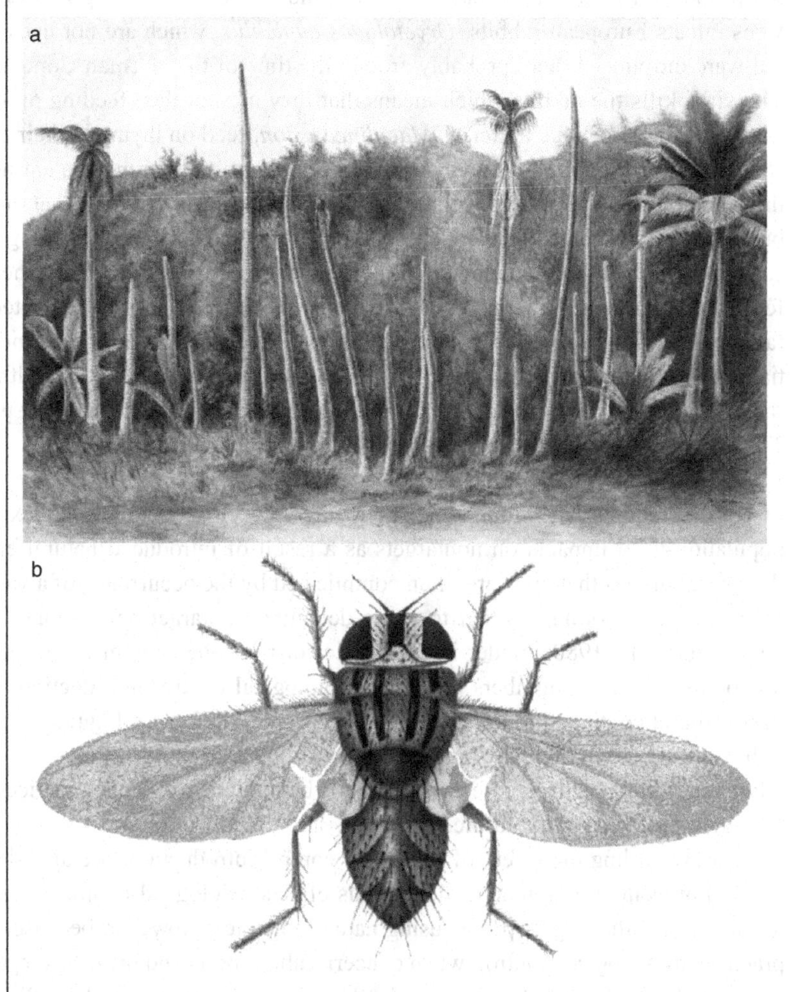

Figure Box 18.3.1 (a) Appearance of failing and dead coconut palms in Fiji after two or sometimes three successive outbreaks of the coconut moth, *Levuana iridescens*. (b) The introduced tachinid *Bessa remota*, which lays its eggs externally on larvae of the coconut moth (Tothill et al., 1930).

confirmed population-level declines in nontarget species, while an additional four examples are suspect. Whether the number is eight or twelve, this is still an extremely low percentage of the total. For 512 insects introduced for biological control of weeds, only four (0.8 percent) caused population-level nontarget effects and, of these, all were thistles or cacti and within the same genera as the intended hosts (so that host switching would have been tested and predicted).

Indirect Effects

The third concern is about indirect effects, which are very difficult to predict. One example that is often cited is of the myxoma virus (see also Chapter 11) which was not purposefully introduced but was first found in the United Kingdom in 1953. This virus infects European rabbits, *Oryctolagus cuniculus*, which are not native to Britain but were introduced there probably around the time of the Norman Conquest in 1066. The virus kills the rabbits which means that they are not then feeding on the grasses. Larvae of the large blue butterfly, *Maculinea arion*, feed on thyme in their early instars and then are transported by ants, *Myrmica sabulei*, to live within the ant nests, where they provide some food for ants but they also eat ant larvae. It seems that the large blue butterflies are completely dependent on the ants and the ants need shorter grass (in this case trimmed by rabbits) to construct their nests. So without the rabbits, there were too few ant colonies and the large blue butterfly is now extirpated from the United Kingdom (although it still occurs in other parts of Europe). The virus is considered indirectly partially to blame for extirpation of this butterfly from the United Kingdom, although other factors had an impact, including land-use practices that did not favor ant populations. This butterfly has also now been successfully reintroduced.

Overall, scientists have found that there are few clear examples of extinctions or population-level impacts on nontargets as a result of introduced natural enemies and that the examples that exist are often complicated by the occurrence of a variety of different factors working together to cause declining nontarget populations. Regardless, beginning in the 1980s, under intense criticism from the ecological and environmental communities, the numbers of classical biological control introductions per decade declined and continued to decline in the 1990s and 2000s (see Figure 3.2). Below, we will continue discussing why biological control in some cases has received a bad reputation regarding nontarget effects and what is being done to change practices to prevent nontarget problems in the future.

Issues regarding the safety of biological control introductions are aimed at identifying and preventing unintended, deleterious effects. Biological control programs strive to impact specific target species using natural enemies. However, because of previous practices in biological control, when concerns about pests and priorities regarding their control differed from today, some natural enemies were released that affected nontarget native invertebrates and plants. The controversy about the environmental safety of biological control that began in the 1980s was principally fueled by concerns about protecting biodiversity. In fact, in this era of extensive reliance on chemical pesticides, a goal of biological control has been in helping the environment by developing methods for an environmentally safer type of pest control that avoided use of synthetic chemical pesticides.

During earlier periods when biological control was practiced up until the 1980s, considerations for preserving all biodiversity were not a high priority. However, that view changed and the field of biological control needed to realize this and make changes. Now, biological control has changed to include practices to preserve biodiversity, not only through reduction of reliance on synthetic chemical pesticides but also through using natural enemies that will not (or only minimally) impact populations of native species in the flora and fauna. But first, we will discuss major reasons that nontarget effects occurred.

18.2 Reasons Nontarget Effects Have Occurred

Relative to the total numbers of classical biological control releases, there are few examples of significant nontarget effects caused by biological control agents. Above we presented the very low percentages of classical biological control releases having population-level impacts on nontarget species. However, some critics say that these low levels of significant nontarget effects are solely low because of a lack of documentation. Although many classical biological control programs are evaluated directly after release to document establishment and efficacy of natural enemies that were released, later post-release monitoring has been minimal, if occurring at all. In addition, after releases, only the natural enemy and pest are generally monitored and only rarely have nontarget species been included. To make before and after comparisons more difficult, there is rarely much objective information about the abundance patterns for many species in the native flora and fauna before the pest or introduced natural enemies arrived (e.g., Box 18.3), a lack of so-called baseline data. Therefore, if native species increase or decrease in abundance after releasing a natural enemy, we often cannot say whether yearly changes in the densities of native species are normal or not, as it is not atypical for native populations to vary from year to year. For these reasons, there are few well-documented examples that prove the extent that nontarget species have been affected by biological control agents.

Among the examples where nontargets have been affected by natural enemies, there are several general scenarios to explain why natural enemies that would later cause nontarget effects were released.

18.2.1 Changing Priorities

Today, we are witnesses to results from biological control programs made in previous eras. In some cases, decisions to introduce natural enemies were made when any concern regarding effects on the native flora and fauna was overshadowed by the problems caused by the pest. This certainly occurred in earlier years when vertebrate predators were released for pest control. Although vertebrate predators, with the exception of a few fish species, are no longer released, these early introductions remain, serving as examples of classical biological control introductions gone wrong.

In the early 1900s, gypsy moth caterpillars were causing extensive defoliation year after year in New England, the northeastern states of the United States. Controlling this pest of hardwood forests and orchards was very difficult and numerous classical biological control introductions were made. One of the natural enemies that was introduced from Europe, a parasitic fly, *Compsilura concinnata*, that attacks larger gypsy moth larvae, became established (Figure 18.1). This is a species of tachinid fly with adult females that pierce the host cuticle and inject living larvae into hosts; this is

Figure 18.1 Adult female of the tachinid parasitoid *Compsilura concinnata*, first introduced to North America for control of gypsy moth, *Lymantria dispar*, in 1906. Approximately 7.5 mm long (photo by Joyce Gross, UCB; Bugwood.org).

possible because the egg hatches within the mother. Researchers knew that in Europe this fly was highly polyphagous. At that time, the urgency for control of gypsy moth was such that polyphagy was not considered a reason to prohibit release of this parasitoid. After establishment, this fly regularly attacked gypsy moth larvae but, although it can have an influence, *C. concinnata* has never had a large enough impact to regulate gypsy moth population densities. In recent years it was noted that the populations of beautiful, large, wild silk moths (Saturniidae) in New England were not as abundant as recorded in the literature prior to the arrival of *C. concinnata*. Jeff Boettner and Joe Elkinton hypothesized that *C. concinnata* could have influenced silk moth densities and they investigated this by placing caterpillars of the native North American cecropia moth (*Hyalophora cecropia*) in the field to see if they were parasitized. Indeed, the cecropia caterpillars were parasitized by *C. concinnata* at high enough levels to strongly suggest that the activity of this parasitoid could be responsible for the present-day low densities of cecropia moths. These researchers also included two other silk moths (*Callosamia promethea* and *Hemileuca maia maia*) in their studies and found the same trends. This is clearly an example of changing priorities over time; maintaining the biodiversity of the endemic moth fauna of New England is a priority for many people today but this certainly was not a priority that carried weight in the early 1900s, although wild silk moths were valued and were commonly collected by hikers and others who reared the moths as a hobby. In fact, *C. concinnata* was released from 1906 to 1986 to control 13 different pest species but recently the tide has turned and it is highly doubtful whether *C. concinnata* would be released today since it is such a generalist.

Another example of changing priorities is the release of the flowerhead weevil, *Rhinocyllus conicus* (Figure 18.2), for control of introduced thistles, but especially the musk thistle, *Cirsium nutans*. It is generally thought that the more specific a natural enemy for use in classical biological control, the better the resulting control and the less chance of nontarget effects. However, this weevil example demonstrates that to prevent nontarget effects, genus-level specificity is not always good enough when there are native species in the same genus that could be impacted. This weevil does not have

Figure 18.2 The flowerhead weevil, *Rhinocyllus conicus* (*c.* 10–15 mm in length), which is native to Eurasia, has been introduced to Argentina, Australia, New Zealand, and North America for control of weedy thistles (drawn by Alison Burke).

broad tastes and it feeds on thistles in the genus *Cirsium* but North America has 90 native species of *Cirsium* that are widespread and several of these species occur in the same areas as the invasive thistle that this weevil was released to control. Based on host specificity testing that had been conducted before releases, researchers predicted that this weevil would feed on the native *Cirsium* species. The weevil has had population-level impacts on the native Platte thistle, *Cirsium canescens*, in the sandhills of Nebraska. Releases of *R. conicus* have resulted in extensive controversy over the choices that were made when releasing this weevil, especially as the weevils feed on at least two native species of *Cirsium* that are considered endangered. However, at the time that releases began in the 1960s, concern for controlling thistles that cattle cannot eat and which outcompeted forage plants on rangeland, was considered a higher priority by the government and land managers than preserving the native thistle flora. Once again, the decision to release these weevils would not be made under today's practices.

18.2.2 Insufficient Information about a Natural Enemy

The cane toad, *Rhinella marina*, provides an excellent example of disastrous problems because of lack of knowledge about the ability of a natural enemy to control a pest. This example occurred long ago and can perhaps be explained by difficulties in communication and in conducting scientific studies during those times. Cane toads were introduced to numerous areas in the 1920s to 1930s. In Australia, these toads were introduced to control greyback (*Dermolepida albohirtum*) and frenchi (*Lepidiota frenchi*) cane beetles (Scarabaeidae) feeding on sugarcane at a time when worries about poor yields in sugarcane controlled such decisions. Insufficient efforts were made to understand the biology of the toad before release. It turned out that *R. marina* is a very effective predator with an incredibly wide host range. The toads would certainly eat the sugarcane beetles if they were in the same locations at the same times, but unfortunately, they are not. When cane toads eat insects, they eat those on the ground but the sugarcane scarabs that the toads were supposed to attack were high on the plants. Also, when the adult scarabs are in the cane fields, there is no cover on the ground and cane toads, avoiding sunlight, aren't normally found in fields without cover. The cane toads therefore provided no control of the target pests. However, these toads readily became established and increased to large numbers. They are toxic when eaten as prey and they also kill and eat many endemic species, with the result that their spread in eastern Australia has radically changed the community-level food webs. Today, the cane toad is regarded as a pest animal itself and huge efforts to control it have been initiated.

Another program based on too little information is the introduction of the predatory snail *Euglandina rosea* to control the giant African snail on Moorea, as described earlier (see Section 18.1.2). This predator had been introduced to Hawaii to control this same pest in 1955 but was not effective and had been implicated as one of several causes of the decline and extinction of the native Hawaiian tree snails that it ate. Unfortunately, this prior experience in Hawaii did not prevent the introduction to Moorea, probably because the Mooreans were not aware of the overall effects of this predator in Hawaii.

18.2.3 Fragile Ecosystems

Nontarget effects certainly would have greater impacts in fragile ecosystems with less complex food webs. In particular, islands often host fewer species of endemic plants and animals than continents, and island endemics are often thought to have fewer preformed defenses against natural enemies. Some of the most strident voices criticizing classical biological control have used examples from Hawaii. This is an extreme example because not only is Hawaii an isolated group of islands, but, because of its location as a stop-over point for shipping, especially when the world relied on sailing ships, many pests were introduced there and many natural enemies were released to control these introduced pests.

More classical biological control introductions have been made in Hawaii than anywhere else in the world. Between 1890 and 1985, 679 species of organisms were introduced to Hawaii for biological control of insects, weeds, and other species and 243 became established. Since 1900, the majority of natural enemies released (71.6 percent) have been parasitoids and predators released to control insects. Relative to this number of releases, the number of negative effects is not far from the worldwide averages. However, few of the reported occurrences of nontarget effects are based on objective data. In fact, in most of the cases it is impossible to tell whether the decline of a nontarget was

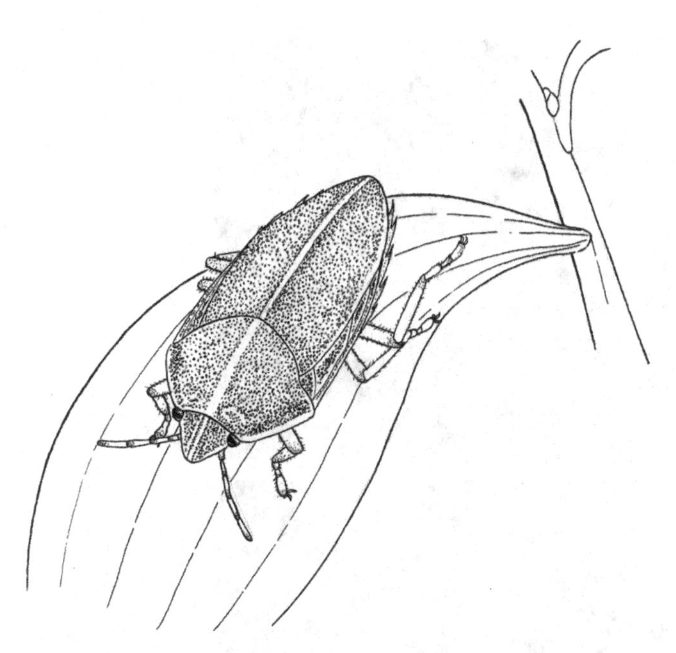

Figure 18.3 The native Hawaiian koa bug, *Coleotichus blackburniae* (*c.* 15 mm in length) (drawn by Alison Burke).

caused by an exotic natural enemy that had been purposefully released, by competition from an invasive species, by some accidentally introduced (invasive) natural enemy, or by some other factor limiting the populations such as habitat depletion or degradation.

The beautiful green koa bug, *Coleotichus blackburniae* (Figure 18.3), that feeds on the native koa trees of Hawaii provides an example of the difficulties in documenting nontarget effects. An agricultural pest, the southern green stink bug (*Nezara viridula*), was accidentally introduced to Hawaii in 1961 and, by 1963, two species of exotic parasitoids were being introduced for its control. One was a parasitic fly (*Trichopoda pilipes*) attacking late-stage nymphs and adults and the second was a tiny egg parasitoid (*Trissolcus basalis*). The southern green stink bug and the koa bug belong to closely related families of true bugs, the Pentatomidae and Scutelleridae, respectively. In 1991, largely based on records from museums, it was stated that the parasitoids introduced to control southern green stink bug had caused the decline of the koa bug. However, an in-depth study by Peter Follett started out by finding that koa bugs were still abundant locally if you went to the field looking for them in the correct places. Life table analyses demonstrated that invasive generalist predators, not introduced for biological control, caused the highest mortality to koa bugs in all habitats on Hawaii. Parasitism of koa bug eggs by the egg parasitoid introduced against *N. viridula*, was virtually nonexistent at 21 of 24 sites. However, at three sites with higher bug densities, egg parasitism levels were higher. In this case, evaluating nontarget effects of biological control introductions is complicated because the releases had been conducted several decades before these data were collected. There is no clear answer regarding whether koa bug populations have declined since the natural enemy releases against *N. viridula* because no one quantified typical koa bug population densities before releases. If koa bug populations

Figure 18.4 (a) Caterpillar of the pyralid *Cactoblastis cactorum* (length 25–30 mm), which attacks prickly pear cactus, is bright orangish-red with black transverse stripes. (b) Adult moth (wing span 2–3.5 cm) and eggs that are laid on top of each other to form long sticks resembling cactus spines (photos by J. Carpenter, USDA ARS).

did decline at some time after the natural enemies were released, the cause of the decline was not documented. Most of the evidence for a decline in koa bug populations comes from Oahu, the most heavily populated island in Hawaii, and low densities of koa bugs on Oahu could easily also result from the current lack of native koa trees and invasions by generalist predators. Since the biological control agents that were introduced in 1963 attacked koa bugs in the laboratory before introductions, this example also shows that prerelease testing predicted that some degree of nontarget impacts would occur, although impacts of predatory invasive species and loss of habitat were not predicted.

18.2.4 Dispersing Natural Enemies

A few examples have shown that natural enemies can disperse unexpectedly. A caterpillar from Argentina, *Cactoblastis cactorum* (Figure 18.4), was introduced in Australia in 1926 to control species of the cactus *Opuntia* that had been introduced and had increased to create impenetrable, thorny thickets over huge areas. Amazing levels of control were recorded in Australia, where *C. cactorum* virtually cleared enormous areas of these weedy cacti. Based on this success, in 1957 and 1960 this caterpillar was introduced to several islands in the Caribbean. This cactus family has no native species in Australia and *C. cactorum* is quite host specific so there were no nontarget concerns in Australia. However, it was a different picture in the Caribbean where there are many species of *Opuntia* that are native. In these Caribbean islands, the target species were primarily weedy native *Opuntia*, but there are also *Opuntia* species in the Caribbean islands, as well as in Mexico and North America, that are valued. No one knows how it traveled, but in 1989, caterpillars of this weed control agent were found in Florida. Closer observation documented *C. cactorum* attacking all six species of *Opuntia* native to Florida and, of special concern, this includes the rare Floridian semaphore cactus (*Opuntia corallicola*). The semaphore cactus is rare because of habitat loss as it lives along the increasingly developed Floridian coast. However, the fact that *C. cactorum* will attack this rare cactus certainly is problematic because the few plants remaining must now be protected. Of more serious concern is the fact that this genus of cactus is diverse in the southwestern United States and Mexico and, in Mexico, the cactus pads (nopales) and fruit are used as food. In fact, *Opuntia* is so well regarded as part of Mexican culture that it is illustrated in the center of the Mexican flag, by an eagle holding a snake in its beak, perched on an *Opuntia* plant. As *C. cactorum* disperses through the southwestern United States and into Mexico, *C. cactorum* may need to be controlled both to maintain biodiversity and ecosystem structure and to protect plants of this genus used as a source of food and cultural identity.

18.3 Direct versus Indirect Effects

Nontarget impacts can potentially take two different forms. (1) Natural enemies can directly impact nontarget as well as target species by using these as hosts or prey. (2) Natural enemies can indirectly impact nontarget species because of their activity influencing other organisms within food webs. An example could be feeding by a natural enemy, which in turn impacts a third species, such as a competing native natural enemy. Indirect effects are especially troublesome because they are as difficult to prove as they are to predict. Another example of an indirect nontarget effect would be competition between exotic and native natural enemies. Generalist lady beetles that have been introduced for aphid control can be so effective that they outcompete native lady beetles. One of these biological control agents, the seven-spotted lady beetle (*Coccinella septempunctata*), eats lady beetles as well as aphids. It is thought that indirect effects of North American introductions of these exotic lady beetles caused decreases in population densities of native North American lady beetles, such as the nine-spotted lady beetle (*Coccinella novemnotata*), after the exotic generalists became established. A similar example including unintended spread as well as both indirect and direct impacts would be the harlequin ladybird or Halloween lady beetle, *Harmonia axyridis*, which has impacted native lady beetle populations in both Europe and North America (Box 18.2).

The introduction of the polyphagous parasitoid *Compsilura concinnata* presents a biological control example where a negative impact on some nontarget species also had a silver lining. This parasitic fly was introduced to North America in 1906, primarily for gypsy moth control, and impacted native silk moth populations (see Section 18.2.1), but this parasitoid is now also held responsible for driving populations of a different European invasive, the browntail moth, *Euproctis chrysorrhoea*, to low densities and reducing its distribution. Browntail moths were introduced in 1897 and populations spread and reached high densities over a large area of northeastern North America. Control of browntail moths is important for forest health but the urticating hairs of these caterpillars are also of significant public health concern, leading to documented human deaths with high levels of exposure, especially when outbreak populations have occurred. However, in 1914, browntail populations began to decline and the distribution began shrinking until this species only remains today in a few very localized areas along the northeastern coast of North America. Experimental work strongly suggests that *C. concinnata* is the primary reason for this population reduction.

18.4 Predicting Nontarget Effects

A meta-analysis based on historical data of classical biological control programs using parasitoids and predators against arthropod pests found that the most successful natural enemies were monophagous or oligophagous, with multiple generations per year. As you can imagine, for any natural enemy that is being considered for release, the host range is a critical factor, along with efficacy in controlling the pest. Of course, predicting nontarget effects is not especially simple and practitioners of biological control have learned that several types of evaluations are required. Researchers studying biological control of weeds have more years of extensive experience with host-range testing than other practitioners of biological control. From early times, programs for biological control of weeds always included nontarget testing to assure that natural enemies being considered for release would not eat crop plants.

It is clear that what we actually want to know before deciding whether a natural enemy should be released is what the host range in the release area will be. How can we predict this before introducing a natural enemy to a new area? One clever approach has been to investigate the host range in the natural enemy's area of origin, including thinking about those species that might be present but are not attacked. In fact, this has been done in biological control of weeds for many years. After the natural enemy has been accurately identified to the species level (an important step; see Chapter 3), a review of the literature will often yield a list of known hosts where it is native. However, such lists are frequently based on random field collections or observations, are often incomplete, and the taxonomy from identifications can be incorrect. An evaluation in the field in the area of endemism can give a first glimpse of how narrow or broad the host range of the natural enemy is in nature. Whether a natural enemy uses plants or pests in the area of origin that are closely related to organisms in the release area can also provide initial information about specificity. This kind of pre-assessment information is used to decide whether a natural enemy is even a candidate for host specificity testing, for example, species found to be polyphagous would not be pursued further.

Host-range testing can require several types of studies and requirements will differ for different types of biological control (Box 18.4). For example, natural enemies that are insects require testing that takes behavior into account, offering choices of different life stages of natural enemies, such as adults versus immatures versus eggs. Natural enemies that are microbes would not require tests where the natural enemies make active choices, although different life stages of hosts could be tested at different doses because pathogenicity and virulence can differ by dose and host stage; estimating and using realistic doses that would be present under field conditions is an important consideration.

Box 18.4 Testing Host Specificity

Some natural enemies feed on only one or a few hosts while others feed on numerous species. This variability in the number of species affected is seen across the diversity of natural enemies feeding on microbes, plants, or other animals. To determine whether nontarget effects could occur, the host range of the natural enemy must be determined. However, with many natural enemies, it is often not so simple to determine host/prey specificity with accuracy. Therefore, specific methods for testing have been developed.

Phytophagous Natural Enemies

Methods based on natural enemy biology have been developed for testing the host specificity of herbivorous arthropods. Different stages of any given insect can react differently to the same plant species and they do not always react in the same way. Both acceptance of plants for oviposition by adults and suitability for development of immatures must be tested. For example, adults of some species are known to lay eggs on plants that are not optimal for development of immatures. Conversely, plant species on which immatures can develop are not always accepted by adults for oviposition. How do you choose which plant species to test? The list of species to test must be short enough to be practical but long enough to answer questions being asked about both economically and environmentally important nontargets. Therefore, it is common that many plant species are tested including, in particular, any beneficial plants that seem like possible hosts. Often, those species closely related to the target pest are chosen for testing first, with successive species for testing being increasingly distantly related. This is called the **centrifugal phylogenetic method** and is based on choosing species for testing that occur in the continent of release, according to their relation to the target. It can work well for natural enemies that only attack species closely related to the host. However, some phytophagous natural enemies decide on acceptability of plant species based on plant chemistry or plant morphology and for such natural enemies, plants to test should instead be chosen based on the chemicals they produce or their morphology.

Another strategy for determining the host specificity of phytophagous natural enemies is called the "reverse-order" method; this method includes separate evaluations of development of immatures and oviposition by adults. First, larvae of the candidate agent are fed plants of interest, including the target weed species as a positive control. The individual insects being tested are each fed species of plants separately, with no choice. If they will not eat or cannot develop successfully, then this

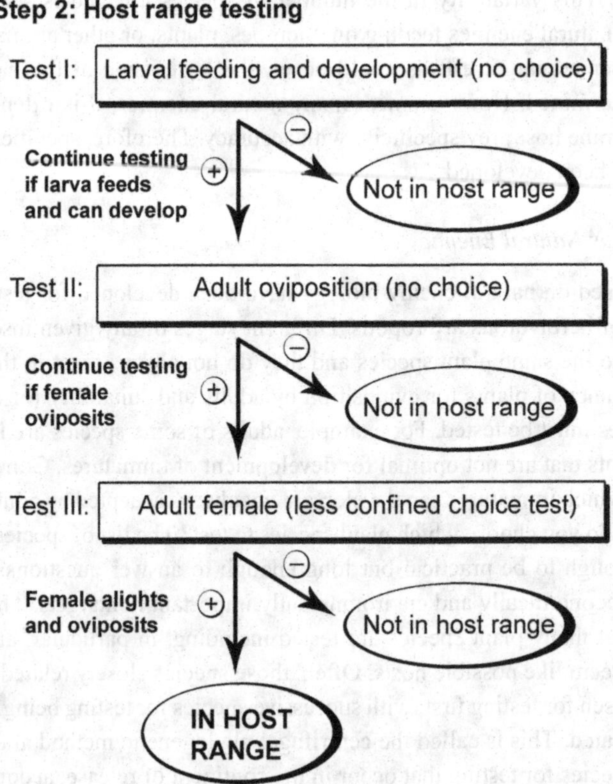

Figure Box 18.4.1 Testing sequence for evaluating the host range of herbivorous arthropods. Referred to as "reverse-order" because ability of larvae to develop when fed on a plant species is tested before adult acceptance of the plant (based on Wapshere, 1989).

phytophagous species could be used. However, plant species on which the insects can develop successfully from egg to reproductive adult next undergo a second tier of testing. Plants are next tested to see whether adults will oviposit on them when not offered a choice. Results will indicate whether adults will ever accept the plant in question, an important fact if this plant species is scarce or any injury to it is not acceptable, as would be the case for rare and endangered plants. For the majority of plant species, the major concern is whether both oviposition and development would occur regularly. Only plants on which adults will lay eggs remain in the study for the third tier of testing when oviposition by adults is tested against a choice of numerous plant species. This entire test therefore identifies both whether immatures can develop and whether adults will lay eggs on plant species of concern.

Once scientists feel certain that they understand the host specificity of a natural enemy and that it seems to have a good chance of efficacy, a decision must be made regarding whether that agent should be released. Here, we should mention an example where a decision was not straightforward, with an interesting twist that impacted arriving at a decision. A plant named *Echium plantagineum* was introduced in the 1800s to Australia where it grew too abundantly and was designated a noxious weed in 1909. This plant is especially problematic because it has been linked with deaths of horses and sheep that have eaten it extensively; the pathology is as a result of liver damage from the pyrrolizidine alkaloids in this plant. Controlling this weed is difficult and classical biological control began to be investigated in 1972. A leaf-mining caterpillar (*Dialectica scalariella*) was introduced in 1980 and a stem-boring beetle (*Phytoecia coerulescens*) and two leaf beetles with root-mining larvae (*Longitarsus aeneus* and *L. echii*) were identified as promising control agents. However, biological control of this plant became a source of controversy. Although the common name used by ranchers for this plant was Paterson's curse, the common name used by beekeepers was Salvation Jane. The abundant flowers of this plant were considered a major source of nectar and pollen for honey bees. Whether to continue introductions of classical biological control agents or not was eventually decided by a panel including scientists, farmers, and beekeepers as well as the general public. In the end, two inquiries found that control of the weed was in the national interest before the natural enemies were released. Perhaps this was the start toward the current processes in some countries of including scientists, stakeholders, and the public in decisions on whether to release biological control agents.

Predators and Parasitoids

For testing the host specificity of predators and parasitoids, once again evaluations start under highly controlled situations. Understanding the behavior of these natural enemies that kill their hosts or prey requires attention to detail. We know that factors like the age and egg load of the natural enemy, densities of natural enemy and host or prey, length of exposure, level of host deprivation prior to testing, and type of testing arena can all affect whether a natural enemy will attack or not. A list of species to test is usually chosen based on taxonomic/phylogenetic affinity and sometimes ecological similarity with the target, along with what has been called "safeguard considerations," which include valued organisms, such as organisms providing ecosystem services.

Taking into account such factors, once a micro-environment for testing has been developed and natural enemies and hosts/prey are available, the first type of study usually undertaken exposes appropriate stages of nontargets to natural enemies without offering a choice. Concurrently, to document that the natural enemy was healthy, a subset of individuals should be provided with hosts/prey that they are known to attack successfully. If the nontarget hosts/prey are attacked, the next stage is to expose nontargets to natural enemies in a small arena, giving the natural enemies a choice of the nontarget versus the target. The last stage requires more effort but can potentially be more valuable, with studies conducted in a larger and more realistic arena, providing numerous nontargets and targets to try to simulate conditions that would occur in nature.

Controlled specificity and efficacy tests under optimal conditions are usually conducted in quarantines, laboratories, or caged areas, depending on the types of tests and natural enemies. Sometimes, some of the testing can be performed in the native area where quarantine might not be needed. Data from controlled exposures will yield the physiological host range, or those hosts that could potentially be used by the natural enemy for development under optimal conditions. This can also be thought of as the "maximal host range."

Concerning macro-natural enemies it is appropriate to also include behavioral components in studies to make sure that the natural enemies will accept the plant, host, or prey under normal conditions and successfully develop on it; in this case, the terminology for trials including behavioral components would be fundamental host range. However, in controlled studies where natural enemies are not given a choice, this can be more like forcing the natural enemy to use that host/prey, which they would seldom or never choose naturally, given a choice. In order to learn the fundamental host range, studies should include natural conditions and should provide choices to see whether the natural enemies would and could choose to utilize various targets and nontargets. However, using no-choice studies is considered the more conservative and safer way to estimate host range.

For many species of natural enemies, all of the prey or hosts that might be attacked under laboratory conditions will not be attacked under natural conditions because of behavior, timing, or habitat. This is referred to as the ecological host range (or, occasionally, as the realized host range) and represents the range of hosts actually attacked under natural conditions. The gypsy moth fungal pathogen *Entomophaga maimaiga* could infect 28 out of 78 species of nontarget caterpillars in the laboratory but only three of these were found infected in the field and infection was at very low levels. This is in part because the gypsy moth has unusual behavior for a caterpillar. Larger larvae rest in the leaf litter and, in doing so, become exposed to the abundant spores of this fungal pathogen in the surface layers of the soil. Few of the other caterpillar species that are potentially susceptible to this fungus ever venture to the ground for long periods of time, let alone spend all day long there as gypsy moth caterpillars do. The ecological host range can therefore be quite different from the physiological or fundamental host ranges, but it is the ecological host range that is important for biological control purposes.

Of course, we are trying to predict the species that a natural enemy will attack in a new area. Information on the flora and fauna of the area for release, as well as contiguous areas where the natural enemy might disperse, are important for consideration. Concerns about nontarget effects of agents for the biological control of weeds have been alleviated when herbivores specializing on a plant family that is not native to the release area are considered for release against an invasive weed belonging to that plant family. We have to make the best estimates based on appropriate information before making decisions about the potential nontarget effects of natural enemies. Adding information about host range from the laboratory and from the area of endemism to knowledge of the ecology and behavior of the natural enemy and the species that inhabit the area of release and could be affected will help to arrive at the best prediction regarding nontarget impact. Once this information is in hand, practitioners of biological control can make informed decisions about whether specific biological control agents should be considered for release. After release, field evaluations to determine nontarget effects are critical so that we can match predictions with results and gain more information toward preventing nontarget effects of biological control. Because natural enemy population densities can vary significantly from year to year, evaluations encompassing several years are preferable.

Indirect effects are much more difficult to predict and these involve ecological investigations at the community level. Information about the communities in which releases will be made may help to identify potential indirect effects. For example, natural enemies to be introduced could compete with native natural enemies or they could attack native natural enemies. Alternatively, introduced natural enemies could hybridize with natives, leading to changes in the genotypes present. A general trend though seems to be that indirect effects can be avoided or decreased if the natural enemy population decreases to low levels after responding to damaging densities of pests. In such cases, there are fewer natural enemies in the environment to impact other species.

18.4.1 Will Host Specificity Change?

Concern has been voiced regarding whether the host ranges of natural enemies will change after release. Biological control of weeds has undergone quite a bit of scrutiny regarding host ranges of natural enemies for potential introduction, but an in-depth analysis of different programs found no evidence for changes in host ranges after introductions. Historically, any changes that have been reported were because of new associations (e.g., exotic herbivores were introduced that had not encountered certain host plants previously), errors in sampling, or changes in the ecology of the biotic community, such as changes in host densities. There have been a few instances when a natural enemy population was very abundant and the regularly used hosts (the target pests) became rare because of overuse by the natural enemy, resulting in the utilization of a previously nonattacked host: a spillover event. This situation has been found to be transient, occurring only with abnormally high natural enemy densities and usually not long after natural enemies had become established. The natural enemies increase to large numbers when using an overabundance of hosts but then run out of hosts through their own efficiency. This kind of spillover situation occurred when the lace bug *Teleonemia scrupulosa* was released against the weed *Lantana camara* in Uganda. When bug populations became very abundant and little lantana remained, the bugs began eating sesame, but this had not been predicted. However, feeding on sesame by this bug was transient and *T. scrupulosa* has now been introduced to a total of 27 countries and islands and has not been reported as a pest of sesame in any other instance.

Projects for the biological control of weeds in the United States, the Caribbean, and Hawaii, including 112 insects, 3 fungi, 1 mite, and 1 nematode, were evaluated by Robert Pemberton (2000). He found that all attacks on nontargets could have been predicted based on test results before organisms were released. Nontarget effects were not due to changes in the host ranges of exotic natural enemies but instead, nontarget plants that were attacked were all closely related to the target weed and occurred in the same habitats. Therefore, nontarget effects could have been prevented by only targeting weed species having few to no closely related species native to the release area. For this analysis, no studies documented unpredictable changes in host range by natural enemies after their introduction. The key seems to be that excellent knowledge of the host range of the natural enemy is required as well as the breadth of species occurring in the release area.

While a true change in absolute host acceptability and preference for a natural enemy would require genetic changes, which are less likely, a change in the intensity of an association already in place is more possible. Generally, this could be a decrease in effectiveness of a natural enemy against a pest or an increase in resistance in that pest against the natural enemy. Reciprocal changes like this would be called coevolution. A major example of coevolution in biological control has been the coevolution documented between the myxoma virus and its rabbit host in Australia (Chapter 11). After release, the virulence of the myxoma virus decreased while rabbit resistance to the virus increased. This change is believed to have been partially driven by natural selection in the pathogen to enhance virus transmission and therefore persistence. The mosquito-transmitted virus is only transferred from living infected rabbits to healthy rabbits so, if rabbits die quickly after infection, the virus had less chance of being transmitted. When the virus became less virulent, rabbits survived longer after infection so there was a longer window for the pathogen to be transmitted to a new host. In examples of a rust fungus attacking a weed and a parasitoid attacking an aphid, natural enemies have become less effective after release, along the lines of what was documented with the myxoma virus (although this has rarely been reported).

Another possible way for populations of natural enemies to change genetically could occur if a very small number of natural enemies are present, causing a genetic bottleneck. In fact small numbers of natural enemies are released very frequently, as they are not always so easy to rear or keep alive before they are released, and then few are available for release. Yet, changes caused by a bottleneck have never been reported. A third possible way for genetic changes to occur is if an exotic natural enemy being released is closely related to a native species and hybridization occurs. To date, this has been reported rarely and, even then, levels of hybridization were low and no changes in the effect of the natural enemies on the hosts were reported.

Fundamental or ecological host range is what is important regarding changes in host specificity. This can be determined by a single behavior or trait that is thought to occasionally be coded by a single gene, especially in cases where individual chemicals act as repellents or deterrents. However, it is much more common that different aspects of host use are determined by a diversity of genes and, in these cases, all of these genes would have to change coincidentally in order to change the range of hosts that are used. It has been argued that there is less chance of such a change for an introduced species (like an exotic biological control agent) than for a native species and phylogenetic studies of these changes in native species show that this might take place over hundreds of thousands of years.

Therefore, as previously stated, there are no examples of changes by natural enemies to use new hosts that could not have been predicted based on the known host range of a natural enemy, either through investigating the activity in the area of origin of the natural enemy or through prerelease studies of host specificity. In addition, prior to releases, researchers should be aware of the relatedness of natural enemies being released to native species. However, biological control researchers should also remain aware of the potential for changes and should conduct studies when appropriate.

18.5 Preventing Nontarget Effects

As described above, nontarget effects of natural enemies have been documented in the literature but these were few and far between, until criticism about adverse nontarget effects of classical biological control releases began in 1983. In 1983, and then 1991, the effects of exotic natural enemies on the native biota in Hawaii were summarized by Howarth, which started reviews of the issues of nontarget effects of biological control. Then, the effects of the weevil *Rhinocyllus conicus* on thistles on the North American Great Plains began being reported and this example drew further attention to the occurrence of nontarget impacts. Although the releases made in previous years (i.e., when decisions regarding release of natural enemies were not based on preserving native biodiversity) remain, the scientific community has been addressing the concerns raised.

18.5.1 Environmental Risk Assessments

Countries each have their own methods for determining the risks of releasing exotic natural enemies for classical or augmentation biological control and their own legislation about how this is regulated. However, biological control using exotic species really requires international collaboration as releases can have international repercussions. For countries on the same continent, introductions of biological control agents could result in establishment and subsequent spread across several countries. Therefore, the Food

Table 18.1 The code of best practices for classical biological control of weeds.

1	Ensure target weed's potential impact justifies release of exotic agents
2	Obtain multi-agency approval for developing a program against the target
3	Select agents with potential to control the target
4	Release safe and approved agents
5	Ensure only the intended agent is released
6	Use appropriate protocols for release and documentation
7	Monitor impact on the target
8	Stop releases of ineffective agents or when control is achieved
9	Monitor impacts on potential non-target organisms
10	Encourage assessment of changes in plant and animal communities
11	Monitor interactions among species of agents
12	Communicate results to the public

Source: based on Balciunas (2000).

and Agriculture Organization of the United Nations developed a Code of Conduct for the Import and Release of Biocontrol Agents to provide guidelines for steps necessary toward releases of exotic natural enemies.

The code was developed to provide a thorough framework for making decisions to promote safe use of exotic natural enemies for biological control. First, requests to be able to import exotic natural enemies into quarantines must be submitted to governmental agencies, regardless of whether use would potentially be for classical or augmentation biological control. The guidelines require a list of the three main groups involved in a biological control release: the national authorities, the importers, and the exporters, and their respective responsibilities during different phases of an introduction program. One major initial responsibility is that the biology of the natural enemy and its potential risks must be summarized. This must also be accompanied by communicating information and results to the scientific community and the public as well as encouraging ecological assessments of the effect of the natural enemy on ecosystems in the areas where it will be released.

Following the establishment of the general code, biological control of weeds researchers recognized the need for establishing professional standards for introductions in their subdiscipline. Therefore, a "Code of Best Practices for Biological Control of Weeds" that includes 12 steps that should be taken for classical biological control of weeds introductions was ratified by members of this subdiscipline (biological control of weeds) at an international symposium in 2000 (Table 18.1). This code ends by including study of plant and animal communities after releases and sharing all results with the public.

We have discussed methods for evaluating host specificity of agents, information central to any decisions, but how are results translated into decisions? Polyphagous agents or those requiring nontargets as part of their life cycle are rarely if ever considered for release today. In this category, an important native parasitic wasp attacking gypsy moth larvae in Japan, *Glyptapanteles liparidis*, has several generations per year and overwinters in larval hosts. Gypsy moth has only one generation per year and overwinters as eggs. Therefore, after its first generation each year in Japan *G. liparidis* parasitizes caterpillars other than gypsy moth. For this reason, this species has not been considered for importation to, and release in, North America; although it is an important natural enemy controlling gypsy moth in Japan, it is thought that this parasitoid would certainly affect nontarget caterpillars in North America. Agents known to attack close relatives of native species that naturally occur in the release areas would also be poor choices for introductions. Attention should also be paid to whether the natural enemy could affect other species in the community in some indirect way. A parasitic wasp (*Diachasmimorpha tryoni*) was introduced to Hawaii early in the twentieth century for control of the Mediterranean fruit fly, also called medfly (*Ceratitis capitata*). Later, a gall-forming fly (*Eutreta xanthochaeta*) in the same family as medfly (Tephritidae) was introduced for biological control of the weed *Lantana camara* in Hawaii. Unfortunately, this parasitic wasp now attacks the agent introduced for biological control of the weed. Once again, practitioners are learning from past experiences and we've learned that a broader vision of possible ecosystem effects is important and must be adopted when planning biological control programs.

How Do You Quantify Risk?

One question critical to evaluating risk is how much of an effect by a natural enemy is considered "adverse." This is a very subjective judgment. For ardent conservationists, an unacceptable risk could be one nontarget individual being affected. However, what most can agree is of concern is when the natural enemy has an impact on population densities of native species, especially including valued species.

Risk is considered the interaction between the potential magnitude of an impact (often referred to as hazard) versus the chance that the interaction will occur (the exposure). Numerous methods for analysis of the risk of biological control introductions have been explored. We will include a stepwise method created by Joop van Lenteren and colleagues for making objective decisions about the environmental risk of biological control agents (Figure 18.5). The major criteria that are of concern are establishment and permanence, host range, dispersal, and direct and indirect effects. At each step, the magnitude of the potential impact as well as the chance of this occurring are considered and the risk for that step is quantified. First, whether the agent can become permanently established is of concern for augmentation but not classical biological control, where establishment is the goal. Next, data on physiological and fundamental host ranges are used to evaluate the potential of the natural enemy to impact related, unrelated, and valued nontarget species. This is followed by evaluations of the extent to which the natural enemy will spread and then the possibilities for direct unintended effects (for example, intraguild predation) or indirect effects. Based on these data, the overall risk can be evaluated, followed by discussions of whether these risks could be avoided or reduced.

Finally, the risk of introducing natural enemies should also be weighed against the pest-control alternatives. No discussion of nontarget effects is complete without mentioning that nontarget effects caused by a natural enemy cannot be considered alone. If decisions are made to control a pest and biological control is not chosen, it is likely that a different form of control, often use of synthetic chemical pesticides, will be used

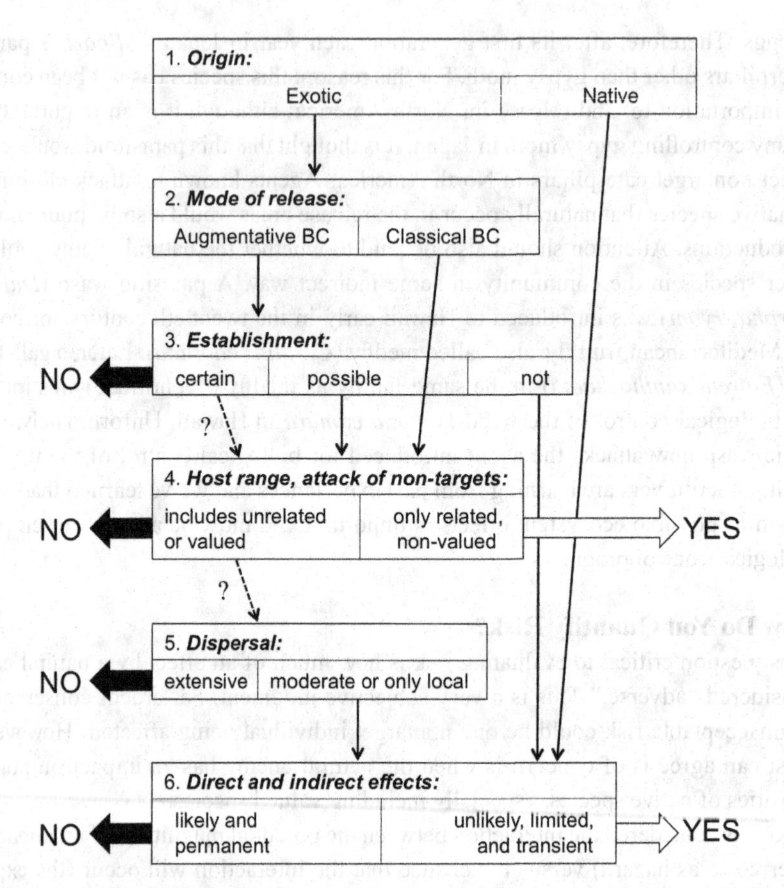

Figure 18.5 A stepwise environmental risk assessment scheme for potential release of an arthropod agent for biological control (BC). YES = release recommended; NO = not recommended. "?" indicates that if desired, information on additional levels in the scheme may be assessed to allow reconsideration of a decision to not release an agent (based on van Lenteren et al., 2006).

instead. In such an instance, the range of flora and/or fauna affected could be much greater than after release of an exotic natural enemy. While the pesticides would be present for a shorter time than a permanently established exotic natural enemy, pesticides might need to be applied repeatedly. Of course, another alternative would be to not control the pest at all. In such a case, an outbreak of the pest could have a strong impact on the ecological community, affecting nontargets but in different ways. For example, an invasive weed not being controlled could outcompete the native plant species and few endemic plants would remain. Along the same lines, if a pestiferous herbivore was eating all of the foliage of endemic plants, normal inhabitants of that ecosystem might not be able to survive under the altered conditions, either through loss of host plants or through physical changes to the microhabitats they require. Therefore, in deciding about using exotic natural enemies for pest control, the effects of the different alternatives on populations of nontargets must also be considered. As an example of trust in the safety of biological control, host-specific biological control agents are usually considered first for pest control in areas where rare and endangered species occur. For example, the highly specific gypsy moth nucleopolyhedrovirus is used against gypsy moth, instead of Bt, in areas hosting threatened Karner blue butterflies (*Plebejus melissa samuelis*).

As a caveat, one of the major types of data used for evaluating risk is data on host range, usually predominantly based on controlled laboratory studies; these data are thought to provide a conservative view in order to prevent any risk of nontarget effects. A recent study investigated the host-range data for five species of insects that had already been released for biological control of weeds. Based on present-day evaluation of these data, none of these five species would have been released today; all would have been considered too risky. However, all of these species had contributed significantly to control of major weeds and none had displayed any environmental risks. The authors suggested that appropriate risk and benefit should be considered at the habitat level and natural enemy behavior should also be included in evaluations. In addition, as stated earlier, procedures should always incorporate post-release monitoring so that we learn more about the potential for direct and indirect nontarget impacts and compare these with the priorities in place for making releases.

18.5.2 Regulations

Use of organisms for pest control is generally regulated by governmental organizations. For exotic macro-natural enemies, environmental risks, as described above, are generally at issue, but rules and regulations differ by country. A major concern is whether the natural enemy is exotic or not, although information about the biology and host specificity of the natural enemy will be required regardless. Organisms for classical biological control are always evaluated on a case by case basis. For augmentative biological control in the United States and Europe, macro-natural enemies and entomopathogenic nematodes are not regulated to the same extent as microorganisms. Microorganisms being developed as pest control products must undergo mammalian safety testing. This is based on mammalian testing required for evaluating new chemical pesticides, although much less testing is required for microbes than chemicals. The cost of registering a microbial biological control agent is therefore much lower than a chemical pesticide. However, the registration requirements will involve registrations of the same product separately in different countries, which can make registration of microbials cost-prohibitive. Particularly when the biological control agent is targeting a niche market, such as a crop that is not grown over huge areas, registration can simply cost too much to justify investment by industry toward developing and producing the beneficial. Some microorganisms that are known to act as antagonists are also known to increase plant resistance to stress and nutrient uptake. They can be sold as plant growth promoters, soil conditioners, plant strengtheners, or wound protectants, thus avoiding the testing needed for registration as a biological control agent. Yet, for microorganisms used to control arthropods or weeds, regulatory requirements can constitute a hurdle for industry considering development of microbial biopesticides.

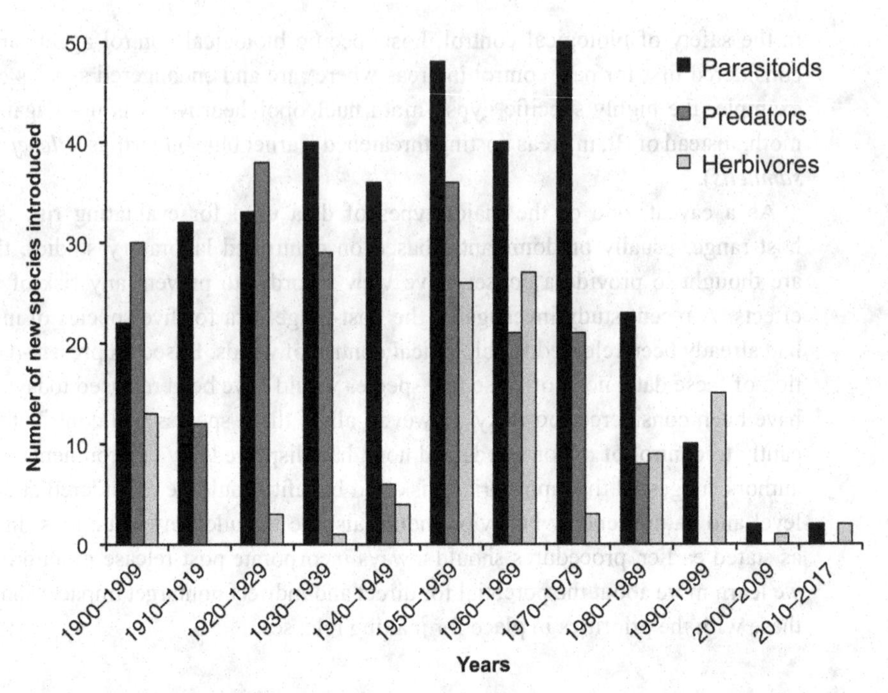

Figure 18.6 Changes in the use of classical biological control in Hawaii, showing the numbers of new natural enemies introduced per decade (from Follett et al., 2000; updated by Follett, Bautista, and Wright).

18.5.3 Some of the Changes to Increase Environmental Safety

Based on the differences in nontarget issues for different biological control strategies, classical and augmentation biological control have responded in different ways.

Classical Biological Control

In general, practices have changed so that more effort and time must be spent evaluating specificity of natural enemies before releases are possible. In addition, once the host specificity testing is completed, it can be slow for regulatory agencies to come to decisions. This has resulted in a decline in the rate of new natural enemy introductions, which began in the 1980s (see Figure 3.2). As an example, for classical biological control in Hawaii, between 1900 and 1980, 3.8 species were released per year (Figure 18.6). This release rate slowed to 2.3 species per year from 1980 to 1989, while between 2000 and 2017, a span of 17 years, 3 agents for biological control of weeds and 4 parasitoids for control of a diversity of invertebrate pests were released and no predators have been released since 1987. Therefore, the releases have decreased to 0.4 per year since 2000.

Augmentation Biological Control

Augmentation biological control is not principally based on use of exotic species but instead on releasing effective natural enemies, either with inundative or inoculative strategies. If exotic species are used, in many countries they now must undergo

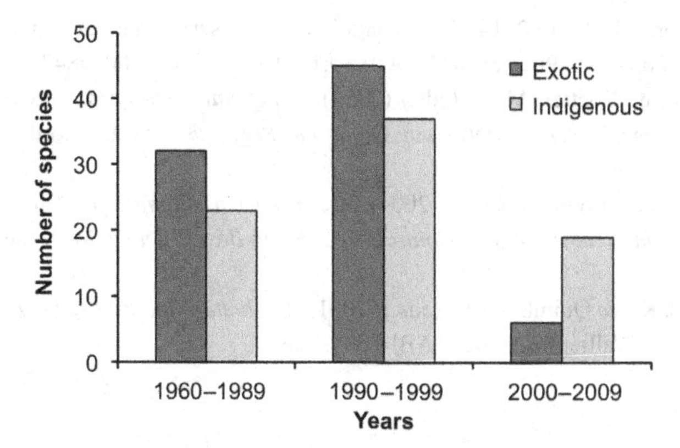

Figure 18.7 Numbers of exotic and indigenous invertebrate natural enemies introduced to the European market over time (van Lenteren, 2012).

environmental risk assessments and subsequent registration with governmental agencies. In 1970, the two species of macro-beneficials commercially available in Europe were exotics. From 1960 to 1999, new species for augmentation biological control were partially exotics and partially native species but from 2000 to 2009, new biological control agents for sale in Europe were predominantly native species (Figure 18.7). Elsewhere in the world, by 2012, some countries were using many native species while others still depended predominantly on exotics for augmentation. In Europe, some commonly used exotic species have been replaced with native species that are thought to be equally effective. We would not be surprised to see this trend in using natives instead of exotics for augmentation continue.

Further Reading

Barratt, B. I. P. (2011). Assessing safety of biological control introductions. *CAB Reviews: Perspectives in Agriculture, Veterinary Science, Nutrition and Natural Resources*, **6**(042), 1–12.

Ehlers, R.-U. (2011). *Regulation of Biocontrol Agents*. Dordrecht, NL: CABI.

Hajek, A. E., Hurley, B. P., Kenis, M., Garnas, J. R., Bush, S. J., Wingfield, M. J., van Lenteren, J. C., & Cock, M. J. W. (2016). Exotic biological control agents: A solution or contribution to arthropod invasions? *Biological Invasions*, **18**, 953–969.

Heimpel, G. E. & Mills, N. J. (2017). *Biological Control: Ecology and Applications*. Cambridge: Cambridge University Press.

Hinz, H. L., Schwarzländer, M., Gassmann, A., & Bourchier, R. S. (2014). Successes we may not have had: A retrospective analysis of selected weed biological control agents in the United States. *Invasive Plant Science and Management*, **7**, 565–579.

Hokkanen, H. & Hajek, A. E. (eds.) (2003). *Environmental Impacts of Microbial Insecticides*. Dordrecht, Netherlands: Kluwer Academic Publications.

Louda, S. M., Pemberton, R. W., Johnson, M. T., & Follett, P. A. (2003). Non-target effects – the Achilles' heel of biological control? *Annual Review of Entomology*, **48**, 365–396.

Suckling, D. M. & Sforza, R. F. H. (2014). What magnitude are observed non-target impacts from weed biocontrol? *PLoS ONE*, **9**(1), e84847. DOI: 10.1371/journal.pone.0084847.

Sundh, I., Wilcks, A., & Goettel, M. S. (eds.) (2012). *Beneficial Microorganisms in Agriculture, Food and the Environment: Safety Assessment and Regulation*. Wallingford, UK: CABI Publishing.

Van Driesche, R. G. & Reardon, R. (eds.) (2004). *Assessing Host Ranges for Parasitoids and Predators used for Classical Biological Control: A Guide to Best Practice*. Morgantown, WV: USDA Forest Service.

Wajnberg, E., Scott, J. K., & Quimby, P. C. (eds.) (2001). *Evaluating Indirect Ecological Effects of Biological Control*. Wallingford, UK: CABI Publishing.

19 Biological Control as Part of Integrated Pest Management

19.1 Using Natural Enemies as "Stand Alone" Strategies

Sometimes, biological control is a "stand alone" method and does not have to be used in conjunction with other methods. This is especially true when effective natural enemies are introduced against pests in uncultivated areas, such as for control of aquatic weeds, rangeland weeds, or arthropod pests in wild lands (i.e., forests, wetlands, etc.), all of these being persistent habitats usually requiring lower levels of management. For example, a classical biological control introduction to combat a forest pest, if effective, would not need further attention unless the new balance established between the introduced natural enemy and pest was disrupted.

Another main benefit of classical biological control is in the developing world where access to pesticides is often restricted for subsistence farmers because of cost. In rural areas in developing countries, biological control can be beneficial, as seen with two African examples of classical biological control: control of mealybugs in staple cassava crops (see Box 8.4) and control of water hyacinth to provide access to waterways (see Box 20.1). The resulting benefits have been food security, biodiversity conservation, reduction in use of pesticides, and sustainability.

If use of natural enemies is intended as a stand alone strategy, generally the focus is on only one pest, as would be typical of a classical biological control introduction. However, use of natural enemies as a stand alone strategy is only possible if the target pest can be maintained below population densities that cause unacceptable damage and in cases where the natural enemies will not be disrupted by controls for other pests in the system.

19.2 Integrated Pest Management

In pest control, more than just one control method is commonly used against one pest or pests that are part of a pest complex requiring various types of control. In such cases, the various types of control that are used must be melded so that they are used in harmony and not at cross purposes. When chemical pesticides must be applied to combat pests, they should be applied so that they do not kill significant amounts of natural enemies controlling other pests in the same system. For example, while spraying fungicides on potatoes might control the fungal pathogen causing late blight (*Phytophthora infestans*), these fungicides also have the potential to kill the entomophthoralean pathogens that are controlling green peach aphid (*Myzus persicae*) populations on potato plants in the same field. To minimize negative impacts on natural enemies, selective pesticides can be sprayed in specific locations or at specific times. A management strategy has been developed to address the complexities that arise in trying to use such an integrated approach and this has been called integrated pest management (IPM). The definition of IPM has evolved since its inception in the 1960s and early 1970s, and numerous definitions are in existence. One of these recent definitions with an ecological emphasis is provided below.

> The careful consideration of all available pest control techniques and subsequent integration of appropriate measures that discourage the development of pest populations and keep pesticides and other interventions to levels that are economically justified and reduce or minimize risks to human health and the environment. IPM emphasizes the growth of a healthy crop with the least possible disruption to agro-ecosystems and encourages natural pest control mechanisms. (Food and Agriculture Organization, 2018)

IPM has been very successful in balancing ecological and economic concerns in a practical manner. The two entomologists who developed this concept and started spreading it around the world, Perry Atkinsson and Ray F. Smith, received the 1997 World Food Prize. Although the roots of IPM were agricultural, uses today have spread throughout pest control of different types of pests in different systems. For example, in New York State, particular attention has been paid to urban IPM programs, because pesticides used in buildings, lawns, and landscapes have a high potential for human exposure. Regardless, most of our discussion of IPM will use agricultural systems as examples.

While this definition for IPM is somewhat long and complicated, the basic goal is to use pest-control tactics only when necessary and, when they must be used, tactics should cause the least collateral harm, especially to other pest controls. This approach is thus called management because sometimes the decision is to employ no control tactics at all and then the term "control" is somewhat inappropriate. IPM suggests using methods for control only if the pest population is causing, or will soon cause, damage above the economic injury level (see Chapter 2). IPM thus always requires a good biological and ecological understanding of the pest system. To use an IPM approach, methods for determining pest densities and knowledge of pest densities that cause economic damage are considered optimal. IPM programs are composed of six basic elements, with biological control fitting in as one of the different control tactics that can be used if control is necessary (Table 19.1).

A critical part of IPM is monitoring the pest population, which is often referred to as sampling or scouting. There are numerous methods for sampling pest populations and these differ based on the pest and the resource it is affecting. For agricultural field crops, a sampling unit could be a leaf or a bud or a specific amount of soil at the base of a plant. For example, to sample tomato fruitworm (*Helicoverpa zea*) eggs, the leaf below the highest open flower is checked because that is where most eggs are laid by the female moths. Traps baited with sex pheromones, the chemicals used by insects for communication between sexes of the same species, are often used to detect or

Table 19.1 Requirements for using integrated pest management (IPM)

1	Pest manager knowledgeable about the system and management strategies
2	Information available about the system being protected, e.g., major pests and when they are active
3	Monitoring the numbers and state of the ecosystem elements, e.g., pest, weather, resource being protected, natural enemies of the pest
4	Economic injury levels for the system
5	Types of control methods
6	Agents and materials to be applied for control, their availability and efficacy

Source: based on Flint & van den Bosch (1981).

quantify densities of moth species. Quantifying the areas on individual plants that display symptoms of infection, or counting the numbers of diseased plants, are used to monitor plant pathogens. Weeds per unit area can be counted to quantify weed populations. The number of samples necessary to arrive at an estimate of the pest population density can be precisely determined for each system based on desired levels of accuracy. During the time that pests are active, sampling is often done on a regular basis, with the sampling frequency dependent on the ability of the pest to increase rapidly to damaging levels. Faster growing populations require more frequent sampling.

Another critical component of IPM is knowledge of the threshold pest density above which damage will lead to economic losses (see Figure 2.1). The concepts of IPM and economic injury levels were originated by entomologists concerned with problems caused in crops by arthropods. However, threshold levels for damaging densities caused by plant pathogens and weeds have also been developed. This concept is less quantifiable when applied to pests that are not causing damage to crops, because it is difficult to assign monetary values to damage or losses. For example, the threshold for "nuisance" pests such as fire ants (*Solenopsis*) living in your backyard or stink bugs overwintering in your attic depends on what is acceptable or tolerable to an individual.

When IPM was first adopted, the initial stage of the transition from scheduled pesticide applications was to monitor pest populations to determine whether pest densities were above damaging levels and to only use pesticides when needed. This was a big change for growers but these initial changes did not incorporate many of the aspects of IPM that researchers were suggesting. With time, as researchers and agricultural advisors demonstrated that alternative control strategies could be used either along with or instead of synthetic pesticides, growers began to trust this new integrated approach. While the concept of IPM is now well known to many, it is used to different extents by different users and according to different systems and circumstances. In the EU, the IPM concept is now considered the standard for all future crop protection and projects to be supported by EU funding must be based on IPM.

Today, adoption of IPM has different levels that can be represented by varying degrees of reliance on pesticides versus natural enemies. Four tiers can be interpreted by their levels of reliance on natural enemies, ranging from complete reliance on pesticides to IPM that completely integrates use of natural enemies (the latter has been referred to

Table 19.2 Levels of use of natural enemies for controlling pests with IPM.

IPM Type	Treatment thresholds
No IPM	Pesticide treatments based on calendar, crop stage, or pest detection but not pest quantification
Low level IPM	Pesticides applied according to thresholds associated with monitoring and timed to minimize impacts on natural enemies and beneficials
Medium level IPM	Thresholds for pesticide treatments adjusted to preserve natural enemy populations
Biointensive IPM	Development of thresholds for releasing natural enemies as alternatives to chemical pesticides

Source: based on Benbrook (1996).

as biointensive IPM) (Table 19.2). An example of biointensive IPM is the control of nuisance flies in dairies in the northeastern United States by integrating sanitation and use of insecticides with a primary focus on use of parasitic wasps that kill flies in the pupal stage, before they become nuisances as adults (Box 19.1).

Box 19.1 Cows without Flies

House flies (*Musca domestica*) and stable flies (*Stomoxys calcitrans*) are common nuisance pests of dairy cattle. The immature stages of these flies, maggots, develop quickly in warm cow manure. After the advent of DDT, many dairy farmers used synthetic chemical pesticides as their only means for fly management. This practice of course resulted in development of insecticide resistance, with resistance found first in house flies. Dairy farmers were probably not well aware that along with developing insecticide-resistant flies, they were killing the natural enemies that previously had been present, resulting in fly populations that were completely out of control.

An assessment of the levels of resistance to insecticides among flies in New York State dairies in 1987 suggested that the usefulness of registered pesticides for fly control was extremely limited. Don Rutz and his colleagues at Cornell University decided that there really must be some natural enemies of flies lurking somewhere in the environment. Once they set about it, 10 species of small parasitoids were found. The most common was *Muscidifurax raptor*, a dark, shiny species of wasp, with adults about 1–2 mm long, that lays eggs after drilling into fly pupae with its ovipositor. *Muscidifurax raptor* has the benefit of being the only parasitoid abundant in both indoor and outdoor fly-breeding microhabitats.

Researchers needed to develop an integrated pest-management strategy that did not rely principally on pesticides, since the registered pesticides were largely ineffective. As is typical of IPM programs, a system was developed with several different types of controls that could be used simultaneously and would complement each other. First, manure had to be removed frequently because *M. raptor* could not respond quickly enough if fly populations were huge. If fly populations were exceedingly high, a pyrethrin insecticide was applied to spaces with abundant flies to knock down the adults. However, central to this program were the inundative applications of *M. raptor*. Fly pupae parasitized by *M. raptor* were mass-produced by a regional insectary and were purchased and released at a rate of 200 per cow in the herd. To release parasitoids, cheesecloth bags containing parasitized fly pupae were hung in dairy barns. Releases were made weekly for approximately 4 months during the spring and summer. Soon after releases, the wasps emerged and flew to search for unparasitized fly pupae on the top of the manure and along the edges of walls. Additional attributes of this natural enemy are that it can search for hosts in areas that sprays cannot easily reach and it can kill pest stages that would not be susceptible to pesticides – such as the well-defended, thick-walled fly pupa.

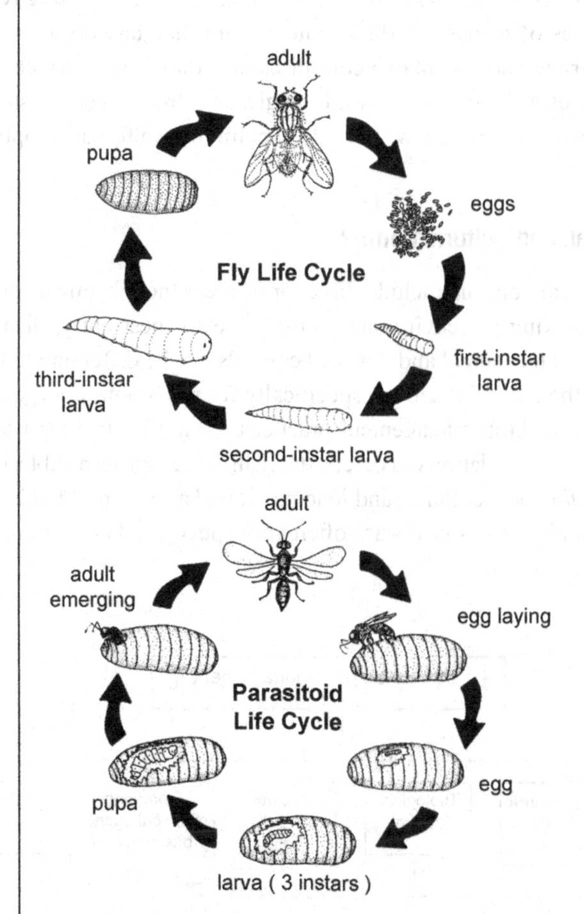

Figure Box 19.1.1 Life cycles of the house fly, *Musca domestica*, and its pupal parasitoid, *Muscidifurax raptor* (Axtell, 1986; © Novartis, reprinted with permission).

For those farms where this IPM program was implemented, fly populations were cut in half and 80 percent fewer insecticide treatments were needed. In summary, this IPM program reduced the costs of pest control for the farms and lowered amounts of insecticides that had to be used – thereby reducing the development of resistance in flies. Last, but certainly not least, the farm was a more pleasant place to work with fewer flies buzzing around and having fewer flies was also greatly appreciated by neighboring homes and businesses.

When pest control is necessary, the alternative control strategies that can be used as part of an IPM system are many and varied. Numerous authors have presented these alternative controls using different groupings and terminologies, but Figure 19.1 provides one example of the relationships between biological control and other options for pest management. Not all of these strategies are acceptable or available to all situations because of differences among individual systems, policies, existing regulations, or personal preferences of managers. IPM strategies are then based on integration of a diversity of appropriate and acceptable control tactics, only one of which is biological control. However, users of IPM often see biological control as a cornerstone among these alternative controls. Below, we will briefly explain these alternative options.

19.2.1 Mechanical, Physical, and Cultural Control

Mechanical and physical controls include direct or indirect (nonchemical) methods for destroying pests or making the environment unsuitable for pest entry, dispersal, survival, or reproduction. Mechanical and physical controls can be differentiated from cultural controls in that the actions taken are specifically for pest control purposes and are not merely extensions of crop management practices. Examples include use of metal barriers around building foundations to deter establishment of eastern subterranean termite (*Reticulotermes flavipes*) colonies and fencing around agricultural fields to exclude deer. Mechanical and physical controls are often very specialized and can require considerable outlays of resources.

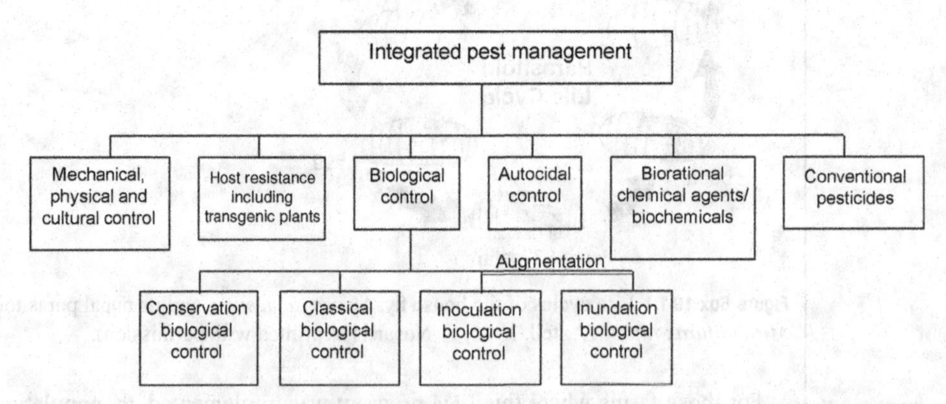

Figure 19.1 Relationship between biological control and other strategies used for integrated pest management. The order of boxes as shown draws parallels between related strategies, such as cultural control and conservation biological control (Eilenberg et al., 2001; with kind permission of Kluwer Academic Publishers).

Cultural controls are modifications of management practices that make the environment less favorable for pest reproduction, dispersal, and survival. Cultural controls include tactics such as improved sanitation, crop rotation, cultivation, trap crops, or adjustment of planting dates. Crop rotation, in particular, has been used extensively for controlling plant pathogens, plant parasitic nematodes, and weeds, as well as insects. If fields are replanted with the same crop, especially year after year, plant pathogens and plant parasitic nematodes can increase to high levels. Rotating crops can break the pest life cycle that depends on presence of the same crop year after year. Conservation biological control might use the same types of techniques as cultural control, involving altering the environment, but for conservation biological control the goal is to increase the natural enemy populations while the goal of cultural control would be to decrease the pest population. Thus, the same changes to the environment could yield the advantages of both conservation biological control and cultural control if, for example, a trap crop planted near a primary crop lures the pest out of the primary crop (cultural control) and flowers produced by the trap crop provide nectar for parasitic wasps that would then remain in the area and live longer to attack more pests within the crop (conservation biological control).

19.2.2 Host Plant Resistance

This strategy was developed based on the fact that different genotypes of plants can vary significantly in their acceptability to pests and in the damage caused by pests. For centuries, humans have selected plants to propagate and grow with positive attributes such as hardiness, productivity, color, and flavor. Past efforts in plant breeding have principally focused on increasing yields of crops and not on increasing resistance to pests. Significant reductions in pest damage to plants in agriculture can be attributed to the efforts over the past few decades of plant breeders who have shifted their focus to developing pest-resistant plant cultivars.

One kind of resistance is inducible; plants can be induced by the activity of pests to produce substances that deter the pest. An example would be the antimicrobial phytoalexins such as pisatin produced in pea plants to inhibit development of plant pathogens. Alternatively, some plants have characteristics that are expressed constantly to resist pests. Such constant, or constitutive, characteristics can include (a) chemical or morphological features that impact pest behavior (antixenosis), or (b) constituents of the plant that disrupt normal growth, development or reproduction of the pest, such as toxins or digestibility reducing factors (antibiosis). Alternatively, some plant genotypes are tolerant to damage by the pests, using mechanisms such as regrowing damaged tissues or producing additional stems or branches to compensate for damage.

Variability in plants can impact organisms that feed on them in many different ways which in turn can impact biological control. As some examples, herbivore development can be altered and with slower development, pests could be exposed to natural enemies for longer. Some herbivores sequester toxic plant compounds gained from plants that help to defend them against natural enemies. Induced plant resistance can cause herbivores to move around more in order to find more acceptable locations for feeding and, in doing so, they could become more exposed to natural enemies. Natural enemies can be attracted to herbivore-induced plant cues (see Chapters 5, 7, and 8) and plant structures can be used, as when mites hide in leaf hairs (see Section 5.3.2).

Advances in genetic engineering have greatly enhanced the possibilities in the area of host plant resistance by enabling the transfer of genes between unrelated organisms to confer resistance to pests. Of particular significance, numerous crop plants have been genetically engineered to either produce the toxins found in *Bacillus thuringiensis* (Bt) for control of insect pests or to be resistant to the active ingredient in the herbicide glyphosate. This technology has been eagerly adopted by growers in some countries, but is also controversial in other parts of the world. A number of important issues remain unresolved, such as potential transfer of engineered genes to closely related weedy species, appropriate marketing and regulation of products from engineered organisms (since some countries do not allow these), and impact of genetically engineered plants on nontarget organisms. Perhaps the greatest long-term challenge for transgenic crops is the potential ability of pests to evolve resistance to transgenics under strong selection pressure. For this example, pests resistant to Bt-engineered plants then could also be resistant to Bt sprays, a mainstay for many growers producing organic crops. In response, intensive studies are being conducted to develop methods to prevent development of resistance to Bt, and US growers planting Bt-engineered crops sign contracts stating that they will use resistance-preventing practices such as planting Bt-free refuges (see Chapter 10). Transgenic crops can be part of IPM programs, if acceptable to the grower, local regulations, and consumers.

19.2.3 Autocidal Control

Using this approach (also called sterile insect technique [SIT] or sterile insect release method [SIRM]), individuals of the pest species are genetically altered to sterilize them and they are then released into the pest population in large numbers to overwhelm the resident population. This strategy is mainly aimed at controlling arthropods and its principal use today is in release of sterile males for insect control, in which case males of the pest insect are sterilized, often using irradiation. Following release in large numbers, they compete with fertile males for female mates, thereby reducing the number of matings that successfully produce offspring. The result is a decrease in the size of the pest population. This strategy has been particularly successful for control of the screwworm fly (*Cochliomyia hominivorax*), a serious pest of livestock. These flies lay their eggs in wounds and feeding by the maggots enlarges the wounds, which then attracts more flies to lay eggs. For this species, each female mates only one time, which makes this method particularly effective. After SIT releases, sterile males are much more abundant than fertile males and when a female mates with one there is no chance that she will reproduce as she will not mate again. Large-scale releases of sterile male screwworm flies started in the 1950s in the southern United States. As screwworm was eliminated, sterile males were released in areas progressively further south, effectively creating barrier zones. This pest was considered eradicated from the United States in 1966 and from Mexico in 1991 and is now absent from much of Central America as successful eradication programs progressively moved further south. Sterile male screwworm flies continue being released in Panama to create a biological barrier so that northward movement does not occur again. The two scientists who developed this method, Raymond Bushland and Edward Knipling, won the World Food Prize in 1992. This basic method has been used successfully against numerous other fly and moth pests, ranging from eradicating the tsetse fly (*Glossina austeni*) that vectors the pathogen causing sleeping sickness in Zanzibar to suppressing the codling moth (*Cydia pomonella*) in an apple-growing region of British Columbia.

19.2.4 Biorational Chemical Agents/Biochemicals

There are numerous chemicals associated with pest control that have biological origins and these are often called biorational pesticides, biorationals, or biochemicals. Perhaps the most commonly used of these are the pests' own behavior-modifying chemicals. Many organisms emit volatile chemical cues that evoke specific behaviors from other individuals of the same or a different species. Pheromones are one category of these chemicals that currently have extensive applications in pest management. Farmers, foresters, homeowners, and government agencies all rely on commercially produced pheromone products. Although the primary use of most pheromones is for detection and monitoring pest abundance and distribution, some are sold commercially for pest control. Mate-attraction pheromones are often used in pest lures or in traps laced with synthetic insecticides or biopesticides. The pheromone-based control method in greatest use is widespread application of pheromones to flood the environment – which disrupts pests' ability to find mates, thereby reducing successful reproduction by the pest. This method, called mating disruption, is presently being used very successfully to slow the westward and southern spread of the invasive gypsy moth (*Lymantria dispar*) in the United States.

Insect growth regulators (IGRs) are naturally occurring hormones or similar synthesized compounds that influence insect growth and are used extensively to control arthropods. IGRs regulate the process of molting, whereby insects repeatedly shed and then form a new cuticle as they grow. When used for control, IGRs kill insects by affecting growth processes, in particular interfering with molting. These insecticides have low toxicity to mammals, but some IGRs affect crabs, shrimp, and other nontarget invertebrates that molt. Concerns about impacts on these nontarget species, some of which are considered beneficial and are economically important, have led to stringent restrictions on allowed uses of IGRs. IGRs are now being examined with renewed interest for use in environments where nontarget impacts are highly unlikely, such as in homes or grain storage elevators. More specific IGRs might be developed for pests affecting high-value resources; however, no species-specific IGRs are presently on the market.

"Botanical" pesticides are derived from plants and are used in the same way as synthetic pesticides but are naturally occurring. Examples include pyrethrins from chrysanthemum flowers, sabadilla from sabadilla lilies (*Schoenocaulon officinale*), and neem from neem trees (*Azadirachta indica*). Naturally occurring botanicals are popular among organic farmers and gardeners because they are derived from "natural" sources. However, scientists point out that some botanicals are no safer as a group than synthetic pesticides and can pose the same questions of mammalian toxicity, carcinogenicity, and environmental impact – so botanicals should be evaluated individually before use.

19.2.5 Synthetic Chemical Pesticides

There is an immense body of literature on synthetic chemical pesticides and only a few aspects relevant to pest management and biological control will be mentioned here. Companies producing synthetic pesticides are continually working to discover new pesticidal chemicals because pests continually become resistant to the pesticides already in use. There is a trend toward discovering chemicals with pesticidal properties that are more specific to the target species to avoid nontarget effects and associated problems, although the trade-off for industry is that their market for each product is then smaller. There is also a trend toward developing less-persistent pesticides to avoid issues regarding long-lived pesticide residues. This can result in the need for more frequent applications, and therefore greater expense for the grower. Users of biological control should be aware of the compatibility of any pesticides they consider using with natural enemies that are also being used.

19.3 Adding an Ecological Understanding to IPM

Although IPM was developed to decrease reliance on synthetic pesticides, these remain the major type of pest control used worldwide today. However, among the scientific community and many members of the public, there remains the desire to place more emphasis on increasing the use of natural enemies. Scientists have presented the challenge that even when the IPM paradigm is being followed, the result is often still short-term single-technology intervention for pest control, especially for simplified landscapes such as monocultural agroecosystems. Thus, often only one type of biological control or host plant resistance may be used to replace pesticides. These arguments state that such therapeutic approaches are only treating the symptoms with the fastest-acting remedy possible, searching for the one so-called "silver bullet" that will solve the problem. Scientists argue that we should remember ecological principles and the natural balance of systems that demonstrate that such single-technology, therapeutic interventions might readily be countered by responses that would neutralize their effectiveness.

Scientists have recommended a more truly integrative approach that simultaneously draws on numerous methods to prevent pest problems that might occur, instead of treating pests only when they become abundant. Such a "total system approach" provides a long-term, more sustainable solution. Ecologically based IPM requires a total change in how pest problems are viewed. Instead of asking how to control one specific pest, we would be asking why the pest is a pest. Such an approach identifies weaknesses in ecosystems themselves or in how systems are managed that allow species to become pestiferous. The IPM goal would not be to eliminate the pest but to bring pest densities within acceptable limits. This approach toward sustainability may be more complicated and certainly would require an in-depth understanding of the ecological interactions in each managed system. Because a "systems" approach is being suggested, this could require scientists from numerous disciplines to work together. For example, use of non-crop plant species as cover crops has been investigated with the goal of providing habitat and food for aboveground natural enemies or controlling soil-borne plant pathogens or plant parasitic nematodes through exposure to deleterious root exudates. These same cover crops could also be effective in reducing erosion, improving soil organic matter, outcompeting weeds, and providing nitrogen for subsequent crops. In this scenario, scientists knowledgeable about insects, plant pathogens, soils, weeds, and crops would need to work together to help develop a sustainable approach.

Specifically relative to biological control, a sustainable approach to pest management requires knowing the resident natural enemies and those factors enhancing their abundance and activity. With this knowledge would come the ability to manipulate these natural enemies, not only for short-term interventions but preferably for long-term, sustainable solutions. Inundative and inoculative strategies are more typical of short-term interventions, while long-term sustainable approaches might depend more on classical biological control and/or conservation. In particular, sustainable agriculture is aimed toward making use of resident natural enemy populations and enhancing their effectiveness by providing habitats and food for their survival and increase.

Understanding the interactions within a system is critical to adopting this type of more ecologically based strategy. An example is the fungal pathogen *Metarhizium acridum* that has been used against African locusts and can be considered a rather slow-acting biopesticide for killing the target pests. So, one might consider how useful this slow-acting pathogen is. However, a different perspective takes the biology and ecology of the system into account: (1) fungal spores produced from the bodies of the first group of locusts that die can infect a second set of locusts and a second application of the fungus is thus not needed; (2) soon after infection, feeding by locusts decreases dramatically so that damage declines long before locusts are dead; (3) infected locusts are more readily eaten by predators, thus potentially increasing the predator population, but they are avoided by scavengers so that some cadavers are left to produce spores for disease transmission; and (4) periods of cloudy weather and rain will lead to higher levels of infection and more transmission of the fungal pathogen. This in-depth knowledge of the pathogen and ecosystem helps with understanding the utility of this pathogen and best methods for application by taking into account the long-term effects of the fungus and not only its immediate effects.

Will more sustainable approaches such as these work? As one example, an experimental farm in the Netherlands used multidisciplinary approaches to farming, following the basic principles of sustainability. Over a 15-year period, pesticide use on the farm was reduced over 90 percent and, while yields were lower than other farms in that area, the reduction in pesticides by using "alternative" methods meant that profits were equivalent.

Table 19.3 Examples of adaptations to biological control strategies for integrated pest management in agroecosystems.

1. Use of synthetic chemical pesticides or other materials affecting natural enemy activity	Only use when economic damage will occur Use materials less toxic or nontoxic to natural enemies Apply lower rates Separate applications from periods of natural enemy activity in space and time
2. Cultural and mechanical controls	Provide resources that natural enemies need Reduce tillage Make sure that traps or barriers that are in place do not impact natural enemy activity
3. Combine use of natural enemies with use of pest-resistant plant varieties	

Source: based on Orr (2009).

19.4 Use of Natural Enemies within IPM Systems

Use of natural enemies has been a central theme in IPM, especially in IPM programs that really take an integrated approach (see Table 19.2). But how are the different types of control integrated? Examples of pest-control practices within IPM systems that can be integrated with natural enemies are presented in Table 19.3. You will recognize that many of these issues are largely drawn from conservation biological control, although the natural enemies being conserved or enhanced could be part of the native biodiversity of the system or may have been applied via augmentation or classical biological control. Of greatest concern impacting all types of natural enemies is the use of synthetic pesticides. In addition to not killing natural enemies, it is important to prevent exposure to sublethal levels of pesticides that can impact the activity of natural enemies. Best practices include spraying pesticides or implementing other controls that would interfere with natural enemies only when pest populations or damage are near or over the economic threshold. Other approaches include using materials less toxic to natural enemies, applying lower rates of pesticides, and separating applications from periods when natural enemies are active, both in space and time. Additionally, cultural and mechanical controls should retain resources used by natural enemies. One problem in agroecosystems has always been that there are often too few resources for natural enemies to survive and proliferate within agricultural fields. The best example is that many invertebrate natural enemies are more abundant and effective when alternate food sources such as nectar and pollen are present. Alternatives suggested for addressing these issues include intercropping (planting different crops in the same field), planting cover crops (planted for periods when crops are not in the field), and maintaining vegetation on the edges of fields, to provide resources for natural enemies. Providing resources also extends to making sure that there is a consistent habitat for important natural enemies so reducing tillage can be important, as can attention to whether traps or barriers that might be used will deter natural enemies. Use of pest-resistant plants also can impact natural enemies either positively or negatively. Of course, each system differs in types of pests and natural enemies that are important and which of the above conditions will impact natural enemies.

Augmentation has been used extensively in IPM programs and inundative, rather than inoculative, biological control has been the major focus. The emphasis on inundation makes this method for using natural enemies in some ways similar to using pesticides. Unfortunately, growers and crop protection specialists still rely principally on synthetic pesticides rather than inundatively applied biological control agents. However, a change is under way. The biological control market has been increasing, with annual increases in sales of 10 percent before 2005 and more than 15 percent increases in sales per year since 2005. In comparison, growth in the synthetic pesticide market is estimated to be between 5 and 6 percent from 2016 to 2021. Therefore, purchases of biological control agents for augmentative use are growing significantly.

Much of this growth in sales of biological control agents is a result of IPM strategies being used in a diversity of systems. Joop van Lenteren in the Netherlands has been very influential in initially developing and promoting use of natural enemies as key among other control strategies in greenhouses and then continuing to foster the growth of this industry with frequent use of biological control agents (Box 19.2). In many countries, IPM systems for greenhouses primarily rely on a mix of natural enemy species for use against different pests, either via inundation for short-term crops or seasonal inoculative releases for longer-term crops. In 2017 it was estimated that 80 percent of the value of arthropod natural enemies sold commercially was used for protected crops and other high-value crops, such as strawberries and grapes.

Box 19.2 Joop van Lenteren: Profile of a Leader in Biological Control

After synthetic chemical pesticides began to be used for agriculture, they were adopted widely by the greenhouse industry. The approach of spraying pesticides was easier for growers compared with the previous, more complicated mix of practices for pest control that included cultural practices, using host plants that were resistant to pests, and earlier uses of biological pesticides. After the switch, synthetic pesticides were used extensively as, in many ways, they were so easy to use. However, pests quickly began to develop resistance, there were unwanted side effects (health risks for workers, environmental pollution, residues on food) and it became clear to some that alternatives were needed. As problems with the new technology began to occur, people who knew a lot about the methods used before synthetic chemical pesticides worried that no one would remember how to do anything except spray pesticides. Also, ecologists and entomologists were concerned about the large-scale use of pesticides. Therefore, in 1955 the International Organization for Biological Control (IOBC) was formed to try to promote safer, nonchemical pest-control methods.

Figure Box 19.2.1 Joop van Lenteren (photo by F. I. C. van Lenteren).

In 1968, commercial biological control in greenhouses started in Europe and Dr. van Lenteren began working as a professor in the Netherlands soon after. He grew up in the middle of the then-largest greenhouse area in the world, where his family was involved in production of vegetables and ornamentals. After having worked as a part-time laborer in greenhouses during most of his holidays, he went to the university and decided that he would never work in a greenhouse again. He studied experimental biology and earned a PhD on behavior and population dynamics of a parasitoid. During his PhD research, he read many papers about biological control, which can be interpreted as applied population dynamics, and he became fascinated by this way of managing pests. During conversations with family members, he realized how little growers knew about nonchemical alternatives for pest control and how problematic some of the chemical control problems already were. He quickly focused on the fact that the greenhouse industry needed help with pest control. To do this, they needed to understand the behavior and ecology of parasitoid species in order to use these natural enemies successfully for biological control. Dr. van Lenteren went about doing just that in his early career. Perhaps one of the reasons that biological control is used so extensively today in greenhouses is because as natural enemies began to be used, academia focused on what was effective and why, as well as troubleshooting any problems that developed. Academicians, industry, and growers worked together to make sure that growers were handling and using natural enemies optimally.

As his career developed, Dr. van Lenteren realized that one stumbling block was getting effective natural enemies to growers. He put together a book providing information so that natural enemies being sold are still of high quality when they are received by growers. The methods he taught ensure that natural enemies had not lost efficacy through mass production or inbreeding and were not damaged during storage or shipping. Another stumbling block was the increasing resistance to the import and release of exotic natural enemies by different countries. This problem was solved by an initiative led by van Lenteren to stimulate biological control researchers, together with the International Organization for Biological Control (IOBC), industry, and governmental representatives to design and publish an environmental risk assessment methodology which is now used by several governments.

Today there are hundreds of commercial natural enemy producers raising invertebrate natural enemies as well as microorganisms for control of invertebrates or plant pathogens in greenhouses as well as outdoors. The growth in greenhouse use has mainly been possible because industry, growers, and researchers all worked together with the goal of making this safer means for pest control an effective alternative to chemical pesticides. Largely owing to van Lenteren, it is no longer a matter of "should we use chemical or biological control?" in the major greenhouse areas, but rather "which natural enemies are we going to apply this season?"

Another example is the use of parasitic wasps to control flies in dairies in the northeastern United States, resulting in decreased pesticide use as well as decreased fly populations (Box 19.1). While this example pertains to individual dairy farmers making decisions about fly control, IPM has also been adopted on much larger scales. Growing cotton in the arid southwestern United States has been a challenge because of three major pests that require very different types of control. Caterpillars of the invasive pink bollworm, *Pectinophora gossypiella*, feed on reproductive structures, damaging the production of lint. Bt-engineered cotton was introduced in the United States in 1996 to control pink bollworm and by 2007, 98 percent of growers had adopted this strategy. Between the Bt-cotton, mating disruption using pheromones, and sterile insect release, the pink bollworm was eradicated from the United States – although growers remain watchful so that it does not return. The second pest is the silverleaf whitefly, *Bemisia tabaci*, which produces honeydew, causing the cotton lint to be sticky and unsaleable. When populations were high, use of broad-spectrum insecticides resulted in the need for multiple sprays because high-density whitefly populations would decline but then resurge. A better alternative was selective insecticides, insect growth regulators that also were introduced in 1996, along with sampling and threshold decisions about when control was needed. This combination controlled an outbreak population but did not kill nontargets like parasitoids and resident generalist predators. So, instead of rebounding, the whitefly population would remain low after crashing as a result of an IGR application. Finally, in 2006, selective insecticides were also introduced for the third key pest, western tarnished plant bugs, *Lygus hesperus*. This allowed previous sampling and decision aids to be more effectively implemented. All of these strategies allowed biological

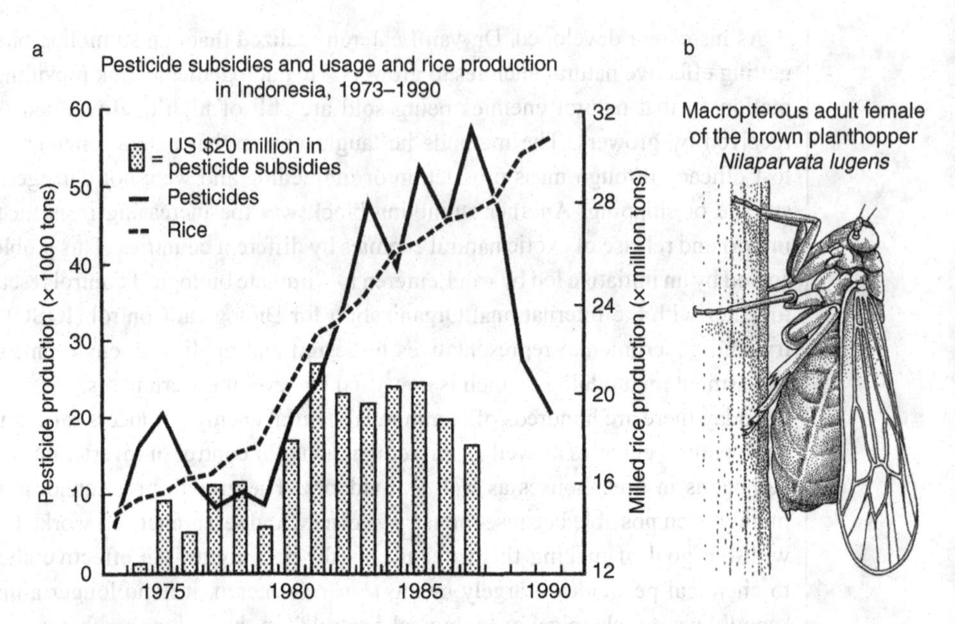

Figure 19.2 Integrated pest management of rice in Indonesia. (a) Once pesticide subsidies were eliminated, insecticide use plummeted but rice production increased. (b) Adult female brown planthopper, *Nilaparvata lugens* (4–5 mm long) (Gullan & Cranston, 2000; illustration by Karina H. McInnes).

control to operate at a higher level in the cotton system. This approach has been called "bioresidual" because with use of selective insecticides to control economic densities of pests, the biological control agents are retained in the area. Researchers who developed this IPM system estimated an 80 percent reduction in foliar insecticides, more than US$500 million in savings on the costs for control, and enhanced ecosystem services since 1996.

The Food and Agriculture Organization (FAO) was influential in dispersing information internationally as the IPM strategy was developed. A stellar example of an FAO-supported IPM project was developed in rice in Southeast Asia. Key to the adoption of IPM was demonstrating that outbreaks of the brown planthopper, *Nilaparvata lugens*, a major pest of rice, were associated with overuse of broad-spectrum insecticides (Figure 19.2). The farmers had been counseled to overspray as insurance against pest problems, but these excessive sprays led to planthopper resistance and lost profits for the farmers. Before insecticide applications were used, the natural enemies occurring in rice paddies had generally kept brown planthopper populations below the economic injury level. The increase in problems as a result of brown planthopper also coincided with the widespread cultivation of modern, high-yielding rice varieties along with increased use of nitrogenous fertilizers and insecticides. Until 1986, the major noninsecticidal method for control of brown planthopper was use of resistant cultivars of rice, but this was not enough. With the threat of the collapse of the rice industry, 57 broad-spectrum insecticides were banned in Indonesia in 1987 by presidential decree following an outbreak of brown planthoppers. IPM was instituted by creating farmer field schools in which farmers learned about the ecology of their crop and its pests and the different control measures available. This built the trust of farmers in the IPM techniques they were learning to use. The Indonesian government took funds from pesticide subsidies and used these to fund the farmer field schools. By the end of 1995, 35,000 trainers and 1.2 million farmers had been exposed to rice IPM through this program.

Another excellent example of integration of control methods including natural enemies is a strategy called "push-pull," basically meaning that the grower tries to repel ("push") the pest away from the crops and in the direction of the "pull plants" at the edges of the field (not the crop itself). This strategy, also named "stimulo-deterrent diversion," has been used very successfully by resource-poor, small-scale farmers for pest control in corn and sorghum, which are main food and cash crops in eastern African countries like Kenya. Some of the major pests are several species of stem-boring moths that are very difficult to control. When the repellent molasses grass (*Melinis minutiflora*) is planted among crop plants, the number of stem borers laying eggs decreases and these same plants are also associated with increased parasitism of corn stemborers by the wasp *Cotesia sesamiae*. The repellent plants are also not a complete loss because they can be fed to livestock. Then at the edges of the fields are the "pull" plants, for example, Sudan grass (*Sorghum vulgare sudanense*), that attract the stem borers that have been repelled. Fortunately, presence of Sudan grass was also associated with greater parasitism levels. The other major pest is witchweed (*Striga*), an obligate root parasite of cereal crops. If corn or sorghum are interplanted with *Desmodium* plants, the stem borers are repelled. Also, exudates released by *Desmodium* cause *Striga* seeds to germinate but inhibit proper growth, a combination called "suicidal germination," resulting in one plant destroying the seed bank of another. This system can increase farmers' earnings compared with use of pesticides and has been adapted to also address climate change for this semi-arid region.

Further Reading

Abrol, D. (ed.) (2013). *Integrated Pest Management: Current Concepts and Ecological Perspective*. San Diego, CA: Academic Press.

Chandler, D., Bailey, A. S., Tatchell, G. M., Davidson, G., Greaves, J., & Grant, W. P. (2011). The development, regulation and use of biopesticides for integrated pest management. *Philosophical Transactions of the Royal Society B*, **366**, 1987–1998.

Flint, M. L. (2012). *IPM in Practice: Principles and Methods of Integrated Pest Management*, 2nd edn. Oakland: University of California Agriculture & Natural Resources Publication.

Radcliffe, E. B., Hutchison, W. D., & Cancelado, R. E. (eds.) (2009). *Integrated Pest Management: Concepts, Tactics, Strategies and Case Studies*. Cambridge: Cambridge University Press.

Sivinski, J. (2013). Augmentative biological control: Research and methods to help make it work. *CAB Reviews*, **8**, 26.

van Lenteren, J. C., Bolckmans, K., Köhl, J., Ravensberg, W. J., & Urbaneja, A. (2018). Biological control using invertebrates and microorganisms: Plenty of new opportunities. *BioControl*, **63**(1), 39–59. DOI: 10.1007/s10526-017-9801-4.

20 Our Changing World: Moving Forward

We began this book talking about the growing human population, growth that will be accompanied by the need for increasing resources as well as the increasing pressures that will be placed on the environment. This increasing pressure is reflected as a diversity of global changes which are often not purposeful. The causes for changes already occurring on our planet are diverse, including harvesting and exploiting natural resources in unsustainable ways, polluting, introducing invasive species, loss and fragmentation of native ecosystems, emerging diseases, and, of course, the changing climate. Based on the source, one can find that each of the preceding is the most critically important driver for changes.

Agriculture, forestry, and public health have principally relied on the use of chemical pesticides for pest control for many decades, developing a "pesticide mindset" among practitioners and the public alike. For multitudes of reasons, ranging from loss of efficacy of pesticides, decreasing availability of effective pesticides, government regulations, and environmental and public scrutiny to increasing desire for pesticide-free products (see Chapters 1 and 19), pest-control alternatives are necessary and more sustainable methods for pest control will be increasingly needed. Therefore, we will finish this book by discussing some of the causes for global change and how biological control will interface with or be impacted by them and/or how biological control can be part of mitigations or solutions.

Relative to pest-control issues, the pressure from an increasing human population can be experienced as humans struggling to be able to produce and distribute enough food and fiber and provide enough shelter, as well as prevent human diseases in a sustainable way. In addition, there will also be increasing encroachment into native ecosystems and disruption of native biodiversity. Potential changes caused by the changing climate at this time are unpredictable but will likely have far-reaching impacts, especially because many of the pests that we have been discussing do not regulate their own temperatures but are dependent on environmental conditions (and can be more active under warmer conditions). Thus, moving forward, our planet will be changing in many ways and we predict that environmentally safe methods for pest control will be critical for keeping up with these challenges.

20.1 Major Challenges

20.1.1 Invasive Species

Invasive species are introduced species with significant impacts on the economy, environment, or public health (see Chapter 1). Invasive species are now known as principally being introduced to new areas by global trade and travel, often referred to as globalization, which breaks down historical barriers to dispersal. Increasing globalization has meant that organisms can be moved quickly around the world and introduced to new areas where they are not native and become pests. With the present onslaught of invasive species around the world, scientists have been asking whether the major or most damaging species have not already been introduced everywhere. However, regardless of the source area and the area of introduction, unfortunately there are many more invasive species from around the world that have not been introduced already and estimates for rates of introductions of most kinds of invasives do not suggest a plateau any time soon; therefore, there are many more species that have yet to be introduced. Unfortunately, very few countries have instituted the types of precautions to prevent further introductions of invasives. An exception is New Zealand, an isolated, ocean-bound nation with excellent surveillance and regulations regarding invasions, where every non-native species found in shipping containers, luggage, and so on, is targeted for eradication. The present levels of global homogenization of biodiversity led experts to predict that without more controls of introductions of invasives, we are moving toward a world with fewer endemics in individual areas and more established alien species, many of which will be common to many regions.

Even if we could know which new invasive species will be introduced (which we often do not), which species will become established and will become invasive is unpredictable. For example, the emerald ash borer, *Agrilus planipennis*, in North America was probably introduced from China, where it is a rare native species. In contrast, in North America this wood borer is at outbreak densities and the worry is that numerous of the North American species of ash trees (*Fraxinus* spp.) will become rare or extinct, just as North America's native American chestnut trees (*Castanea dentata*) became rare early in the twentieth century as the result of an introduced fungal disease (see Box 17.1). It has been estimated that the 98 species of native ash-specific herbivores in North America would be at risk if the 16 species of *Fraxinus* in North America were eliminated by the emerald ash borer. So, as a result of damage by emerald ash borers, not only the tree species will become extinct, but any of the associated native herbivores that are tightly linked with ash trees are also in jeopardy.

Historically, the only way that invasive species have been controlled so that continual manipulations (such as repeated applications of chemical insecticides) are not needed is by eradication or classical biological control. Eradication is only possible very early in infestations and under specific conditions. As you have read, early introductions of natural enemies for classical biological control at times were short of host specificity data and priorities often focused primarily on pest reduction. Such programs at times resulted in impacts on native species by natural enemies that were released. Armed with an appreciation and resources for preservation of native biodiversity, attention has been drawn to this issue and the number of classical biological control introductions plummeted from the 1980s to the 2000s. After the realizations that vertebrates and other generalists should no longer be used (except perhaps in a few special and isolated circumstances) further studies demonstrated that "drifting" in specificity of biological control agents to hosts/prey rarely had a strong enough impact to cause changes in population densities of nontargets. In addition, today much more extensive testing is required before decisions can be made to introduce a natural enemy. Biological control practitioners are now aware of the balance that is necessary between controlling an invasive and impacting nontargets; risk assessments are necessary to make sure that the pests being targeted are in fact causing significant damage and precautions are in place so that natural enemies to be released will not themselves impact biodiversity significantly. Classical biological control is still being used but programs are slow to move forward, in part because of the time needed to conduct increased host specificity testing and the time required for regulatory groups to make decisions about releases. Classical biological control is especially being used for very impactful invasive species with few other options for control. Researchers working to control invasives on wildlands have documented that classical biological control of both invasive arthropods and weeds has provided valuable protection for biodiversity and ecosystem services, as well as natural products that are harvested. It remains for the community involved with classical biological control to overcome the bad reputation that persists, largely as a legacy of introductions conducted before protection of biodiversity became a higher priority.

20.1.2 Climate Change

Although exact predictions about temporal and spatial changes are not possible, the Earth's climate is changing. In North America, the mean annual temperatures have increased by up to 2°C between 1901 and 2006 and temperatures are projected to increase an additional 1–1.5°C by 2050. It is thought that changes will be greater along mid-latitude coasts and in northern latitudes. The major changes that are expected to impact organisms are rising temperatures, increased atmospheric CO_2 concentrations, and increasingly variable precipitation. However, how will these changes impact pest populations and biological control?

First, changes in environmental conditions could directly impact the physiology and behavior of biological control agents or the pests they are attacking. None of the major players in biological control (invertebrates, with emphasis on arthropods, plants, and microorganisms) thermoregulate or have strong enough dispersal abilities to effectively track favorable environmental conditions across broad landscapes, which makes them quite sensitive to environmental changes. Second, the life history (i.e., the timing of feeding, mating, and dormancy) of these organisms is strongly linked to the season so changes in the timing of important seasonal events could have strong impacts. Third, changing the climate could change the interactions among species involved in biological control. An example of this last category could include environmental changes altering the levels and effects of secondary plant compounds on biological systems. In cotton, increased CO_2 caused increased levels of the phenolic aldehyde, gossypol, which had a negative impact on the cotton aphid, *Aphis gossypii*, corresponding with improved fitness for the coccinellid predator, *Propylaea japonica*, when it was feeding on these aphids. Alternatively, increased CO_2 and temperatures in alfalfa decreased biological control efficacy against beet armyworm, *Spodoptera exigua*, caterpillars by parasitoids, based largely on decreased nutritional quality of the plants and faster development of the caterpillars, so that the parasitoids were unable to complete development fast enough for optimal efficacy. In Sitka spruce trees in the United Kingdom, the green spruce aphid, *Elatobium abietinum*, is preyed on by two species of coccinellids that naturally occur in that ecosystem. Under drought conditions, the coccinellids ate more aphids although moderate levels of drought that were intermittent resulted in the opposite effect, reduced feeding by larval coccinellids. These examples suggest that climatic feedbacks are likely to be complex and/or difficult to predict.

Biological control, regardless of whether this involves classical biological control, augmentation, or conservation, is dependent on interactions between pests and natural enemies. In the past several decades, highly specific associations have been emphasized, especially for classical biological control. As suggested above, changes in environmental conditions because of climate change can sometimes result in improved biological control while at other times such changes could result in decreased efficacy of biological control. At present, the major opinion is that whether the impact of climate change on biological control will be positive, negative, or neutral, the results will be system dependent and increased studies are necessary to be able to predict results of climate change ahead of time.

20.1.3 Sustainability

Sustainability can mean avoiding depletion of natural resources in order to maintain ecological balance, a theme that is very relevant to the increasing human population. In fact, sustainability is often characterized as the intersection of economic development, social development, and environmental protection. Sustainability is usually mentioned as a goal of agriculture and forestry, both being practices where humans manipulate environments and expect planned results, such as high yields from crops that they want to plant again and again for equal or better yields. Biological control has been "considered among the most promising technologies for sustainable agriculture" as it reduces use of synthetic chemical pesticides, has less negative impact on the environment, and is safer for workers (Tracy, 2014). As an example, we will discuss augmentation as one type of biological control used in this context and will refer to agents being used as biopesticides. At present, biopesticides are only a small share of the total products being used for pest control. Although use of biopesticides is increasing, still the use of synthetic pesticides is far more common.

Can use of biopesticides (and for that matter also classical and conservation biological control) be scaled up to meet the goals of sustainability? There is presently significant growth in use of biopesticides, driven by issues such as development of resistance to chemical pesticides and pesticides being banned from use, consumer preferences for sustainably produced food, eco-labeling foods produced in this way, and health and safety issues for workers, which all favor biopesticides over conventional chemical controls. Yet, using biological control agents is often more costly than using synthetic pesticides and their use requires increased training for workers. Because living organisms are being used in biological control, research and development of new products is slow and regulatory issues can be encountered at federal, regional, and even municipal levels. An increased focus of research on calculating the overall costs of using different control strategies could enhance adoption of biopesticides. For this to be possible researchers must quantify crop yields based on different control treatments in order to report savings gained through use of different control strategies (although it can be difficult to calculate savings for preserving biodiversity).

Reasons that biopesticides are not used more extensively are different in the two major systems where they are used: greenhouses versus open fields. In some countries biological control agents are used extensively in greenhouses. For example, as of 2009, 90 percent of Dutch vegetable production in greenhouses was under integrated pest management which included the use of biological control agents. However, the estimated global average of use of biological control in greenhouses across all countries was much less than the Netherlands. Use of biological control agents in greenhouses is preferred because pesticides may no longer work or may no longer be available and exposure of workers to pesticides in an enclosed greenhouse can be more dangerous than in open fields.

Biopesticides are used less for growing crops in open fields because results may seem less predictable to growers in this more variable environment. There are, however, several examples of important use of biopesticides outdoors, for example, CpGV for control of codling moth, *Cydia pomonella*, in apple orchards (see Box 11.2), and use of the fungus *Metarhizium anisopliae* for control of pests in sugarcane (see Box 12.2). One major reason for the lack of trust in this method can be ascribed to pesticide drift from neighboring farms which can kill arthropod or fungal natural enemies. If growers in the same district coordinate, this problem could be ameliorated but it would require a new approach: growers would need to plan and communicate about practices with neighbors and be ready to compromise to address community-wide concerns. Sustainable use of biological control can also require continuing research and development to understand what is necessary for control that reaps economic benefits. In the case of cotton in the southwestern US, reduced pesticide use allowed native predator populations to provide control (see Chapter 19). In New Zealand, in high-intensity agriculture a loss of biological control efficacy was found 7 years after introduction of a parasitoid to control Argentine stem weevils (*Listronotus bonariensis*) in pastures. Researchers concluded that the low plant and enemy diversity in the simplified landscape of intensive large-scale agriculture may have facilitated evolution of resistance by the weevils to the parasitoids.

So, at least for now, some regions provide excellent examples where biological control enhances sustainability, although we have a long way to go to institute these methods more in open field agriculture where many interacting agents occur under less predictable conditions. Research from conservation biological control at the landscape level has been demonstrating that landscape design to reduce simplification promotes

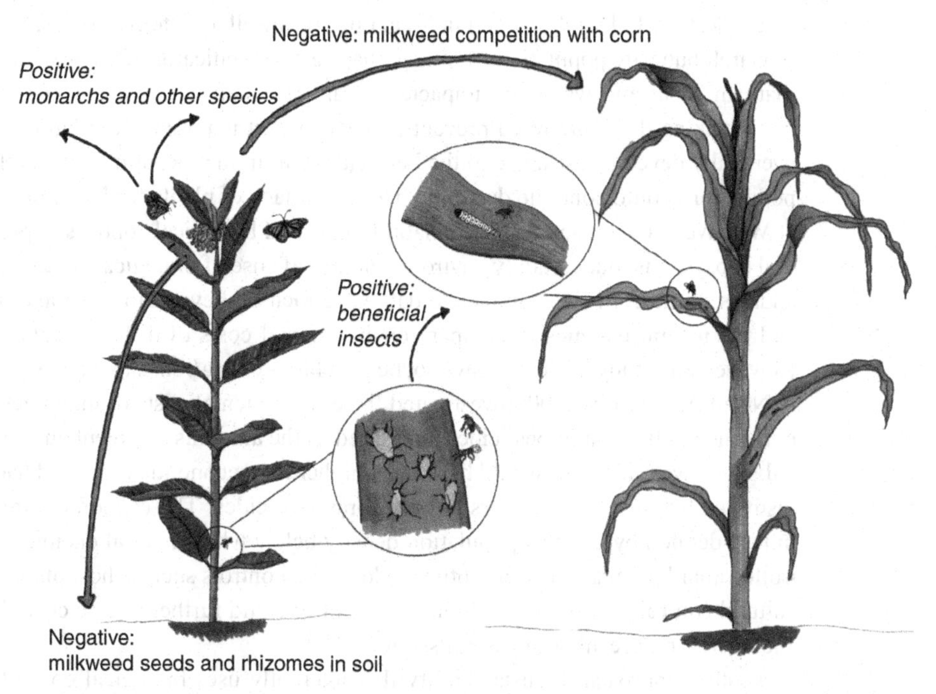

Figure 20.1 Complex effects of the presence of milkweed on corn, with both positive and negative impacts. For example, beneficial insects such as *Trichogramma* wasp populations benefit from honeydew produced by aphids feeding on milkweed, leading to greater parasitization of eggs of insect pests such as the European corn borer. Milkweed plants also directly support monarch butterflies and species diversity. Negative effects include competition for water, light, and nutrients and an increase of milkweed seeds and rhizomes in the soil which can become weeds in the following years' crops of other kinds (based on DiTommaso et al., 2016).

sustainability. One example demonstrates how control activities can echo through ecosystems, affecting numerous different agents, including native species. In recent years in the United States, corn that is being grown is engineered to be tolerant to glyphosate so that weeds in corn fields can be controlled using this herbicide. The major weeds that this herbicide kills include milkweed, *Asclepias* spp., well known as the only genus of food plants eaten by caterpillars of the iconic migratory monarch butterfly, *Danaus plexippus*. Lack of milkweeds on the migration route of these butterflies is one of the proposed causes of the declining populations of these butterflies in North America. Milkweed also hosts populations of three aphid species. A major pest of corn is the European corn borer, *Ostrinia nubilalis*, and if the aphids are present in the crop, *Trichogramma* egg parasitoids provide some control as the adult parasitoids persist in the field because they can feed on aphid honeydew. So, growers must balance the cost of having milkweed in their corn fields as a competitor of corn, against its value as a resource for minor pests (aphids) supporting natural enemies (*Trichogramma*) killing the major pest (corn borers) as well as serving as food for monarch butterfly caterpillars (Figure 20.1). This system is complicated relative to control of a weed and an insect pest, including direct and indirect effects as well as external forces, for example, monarch butterfly populations, but is perhaps a good indicator of the many interactions that can occur and which are impacted by any decision that is made. If growers continue making the same weed prevention decisions, it is possible that milkweeds could eventually develop resistance to the herbicide, but, in the meantime, monarch butterfly populations could continue declining because of lack of plants for larval development.

We have not yet discussed sustainability in use of biological controls to preserve natural ecosystems, but certainly environmentally safe use of biological control can help to address this need. Few examples to date have focused beyond the pest and its damage and the natural enemies to compare environmental costs of different control options. However, one study of the invasive soybean aphid, *Aphis glycines*, which was first found in North America in 2000, investigated the environmental costs of manufacture, transport, and application of pesticides, measured as the amounts of greenhouse gases emitted. Costs were high but would have been higher if an economic threshold had not been instituted below which growers did not spray pesticides. The economic threshold was in part defined by the pest population density below which natural enemies often controlled aphid populations. The authors added that controls such as host plant resistance, cultural controls, and classical biological control could further reduce costs for control and, therefore, greenhouse gas emissions.

Another approach to sustainability that integrally uses biological control agents is a multifaceted program called Working for Water that aims to reduce populations of invasive plants in South Africa and thereby conserve water, with the side effect of also protecting native biodiversity in the context of a program that educates and involves the public (Box 20.1).

Box 20.1 Working for Water

Overall, South Africa is a water-stressed country, dominated by regions that are semi-arid to arid, and the native vegetation in these areas is adapted to low rainfall. However, many non-native plants that have been introduced have invaded natural vegetation communities in South Africa. Many of these invasive plants grow very quickly and, as a consequence of this fast growth, use much more water than the native vegetation. Some invasive plants are aquatic weeds that can block waterways and change aquatic ecosystems. Of the approximately 9,000 species of plants introduced to South Africa, 379 are considered problematic (defined as covering at least 10 percent of the country's land area). In 1995, the government of South Africa began a program toward clearing alien invasive plants in order to free South Africa's water resources for humans and native biodiversity. This program hires and trains unemployed members of marginalized communities as well as providing social services. A variety of methods for control are used and different controls are usually integrated for any one problem. Physical/mechanical control including physically removing plants, controlled burning, and applications of herbicides are used along with classical biological control.

Figure Box 20.1.1 Working for Water team manually removing water hyacinth from the Vaal River in South Africa as part of an integrated management program against this weed that also relies on biological control (photo by Julie Coetzee).

In this program, for classical biological control, a total of 93 species have been introduced against 59 species of invasive plants. However, some of the invasive plants, such as eucalyptus, acacias, and pines, are planted by the forestry industry and smaller-scale farmers and have byproducts (e.g., wood) used by households for cooking. For such plants that are valued by some and not wanted by others, biological control using natural enemies that attack only the seeds has been employed, thereby targeting only the spread but not the growth of these trees. Since it began, the program has cleared invasive plants from more than 1 million ha and hired and trained over 20,000 people of which about 60 percent were women. This program is known worldwide as a success story of a sustainable environmental conservation initiative where South Africans say they are paying back the environment for ecosystem services provided.

20.1.4 The need to Increase Use of Biological Controls in Public Health

Human populations worldwide face major problems because of infectious diseases, especially diseases caused by pathogens vectored by arthropods. Some of the vectors and pathogens responsible have been known for many years in specific areas, for example, tsetse flies vectoring sleeping sickness in tropical Africa and mosquitoes vectoring numerous viruses worldwide. However, pathogens are now often being moved globally or new strains are evolving. These new diseases are called emerging and some very important vectored pathogens fall in this category of new diseases, for example, the Zika virus. On top of that, several known human diseases vectored by insects and mites are spreading, for example, bacterial and virus diseases transmitted by *Ixodes* ticks. Prevalence of tick-transmitted human diseases are increasing in North America and in parts of Europe. If we focus on vectored pathogens of humans, use of synthetic chemical pesticides has dominated control efforts but now, with abundant repeated applications each year, insecticide resistance is a recurring problem. Development of new pesticides is not meeting demands and there are numerous reasons not to continue relying only on chemical pesticides (see Chapters 1 and 19), so alternative controls are necessary. As one example, applying synthetic neurotoxins (present in some chemical pesticides) to kill adult mosquitoes within houses is a common approach leading to human contact with these materials, which could lead to negative health outcomes in the short and long term.

At the International Congress of Entomology in Orlando, Florida, in 2016, a symposium was organized by Matt Thomas to discuss why biological controls are not being used more to help control disease vectors. Biological control of invertebrates has historically focused most on herbivorous arthropods and weeds impacting agriculture, forestry, and wildlands and biological control of vectors of human diseases using biological means has lagged. While this symposium was directed toward insect vectors and public health, and therefore was about human diseases, discussions were also at times broadly relevant to considerations about other pests associated with public health, such as bed bugs, or vectored pathogens of other vertebrates or plants.

Everyone agrees that biological control would be a great way to control disease vectors. Major methods for biological control in public health have been the use of mosquito fish (*Gambusia*) that feed on mosquito larvae (see Box 7.1), with variable success but also causing nontarget effects. Augmentation biological control using microorganisms has shown greatest promise, especially for mosquito control. Bacteria have been the most successful agents, particularly *B. thuringiensis israelensis*, but in this case only larvae of mosquitoes will be controlled, with attendant issues about finding their breeding sites, which can be numerous, hard to find, and transient. Control strategies using insecticides are generally focused on killing adults, especially because it can be difficult to locate all areas where larvae are developing and the adult stage does the vectoring. Mosquito-pathogenic fungi are being investigated for control of adult mosquitoes while viruses and nematodes have been explored for larval control. Of course we must mention the new approach being investigated of introducing mosquitoes containing an adapted bacterial symbiont (*Wolbachia*). The symbiont makes *Aedes aegypti* (a principal vector of dengue) more resistant to infection with the dengue pathogen (so they will not vector it) and shortens the lives of female mosquitoes. The goal is for this bacterial symbiont to be transferred into the native mosquito populations so that dengue will not be vectored. This new, exciting approach for control is still under evaluation but appears to be promising. However, this strategy is presently specific to one host/pathogen system. Also, efforts to control the vectors must proceed hand in hand with medical efforts like vaccination programs.

Especially directed at control of mosquito vectors, the field of integrated vector management (IVM) is working toward vector control that is more efficient and cost effective along with being ecologically safe and sustainable. The main control tactics presently consist of combinations of pesticide use and cultural controls, but a growing community of researchers are working to add viable biological control options to the mix.

20.2 Acceptance by the Public, Scientists, and Governments

People must be willing to use biological controls and this once more brings into question what is meant by this term biological control. Classical, augmentation, and conservation are completely different strategies, though usually when people mention biological control, they are referring to only one of the strategies but the strategy being discussed is not specifically mentioned. As explained in Chapter 18, opinions of some people have been anti-"biological control" after the 1980–2000 controversy about effects on biodiversity due to continuing results from older practices that were not based on preserving biodiversity. This controversy was initially centered principally on classical biological control, although augmentation also became implicated. Moving forward, this controversy is being solved by intensive evaluations of past examples helping to guide changes in practices for any new introductions. It now behooves the biological control community to disseminate the information about these changes as well as information about protocols that have been changed to enhance environmental safety. Classical biological control has been a major way to try to control invasive species and, as you have read, numbers of new invasive species are not projected to decrease any time soon.

Relative to pest-control methods used on food and horticultural crops, biological control strategies increasingly include augmentation and conservation. In one Canadian survey, people stated that they consider use of biological control agents safer than synthetic insecticides. However, we know that biological control is not used as much as it could be, although it is used extensively in the greenhouse industry in Europe and northern North America. At present, governmental regulations can create barriers to being able to use biological controls, in particular new biopesticides. Costs for developing a new chemical pesticide have been estimated at $250 million while costs for developing a new biopesticide are $3–7 million and this requires approximately three years. Costs are partly due to screening many new microbes or chemicals to find new effective and safe agents. Although testing requirements are generally less stringent for microbes than for chemical pesticides, these data are still expensive to generate, both in money and time. Especially when these products are being developed by small companies for smaller crops, the required testing and time for regulators to reach decisions can prohibit development of new materials. Regulators provide no allowance for the beneficial aspects of these biopesticides, such as persistence and continued reproduction or side effects of plant growth promotion. As there are fewer chemical pesticides and as alternative controls are increasingly necessary (see Chapter 1), hopefully regulations can be adapted to provide more support for biopesticides so that development of new, safe products will be more possible.

Conservation biological control can escape notice by the public at present but more examples of successful use could go a long way toward bringing its value to the public eye. The different types of biological control all need acceptance and recognition by users and the public, but all could benefit from more outreach to share developments and results and to demonstrate that these methods are safe.

Regardless of the biological control strategy being used, the best way to promote adoption is to document results and benefits as clearly as possible. Control of pests in agricultural crops should document not only decreased pest populations but also increased yields and costs for control as well as increased profits. For pests in other systems where economic value is not measured directly in yield and profit (such as rangelands and natural areas) cost:benefit ratios can be useful. For classical biological control, it is also important to conduct post-release monitoring so that we learn more about the potential for direct and indirect impacts.

20.3 International Cooperation Is Necessary

Biological control practices are used by an international community of university and governmental scientists and students as well as by industry. This community gains by sharing information about pests and biological control agents, methods, and successes as well as information about host specificity and direct and indirect nontarget impacts. The International Organisation for Biological and Integrated Control (IOBC) is a good example of international collaboration among scientists. Workshops are held at regular intervals with focus on, for example, a crop ecosystem (e.g., greenhouse crops) or an organismal group for biological control (e.g., antagonists to plant diseases).

Biological control methods can be more difficult to use compared with continuing to use preexisting pesticide spray technology that uses long-standing methods for application of nonliving materials. Biological control is not run by large multinational companies as is the chemical pesticide industry, and international willingness and ability to communicate and share information is proving important to moving forward. At present, an important organization for augmentation biological control is the International Biocontrol Manufacturers Association (IBMA) which includes representatives from many of the individual, smaller companies producing biological control agents, and this larger organization facilitates communication. This organization also is important in providing a means for smaller companies to have a larger impact since they are acting together, for example, through representation on governmental committees deciding rules and regulations about biological control.

An issue has arisen that prevents a means for international cooperation that existed previously. In 2009, the Convention on Biological Diversity (a global program initiated by the United Nations Environment Programme) instituted an agreement regarding sharing of benefits arising from the use of genetic resources. This came about to prevent biopiracy, where the profits from chemical and drug development were not being seen by the countries that were the sources of the original biological materials. Unfortunately, this new legislation also affected collection of new biological control agents in other countries. This principally impacts classical biological control of invasive species but also could prevent finding new biological control agents for augmentation. Classical biological control is conducted by governmental (national or international) agencies and universities and is not conducted by companies that will profit from successful controls. Biological control agents for augmentation are produced and sold by industries but these are smaller industries that do not reap the larger profits that international chemical and drug companies do. Therefore, use for biological control falls outside of the policy's stated goals of preventing bioprospecting, as no large profits are gained for natural enemies that are found, and an exception is being sought.

Many of the problems and issues described in this chapter are, in fact, global in magnitude, such as climate change, or the same issues are being experienced in different regions. Sharing information and resources globally toward sustainable solutions to pest control to feed the growing human population while preserving biodiversity and ecosystem functioning is the key to moving forward. Seventeen sustainable development goals were formulated by the United Nations in 2015 in order to address such issues. Numbers 2 and 3 are to end hunger and promote food security and to ensure health and well-being and additional goals concern conservation of aquatic and terrestrial ecosystems. Biological control can clearly play a part in working toward these global goals.

Further Reading

Bale, J. S., van Lenteren, J. C., & Bigler, F. (2008). Biological control and sustainable food production. *Philosophical Transactions of the Royal Society B*, **363**, 761–776.

Cock, M. J. W., van Lenteren, J. C., Brodeur, J., Barratt, B. I. P., Bigler, F., Bolckmans, K., Cônsoli, F. L., Haas, F., Mason, P. G. & Parra, J. I. P. (2009). Do new Access and Benefit sharing procedures under the Convention on Biological Diversity threaten the future of biological control? *BioControl*, **55**, 199–218.

DiTommaso, A., Averill, K. M., Hoffmann, M. P., Fuchsberg, J. R., & Losey, J. E. (2016). Integrating insect, resistance, and floral resource management in weed control decision-making. *Weed Science*, **64**, 743–756.

Ehlers, R.-U. (ed.) (2011). *Regulation of Biological Control Agents*. Wallingford, UK: CABI.

Eigenbrode, S. D., Davis, T. S., & Crowder, D. W. (2015) Climate change and biological control in agricultural systems: Principles and examples from North America. In *Climate Change and Insect Pests*, ed. C. Björkman & P. Niemelä, pp. 119–135. Wallingford, UK: CABI.

Garnas, J. R., Hurley, B. P., Slippers, B., & Wingfield, M. J. (2012). Biological control of forest plantation pests in an interconnected world requires greater international focus. *International Journal of Pest Management*, **58**, 211–223.

Heimpel, G. E., Yang, Y., Hill, J. D., & Ragsdale, J. W. (2013). Environmental consequences of invasive species: Greenhouse gas emissions of insecticide use and the role of biological control in reducing emissions. *PLoS ONE*, **8**(8), e72293.

Kremen, C. & Miles, A. (2012). Ecosystem services in biologically diversified versus conventional farming systems: Benefits, externalities, and trade-offs. *Ecology and Society*, **17**(4), 40.

Seebens, H., Blackburn, T. M., Dyer, E. E., Genovesi, P., Hulme, P. E., Jeschke, J. M., Pagad, S., Pyšek, P., Winter, M., Arianoutsou, M., et al. (2017). No saturation in the accumulation of alien species worldwide. *Nature Communications*, **8**, 14435.

Thomas, M. B., Godfray, H. C. J., Read, A. F., van den Berg, H., Tabashnik, B. E., van Lenteren, J. C., Waage, J. K. & Takken, W. (2012). Lessons from agriculture for the sustainable management of malaria vectors. *PLOS Medicine*, **9**(7), e1001262.

Tracy, E. F. (2014). The promise of biological control for sustainable agriculture: A stakeholder-based analysis. *Journal of Science Policy & Governance*, **5**(1).

Van Driesche, R., Simberloff, D., Blossey, B., Causton, C., Hoddle, M., Marks, C., Heinz, K., Wagner, D., & Warner, K. (eds.) (2016). *Integrating Biological Control into Conservation Practice*. Oxford: Wiley Blackwell.

Glossary

Definitions in this glossary are derived from a variety of sources, most of which are included in the general references.

Abiotic Not associated with living organisms.

Acclimatization societies Societies organized in the eighteenth and nineteenth centuries, during the period of colonialism, with goals of introducing non-native species (usually from Europe) to enrich the flora and fauna of newly colonized regions.

Ambushers Entomopathogenic nematodes that raise nearly all of their body off the substrate (called nictating) to attach to passing hosts. Foraging behaviors are species specific.

Antagonists The term used in biological control of plant disease to refer to natural enemies of plant pathogens; in this context the term refers to microorganisms employing several strategies, including competition.

Antibiosis (a) Biological control of plant disease: inhibition of one organism by a metabolite produced by another microorganism. (b) Plant resistance: plant characteristics that affect phytophagous insects in a negative manner (e.g., reduced fecundity, increased mortality).

Antibiotic A substance produced by a microorganism that is damaging to another at low concentrations.

Antixenosis (nonpreference) Plant characteristics negatively impacting behavior of a phytophagous insect, for example, causing an insect to leave a plant.

Apparent competition Negative interactions between two individuals that interact indirectly by sharing a common resource.

Appressoria (singlular = appressorium) A specialized fungal cell used to penetrate a host.

Augmentation Biological control strategies where natural enemies are released without the expectation of permanent establishment and autonomous regulation of the host; this term includes both inoculative and inundative biological control.

Autoecious Completing its entire life cycle on the same host as compared with species that require two different hosts (pertaining to rust fungi).

Banker plants Plants inhabited by both pest and natural enemy that are introduced to an area as a source of natural enemies. The natural enemies persist on the plant and disperse to find pests on crops; used primarily in biological control in greenhouses.

Beetle banks Banks of soil within crop fields where native vegetation is allowed to establish over time thus providing habitat for predatory arthropods that disperse into the crop and control pests.

Beneficial organism In biological control, living organisms that aid in pest control (i.e., natural enemies).

Bioherbicides Biologically based control agents applied to control weeds using augmentation biological control. This term is used for living microorganisms (e.g., plant pathogens, see Chapter 15) but also refers to compounds and secondary metabolites derived from organisms.

Biological control (or Biocontrol) The use of living organisms to suppress the population of a specific pest organism, making it less abundant or less damaging than it would otherwise be.

Biologically based pest management Managing pests using biologically based technologies when control is needed.

Biologically based technologies This umbrella term has been defined as including biological control (including predators, parasites, competitors, and pathogens that spread on their own and genetic engineering applied to this approach), microbial pesticides (microbes applied to suppress pest populations and genetic engineering applied to this approach), pest-modifying chemicals, genetic manipulations of pest populations, and plant immunization (OTA, 1995).

Biopesticides Biologically based control agents applied for augmentation biological control. This term is often used for living microorganisms (e.g., pathogens of arthropod pests) but also refers to compounds and secondary metabolites derived from organisms.

Bioprotectant Natural enemies released when the pest is not present so the pest will not cause damage, used in plant pathology when antagonists are applied to healthy plants or seeds so that pathogens will not be able to infect.

Biorational pest management Pest management techniques that are based on biological and ecological interactions that reduce the negative effects of pest populations without harmful side effects to man or the environment. Included in this term are cultural control, host resistance, biological control, autocidal control, and biorational chemical agents.

Biotroph A pathogen that requires living host tissue for its nutrition; this term is generally used in plant pathology.

Biotype A strain of a species, morphologically indistinguishable from other members but exhibiting distinct physiological characteristics; especially with regard to its adaptation to climatic conditions.

Blind release Greenhouse application of macro-natural enemies on a regular basis, regardless of finding pests.

Bottom-up control Regulation of ecosystems by nutrient supply or organisms at the base of a food chain.

Cambium A layer of partially undifferentiated and actively dividing plant cells from which vascular plants produce new tissue.

Canker An area of a stem, branch, or twig dead and sunken as a result of an activity of a plant pathogen.

Chronic Occurring over a long period of time.

Classical biological control The intentional introduction of an exotic biological control agent for permanent establishment and long-term pest control.

Coevolution Evolutionary interactions between two closely associated organisms, for example, a parasitoid and its phytophagous host; usually referring to changes taking place in which the evolution of a specific trait in one organism leads to reciprocal development of a trait in the second organism, directly in response to the initial trait change in the first.

Coevolutionary arms race Responses of natural enemies to hosts/prey through evolutionary time and vice versa; for example, over time natural enemies are better able to attack and utilize hosts/prey while hosts/prey become more resistant.

Community The group of species occurring at a specific habitat and time and interacting with each other.

Competitive displacement Displacement of one species by another that is competitively superior; in general this occurs when a better competitor is introduced to a new area.

Competitors Organisms interfering with each other because they need the same limiting resource.

Conducive soil A soil that allows development of plant disease.

Conidia (singular = conidium) Asexual fungal spores formed on the ends of specialized hyphae called conidiophores.

Conservation biological control Modification of the environment or existing practices to protect and enhance specific natural enemies or other organisms to reduce the effect of pests.

Constitutive (characteristics) Something always present (opposite of inducible).

Consumer An organism that feeds on a resource or another consumer.

Consumer–resource interaction An interaction in which one organism depends on another for nourishment. In biological control, this could refer to predators and prey.

Cross protection A plant inoculated with a mild strain of a virus, and this then protects the plant from infection by more virulent strains of the same virus.

Cruisers Entomopathogenic nematodes that do not raise their bodies off the substrate to forage (as opposed to ambushers) and display more movement than ambushers. Foraging behaviors are species specific.

Cryptogenic Of uncertain origin, for example, referring to a pest on a crop that has been moved around the world so extensively so that no one knows where the pest is native.

Cuticle The outermost, protective layer covering certain organisms. This term is often used in relation to insects or plants.

Damping off Mortality of seedlings near the soil surface, often evidenced by wilted and decaying seedlings lying on the soil.

Defoliation Removing leaves from a plant, often used in entomology to refer to insects feeding on trees.

Delayed density dependence In population ecology, the magnitude of an effect determined by the density of a population at some time in the past.

Delta (δ) endotoxin Toxic protein occurring in a crystalline form, within sporangia (spore-bearing cells) of *Bacillus thuringiensis*.

Density-dependent mortality Mortality of members of a population in relation to the density of the prey or host population.

Density-independent mortality Mortality of members of a population with no relation to prey or host density, for example, in response to severe weather.

Disease Invertebrate pathology: state of not being healthy. Plant pathology: malfunctioning of host cells and tissues leading to development of external or internal reactions or alterations of the plant.

Domatia (singular = domatium) Small areas, often located where primary veins converge on leaves in some woody plants, often consisting of a depression that can be enclosed by leaf tissues or hairs.

Ecological host range The host range of a natural enemy under natural conditions in the field.

Ecologically based pest management Pest management based on facilitating the natural processes controlling pests but supplemented by biological control organisms, biological control products (genes or gene products derived from living organisms), narrow-spectrum synthetic chemicals, and resistant plants (National Research Council, 1996).

Economic injury level The density of a pest population at which the cost of control measures is justified. At these pest densities, the cost of control is equal to the loss suffered if control action is not taken.

Ecosystem Biological community of interacting organisms and their physical environment.

Ecosystem services Benefits that organisms obtain from ecosystems. Biological control is generally considered a "regulating" ecosystem service.

Ecotype A specific strain of a species adapted to a certain suite of environmental conditions.

Ectoparasitoid A parasitoid that develops externally on the body of the host.

Emerging disease An infectious disease whose prevalence has increased in the past 20 years. The responsible pathogens have generally evolved or have spread from another area.

Encapsulation An immune response by arthropods where an invader is surrounded by blood cells that eventually form a capsule.

Endemic Native, originating in, and limited to, that geographic area.

Endoparasitoid A parasitoid that develops internally, within the body of the host.

Endophyte A symbiont, often fungal or bacterial, that lives within a plant for at least part of its life cycle but is not a plant pathogen.

Entomopathogenic Adjective describing pathogens infecting insects.

Environmental risk Actual or potential threat of adverse effects on living organisms or the environment.

Epiphytic Occurring on the surface of a plant or plant organ without causing infection.

Epizootic The widespread occurrence of a disease in a large proportion of a population of a nonhuman animal species. This term is analogous to the term epidemic, which refers to disease in human populations.

Equilibrium The density at which a population will remain, if not disturbed.

Eradication Total elimination of an organism from an area.

Eukaryote Organisms with cells containing genetic material within a nucleus. This excludes viruses, prions, bacteria, and other microorganisms lacking a nucleus.

Exotic Introduced from another geographic area.

Exploitative (resource) competition Individuals share a common but limiting resource but there is an absence of direct interaction among competitors and negative interactions are mediated through the limiting resource.

Extinction No longer in existence, globally.

Extirpation Local extinction (the species would still occur in other areas).

Extrinsic (i.e., exogenous) Coming from outside of something.

Facultative pathogen An organism having the ability to live as a pathogen but living as a pathogen is not a requirement.

Facultative saprotroph An organism having the ability to live as a saprotroph, using dead organic matter for food, but living as a saprotroph is not a requirement.

Fitness The reproductive contribution of an individual to the next generation, usually measured as the number of offspring or close kin that survive to reproductive age.

Food chain An ecological system described by feeding relationships and energy and material flow, for example, including plants, herbivores, decomposers, and predators.

Food web The network of feeding relationships among organisms in a localized area and time.

Formulation Relative to microorganisms for use in biological control, this term includes aids for preserving organisms, delivering them to targets, and improving their activity.

Frass Solid excreta from insects, especially larvae.

Functional redundancy When different species contribute in equivalent ways to ecosystem function.

Functional response Response of predators to prey in which the number of prey eaten by predators changes in response to prey density, that is, more prey are eaten when prey density is greater.

Fundamental host range The set of species that can support development of a natural enemy and would be chosen for use by the natural enemy.

Gall A swelling or overgrowth on a plant that results from stimulation by fungi, bacteria, or arthropods.

Genetically modified Organisms whose genetic material (DNA) has been modified in a way that does not naturally occur.

Glasshouse A building for growing plants in protected or controlled conditions. In North America, the word "greenhouse" is equivalent.

Gram negative Bacteria that appear red when stained using the Gram stain procedure, a method for preliminary identification of bacteria.

Gram positive Bacteria that appear violet when stained using the Gram stain procedure, a method for preliminary identification of bacteria.

Greenhouse A building for growing plants in protected or controlled conditions. The word "greenhouse" is often equivalent to "glasshouse."

Gregarious parasitoid A parasitoid for which more than one individual can develop in or on one host individual.

Guild Group of species that use a common resource base in a similar way.

Habitat displacement Spatial segregation of one species forced by the presence of a stronger competitor.

Habitat enhancement One major type of conservation biological control focusing on enhancing habitats to increase natural enemy populations or activity.

Hemimetabola Insects with body forms gradually changing at each molt, with wing buds increasing gradually to the fully winged adult state. The insects have incomplete metamorphosis.

Hemocoel The main body cavity of many invertebrates, including insects and mites, equivalent to a circulatory system and filled with "blood" or hemolymph.

Herbivore Animal feeding on living plants.

Herbivore-induced plant volatiles (HIPVs) Specific volatile organic compounds that plants produce in response to herbivory and which are thought to be used for communication with other organisms.

Holometabola Insects with body forms abruptly changing at the molt to pupae and again at the molt to adults. These insects have complete metamorphosis.

Honeydew Watery fluid containing sugars excreted from the anus of some Homoptera (e.g., aphids, scale insects, mealybugs, whiteflies).

Horizontal transmission Transmission of an infectious agent from one individual host to another, except direct transmission from parent to offspring.

Host An organism that harbors another organism, such as a pathogen, parasite, or parasitoid, either internally or externally.

Host feeding Activity by females of some parasitoid species of piercing a host with the ovipositor and then drinking host fluids that ooze from the wound. This is not always fatal to the host.

Host specificity The degree of limitation in the number of different plant or animal species that can be used for food by herbivorous or carnivorous species.

Hyperparasite A parasite that lives at the expense of another parasite (used with reference to both plant pathogens and insects).

Hyperparasitoid A parasitoid that lives at the expense of another parasitoid (used with reference to insects only).

Hypersensitive Excessively sensitive.

Hypha (plural = hyphae) A singular, usually tubular, branch of the body of a fungus.

Hypovirulence Reduced virulence.

Idiobiont A parasitoid attacking hosts that do not develop further after parasitization.

Induced When a specific action occurs in response to a stimulus.

Induced resistance A resistance process that is activated in response to an attack.

Infection Invertebrate pathology: entry of a pathogen into the body of a susceptible host. Plant pathology: establishment of a pathogen within a host plant.

Infective juvenile Third stage entomopathogenic nematodes that disperse and infect a new host.

Inoculative biological control (i.e., inoculative augmentation) The intentional release of a living organism as a biological control agent with the expectation that it will multiply and control the pest for an extended period, but not that it will do so permanently.

Instar For insects, the growth stage between two successive molts.

Interference competition Negative interactions as a result of direct interactions among competitors or via modification of a shared resource by one competitor so that it cannot be used by the other competitor.

Intraguild predation Predation by one member of a guild on another member of the same guild, for example, for two predators that feed on the same prey, one predator eats the second.

Intrinsic Naturally belonging (to an organism) and not because of external or environmental factors.

Introduced Introduced from another geographic area.

Inundative biological control (i.e., inundative augmentation) The use of living organisms to control pests when control is achieved exclusively by the organisms themselves that have been released. Therefore reproduction by the organisms released is not expected for control.

Invasive Not native to that region and having an ecological and/or economic impact.

Invasive species treadmill If invasives are removed from an environment, other invasives move in to use the vacant niche instead of native species (term is especially used for invasive weeds).

Inverse density dependence A relationship where an effect increases with decreasing densities of an organism.

Invertebrate An animal without a backbone.

Keystone predator Dominant predator whose removal results in large changes throughout a food web.

Koinobiont A parasitoid attacking hosts that continue developing after parasitization.

***K*-selection** Selection characteristic of populations at or near the carrying capacity of an environment that is usually more stable. K-selected species are often known for competition for resources and slower development of young.

Larva (plural = larvae) An immature insect after it has emerged from the egg and before pupation; usually refers to holometabolous insects.

Lesion An injury or hurt, commonly used to refer to injuries to plants caused by growth of pathogens.

Life table A table documenting complete data on the mortality of a population through time. For insects, this is often organized as mortality by instar.

Macro-beneficial Beneficial organisms (e.g., natural enemies, bees) that are large enough to be seen by an unaided eye.

Macro-organism Organisms that are large enough to be seen by an unaided eye. In the context of biological control this term usually refers to invertebrate natural enemies.

Mating disruption Mass application of pheromones so that mates cannot find each other.

Melanization The process through which invertebrates darken through deposition of melanin. This commonly occurs after an organism molts or during immune responses.

Meta-analysis Use of statistics to analyze data sets that are composed of many different data sets on a common subject, with the goal of increasing statistical power.

Metapopulation A set of local populations connected to each other through dispersal.

Microbial control The use of microorganisms as biological control agents. This should be considered a subset of biological control.

Microbial pesticides Microorganisms used for inoculative or inundative biological control.

Microbiome The community of microorganisms that are present in a particular environment.

Microhabitat A habitat that is small or of limited extent, differing from surrounding larger habitats.

Microorganism (i.e., microbe) An organism visible only when using a microscope.

Molting The periodic process of shedding the outer covering; for arthropods this is necessary during larval or nymphal growth because the hardened exoskeleton does not allow further growth.

Monoculture Cultivation with a single plant species.

Monophagous Feeding on only one species of organism.

Morphology Study of the form of organisms.

Multiple parasitism More than one parasitoid species occurs simultaneously in or on the body of a host.

Mutualism Symbiotic relationship where both members of the relationship benefit from the association.

Mycetocyte A cell within insects containing symbiotic microorganisms. Such cells can be scattered throughout the body (particularly found in the fat body) or aggregated in an organ called a mycetome.

Mycoherbicide Fungal pathogens of weeds used for inundative or inoculative biological control.

Mycoparasite Plant pathology: term specifically meaning a fungus that attacks other fungi.

Mycorrhiza (plural = mycorrhizae) An association between a fungus and the roots of a plant that is symbiotic or mildly pathogenic.

Native Originating in the area; part of the naturally occurring flora and fauna.

Natural control Naturally occurring control of organisms, often referring to control of one species by another species.

Natural enemies In a general sense, the parasites, predators, pathogens, and competitors associated with a species of animal or plant that cause debility or mortality.

Necrosis When a living organism's tissues die or degenerate; this term describes a symptom and is often used to describe darkened or wilted plant parts.

Necrotroph A microorganism feeding only on dead organic tissues; this term is generally used in plant pathology.

Neoclassical biological control Classical biological control using a natural enemy that did not coevolve with the pest (this term is the same as "new association" classical biological control).

New association Classical biological control term referring to situations where a natural enemy is used against a host with which it did not coevolve, that is, the natural enemy and host did not originate from the same area.

Niche The set of ecological conditions under which a species can live and eat such that it is able to reproduce and exploit similar conditions elsewhere.

Nonconsumptive effects Reduction in likelihood of use of a site by a pest as a result of the possibility of predation by a natural enemy; absence from a site is not due to predation itself.

Nontarget organism In the context of biological control, all species that are not the pest(s) that are the focus of control programs.

Numerical response Response by predators to prey in which density of predators in a given area increases because of reproduction in relation to increasing prey density.

Nymph An immature insect after emerging from an egg and before becoming an adult; usually refers to hemimetabolous insects.

Obligate pathogen A pathogen that, in nature, can only grow and reproduce in or on another living organism.

Occlusion bodies (OBs) Produced by certain viruses, a protein matrix containing virions.

Oligophagous Feeding on a limited number of species.

Outbreak population Explosive increase in the abundance of a species over a relatively short time period.

Ovipositor Structure in parasitic Hymenoptera at the distal end of the body, through which eggs pass during egg laying, or oviposition. The ovipositor can be used to lay eggs on surfaces or can be used to deposit eggs within hosts or in structures in which hosts live.

Parasite An organism that uses another organism for nourishment, living on or in a host and benefiting at the host's expense.

Parasitoid An insect that, when immature, parasitizes another insect, killing its host, and is subsequently free-living as an adult.

Parasporal body (or parasporal crystal) In *Bacillus thuringiensis*, the crystal within the sporangium (spore-bearing cell) that contains the delta (δ) endotoxin.

Pathogen A microorganism that lives as a parasite on or in a larger host organism, causing debility or mortality.

Pathogenicity Potential ability to cause disease.

Pest A species that inflicts harm on humans, domesticated animals or cultivated crops; basically, organisms that compete with man and his needs.

Pesticide Any substance used for controlling, preventing, destroying, disabling, or repelling a pest.

Pesticide treadmill A syndrome through which use of pesticides leads to pesticide resistance, pest resurgence, and secondary pest outbreaks which then results in increased pesticide use and this only leads to repeated occurrence of these effects, resulting in another increase in pesticide use.

Phagostimulant A compound that stimulates feeding.

Pheromone A substance secreted by one animal that causes a specific reaction when received by another animal of the same species.

Phloem-feeding Feeding on the liquid running through the nutrient-conducting tissues of plants.

Phoresy An association between two organisms where one travels on the body of another, without being parasitic.

Physiological host range The range of hosts that a natural enemy can utilize for growth and reproduction. This differs from the ecological host range because these hosts might not normally be encountered by the natural enemy in nature.

Phytoalexins Substances produced in higher plants in response to chemical, physical, or biological stimuli that inhibit the growth of certain microorganisms.

Phytophagous insect An insect that feeds on plants.

Plasmid An extrachromosomal piece of DNA that can replicate independently and can be transferred between organisms.

Polar filament When a microsporidian spore germinates, the tube that everts and extends to puncture a host cell. The cytoplasm of the microsporidian spore then flows through the polar filament and into the pierced host cell.

Polyculture Simultaneously growing two or more species or cultivars of plants at agronomically close distances.

Polyphagous Feeding on a diversity of species.

Population dynamics Changes in size and structure of populations over time.

Population regulation Relative to population dynamics, the control of a population; often refers to the return of a population to an equilibrium density as a result of density-dependent processes.

Positive controls Insect pathology: Groups of organisms included in bioassays to confirm a known response, such as susceptibility to a pathogen. These are included in studies to compare with the unknown response of a treatment being tested.

Predator An animal that eats more than one other animal during its life.

Preemptive exclusion Colonization of a resource by a microorganism before other microorganisms, thus excluding newcomers. These pioneer colonists can alter the habitat so later-arriving microorganisms cannot compete well.

Prevalence (of a disease) Total number of cases of a particular disease at one time, in a population.

Prey A food item of a predator.

Primary infections The first infections of the season by the persistent stages of a pathogen.

Primary parasitoid A parasitoid that develops in or on a host that is not a parasitoid.

Prokaryote Organisms whose genetic material is not contained within a nucleus. This includes the bacteria, viruses, and their relatives such as prions.

Proovigenic Parasitoids that emerge as adults with their full lifetime complement of mature eggs.

Pustule Plant pathology: Blister-like growth of a pathogen (usually referring to fungi) as it breaks through the plant surface.

Recombinant In which segments of two DNA molecules have been exchanged. If recombination occurs, progeny from a cross between two parents would have combinations of alleles not displayed by either parent.

Replication Reproduction, copying.

Reproductive parasite Parasites that influence the ability of hosts to reproduce.

Resistance The ability of an organism to exclude or overcome, completely or partially, the effect of a deleterious agent. Often used to describe the ability of previously susceptible pests to now survive exposure to specific synthetic chemical pesticides.

Resource An aspect of the environment that is consumed by one individual and is then not available to another. Relative to biological control, this could be an herbivorous pest eaten by a predator.

Rhizobacteria Bacteria living in the root zone.

Rhizosphere The soil around a living plant root.

Rhizosphere competent Microorganisms surviving and remaining active around roots as the roots develop and grow.

Risk Relative to a situation, the extent of problems that could be caused (the magnitude or hazard) versus the chance that this situation will occur.

***r*-selection** Characteristic of species with populations capable of increasing rapidly, usually more characteristic of unstable environments.

Sampling The process of taking a small part of something as a sample to provide estimation about the whole.

Saprophyte An organism using dead organic matter as food.

Sclerotization Stiffening of the cuticle of an invertebrate by cross-linkage of protein chains. This process commonly occurs after a molt.

Scouting Sampling to detect pests.

Seasonal inoculative release When a natural enemy does not persist after release as a result of seasonal effects and is thus released inoculatively each season.

Secondary cycling Cycles from infection to production of inoculum and infection of another host that occur after the first cycle during one season; the first cycle is initiated by primary infections and later cycles are caused by secondary infections. This term is used in biological control of arthropods using pathogens, biological control of weeds using plant pathogens, and in plant pathology.

Secondary infections Infections caused by inoculum produced as a result of primary or a subsequent infection.

Secondary metabolites (produced by microorganisms) Organic compounds produced by microorganisms that are not required for normal growth, development, and reproduction.

Secondary pest outbreak Rapid numerical response (increase by reproduction) to pest status of an organism that is typically not pestiferous. Commonly occurs after use of a broad-spectrum pesticide for control of another pest resulting in mortality of the natural enemies that normally regulate populations of the secondary pest.

Secondary plant compounds Plant chemicals that function in defense against herbivores or plant pathogens.

Selection pressure Effect of natural selection causing decreases in species that produce fewer surviving offspring.

Sequester In the context of interactions between herbivores and secondary plant compounds: when a compound is separated from the ingested plant material and stored within the herbivore, often considered as subsequently providing defense.

Siderophore A chemical produced by an organism that binds cations, especially Fe^{+2}, and helps to transport it into the organism, especially in areas where that cation is limiting.

Solitary parasitoid Parasitoid species for which only one individual can develop in or on one host individual.

Spillover This term has broad usage, ranging from host specificity of pathogens and parasitoids to changes in habitat use by generalist predators but the general meaning is similar: an unexpected event resulting in an expansion of resources being used. As an example, for pathogens: when a pathogen usually infects one host species but then, as when disease prevalence is high or the primary host population is low, expands host use to include additional susceptible species.

Spiracle For arthropods, external opening of the gas exchange system, which is composed of series of ever-smaller tubes called tracheae and tracheoles.

Sporangium A structure containing one or more asexual spores, used in reference to bacteria and fungi.

Spore Reproductive unit of most fungi and some bacteria; analogous to a seed of a green plant.

Superorganism A social unit of social animals that work together (such as ants or bees).

Superparasitism Situation where more parasitic individuals occur in or on a host than can survive.

Suppressive soil A soil in which plant disease is reduced or absent although a pathogen and a susceptible host are both present.

Sustainable Ability of being maintained over time, based on current conditions and practices; relating to ecosystems or renewable resources.

Sustainable agriculture Farming systems that are capable of maintaining their productivity and usefulness to society indefinitely; they should be resource-conserving, socially supportive, commercially competitive, and environmentally sound.

Symbiosis Living together of two or more different kinds of organisms.

Synergism Interaction of agents or conditions so that the total effect is greater than the sum of the individual effects.

Synovigenic Parasitoid species with adult females that mature eggs through part or all of their adult lives.

Systemic response A response occurring throughout an organism's body and not only in a particular part.

Target pest resurgence Rapid increase of a pest population after use of a broad-spectrum pesticide that results in destruction of the natural enemies that would normally help to control populations of the pest.

Tissue specific Restricted to specific tissues, as is characteristic for some microsporidia and viruses.

Tolerance In plant resistance, the ability of a host plant to withstand injury by pests.

Top-down control Regulation of ecosystems by organisms higher in food chains (often predators/parasites), rather than factors like soil nutrients or primary productivity at the bottom of the food chain.

Transconjugate In prokaryotes, transfer of DNA from one cell to another.

Transgenic Used to describe genetically modified organisms containing foreign genes inserted by means of recombinant DNA.

Transmission Passage of microorganisms (in this case, disease-causing pathogens) from an infected host or reservoir to another potential host.

Trichome Hair-like outgrowth from the epidermal cell of a plant.

Triungulin An active first instar larva of insects that usually disperses before changing into a less dispersive stage in later larval instars.

Trophic cascade Natural enemy-host/prey interactions that alter the abundance, biomass, or productivity of a community across more than one trophic level.

Vector A bearer. An organism that transmits a disease-producing organism from one host to another.

Vertebrate Animal with a backbone.

Vertical transmission Direct transmission of an infectious agent from parent to offspring.

Virion Viral DNA or RNA and the protein capsule surrounding it and sometimes an external envelope.

Virulence The relative, and quantifiable, capacity of a pathogen to cause disease; the disease-producing power of a pathogen.

Zoophytophagy Opportunistic omnivory, usually by carnivorous species that consume plants when prey is scarce or as a source of essential nutrients that are absent in prey populations.

Bibliography

Agrawal, A. A. (2000). Mechanisms, ecological consequences and agricultural implications of tri-trophic interactions. *Current Opinion in Plant Biology*, **3**, 329–335.

Agrios, G. N. (2005). *Plant Pathology*, 5th edn. San Diego, CA: Academic Press.

Albajes, R., Gullino, M. L., van Lenteren, J. C., & Elad, Y. (eds.) (1999). *Integrated Pest and Disease Management in Greenhouse Crops*. Dordrecht, Netherlands: Kluwer Academic.

Andow, D. A. (1991). Vegetational diversity and arthropod population response. *Annual Review of Entomology*, **36**, 561–586.

Andrews, J. H. & Harris, R. F. (1986). *r*- and *K*-selection and microbial ecology. *Advances in Microbial Ecology*, **9**, 99–147.

Annecke, D. P., Karny, M., & Burger, W. A. (1969). Improved biological control of the prickly pear, *Opuntia megacantha* Salm-Dyck, in South Africa through use of an insecticide. *Phytophylactica*, **1**, 9–13.

Aquilino, K. M., Cardinale, B. J., & Ives, A. R (2005). Reciprocal effects of host plant and natural enemy diversity on herbivore suppression: An empirical study of a model tritrophic system. *Oikos*, **108**, 275–282.

Askew, R. R. (1971). *Parasitic Insects*. London: Heinemann Educational Books.

Askew, R. R. (1975). The organisation of chalcid-dominated parasitoid communities centered upon endophytic hosts. In *Evolutionary Strategies of Parasitoids*, ed. P. W. Price, pp. 130–153. New York: Plenum Press.

Axtell, R. C. (1986). *Fly Control in Confined Livestock and Poultry Production*. Greensboro, NC: CIBA-GEIGY Corp.

Bajwa, W. I. & Kogan, M. (2002). Compendium of IPM definitions (CID) – What is IPM and how is it defined in the worldwide literature? IPPC Publication No. 998. Integrated Plant Protection Center (IPPC), Oregon State University, Corvallis, OR.

Baker, K. F. & Griffin, G. J. (1995). Molecular strategies for biological control of fungal plant pathogens. In *Novel Approaches to Integrated Pest Management*, ed. R. Reuveni, pp. 153–182. Boca Raton, FL: CRC Press.

Balciunas, J. K. (2000). A proposed code of best practices for classical biological control of weeds. In *Proceedings of the X International Symposium on Biological Control of Weeds, 5–9 July 1999, Bozeman, Montana*, ed. N. R. Spencer, pp. 435–436. Bozeman: Montana State University.

Banfield-Zanin, J. A. & Leather, S. R. (2016). Prey-mediated effects of drought on the consumption rates of coccinellid predators of *Elatobium abietinum*. *Insects*, **7** (4), 49.

Barbosa, P. (ed.) (1998). *Conservation Biological Control*. San Diego, CA: Academic Press.

Barbosa, P., Gross, P., & Kemper, J. (1991). Influence of plant allelochemicals on the tobacco hornworm and its parasitoid, *Cotesia congregata*. *Ecology*, **72**, 1567–1575.

Barbosa, P., Saunders, J. A., Kemper, J., Trumbule, R., Olechno, J., & Martinat, P. (1986). Plant allelochemicals and insect parasitoids: Effects of nicotine on *Cotesia congregata* (Say) (Hymenoptera: Braconidae) and *Hyposoter annulipes* (Cresson) (Hymenoptera: Ichneumonidae). *Journal of Chemical Ecology*, **12**, 1319–1328.

Barron, G. L. (1979). Observations on predatory fungi. *Canadian Journal of Botany*, **57**, 187–193.

Bateman, R. P. (2000). Rational pesticide use: An alternative escape from the treadmill? *Biocontrol News & Information*, **21**, 96–100.

Baum, J. A., Johnson, T. B., & Carlton, B. C. (1998). *Bacillus thuringiensis*, natural and recombinant bioinsecticide products. In *Biopesticides, Use and Delivery*, ed. F. R. Hall & J. J. Menn, pp. 189–209. Totowa, NJ: Humana Press.

Becker, N. (2000). Bacterial control of vector-mosquitoes and black flies. In *Entomopathogenic Bacteria: From Laboratory to Field Application*, ed. J.-F. Charles, A. Delécluse, & C. Nielsen-LeRoux, pp. 383–398. Dordrecht, Netherlands: Kluwer Academic Publishers.

Bedding, R. A. (1993). Biological control of *Sirex noctilio* using the nematode *Deladenus siricidicola*. In *Nematodes and the Biological Control of Insect Pests*, ed. R. Bedding, R. Akhurst, & H. Kaya, pp. 11–20. East Melbourne, Australia: CSIRO.

Bellows, T. S. & Fisher, T. W. (eds.) (1999). *Handbook of Biological Control*. San Diego, CA: Academic Press.

Bellows, T. S., Paine, T. D., Gould, J. R., Bezark, L. G., & Ball, J. C. (1992). Biological control of ash whitefly: A success in progress. *California Agriculture*, **46** (1), 24–28.

Benbrook, C. M. (1996). *Pest Management at the Crossroads*. Yonkers, NY: Consumers Union.

Berenbaum, M. R. (1995). Turnabout is fair play: Secondary roles for primary compounds. *Journal of Chemical Ecology*, **21**, 925–940.

Bergelson, J. & Crawley, M. J. (1989). The theory and practice of biological control. *Comments on Modern Biology, C, Comments on Theoretical Biology*, **1**, 197–215.

Berner, D. K., Bruckart, W. L., Cavin, C. A., Michael, J. L., Carter, M. L., & Luster, D. G. (2009). Best linear unbiased prediction of host-range of the facultative parasite *Colletotrichum gloeosporioides* f. sp. *salsolae*, a potential biological control agent of Russian thistle. *Biological Control*, **51**, 158–168.

Biddinger, D. & Butler, B. (2013). Biological control of mites in Pennsylvania and Maryland apple orchards. *Fruit Times*, July 26. Retrieved from http://extension.psu.edu/plants/tree-fruit/news/2013/biological-control-of-mites-in-pennsylvania-and-maryland-apple-orchards.

Blossey, B. & Hunt-Joshi, T. R. (2003). Belowground herbivory by insects: Influence on plants and aboveground herbivores. *Annual Review of Entomology*, **48**, 521–547.

Blossey, B., Skinner, L. C., & Taylor, J. (2001). Impact and management of purple loosestrife (*Lythrum salicaria*) in North America. *Biodiversity and Conservation*, **10**, 1787–1807.

Boettner, G. H., Elkinton, J. S., & Boettner, C. J. (2000). Effects of a biological control introduction on three nontarget native species of saturniid moths. *Conservation Biology*, **14**, 1798–1806.

Borriss, R., Chen, X.-H., Rueckert, C., Blom, J., Becker, A., Baumgarth, B., Fan, B., Pukall, R., Schumann, P., Spröer, C., et al. (2011). Relationship of *Bacillus amyloliquefaciens* clades associated with strains DSM 7T and FZB42T: A proposal for *Bacillus amyloliquefaciens* subsp. *amyloliquefaciens* subsp. nov. and *Bacillus amyloliquefaciens* subsp. *plantarum* subsp. nov. based on complete genome sequence comparisons. *International Journal of Systematic and Evolutionary Microbiology*, **61**, 1786–1801.

Boucias, D. G. & Pendland, J. C. (1998). *Principles of Insect Pathology*. Boston, MA: Kluwer Academic Publishers.

Bourchier, R. S. & Van Hezewijk, B. H. (2013). *Euphorbia esula* L., Leafy spurge (Euphorbiaceae). In *Biological Control Programmes in Canada 2001–2012*, ed. P. G. Mason & D. R. Gillespie., pp. 315–320. Wallingford, UK: CABI Publishing.

Brennan, E. B. (2013). Agronomic aspects of strip intercropping lettuce with alyssum for biological control of aphids. *Biological Control*, **65**, 302–311.

Briese, D. T. & McLaren, D. A. (1997). Community involvement in the distribution and evaluation of biological control agents: Landcare and similar groups in Australia. *Biocontrol News & Information*, **18**, 39N-49N.

Brown, G. C. (1987). Modeling. In *Epizootiology of Insect Diseases*, ed. J. R. Fuxa & Y. Tanada, pp. 43–68. New York: John Wiley & Sons.

Bugg, R. L. & Pickett, C. H. (1998). Introduction: Enhancing biological control – habitat management to promote natural enemies of agricultural pests. In *Enhancing Biological Control: Habitat Management to Promote Natural Enemies of Agricultural Pests*, ed. C. H. Pickett & R. L. Bugg, pp. 1–23. Berkeley, CA: University of California Press.

Bull, D. L. & Menn, J. J. (1990). Strategies for managing resistance to insecticides in *Heliothis* pests of cotton. In *Managing Resistance to Agrochemicals*, ed. M. B. Green, H. M. LeBaron, & W. K. Moberg, pp. 118–133. Washington, DC: American Chemical Society.

Burdon, J. J. & Marshall, D. R. (1981). Biological control and the reproductive mode of weeds. *Journal of Applied Ecology*, **18**, 649–658.

Burges, H. D. (ed.) (1998). *Formulation of Microbial Biopesticides*. Dordrecht, Netherlands: Kluwer Academic Publishers.

Butler, L., Zivkovich, C., & Sample, B. E. (1995). Richness and abundance of arthropods in the oak canopy of West Virginia's Eastern Ridge and Valley Section during a study of impact of *Bacillus thuringiensis* with emphasis on macrolepidoptera larvae. *Bulletin of the Agricultural Forest Experiment Station, West Virginia University*, **711**, 1–19.

Cade, W. (1975). Acoustically orienting parasitoids: Fly phonotaxis to cricket song. *Science*, **190**, 1312–1313.

Caltagirone, L. E. (1981). Landmark examples in classical biological control. *Annual Review of Entomology*, **26**, 213–232.

Caltagirone, L. E., Shea, K. P., & Finney, G. L. (1964). Parasites to aid control of navel orangeworm. *California Agriculture*, **18(1)**, 10–12.

Campbell, R. (1989). *Biological Control of Microbial Plant Pathogens*. Cambridge: Cambridge University Press.

Cartwright, B. & Kok, L. T. (1990). Feeding by *Cassida rubiginosa* (Coleoptera: Chrysomelidae) and the effects of defoliation on growth of musk thistles. *Journal of Entomological Science*, **25**, 538–547.

Center, T. D., Hill, M. P., Cordo, H., & Julien, M. H. (2002). Waterhyacinth. In *Biological Control of Invasive Plants in the Eastern United States*, ed. R. Van Driesche, B. Blossey, M. Hoddle, S. Lyon, & R. Reardon, pp. 41–64. Morgantown, WV: USDA Forest Service Publication, FHTET-2002-04.

Chandler, J. M. (1980). Assessing losses caused by weeds. In *Proceedings of the E. C. Stakman Communication Symposium*, ed. R. J. Aldrich. University of Minnesota Agricultural Experiment Station, Miscellaneous Publication 7.

Chaney, W. E. (1998). Biological control of aphids in lettuce using in-field insectaries. In *Enhancing Biological Control: Habitat Management to Promote Natural Enemies of Agricultural Pests*, ed. C. H. Pickett & R. L. Bugg, pp. 73–83. Berkeley: University of California Press.

Charudattan, R. (2005). Ecological, practical, and political inputs into selection of weed targets: What makes a good biological control target? *Biological Control*, **35**, 183–196.

Charudattan, R. (2010). A reflection on my research in weed biological control: Using what we have learned for future applications. *Weed Technology*, **24**, 208–217.

Charudattan, R. (2015). Weed control with microbial bioherbicides. In *Weed Science for Sustainable Agriculture, Environment and Biodiversity*, ed. A. N. Rao & N. T. Yaduraju, Vol. 1, pp. 79–96. Proceedings of the Plenary and Lead Papers of the 25th Asian-Pacific Weed Science Society Conference, Hyderabad, India.

Clarke, A. R. & Walter, G. H. (1995). "Strains" and the classical biological control of insect pests. *Canadian Journal of Zoology*, **73**, 1777–1790.

Cloyd, R. A. (2012). Indirect effects of pesticides on natural enemies. In *Pesticides – Advances in Chemical and Botanical Pesticides*, ed. R. P. Soundararajan, pp. 127–150. Rijeka, Croatia: InTech.

Cock, M. J. W., Murphy, S. T., Kairo, M. T. K., Thompson, E., Murphy, R. J., & Francis, A. W. (2016). Trends in the classical biological control of insect pests by insects: An update of the BIOCAT database. *BioControl*, **61**, 349–363.

Cock, M. J. W., van Lenteren, J. C., Brodeur, J., Barratt, B. I. P., Bigler, F., Bolckmans, K., Cônsoli, F. L., Haas, F., Mason, P. G., & Parra, J. R. P. (2010). Do new Access and Benefit Sharing procedures under the Convention on Biological Diversity threaten the future of biological control? *BioControl*, **55**, 199–218.

Coll, M. (1998). Parasitoid activity and plant species composition in intercropped systems. In *Enhancing Biological Control: Habitat Management to Promote Natural Enemies of Agricultural Pests*, ed. C. H. Pickett & R. L. Bugg, pp. 85–120. Berkeley: University of California Press.

Cook, R. J. (1993). Making greater use of introduced microorganisms for biological control of plant pathogens. *Annual Review of Phytopathology*, **31**, 53–80.

Coppel, H. C. & Mertins, J. W. (1977). *Biological Insect Pest Suppression*. Berlin: Springer-Verlag.

Costa, A. S. & Müller, G. W. (1980). Tristeza control by cross protection. *Plant Disease*, **64**, 538–541.

Crawley, M. J. (1989). The successes and failures of weed biocontrol using insects. *Biocontrol News & Information*, **10**, 213–223.

Cruz, Y. P. (1981). A sterile defender morph in a polyembryonic hymenopterous parasite. *Nature*, **294**, 446–447.

Culliney, T. W. (2005). Benefits of classical biological control for managing invasive plants. *Critical Reviews in Plant Sciences*, **24**, 131–150.

Dahlsten, D. L. & Garcia, R. (eds.) (1989). *Eradication of Exotic Pests*. New Haven, CT: Yale University Press.

Darwin, C. (1859). *On the Origin of Species*. London: John Murray & Sons.

Davies, K. G., Flynn, C. A., Laird, V., & Kerry, B. R. (1990). The life-cycle, population dynamics and host specificity of a parasite of *Heterodera avenae*, similar to *Pasteuria penetrans*. *Revue de Nématologie*, **13**, 303–309.

Deacon, J. W. (1991). Significance of ecology in the development of biocontrol agents against soil-borne plant pathogens. *Biocontrol Science and Technology*, **1**, 5–20.

DeBach, P. H. (1964). *Biological Control of Insect Pests and Weeds*. New York: Reinhold Publishing Corporation.

DeBach, P. (1974). *Biological Control by Natural Enemies*. London: Cambridge University Press.

DeBach, P. & Rosen, D. (1991). *Biological Control by Natural Enemies*. Cambridge: Cambridge University Press.

DeBach, P., Rosen, D., & Kennett, C. E. (1971). Biological control of coccids by introduced natural enemies. In *Biological Control*, ed. C.B. Huffaker, pp. 165–194. New York: Plenum Press.

DeBach, P. & Sundby, R. A. (1963). Competitive displacement between ecological homologues. *Hilgardia*, **34**, 105–166.

De Barro, P. J. & Coombs, M. T. (2009). Post-release evaluation of *Eretmocerus hayati* Zolnerowich and Rose in Australia. *Bulletin of Entomological Research*, **99**, 193–206.

De Barro, P. J., Liu, S.-S., Boykin, L. M., & Dinsdale, A. B. (2011). *Bemisia tabaci*: A statement of species status. *Annual Review of Entomology*, **56**, 1–19.

Delfosse, E. S. & Cullen, J. M. (1980). New activities in biological control of weeds in Australia. II. *Echium plantagineum*: Curse or salvation? In *Proceedings of the Fifth International Symposium on Biological Control of Weeds*, Brisbane, Australia, pp. 563–574.

De Moraes, C. M., Lewis, W. J., Paré, P. W., Alborn, H. T., & Tumlinson, J. H. (1998). Herbivore-infested plants selectively attract parasitoids. *Science*, **393**, 570–573.

Dennill, G. B. (1985). The effect of the gall wasp *Trichilogaster acaciaelongifoliae* (Hymenoptera: Pteromalidae) on reproductive potential and vegetative growth of the weed *Acacia longifolia*. *Agriculture, Ecosystems & Environment*, **14**, 53–61.

Denoth, M., Frid, L., & Myers, J. H. (2002). Multiple agents in biological control: improving the odds? *Biological Control*, **24**, 20–30.

de Oliviera, M. R. V. (1998). South America. In *Insect Viruses and Pest Management*, ed. F. R. Hunter-Fujita, P. F. Entwistle, H. F. Evans, & N. E. Crook, pp. 339–355. Chichester, UK: John Wiley & Sons.

Dietrick, E. J., Schlinger, E. I., & Garber, M. J. (1960). Vacuum cleaner principle applied in sampling insect populations in alfalfa fields by new machine method. *California Agriculture*, **14** (1), 9–11.

Di Giallonardo, F. & Holmes, E. C. (2015). Exploring host-pathogen interactions through biological control. *PLOS Pathogens*, **11(6)**, e1004865.

DiTommaso, A., Averill, K. M., Hoffmann, M. P., Fuchsberg, J. R., & Losey, J. E. (2016). Integrating insect, resistance, and floral resource management in weed control decision-making. *Weed Science*, **64**, 743–756.

Dixon, A. F. G. (2000). *Insect Predator–Prey Dynamics: Ladybird Beetles and Biological Control*. Cambridge: Cambridge University Press.

Dixon, B. (1994). Keeping an eye on *B. thuringiensis*. *BioTechnology*, **12**, 435.

Dobson, A. P. (1988). Restoring island ecosystems: The potential of parasites to control introduced mammals. *Conservation Biology*, **2**, 31–39.

Douglas, A. E. (1998). Nutritional interactions in insect-microbial symbioses: Aphids and their symbiotic bacteria *Buchnera*. *Annual Review of Entomology*, **43**, 17–37.

Duetting, P. S. (2002). Effect of field pea surface wax variation on infection of the pea aphid by the fungal pathogen, *Pandora neoaphidis*. MS thesis, University of Idaho.

Dyer, L. A. (1995). Tasty generalists and nasty specialists? Antipredator mechanisms in tropical lepidopteran larvae. *Ecology*, **76**, 1483–1496.

Dyer, L. A. & Gentry, G. (1999). Predicting natural-enemy responses to herbivores in natural and managed systems. *Ecological Applications*, **9**, 402–408.

Ehler, L. E. (1998). Conservation biological control: Past, present, and future. In *Conservation Biological Control*, ed. P. Barbosa, pp. 1–8. San Diego, CA: Academic Press.

Eigenbrode, S. D., Castagnola, T., Roux, M.-B., & Steljes, L. (1996). Mobility of three generalist predators is greater on cabbage with glossy leaf wax than on cabbage with a wax bloom. *Entomologia Experimentalis et Applicata*, **81**, 335–343.

Eilenberg, J., Enkegaard, A., Vestergaard, S., & Jensen, B. (2000). Biocontrol of pests on plant crops in Denmark: Present status and future potential. *Biocontrol Science and Technology*, **10**, 703–716.

Eilenberg, J., Hajek, A., & Lomer, C. (2001). Suggestions for unifying the terminology in biological control. *BioControl*, **46**, 387–400.

Elad, Y., Chet, I., Boyle, P., & Henis, Y. (1983). Parasitism of *Trichoderma* spp. on *Rhizoctonia solani* and *Sclerotium rolfsii* – scanning electron microscopy and fluorescence microscopy. *Phytopathology*, **73**, 85–88.

Elkinton, J. S., Healy, W. M., Buonaccorsi, J. P., Boettner, G. H., Hazzard, A. M., & Smith, H. R. (1996). Interactions among gypsy moths, white-footed mice, and acorns. *Ecology*, **77**, 2332–2342.

Embree, D. G. (1966). The role of introduced parasites in the control of the winter moth in Nova Scotia. *Canadian Entomologist*, **98**, 1159–1168.

Engelstädter, J. & Hurst, G. D. D. (2009).The ecology and evolution of microbes that manipulate host reproduction. *Annual Review of Ecology, Evolution and Systematics*, **40**, 127–149.

Eriksson, J., Khortstam, K., & Ryvarden, L. (1981). *The Corticiaceae of North Europe*. Oslo, Norway: Fungiflora.

Essig, E. O. (1942). *College Entomology*. New York: Macmillan Co.

Eubanks, M. D., Blackwell, S. A., Parish, C. J., DeLamar, Z. D., & Hull-Sanders, H. (2002). Intraguild predation of beneficial arthropods by red imported fire ants in cotton. *Environmental Entomology*, **31**, 1168–1174.

Evans, E. W. & England, S. (1996). Indirect interactions in biological control of insects: Pests and natural enemies in alfalfa. *Ecological Applications*, **6**, 920–930.

Federici, B. (1999). *Bacillus thuringiensis* in biological control. In *Handbook of Biological Control*, ed. T. S. Bellows & T. W. Fisher, pp. 575–593. San Diego, CA: Academic Press.

Feitelson, J. S., Payne, J., & Kim, L. (1992). *Bacillus thuringiensis*: Insects and beyond. *Bio/Technology*, **10**, 271–275.

Felske, A. & Akkermans, A. D. L. (1998). Spatial homogeneity of abundant bacterial 16S rRNA molecules in grassland soils. *Microbial Ecology*, **36**, 31–36.

Fenner, F. & Fantini, B. (1999). *Biological Control of Vertebrate Pests: The History of Myxomatosis – An Experiment in Evolution*. Wallingford, UK: CABI Publishing.

Fenner, F. & Myers, K. (1978). Myxoma virus and myxomatosis in retrospect: The first quarter century of a new disease. In *Viruses and Environment*, ed. E. Kurstak & K. Maramorosch, pp. 539–570. New York: Academic Press.

Fields, P. G. & White, N. D. G. (2002). Alternatives to methyl bromide treatments for stored-product and quarantine insects. *Annual Review of Entomology*, **47**, 331–359.

Finch, S. & Collier, R. H. (2000). Host-plant selection by insects – a theory based on "appropriate/inappropriate landings" by pest insects of cruciferous plants. *Entomologia Experimentalis et Applicata*, **96**, 91–102.

Fischer, M. J., Havill, N. P., Brewster, C. C., Davis, G. A., Salom, S. M., & Kok, L. T. (2015). Field assessment of hybridization between *Laricobius nigrinus* and *L. rubidus*, predators of Adelgidae. *Biological Control*, **82**, 1–6.

Flint, M. L. & Dreistadt, S. H. (1998). *Natural Enemies Handbook: The Illustrated Guide to Biological Control*. Berkeley: University of California Statewide IPM Project Publication 3386.

Flint, M. L. & Gouveia, P. (2001). *IPM in Practice: Principles and Methods of Integrated Pest Management*. Oakland: University of California, Division of Agriculture and Natural Resources.

Flint, M. L. & van den Bosch, R. (1981). *Introduction to Integrated Pest Management*. New York: Plenum Press.

Follett, P. A., Duan, J., Messing, R. H., & Jones, V. P. (2000). Parasitoid drift after biological control introductions: Re-examining Pandora's box. *American Entomologist*, **46**, 82–94.

Food and Agriculture Organization (UN). (2018). AGP – Integrated Pest Management. Retrieved from http://www.fao.org/agriculture/crops/thematic-sitemap/theme/pests/ipm/en/.

Fowler, S. V., Syrett, P., & Hill, R. L. (2000). Success and safety in the biological control of environmental weeds in New Zealand. *Austral Ecology*, **25**, 553–562.

Friedman, M. J. (1990). Commercial production and development. In *Entomopathogenic Nematodes for Biological Control*, ed. R. Gaugler & H. K. Kaya, pp. 153–172. Boca Raton, FL: CRC Press.

Frost, R. (1936). *A Further Range*. New York: Henry Hold & Co.

Fuester, R., Hajek, A. E., Schaefer, P., & Elkinton, J. S. (2014). Biological control of *Lymantria dispar*. In *The Use of Classical Biological Control to Preserve Forests in North America*, ed. R. G. Van Driesche & R. Reardon, pp. 49–82. Morgantown, WV: USDA Forest Service, Forest Health Technology Enterprise Team, FHTET-2013-02.

Fulbright, D. W. (1999). Hypovirulence to control fungal pathogenesis. In *Handbook of Biological Control*, ed. T. S. Bellows & T. W. Fisher, pp. 691–712. San Diego, CA: Academic Press.

Funasaki, G. Y., Lai, P.-Y., Nakahara, L. M., Beardsley, J. W., & Ota, A. K. (1988). A review of biological control introductions in Hawaii: 1890 to 1985. *Proceedings of the Hawaiian Entomological Society*, **28**, 105–160.

Fuxa, J. R. (1987). Ecological considerations for the use of entomopathogens in IPM. *Annual Review of Entomology*, **32**, 225–251.

Fuxa, J. R. (1998). Environmental manipulation for microbial control of insects. In *Conservation Biological Control*, ed. P. Barbosa, pp. 255–268. San Diego, CA: Academic Press.

Gage, S. H. & Haynes, D. L. (1975). Emergence under natural and manipulated conditions of *Tetrastichus julis*, an introduced larval parasite of the cereal leaf beetle, with reference to regional population management. *Environmental Entomology*, **4**, 425–434.

Garcia, R., Caltagirone, L. E., & Gutierrez, A. P. (1988). Comments on a redefinition of biological control. *BioScience*, **38**, 692–694.

Garcia, R. & Legner, E. F. (1999). Biological control of medical and veterinary pests. In *Handbook of Biological Control*, ed. T. S. Bellows & T. W. Fisher, pp. 935–953. San Diego, CA: Academic Press.

Gardiner, M. M., Landis, D. A. Gratton, C., DiFonzo, C. D., O'Neal, M., Chacon, J. M., et al. (2009). Landscape diversity enhances biological control of an introduced crop pest in the north-central USA. *Ecological Applications*, **19**, 143–154.

Garnas, J. R., Hurley, B. P., Slippers, B., & Wingfield, M. J. (2012). Biological control of forest plantation pests in an interconnected world requires greater international focus. *International Journal of Pest Management*, **58**, 211–223.

Gause, G. F. (1934). *The Struggle for Existence*. Baltimore, MD: Williams and Wilkins.

Gelernter, W. D. & Lomer, C. J. (2000). Success in biological control of above-ground insects by pathogens. In *Biological Control: Measures of Success*, ed. G. Gurr & S. Wratten, pp. 297–322. Dordrecht, Netherlands: Kluwer Academic Publishers.

Gettys, L. A., Tipping, P. W., Della Torre, C. J., III, Sardes, S. N., & Thayer, K. M. (2014). Can herbicide usage be reduced by practicing IPM for waterhyacinth (*Eichhornia crassipes*) control? *Proceedings of the Florida State Horticultural Society*, **127**, 1–4.

Glare, T., Caradus, J., Gelernter, W., Jackson, T., Keyhani, N., Köhl, J., Marrone, P., Morin, L., & Stewart, A. (2012). Have biopesticides come of age? *Trends in Biotechnology*, **30**, 250–258.

Glare, T. R. & O'Callaghan, M. (2000). *Bacillus thuringiensis: Biology, Ecology and Safety*. Chichester, UK: Wiley & Sons.

Godwin, P. A. & ODell, T. M. (1981). Intensive laboratory and field evaluations of individual species: *Blepharipa pratensis* (Meigen) (Diptera: Tachinidae). In *The Gypsy Moth: Research Toward Integrated Pest Management*, ed. C. C. Doane & M. L. McManus, pp. 375–394. Washington, DC: USDA, Forest Service, Technical Bulletin 1584.

Goeden, R. D. & Andrés, L. A. (1999). Biological control of weeds in terrestrial and aquatic environments. In *Handbook of Biological Control*, ed. T. S. Bellows & T. W. Fisher, pp. 871–890. San Diego, CA: Academic Press.

Goldberg, L. J. & Margalit, J. (1977). A bacterial spore demonstrating rapid larvicidal activity against *Anopheles sergentii, Uranotaenia unguiculata, Culex univitattus, Aedes aegypti,* and *Culex pipiens*. *Mosquito News*, **37**, 355–358.

Gordh, G., Legner, E. F., & Caltagirone, L. E. (1999). Biology of parasitic Hymenoptera. In *Handbook of Biological Control*, ed. T. S. Bellows & T. W. Fisher, pp. 355–381. San Diego, CA: Academic Press.

Gould, J., Hoelmer, K., & Goolsby, J. (eds.) (2008). *Classical Biological Control of* Bemisia tabaci *in the United States – A Review of Interagency Research and Implementation*. Dordrecht, Netherlands: Springer.

Graham, F. (1970). *Since Silent Spring*. Boston, MA: Houghton-Mifflin.

Gray, M. E., Ratcliffe, S. T., & Rice, M. E. (2009). The IPM paradigm: Concepts, strategies and tactics. In *Integrated Pest Management: Concepts, Tactics, Strategies and Case Studies*, ed. E. B. Radcliffe, W. E. Hutchison & R. E. Cancelado, pp. 1–13. Cambridge: Cambridge University Press.

Grbić, M., Ode, P. J., & Strand, M. R. (1992). Sibling rivalry and brood sex ratios in polyembryonic wasps. *Nature*, **360**, 254–255.

Greathead, D. J. (1994). History of biological control. *Antenna*, **18**, 187–199.

Grewal, P. S. (2002). Formulation and application technology. In *Entomopathogenic Nematology*, ed. R. Gaugler, pp. 265–287. Wallingford, UK: CABI Publishing.

Grime, J. P. (1977). Evidence for the existence of three primary strategies in plants and its relevance to ecological and evolutionary theory. *American Naturalist*, **111**, 1169–1194.

Griswold, G. H. (1929). On the bionomics of a primary parasite and of two hyperparasites of the geranium aphid. *Annals of the Entomological Society of America*, **22**, 438–457.

Gross, P. (1991). Influence of target pest feeding niche on success rates in classical biological control. *Environmental Entomology*, **20**, 1217–1227.

Gullan, P. J. & Cranston, P. S. (2000). *The Insects: An Outline of Entomology*, 2nd edn. Oxford: Blackwell Science.

Gurr, G. M. & Wratten, S. D. (eds.) (2000). *Biological Control: Measures of Success*. Dordrecht, Netherlands: Kluwer Academic Publishers.

Gurr, G. M., Wratten, S. D., Landis, D. A., & You, M. (2017). Habitat management to suppress pest populations: Progress and prospects. *Annual Review of Entomology*, **62**, 91–109.

Gwynn, R. L. (ed.) (2014). *A World Compendium – The Manual of Biocontrol Agents*, 5th edn. Alton, UK: British Crop Protection Council.

Hagen, K. S., Sawall, E. F., Jr., & Tassen, R. L. (1970). The use of food sprays to increase effectiveness of entomophagous insects. *Proceedings of the Tall Timbers Conference on Ecological Animal Control by Habitat Management*, **2**, 59–81.

Hajek, A. E. (1997). Fungal and viral epizootics in gypsy moth (Lepidoptera: Lymantriidae) populations in central New York. *Biological Control*, **10**, 58–68.

Hajek, A. E. (1999). Pathology and epizootiology of the lepidoptera-specific mycopathogen *Entomophaga maimaiga*. *Microbiology and Molecular Biology Reviews*, **63**, 814–835.

Hajek, A. E. (2004). *Natural Enemies: An Introduction to Biological Control*. Cambridge: Cambridge University Press.

Hajek, A. E., Delalibera, I., Jr., & Butler, L. (2003). Entomopathogenic fungi as classical biological control agents. In *Environmental Impacts of Microbial Insecticides*, ed. H. M. T. Hokkanen & A. E. Hajek, pp. 15–34. Dordrecht, Netherlands: Kluwer Academic Publishers.

Hajek, A. E., Gardescu, S., & Delalibera, I., Jr. (2016). *Classical Biological Control of Insects and Mites: A Worldwide Catalogue of Pathogen and Nematode Introductions*. USDA Forest Service, FHTET-2016-06.

Hajek, A. E., Hurley, B. P., Kenis, M., Garnas, J. R., Bush, S. J., Wingfield, M. J., van Lenteren, J. C., & Cock, M. J. W. (2016). Exotic biological control agents: a solution or contribution to arthropod invasions? *Biological Invasions*, **18**, 953–969.

Hajek, A. E., McManus, M. L., & Delalibera, I., Jr. (2007). A review of introductions of pathogens and nematodes for classical biological control of insects and mites. *Biological Control*, **41**, 1–13.

Hajek, A. E., Wraight, S. P., & Vandenberg, J. D. (2001). Control of arthropods using pathogenic fungi. In *Bio-Exploitation of Fungi*, ed. S. B. Pointing & K. D. Hyde, pp. 309–347. Hong Kong: Fungal Diversity Press.

Hall, F. R. & Menn, J. J. (1999). *Biopesticides, Use and Delivery*. Totowa, NJ: Humana Press.

Handelsman, J. (2002). Future trends in biocontrol. In *Biological Control of Crop Diseases*, ed. S. S. Gnanamanickam, pp. 443–448. New York: Dekker.

Hansen, R. W., Spencer, N. R., Fornasari, L., Quimby, P. C., Jr., Pemberton, R. W., & Nowierski, R. M. (2004). Leafy spurge. In *Biological Control of Invasive Plants in the United States*, ed. E. M. Coombs, J. K. Clark, G. L. Piper & A. F. Cofrancesco Jr., pp. 233–262. Corvallis: Oregon State University Press.

Harris, J. (2000). *Chemical Pesticide Markets, Health Risks, and Residues*. Wallingford, UK: CABI Publishing.

Harris, P. (1981). Stress as a strategy in the biological control of weeds. In *Biological Control in Plant Production*, ed. G. C. Papavizas, pp. 333–340. Totowa, NJ: Allanheld.

Harris, P. (1986). Biological control of weeds. In *Biological Plant and Health Protection, Biological Control of Plant Pests and of Vectors of Human and Animal Diseases*, ed. J. M. Franz, International Symposium of the Akademie der Wissenschaften und der Literatur, Mainz, Nov. 15–17, 1984, pp. 123–138. Stuttgart: G. Fischer.

Harris, P. (1991). Classical biocontrol of weeds: Its definitions, selection of effective agents, and administrative-political problems. *The Canadian Entomologist*, **123**, 827–849.

Harris, P., Wilkinson, A. T. S., Thompson, L. S., & Neary, M. (1978). Interaction between the cinnabar moth, *Tyria jacobaeae* L. (Lep., Arctiidae) and ragwort, *Senecio jacobaea* L. (Compositae) in Canada. *Proceedings of the Fourth International Symposium on Biological Control of Weeds*, pp. 174–180.

Hartley, S. E. & Jones, C. G. (1997). Plant chemistry and herbivory, or why the world is green. In *Plant Ecology*, ed. M. J. Crawley, pp. 284–324, 2nd edn. Oxford: Blackwell.

Harvey, C. H. & Eubanks, M. D. (2005). Intraguild predation of parasitoids by *Solenopsis invicta*: A non-disruptive interaction. *Entomologia Experimentalis et Applicata*, **114**, 127–135.

Havill, N. P., Davis, G., Mausel, D. I., Klein, J., McDonald, R., Jones, C., Fischer, M., Salom, S., & Caccone, A. (2012). Hybridization between a native and introduced predator of Adelgidae: An unintended result of classical biological control. *Biological Control*, **63**, 359–369.

Hawkins, B. A. & Marino, P. C. (1997). The colonization of native phytophagous insects in North America by exotic parasitoids. *Oecologia*, **112**, 566–571.

Hawkins, B. A., Mills, N. J., Jervis, M. A., & Price, P. W. (1999). Is the biological control of insects a natural phenomenon? *Oikos*, **86**, 493–506.

Heimpel, G. E. & Mills, N. J. (2017). *Biological Control: Ecology and Applications*. Cambridge: Cambridge University Press.

Henry, J. E. & Oma, E. A. (1981). Pest control by *Nosema locustae*, a pathogen of grasshoppers and crickets. In *Microbial Control of Pests and Plant Diseases 1970–1980*, ed. H. D. Burges, pp. 573–586. New York: Academic Press.

Herren, H. R., Bird, T. J., & Nadel, D. J. (1987). Technology for automated aerial release of natural enemies of the cassava mealybug and cassava green mite. *Insect Science and Application*, **8**, 883–885.

Higley, L. G. & Pedigo, L. P. (eds.) (1996). *Economic Thresholds for Integrated Pest Management*. Lincoln: University of Nebraska Press.

Hintz, W. (2007). Development of *Chondrostereum purpureum* as a mycoherbicide for deciduous brush control. In *Biological Control: A Global Perspective*, ed. C. Vincent, M. S. Goettel, & G. Lazarovits, pp. 284–292. Wallingford, UK: CABI Publishing.

Hinz, H. L., Schwarzländer, M., & Gaskin, J. (2008). Does phylogeny explain the host-choice behaviour of potential biological control agents for Brassicaceae weeds? In *Proceedings, XII International Symposium of Biological Control of Weeds*, ed. M. H. Julien, R. Sforza, M. C. Bon, H. C. Evans, P. E. Hatcher, H. L. Hinz, & B. G. Rector, pp. 418–425. Wallingford, UK: CABI Publishing.

Hinz, H. L., Schwarzländer, M., Gassmann, A., & Bourchier, R. S. (2014). Successes we may not have had: A retrospective analysis of selected weed biological control agents in the United States. *Invasive Plant Science and Management*, **7**, 565–579.

Hoch, H. C. & Fuller, M. S. (1977). Mycoparasitic relationships: I. Morphological features of interactions between *Pythium acanthicum* and several fungal hosts. *Archives of Microbiology*, **111**, 207–224.

Hochberg, M. E. (1989). The potential role of pathogens in biological control. *Nature*, **337**, 262–264.

Hoddle, M. S. (1999). Biological control of vertebrate pests. In *Handbook of Biological Control: Principles and Applications of Biological Control*, ed. T. S. Bellows & T. W. Fisher, pp. 955–975. San Diego, CA: Academic Press.

Hoddle, M. S., Van Driesche, R. G., & Sanderson, J. P. (1998). Biology and use of the whitefly parasitoid *Encarsia formosa*. *Annual Review of Entomology*, **43**, 645–669.

Hoffmann, A. A., Montgomery, B. L., Popovici, J., Iturbe-Ormaetxe, I., Johnson, P. H., Muzzi, F., Greenfield, M., Durkan, M., Leong, Y. S., Dong, Y., et al. (2011). Successful establishment of *Wolbachia* in *Aedes* populations to suppress dengue transmission. *Nature*, **476**, 454–457.

Hoffmann, J. H. (1995). Biological control of weeds: The way forward, a South African perspective. *BCPC Symposium Proceedings*, **64**, 77–89.

Hoffmann, J. H. & Moran, V. C. (1992). Oviposition patterns and the supplementary roles of a seed-feeding weevil, *Rhyssomatus marginatus* (Coleoptera: Curculionidae), in the biological control of a perennial leguminous weed, *Sesbania punicea*. *Bulletin of Entomological Research*, **82**, 343–347.

Hoffmann, J. H. & Moran, V. C. (1995). Localized failure of a weed biological control agent attributed to insecticide drift. *Agriculture, Ecosystems & Environment*, **52**, 197–203.

Hoffmann, M. P. & Frodsham, A. C. (1993). *Natural Enemies of Vegetable Insect Pests*. Ithaca, NY: Cornell Cooperative Extension.

Hokkanen, H. & Lynch, J. M. (eds.) (1995). *Biological Control: Benefits and Risks*. Cambridge: Cambridge University Press.

Hokkanen, H. & Pimentel, D. (1984). New approach for selecting biological control agents. *Canadian Entomologist*, **116**, 1109–1121.

Hokkanen, H. & Pimentel, D. (1986). New associations in biological control: Theory and practice. *Canadian Entomologist*, **121**, 829–840.

Holloway, J. K. (1957). Weed control by insect. *Scientific American*, **197** (1), 56–62.

Holm, L. G., Plucknett, D. L., Pancho, J. V., & Herberger, J. P. (1977). *The World's Worst Weeds: Distribution and Biology*. Honolulu: University of Hawaii Press.

Hough-Goldstein, J., Lake, E., & Reardon, R. (2012). Status of an ongoing biological control program for the invasive vine, *Persicaria perfoliata*, in eastern North America. *BioControl*, **57**, 181–189.

Howarth, F. G. (1983). Biological control: Panacea or Pandora's box? *Proceedings of the Hawaiian Entomological Society*, **24**, 239–244.

Howarth, F. G. (1991). Environmental impacts of classical biological control. *Annual Review of Entomology*, **36**, 485–510.

Hoy, M. A. (1993). Biological control in U.S. agriculture: Back to the future. *American Entomologist*, **39**, 140–150.

Hrusa, G. F. & Gaskin, J. F. (2008). The *Salsola* complex in California (Chenopodiaceae): Characterization and status of *Salsola australis* and the autochthonous allopolyploid *Salsola ryanii* sp. nov. *Madroño*, **55**, 113–131.

Huang, N., Enkegaard, A., Osborne, L. S., Ramakers, P. M. J., Messelink, G. J., Pijnakker, J., & Murphy, G. (2011) The Banker Plant Method in biological control. *Critical Reviews in Plant Sciences*, **30**, 3, 259–278

Huffaker, C. B. (1958). Experimental studies on predation: Dispersion factors and predator–prey oscillations. *Hilgardia*, **27**, 343–383.

Huffaker, C. B., Messenger, P. S., & DeBach, P. (1971). The natural enemy component in natural control and the theory of biological control. In *Biological Control*, ed. C. B. Huffaker, pp. 16–67. New York: Plenum Press.

Huffaker, C. B., Shea, K. P., & Herman, S. G. (1963). Experimental studies on predation: Complex dispersion and levels of food in an acarine predator-prey interaction. *Hilgardia*, **34**, 305–330.

Hughes, C. E. (1995). Protocols for plant introductions with particular reference to forestry: Changing perspectives on risks to biodiversity and economic development. *Proceedings Brighton Crop Protection Conference Symposium: Weeds in a Changing World*, **64**, 15–32.

Huigens, M. E., Woelke, J. B., Pashalidou, F. G., Bukovinszky, T., Smid, H. M., & Fatouros, N. E. (2010). Chemical espionage on species-specific butterfly anti-aphrodisiacs by hitchhiking *Trichogramma* wasps. *Behavioral Ecology*, **21**, 470–478.

Hull, L. A., Hickey, K. D., & Kanour, W. W. (1983). Pesticide usage patterns and associated pest damage in commercial apple orchards of Pennsylvania. *Journal of Economic Entomology*, **76**, 577–583.

Hunter-Fujita, F. R., Entwistle, P. F., Evans, H. F., & Crook, N. E. (1998). *Insect Viruses and Pest Management*. Chichester, UK: John Wiley & Sons.

Jackson, T. A., Pearson, J. F., O'Callaghan, M., Mahanty, H. K., & Willocks, M. J. (1992). Pathogen to product-development of *Serratia entomophila* (Enterobacteriaceae) as a commercial biological control agent for the New Zealand grass grub (*Costelytra zealandica*). In *Use of Pathogens in Scarab Pest Management*, ed. T. R. Glare & T. A. Jackson, pp. 191–198. Andover, UK: Intercept Ltd.

James, D. G., Orre-Gordon, S., Reynolds, O. L., & Simpson, M. (2012). Employing chemical ecology to understand and exploit biodiversity for pest management. In *Biodiversity and Insect Pests: Key Issues for Sustainable Management*, ed. G. M. Gurr, S. D. Wratten, W. E. Snyder, & D. M. Y. Read, pp. 185–195. Chichester, UK: Wiley-Blackwell.

Jeschke, J. M., Gómez Aparicio, L., Haider, S., Heger, T., Lortie, C. J., Pyšek, P., & Strayer, D. (2012). Support for major hypotheses in invasion biology is uneven and declining. *Neobiota*, **14**, 1–20.

Johnson, M. T., Follett, P. A., Taylor, A. D., & Jones, V. P. (2005). Impacts of biological control and invasive species on a non-target native Hawaiian insect. *Oecologia*, **142**, 529–540.

Julien, M., Center, T. D., & Tipping, P. W. (2002). Floating fern (*Salvinia*). In *Biological Control of Invasive Plants in the Eastern United States*, ed. R. Van Driesche, B. Blossey, M. Hoddle, S. Lyon, & R. Reardon, pp. 17–32. USDA Forest Service, FHTET-2002-04.

Julien, M. & White, G. (1997). *Biological Control of Weeds: Theory and Practical Application* (ACIAR Monograph 49). Canberra: Australian Centre for International Agricultural Research.

Kairo, M. T. K., Paraiso, O., Das Gautam, R., & Peterkin, D. D. (2013). *Cryptolaemus montrouzieri* (Mulsant) (Coccinellidae: Scymninae): A review of biology, ecology, and use in biological control with particular references to potential impact on non-target organisms. *CAB Reviews*, **8**, 1–20.

Kaplan, I. & Thaler, J. S. (2010). Plant resistance attenuates the consumptive and non-consumptive impacts of predators on prey. *Oikos*, **109**, 1105–1113.

Kenis, M., Auger-Rozenberg, M. A., Roques, A., Timms, L., Péré, C., Cock, M. J. W., Settele, J., Augustin, S., & Lopez-Vaamonde, C. (2009). Ecological effects of invasive alien insects. *Biological Invasions*, **11**, 21–45.

Kenis, M., Hurley, B., Hajek, A. E., & Cock, M. (2017). Classical biological control against insect tree pests: Facts and figures. *Biological Invasions*, **19**, 3401–3417.

Kerr, A. (1980). Biological control of crown gall through production of agrocin 84. *Plant Disease*, **64**, 25–30.

Kerry, B. R. (2000). Rhizosphere interactions and the exploitation of microbial agents for the biological control of plant-parasitic nematodes. *Annual Review of Phytopathology*, **38**, 423–441.

Kerry, B. R. (2001). Exploitation of the nematophagous fungal *Verticillium chlamydosporium* Goddard for the biological control of root-knot nematodes (*Meloidogyne* spp.). In *Fungi as Biocontrol Agents: Progress, Problems and Potential*, ed. T. M. Butt, C. Jackson, & N. Magan, pp. 155–167. Wallingford, UK: CAB International.

Kessler, A. & Baldwin, I. T. (2001). Defensive function of herbivore-induced plant volatile emission in nature. *Science*, **291**, 2141–2144.

Kimberling, D. N. (2004). Lessons from history: Predicting successes and risks of intentional introductions for arthropod biological control. *Biological Invasions* **6**, 301–318.

Kiritani, K., Kawahara, S., Sasaba, T., & Nakasuji, F. (1972). Quantitative evaluation of predation by spiders on the green rice leafhopper, *Nephotettix cincticeps* Uhler, a sight count method. *Researches on Population Ecology*, **13**, 187–200.

Klemola, N., Andersson, T., Ruohomäki, K., & Klemola, T. (2010). Experimental test of parasitism hypothesis for population cycles of a forest lepidopteran. *Ecology*, **91**, 2506–2513.

Kogan, M. (1994). Plant resistance in pest management. In *Introduction to Insect Pest Management*, ed. R. L. Metcalf & W. H. Luckmann, 3rd edn., pp. 73–128. New York: John Wiley & Sons.

Kogan, M. (1998). Integrated pest management: Historical perspectives and contemporary developments. *Annual Review of Entomology*, **43**, 243–270.

Kok, L. T. (2001). Classical biological control of nodding and plumeless thistles. *Biological Control*, **21**, 206–213.

Kondo, A. & Hiramatsu, T. (1999). Resurgence of the peach silver mite, *Aculus fockeui* (Nalepa et Trouessart) (Acari: Eriophyidae), induced by a synthetic pyrethroid fluvalinate. *Applied Entomology and Zoology*, **34**, 531–535.

Krebs, C. J. (2001). *Ecology: The Experimental Analysis of Distribution and Abundance*, 5th edn. San Francisco, CA: Benjamin Cummings.

Krieg, A., Huger, A. M., Langenbruch, G. A., & Schnetter, W. (1983). *Bacillus thuringiensis* var. *tenebrionis*: A new pathotype effective against larvae of Coleoptera. *Zeitschrift für Angewandte Entomologie*, **96**, 500–508.

Kuris, A. M. (2003). Did biological control cause extinction of the coconut moth, *Levuana iridescens*, in Fiji? *Biological Invasions*, **5**, 133–141.

Lafferty, K. D. & Kuris, A. M. (1996). Biological control of marine pests. *Ecology*, **77**, 1989–2000.

Landis, D. A. (2017). Designing agricultural landscapes for biodiversity-based ecosystem services. *Basic and Applied Ecology*, **18**, 1–12.

Landis, D. A., Wratten, S. D., & Gurr, G. M. (2000). Habitat management to conserve natural enemies of arthropod pests in agriculture. *Annual Review of Entomology*, **45**, 175–201.

Langewald, J. & Kooyman, C. (2007). Green Muscle, a fungal biopesticide for control of grasshoppers and locusts in Africa. In *Biological Control: A Global Perspective*, ed. C. Vincent, M. S. Goettel, G. Lazarovits, pp. 311–318. Wallingford, UK: CABI Publishing.

Larson, D. L., James, J. B., Grace, B., & Larson, J. L. (2008). Long-term dynamics of leafy spurge (*Euphorbia esula*) and its biocontrol agent, flea beetles in the genus *Aphthona*. *Biological Control*, **47**, 250–256.

Lawton, J. H. (1990). Biological control of plants: A review of generalisations, rules and principles using insects as agents. In *Alternatives to the Chemical Control of Weeds: Proceedings of International Conference held at Rotorua, New Zealand, July 1989*, ed. C. Bassett, L. J. Whitehouse, & J. A. Zabkiewicz, 3–17. Rotorua, New Zealand: Ministry of Forestry, Forest Research Institute.

Lawton, J. H. & McNeill, S. (1979). Between the devil and the deep blue sea: On the problem of being an herbivore. In *Population Dynamics*, ed. K. Anderson, B. Turner, & L. R. Taylor, pp. 223–245. Oxford: Blackwell.

Leake, D. V, Leake, J. B., & Roeder, M. L. (1993). *Desert and Mountain Plants of the Southwest*. Norman: Oklahoma University Press.

Lennox, C. L., Morris, M. J., Van Rooi, C., Serdani, M., Wood, A. R., Den Breeÿen, A., Markram, J. L., & Samuels, G. (2004). A decade of biological control of *Acacia saligna* in South Africa, using the gall rust fungus, *Uromycladium tepperianum*. *Proceedings of the XI International Symposium on Biological Control of Weeds*, Canberra, Australia, April 27–May 2, 2003, pp. 574–575.

Letourneau, D. K., Jedlicka, J. A., Bothwell, S. B., & Moreno, C. R. (2009). Effects of natural enemy biodiversity on the suppression of arthropod herbivores in terrestrial ecosystems. *Annual Review of Ecology, Evolution and Systematics*, **40**, 573–592.

Lewis, S. (1989). *Cane Toads: An Unnatural History*. New York: Dolphin/Doubleday.

Lewis, W. J., van Lenteren, J. C., Phatak, S. C., & Tumlinson, J. H., III (1997). A total system approach to sustainable pest management. *Proceedings of the National Academy of Sciences, U.S.A.*, **94**, 12,243–12,248.

Li, L.-Y. (1994). Worldwide use of *Trichogramma* for biological control of different crops: A survey. In *Biological Control with Egg Parasitoids*, ed. E. Wajnberg & S. A. Hassan, pp. 37–53. Wallingford, UK: CABI Publishing.

Li, Z., Alves, S. B., Roberts, D. W., Fan, M., Delalibera, I., Jr., Tang, J., et al. (2010). Biological control of insects in Brazil and China: History, current programs and reasons for their successes using entomopathogenic fungi. *Biocontrol Science and Technology*, **20**, 117–136.

Liang, W. & Huang, M. (1994). Influence of citrus orchard ground cover plants on arthropod communities in China: A review. *Agriculture, Ecosystems & Environment*, **45**, 175–201.

Lindow, S. E. & Brandl, M. T. (2003). Microbiology of the phyllosphere. *Applied and Environmental Microbiology*, **69**, 1875–1883.

Lomer, C. J., Bateman, R. P., Johnson, D. L., Langewald, J., & Thomas, M. (2001). Biological control of locusts and grasshoppers. *Annual Review of Entomology*, **46**, 667–702.

Lomer, C. J., Thomas, M. B., Douro-Kpindou, O.-K., Gbongboui, C., Godonou, I., Langewald, J., & Shah, P. A. (1997). Control of grasshoppers, particularly *Hieroglyphus daganensis*, in northern Benin using *Metarhizium anisopliae*. *Memoirs of the Entomological Society of Canada*, **171**, 301–311.

Losey, J. E. & Denno, R. F. (1998). Positive predator–predator interactions: Enhanced predation rates and synergistic suppression of aphid populations. *Ecology*, **79**, 2143–2152.

Louda, S. M., Kendall, D., Connor, J., & Simberloff, D. (1997). Ecological effects of an insect introduced for the biological control of weeds. *Science*, **277**, 1088–1090.

Louda, S. M., Pemberton, R. W., Johnson, M. T., & Follett, P. A. (2003). Non-target effects – The Achilles' heel of biological control? *Annual Review of Entomology*, **48**, 365–396.

Luck, R. F., Shepard, B. M., & Kenmore, P. E. (1988). Experimental methods for evaluating arthropod natural enemies. *Annual Review of Entomology*, **33**, 367–391.

Luckmann, W. H. & Metcalf, R. L. (1994). The pest management concept. In *Introduction to Insect Pest Management*, ed. R. L. Metcalf & W. H. Luckmann, 3rd edn., pp. 1–34. New York: John Wiley & Sons.

Lumsden, R. D., Lewis, J. A., & Locke, J. C. (1993). Managing soilborne plant pathogens with fungal antagonists. In *Pest Management: Biologically Based Technologies*, ed. R. D. Lumsden & J. L. Vaughn, pp. 196–203. Washington, DC: American Chemical Society.

Lym, R. G. & Zollinger, R. K. (1995). Integrated management of leafy spurge. North Dakota State Extension Service Publication W-866.

Lynch, L. D. & Thomas, M. B. (2000). Nontarget effects in the biocontrol of insects with insects, nematodes and microbial agents: The evidence. *Biocontrol News & Information*, **21**, 117N–130N.

Maczey, N., Tanner, R., & Shaw, R. (2012). Understanding and addressing the impact of invasive non-native species in the UK Overseas Territories in the South Atlantic: A review of the potential for biocontrol/ Preliminary results Ascension Island. Report ref # TR10086. Wallingford, UK: CABI Publishing.

Madden, J. L. (1968). Behavioural responses of parasites of the symbiotic fungus associated with *Sirex noctilio* F. *Nature*, **218**, 189–190.

Malcolm, S. B. (1992). Prey defence and predator foraging. In *Natural Enemies: The Population Biology of Predators, Parasites and Diseases*, ed. M. J. Crawley, pp. 458–475. Oxford: Blackwell Scientific Publications.

Marshall, J. D. & Fenner, F. (1958). Studies in the epidemiology of infectious myxomatosis of rabbits: VI. Changes in the innate resistance of Australian wild rabbits exposed to myxomatosis. *Journal of Hygiene*, **56**, 288–302.

Mathre, D. E., Cook, R. J., & Callan, N. W. (1999). From discovery to use: Traversing the world of commercializing biocontrol agents for plant disease control. *Plant Disease*, **83**, 972–983.

McCoy, C. W., Stuart, R. J., Duncan, L. W., & Shapiro-Ilan, D. I. (2007). Application and evaluation of entomopathogens for citrus pest control. In *Field Manual of Techniques in Invertebrate Pathology*, ed. L. A. Lacey & H. K. Kaya, pp. 567–581. Dordrecht, Netherlands: Springer.

McEvoy, P. B. & Coombs, E. M. (1999). Biological control of plant invaders: Regional patterns, field experiments, and structured population models. *Ecological Applications*, **9**, 387–401.

McEvoy, P. B., Rudd, N. T., Cox, C. S., & Huso, M. (1993). Disturbance, competition, and herbivory effects on ragwort *Senecio jacobaea* populations. *Ecological Monographs*, **63**, 55–75.

McFadyen, R. E. C. (1998). Biological control of weeds. *Annual Review of Entomology*, **43**, 369–393.

McFadyen, R. E. C. (2000). Successes in biological control of weeds. In *Proceedings of the X International Symposium on Biological Control of Weeds*, ed. N. R. Spencer, pp. 3–14. Bozeman: Montana State University.

Meehan, T. D., Werling, B. P., Landis, D. A., & Gratton, C. (2011). Agricultural landscape simplification and insecticide use in the Midwestern United States. *Proceedings of the National Academy of Sciences, U.S.A.*, **108**, 11,500–11,505.

Mendes, R., Kruijt, M., De Bruijn, I., Dekkers, E., Van der Voort, M., Schneider, J. H. M., Piceno, Y. M., DeSantis, T. Z., Andersen, G. L., Bakker, P. A., et al. (2011). Deciphering the rhizosphere microbiome for disease-suppressive bacteria. *Science*, **332**, 1097–1100.

Messelink, G. J., Bennison, J., Alomar, O., Ingegno, B. L., Tavella, L., Shipp, L., Palevsky, E., & Wäckers, F. L. (2014). Approaches to conserving natural enemy populations in greenhouse crops: Current methods and future prospects. *BioControl*, **59**, 377–393.

Milbrath, L. R. & Nechols, J. R. (2014). Plant-mediated interactions: Considerations for agent selection in weed biological control programs. *Biological Control*, **72**, 80–90.

Mills, N. J. (2006). Accounting for differential success in the biological control of homopteran and lepidopteran pests. *New Zealand Journal of Ecology*, **30**, 61–72.

Moosavi, M. R. & Zare, R. (2015). Factors affecting commercial success of biocontrol agents of phytonematodes. In *Biocontrol Agents of Phytonematodes*, ed. T. Askary & P. R. P. Martinelli, pp. 423–445. Wallingford, UK: CABI.

Mora, C., Tittensor, D. P., Adl, S., Simpson, A. G. B., & Worm, B. (2011). How many species are there on earth and in the ocean? *PLoS Biology*, **9**, e1001127.

Morrison, L. W., Dall'aglio-Holvercem, C. G., & Gilbert, L. E. (1997). Oviposition behavior and development of *Pseudacteon* flies (Diptera, Phoridae), parasitoids of *Solenopsis* fire ants (Hymenoptera, Formicidae). *Environmental Entomology*, **26**, 716–724.

Moscardi, F. (1999). Assessment of the application of baculoviruses for control of Lepidoptera. *Annual Review of Entomology*, **44**, 257–289.

Mota-Sanchez, D., Bills, P. S., & Whalon, M. E. (2002). Arthropod resistance to pesticides: Status and overview. In *Pesticides in Agriculture and the Environment*, ed. W. B. Wheeler, pp. 241–272. New York: Dekker.

Mota-Sanchez, D. & Wise, J. C. (2017). *Arthropod Pesticide Resistance Database 2016*. Retrieved from www.pesticideresistance.org

Murdoch, W. & Briggs, C. J. (1996). Theory for biological control: Recent developments. *Ecology*, **77**, 2001–2013.

Murdoch, W., Briggs, C. J., & Swarbrick, S. (2005). Host suppression and stability in a parasitoid–host system: Experimental demonstration. *Science*, **309**, 610–613.

Murphy, B. C., Rosenheim, J. A., Granett, J., Pickett, C. H., & Dowell, R. V. (1998). Measuring the impact of a natural enemy refuge: The prune tree/vineyard example. In *Enhancing Biological Control: Habitat Management to Promote Natural Enemies of Agricultural Pests*, ed. C. H. Pickett & R. L. Bugg, pp. 297–309. Berkeley: University of California Press.

Murphy, G. D., Ferguson, G., Frey, K., Lambert, L., Mann, M., & Matteoni, J. (2002). The use of biological control in Canadian greenhouse crops. In *Integrated Control in Protected Crops, Temperate Climate*, ed. S. Enkegaard. *IOBC WPRS Bulletin*, **25(1)**, 193–196.

Murray, E. (1993). The sinister snail. *Endeavour*, **17**, 78–83.

Myers, J. H. (1985). How many insect species are necessary for successful biocontrol of weeds? In *Proceedings of the 6th International Symposium on the Biological Control of Weeds, Agriculture Canada*, ed. E. S. Delfosse, pp. 77–82. Ottawa, Canada: Canadian Govt. Printing Office.

Myers, J. H. & Harris, P. (1980). Distribution of *Urophora* galls in flowers heads of diffuse and spotted knapweed in British Columbia. *Journal of Applied Ecology*, **17**, 359–367.

Myers, J. H. & Risley, C. (2000). Why reduced seed production is not necessarily translated into successful biological weed control. In *Proceedings, X International Symposium on Biological Control of Weeds*, ed. N. R. Spencer, pp. 569–581. Bozeman: Montana State University.

Naranjo, S. E. & Ellsworth, P. C. (2009). Fifty years of the integrated control concept: Moving the model and implementation forward in Arizona. *Pest Management Science*, **65**, 1267–1286.

Naranjo, S. E. & Ellsworth, P. C. (2010). Fourteen years of *Bt* cotton advances IPM in Arizona. *Southwestern Entomologist*, **35**, 437–444.

National Academy of Sciences (US). (1988). *Research Briefings 1987: Report of the Research Panel on Biological Control in Managed Ecosystems*. Washington, DC: National Academy Press.

National Research Council (US), Committee on Pest and Pathogen Control Through Management of Biological Control Agents and Enhanced Cycles and Natural Processes. (1996). *Ecologically Based Pest Management: New Solutions for a New Century*. Washington, DC: National Academy Press.

Neuenschwander, P. & Herren, H. (1988). Biological control of the cassava mealybug, *Phenacoccus manihoti*, by the exotic parasitoid *Epidinocarsis lopezi* in Africa. *Philosophical Transactions of the Royal Society of London B*, **318**, 319–333.

Newhouse, J. R. (1990). Chestnut blight. *Scientific American*, **263(1)**, 106–111.

Newman, R. M., Thompson, D. C., & Richman, D. B. (1998). Conservation strategies for the biological control of weeds. In *Conservation Biological Control*, ed. P. Barbosa, pp. 371–396. San Diego, CA: Academic Press.

Nielsen, C., Milgroom, M. G., & Hajek, A. E. (2005). Genetic diversity in the gypsy moth fungal pathogen *Entomophaga maimaiga* from founder populations in North America and source populations in Asia. *Mycological Research*, **109**, 941–950.

Nilsson, C. (1985). Impact of ploughing on emergence of pollen beetle parasitoids after hibernation. *Zeitschrift für Angewandte Entomologie*, **100**, 302–308.

Nordlund, D. A. (1996). Biological control, integrated pest management and conceptual models. *Biocontrol News & Information*, **17**, 35N–44N.

Nottingham, L. B., Dively, G. P., Schultz, P. B., Herbert, D. A., & Kuhar, T. P. (2016). Natural history, ecology, and management of the Mexican bean beetle (Coleoptera: Coccinellidae) in the United States. *Journal of Integrated Pest Management*, **7**(1), 2.

Nowierski, R. M., Zeng, Z., Schroeder, D., Gassmann, A., FitzGerald, B. C., & Cristofaro, M. (2002). Habitat associations of *Euphorbia* and *Aphthona* species for Europe: Development of predictive models for natural enemy release with ordination analysis. *Biological Control*, **23**, 1–17.

Nyrop, J., English-Loeb, G., & Roda, A. (1998). Conservation biological control of spider mites in perennial cropping systems. In *Conservation Biological Control*, ed. P. Barbosa, pp. 307–333. San Diego, CA: Academic Press.

Oliver, K. M., Noge, K., Huang, E. M., Campos, J. M., Becerra, J. X., & Hunter, M. S. (2012). Parasitic wasp responses to symbiont-based defense in aphids. *BMC Biology*, **10**, 11.

Onstad, D. W., Fuxa, J. R., Humber, R. A., Oestergaard, J., Shapiro-Ilan, D. I., Gouli, V. V., Anderson, R. S., Andreadis, T. G., & Lacey, L. A. (2006). An *Abridged Glossary of Terms Used in Invertebrate Pathology*, 3rd edn. Society for Invertebrate Pathology. Retrieved from www .sipweb.org/resources/glossary.html.

Orr, D. (2009). Biological control and integrated pest management. In *Integrated Pest Management: Innovation-Development Processes*, ed. R. Peshin & A. K. Dhawan, pp. 207–239. Dordrecht, Netherlands: Springer.

OTA (Office of Technology Assessment; US Congress). (1995). *Biologically Based Technologies for Pest Control*, OTA-ENV-636. Washington, DC: US Government Printing Office.

Paine, R. T. (1996). Preface. In *Food Webs: Integration of Patterns and Dynamics*, ed. G. A. Polis & K. O. Winemille, pp. ix–x. New York: Chapman & Hall.

Parra, J. R. P. (2014). Biological control in Brazil: An overview. *Scientia Agricola*, **71**, 345–355.

Parrella, M. P., Hansen, L. S., & van Lenteren, J. (1999). Glasshouse environments. In *Handbook of Biological Control*, ed. T. S. Bellows & T. W. Fisher, pp. 819–839. San Diego, CA: Academic Press.

Pedigo, L. P. (1996). *Entomology and Pest Management*, 2nd edn. Upper Saddle River, NJ: Prentice Hall.

Pemberton, R. W. (2000). Predictable risk to native plants in weed biological control. *Oecologia*, **125**, 489–494.

Perfecto, I. & Castiñeiras, A. (1998). Deployment of the predaceous ants and their conservation in agroecosystems. In *Conservation Biological Control*, ed. P. Barbosa, pp. 269–289. San Diego, CA: Academic Press.

Perkins, J. H. & Garcia, R. (1999). Social and economic factors affecting research and implementation of biological control. In *Handbook of Biological Control*, ed. T. S. Bellows & T. W. Fisher, pp. 993–1009. San Diego, CA: Academic Press.

Perkins, R. C. L. (1897). The introduction of beneficial insects into the Hawaiian Islands. *Nature*, **55**, 499–500.

Pickett, C. H. & Bugg, R. L. (eds.) (1998). *Enhancing Biological Control: Habitat Management to Promote Natural Enemies of Agricultural Pests*. Berkeley: University of California Press.

Pimentel, D. (ed.) (2002). *Biological Invasions: Economic and Environmental Costs of Alien Plant, Animal, and Microbe Species*. Boca Raton, FL: CRC Press.

Pimentel, D., Acquay, H., Biltonen, M., Rice, P., Silva, M., Nelson, J., Lipner, V., Giordano, S., Horowitz, A., & D'Amore, M. (1992). Assessment of environmental and economic impacts of pesticide use. In *The Pesticide Question: Environment, Economics, and Ethics*, ed. D. Pimentel & H. Lehman, pp. 47–84. New York: Chapman & Hall.

Price, P. W. (1984). The concept of the ecosystem. In *Ecological Entomology*, ed. C. B. Huffaker & R. L. Rabb, pp. 19–50. New York: John Wiley & Sons.

Quarles, W. (ed.) (2015). 2015 Directory of least-toxic pest control products. *The IPM Practitioner*, **34** (11/12), 1–41.

Quicke, D. L. J. (1997). *Parasitic Wasps*. London: Chapman & Hall.

Radcliffe, E. B. & Flanders, K. L. (1998). Biological control of alfalfa weevil in North America. *Integrated Pest Management Reviews*, **3**, 225–242.

Ramula, S., Knight, T. M., Burns, J. H., & Buckley, Y. M. (2008). General guidelines for invasive plant management based on comparative demography of invasive and native plant populations. *Journal of Applied Ecology* **45**, 1124–1133.

Reilly, J. R. & Hajek, A. E. (2012). Prey processing by avian predators increases virus transmission in the gypsy moth. *Oikos*, **121**, 1311–1316.

Rezende, J. M., Beatriz, A., Zanardo, R., da Silva Lopes, M., Delalibera, I., Jr., & Rehner, S. A. (2015). Phylogenetic diversity of Brazilian *Metarhizium* associated with sugarcane agriculture. *BioControl*, **60**, 495–505.

Rishbeth, J. R. (1975). Stump inoculation: A biological control of *Fomes annosus*. In *Biology and Control of Soil-borne Pathogens*, ed. G. W. Bruehl, pp. 158–162. St. Paul, MN: American Phytopathological Society.

Root, R. B. (1973). Organization of plant-arthropod association in simple and diverse habitats: The fauna of collards (*Brassica oleracea*). *Ecological Monographs*, **43**, 95–124.

Rosenheim, J. A. (1998). Higher-order predators and the regulation of insect herbivore populations. *Annual Review of Entomology*, **43**, 421–447.

Rosskopf, E. N., Charudattan, R., & Kadir, J. B. (1999). Use of plant pathogens in weed control. In *Handbook of Biological Control*, ed. T. S. Bellows & T. W. Fisher, pp. 891–918. San Diego, CA: Academic Press.

Ruberson, J. R., Kring, T. J., & Elkassabany, N. (1998). Overwintering and the diapause syndrome of predatory Heteroptera. In *Predatory Heteroptera in Agroecosystems: Their Biology and Use in Biological Control*, ed. M. Coll & J. R. Ruberson, pp. 46–69. Lanham, MD: Entomological Society of America.

Rutz, D. A. & Watson, D. W. (1998). Parasitoids as a component in an integrated fly-management program on dairy farms. In *Mass-Reared Natural Enemies: Application, Regulation, and Needs*, ed. R. L. Ridgway, M. P. Hoffmann, M. N. Inscoe, & C. S. Glenister, pp. 185–201. Lanham, MD: Entomological Society of America.

Samuels, G. J. & Hebbar, P. K. (2015). *Trichoderma: Identification and Agricultural Applications*. St. Paul, MN: APS Press.

Sasser, J. N. & Freckman, D. W. (1987). A world perspective on nematology: The role of the society. In *Vistas on Nematology: A Commemoration of the Twenty-fifth Anniversary of the Society of Nematologists*, ed. J. A. Veech & D. W. Dickson, pp. 7–14. Hyattsville, MD: Society of Nematologists.

Saunders, G., Cooke, B., McColl, K., Shine, R., & Peacock, T. (2010). Modern approaches for the biological control of vertebrate pests: An Australian perspective. *Biological Control*, **52**, 288–295.

Schaab, R. (1996). Economy and ecology of biological control activities in Africa: A case study on the cassava mealybug, *Phenacoccus manihoti* Mat. Ferr. Dissertation, University Hohenheim, Germany.

Schellhorn, N. A., Gagic, V., & Bommarco, R. (2015). Time will tell: Resource continuity bolsters ecosystem services. *Trends in Ecology & Evolution*, **30**, 524–530.

Schlinger, E. I. & Dietrick, E. J. (1960). Biological control of insect pests aided by strip-farming alfalfa in experimental program. *California Agriculture*, **14**(1), 8–9, 15.

Schmidt, J. M. (1992). Host recognition and acceptance by *Trichogramma*. In *Biological Control with Egg Parasitoids*, ed. E. Wajnberg & S. A. Hassan, pp. 165–200. Wallingford, UK: CAB International.

Schroeder, D. (1983). Biological control of weeds. In *Recent Advances in Weed Control*, ed. W. E. Fletcher, pp. 41–78. Farnham Royal, UK: Commonwealth Agricultural Bureau.

Seebens, H., Blackburn, T. M., Dyer, E. E., Genovesi, P., Hulme, P. E., Jeschke, J. M., Pagad, S., Pyšek, P., Winter, M., Arianoutsou, M., et al. (2017). No saturation in the accumulation of alien species worldwide. *Nature Communications*, **8**, 14435.

Shapiro-Ilan, D. I., Hazir, S., & Glazer, I. (2017). Basic and applied research: Entomopathogenic nematodes. In *Microbial Agents for Control of Insect Pests: From Discovery to Commercial Development and Use*, ed. L. A. Lacey, pp. 91–105. Amsterdam: Academic Press.

Shaw, S. R. (1993). Observations on the ovipositional behaviour of *Neoneurus mantis*, an ant-associated parasitoid from Wyoming (Hymenoptera: Braconidae). *Journal of Insect Behavior*, **6**, 649–658.

Shu, S., Swedenborg, P. D., & Jones, R. L. (1990). A kairomone for *Trichogramma nubilale* (Hymenoptera: Trichogrammatidae): Isolation, identification, and synthesis. *Journal of Chemical Ecology*, **16**, 521–529.

Shuler, M. L., Hammer, D. A., Granados, R. R., & Wood, H. A. (1995). Overview of baculovirus-insect culture system. In *Baculovirus Expression Systems and Biopesticides*, ed. M. L. Shuler, H. A. Wood, R. R. Granados, & D. A. Hammer, pp. 1–11. New York: John Wiley & Sons.

Silvestri, F. (1906). Contribuzioni alla conoscenza biologica degli imenotteri parassiti: I. Biologica del *Litomastix truncatellus* (Dalm.). *Bollettino del Laboratorio di Zoologia Generale e Agraria della Facolta Agraria in Portici*, **1**, 17–64.

Sivasubramaniam, V. & Subramaniam, S. (2015). Area-wide releases and evaluation of the parasitoid *Eretmocerus hayati* (Hymenoptera:Aphelinidae) for silverleaf whitefly control. *Acta Horticulturae*, **1105**, 81–88.

Slippers, B., de Groot, P., & Wingfield, M. J. (eds.) (2012). *The Sirex Woodwasp and its Fungal Symbiont: Research and Management of a Worldwide Invasive Pest*. Dordrecht, Netherlands: Springer.

Slobodkin, L. B. (1988). Intellectual problems of applied ecology. *BioScience*, **38**, 337–342.

Smart, G. C., Jr. (1995). Entomopathogenic nematodes for the biological control of insects. *Journal of Nematology*, **27**, 529–534.

Smith, H. S. (1919). On some phases of insect control by the biological method. *Journal of Economic Entomology*, **12**, 288–292.

Sprenkel, R. K., Brooks, W. M., Van Duyn, J. W., & Deitz, L. L. (1979). The effects of three cultural variables on the incidence of *Nomuraea rileyi*, a phytophagous Lepidoptera, and their predators on soybeans. *Environmental Entomology*, **3**, 334–339.

Steinhaus, E. A. (1975). *Disease in a Minor Chord*. Columbus: Ohio University Press.

Stern, V. M., Smith, R. F., van den Bosch, R., & Hagen, K. S. (1959). The integration of chemical and biological control of the spotted alfalfa aphid: I. The integrated control concept. *Hilgardia*, **29**, 81–101.

Stiling, P. (1990). Calculating the establishment rates of parasitoids in classical biological control. *American Entomologist*, **36**, 225–230.

Stiling, P. (1993). Why do natural enemies fail in classical biological control programs? *American Entomologist*, **39**, 31–37.

Stiling, P. & Cornelissen, T. (2005). What makes a successful biocontrol agent? A meta-analysis of biological control agent performance. *Biological Control*, **34**, 236–246.

Stokstad, E. (2013). Pesticide planet. *Science*, **341**, 730–731.

Strand, M. R. & Pech, L. L. (1995). Immunological basis for compatibility in parasitoid–host relationships. *Annual Review of Entomology*, **40**, 31–56.

Strong, D. R., Lawton, J. H., & Southwood, T. R. E. (1984). *Insects on Plants*. Oxford: Oxford University Press.

Suckling, D. M. & Sforza, R. F. H. (2014). What magnitude are observed non-target impacts from weed biocontrol? *PLoS ONE*, **9**(1), e84847.

Sutton, J. C. & Peng, G. (1993). Manipulation and vectoring of biocontrol organisms to manage foliage and fruit diseases in cropping systems. *Annual Review of Phytopathology*, **31**, 473–493.

Swezey, O. H. (1943). Biographical sketch of the work of Albert Koebele in Hawaii. *Bulletin of the Hawaiian Sugar Planters' Association Experiment Station, Entomological Series Bulletin*, **22**, 5–8.

Symondson, W. O. C., Sunderland, K. D., & Greenstone, M. H. (2002). Can generalist predators be effective biocontrol agents? *Annual Review of Entomology*, **47**, 561–594.

Takken, W. & Knols, B. G. J. (2009). Malaria vector control: Current and future strategies. *Trends in Parasitology*, **25**, 101–104.

Tanada, Y. & Kaya, H. K. (1993). *Insect Pathology*. San Diego, CA: Academic Press.

Tang, W. H. (1994). Yield-increasing bacteria (YIB) and biocontrol of sheath blight of rice. In *Improving Plant Productivity and Rhizosphere Bacteria*, ed. M. H. Ryder, P. M. Stephens, & G. D. Bowen, pp. 267–273. Adelaide, Australia: CSIRO, Division of Soils.

Tauber, M. J., Tauber, C. A., Daane, K. M., & Hagen, K. S. (2000). Commercialization of predators: Recent lessons from green lacewings (Neuroptera: Chrysopidae: *Chrysoperla*). *American Entomologist*, **46**, 26–38.

Thacker, J. R. M. (2002). *An Introduction to Arthropod Pest Control*. Cambridge: Cambridge University Press.

Thies, C. & Tscharntke, T. (1999). Landscape structure and biological control in agroecosystems. *Science*, **285**, 893–895.

Thomas, M. B. (1999). Ecological approaches and the development of "truly integrated" pest management. *Proceedings of the National Academy of Sciences, U.S.A.*, **96**, 5944–5951.

Thomas, M. B. (2018). Biological control of human disease vectors: A perspective on challenges and opportunities. *BioControl*, **63**(1), 61–69.

Thomas, M. B., Godfray, H. C. J., Read, A. F., van den Berg, H., Tabashnik. B. E., van Lenteren, J. C., Waage, J. K., & Takken, W. (2012). Lessons from agriculture for the sustainable management of malaria vectors. *PLOS Medicine*, **9**(7), e1001262.

Thomas, M. B. & Willis, A. J. (1998). Biocontrol – risky but necessary? *Trends in Ecology & Evolution*, **13**, 325–329.

Thomas, M. B., Wratten, S. D., & Sotherton, N. W. (1991). Creation of "island" habitats in farmland to manipulate populations of beneficial arthropods: Predator densities and emigration. *Journal of Applied Ecology*, **28**, 906–917.

Thomas, M. B., Wratten, S. D., & Sotherton, N. W. (1992). Creation of "island" habitats in farmland to manipulate populations of beneficial arthropods: Predator densities and species composition. *Journal of Applied Ecology*, **29**, 524–531.

Thomas, P. A. & Room, P. M. (1986). Taxonomy and control of *Salvinia molesta*. *Nature*, **320**, 581–584.

Thompson, W. M. O. (ed.) (2011). *The Whitefly, Bemisia tabaci (Homoptera: Aleyrodidae) Interaction with Geminivirus-Infected Host Plants*. Dordrecht, Netherlands: Springer.

Tisdell, C. (1990). Economic impact of biological control of weeds and insects. In *Critical Issues in Biological Control*, ed. M. Mackauer, L. E. Ehler, & J. Roland, pp. 301–316. Andover, UK: Intercept.

Tomasetto, F., Tylianakis, J. M., Reale, M., Wratten, S., & Goldson, S. L. (2017). Intensified agriculture favors evolved resistance to biological control. *Proceedings of the National Academy of Sciences, U.S.A.*, **114**, 3885–3890.

Topham, M. & Beardsley, J. W. (1975). Influence of nectar source plants on the New Guinea sugarcane weevil parasite, *Lixophaga sphenophori* (Villeneuve). *Proceedings of the Hawaiian Entomological Society*, **22**, 145–154.

Tothill, J. D., Taylor, T. H. C., & Paine, R. W. (1930). *The Coconut Moth in Fiji: A History of its Control by Means of Parasites*. London: Imperial Bureau of Entomology.

Tracy, E. F. (2014). The promise of biological control for sustainable agriculture: A stakeholder-based analysis. *Journal of Science Policy & Governance*, **5** (1).

Traveset, A. (1990). Bruchid egg mortality on *Acacia farnesiana* caused by ants and abiotic factors. *Ecological Entomology*, **15**, 463–467.

Trouvelot, B. (1931). Recherches sur les parasites et predateurs attaquant le doryphore en Amerique du Nord. *Annales des Épiphyties*, **17**, 408–445.

Tschumi, M., Albrecht, M., Bärtschi, C., Collatz, J., Entling, M. H., & Jacot, K. (2016). Perennial, species-rich wildflower strips enhance pest control and crop yield. *Agriculture, Ecosystems & Environment*, **220**, 97–103.

Tumlinson, J. H., Lewis, W. J., & Vet, L. E. M. (1993). How parasitic wasps find their hosts. *Scientific American*, **268**, 100–106.

Turner, C. E. & McEvoy, P. B. (1995). Tansy ragwort (*Senecio jacobaea* L., Asteraceae). In *Biological Control in the Western United States*, ed. J. R. Nechols, L. A. Andres, J. W. Beardsley, R. D. Goeden, & C. G. Jackson, pp. 264–269. Oakland: University of California, Publication 3361.

Uka, D., Hiraoka, T., & Iwabuchi, K. (2006). Physiological suppression of the larval parasitoid *Glyptapanteles pallipes* by the polyembryonic parasitoid *Copidosoma floridanum*. *Journal of Insect Physiology*, **52**, 1137–1142.

United Nations. (2015). Sustainable development goals: 17 goals to transform our world. Retrieved from www.un.org/sustainabledevelopment/sustainable-development-goals/

Urbaneja, A., Gonzalez-Cabrera, J., Arno, J., & Gabarra, R. (2012). Prospects for the biological control of *Tuta absoluta* in tomatoes of the Mediterranean basin. *Pest Management Science*, **68**, 1215–1222.

van Alphen, J. J. M. & Jervis, M. A. (1996). Foraging behaviour. In *Insect Natural Enemies: Practical Approaches to their Study and Evaluation*, ed. M. Jervis & N. Kidd, pp. 1–62. London: Chapman & Hall.

van den Bosch, R. & Hagen, K. S. (1966). Predaceous and parasitic arthropods in California cotton fields. *California Agricultural Experiment Station Bulletin*, **860**.

van den Bosch, R., Messenger, P. S., & Gutierrez, A. P. (1982). *An Introduction to Biological Control*. New York: Plenum Press.

van den Bosch, R. & Stern, V. (1969). The effect of harvesting practices on insect populations in alfalfa. *Proceedings of the Tall Timbers Conference on Ecological Animal Control*, **1**, 47–54.

van den Bosch, R. & Telford, A. D. (1964). Environmental modification and biological control. In *Biological Control of Insect Pests and Weeds*, ed. P. H. DeBach, pp. 459–488. New York: Reinhold Publishing Corporation.

Van de Peer, Y., Ben-Ali, A., & Meyer, A. (2000). Microsporidia: Accumulating molecular evidence that a group of amitochondriate and suspectedly primitive eukaryotes are just curious fungi. *Gene*, **246**, 1–8.

Van Driesche, R. G. & Bellows, T. S., Jr. (eds.) (1993). *Steps in Classical Arthropod Biological Control*. Lanham, MD: Entomological Society of America.

Van Driesche, R. G., Carruthers, R. I., Center, T., Hoddle, M. S., Hough-Goldstein, J., Morin, L., Smith, L., Wagner, D. L., Blossey, B., Brancatini, V., et al. (2010). Classical biological control for the protection of natural ecosystems. *Biological Control*, **54**, S2–S33.

Van Driesche, R. G., Hoddle, M., & Center T. E. (2008). *Control of Pests and Weeds by Natural Enemies: An Introduction to Biological Control*. Chichester, UK: Wiley-Blackwell.

Van Driesche, R. G., Pratt, P. D., Center, T. D., Rayamajhi, M. B., Tipping, P. W., Purcell, M., Fowler, S., Causton, C., Hoddle, M. S., Kaufman, L., et al. (2016). Cases of biological control restoring natural systems. In *Integrating Biological Control into Conservation Practice*, ed. R. Van Driesche, D. Simberloff, B. Blossey, C. Causton, M. Hoddle, C. O. Marks, K. Heinz, D. Wagner, & K. Warner, pp. 208–246. Chichester, UK: John Wiley & Sons.

Van Klinken, R. D. & Edwards, O. (2002). Is host specificity of weed biological control agents likely to evolve rapidly following establishment? *Ecology Letters*, **5**, 590–596.

van Lenteren, J. C. (1980). Evaluation of control capabilities of natural enemies: Does art have to become science? *Netherlands Journal of Zoology*, **30**, 369–381.

van Lenteren, J. C. (2000). Success in biological control of arthropods by augmentation of natural enemies. In *Biological Control: Measures of Success*, ed. G. Gurr & S. Wratten, pp. 77–103. Dordrecht, Netherlands: Kluwer Academic Publishers.

van Lenteren, J. C. (2003). Need for quality control of mass-produced biological control agents. In *Quality Control and Production of Biological Control Agents: Theory and Testing Procedures*, ed. J. C. van Lenteren, pp. 1–18. Wallingford, UK: CABI Publishing.

van Lenteren, J. C. (2012). The state of commercial augmentative biological control: Plenty of natural enemies, but a frustrating lack of uptake. *BioControl*, **57**, 1–20.

van Lenteren, J. C., Bale, J., Bigler, F., Hokkanen, H. M. T., & Loomans, A. J. M. (2006). Assessing risks of releasing exotic biological control agents of arthropod pests. *Annual Review of Entomology*, **51**, 609–634.

van Lenteren, J. C., Bolckmans, K., Köhl, J., Ravensberg, W., & Urbaneja, A. (2018). Biological control using invertebrates and microorganisms: Plenty of new opportunities. *BioControl*, **63**(1), 39–59.

van Lenteren, J. C., Hale, A., Klapwijk, J. N., Van Schelt, J., & Steinberg, S. (2003). Guidelines for quality control of commercially produced natural enemies. In *Quality Control and Production of Biological Control Agents: Theory and Testing Procedures*, ed. J. C. van Lenteren, pp. 265–303. Wallingford, UK: CABI Publishing.

van Lenteren, J. C. & Martin, N. A. (1999). Biological control of whiteflies. In *Integrated Pest and Disease Management in Greenhouse Crops*, ed. R. Albajes, M. Lodovica Gullino, J. C. van Lenteren, & Y. Elad, pp. 202–216. Dordrecht, Netherlands: Kluwer Academic Publishers.

van Lenteren, J. C., Roskam, M. M., & Timmer, R. (1997). Commercial mass production and pricing of organisms for biological control of pests in Europe. *Biological Control*, **10**, 143–149.

Vega, F. E. & Kaya, H. K. (2012). *Insect Pathology*, 2nd edn. San Diego, CA: Academic Press.

Viggiani, G. (1964). La specializzazione entomoparassitica in alcuni Eulofidi (Hym., Chalcidoidea). *Entomophaga*, **9**, 111–118.

Vittum, P. J., Villani, M. G., & Tashiro, H. (1999). *Turfgrass Insects of the United States and Canada*, 2nd edn. Ithaca, NY: Comstock Publishing Associates.

Waage, J. (1990). Ecological theory and the selection of biological control agents. In *Critical Issues in Biological Control*, ed. M. Mackauer, L. E. Ehler, & J. Roland, pp. 135–157. Andover, UK: Intercept.

Waage, J. (1995). The use of exotic organisms as biopesticides: Some issues. In *Biological Control, Benefits and Risks*, ed. H. M. T. Hokkanen & J. M. Lynch, pp. 93–100. Cambridge: Cambridge University Press.

Wäckers, F. L. & van Rijn, P. C. J. (2012). Pick and mix: Selecting flowering plants to meet the requirements of target biological control insects. In *Biodiversity and Insect Pests: Key Issues for Sustainable Management*, ed. G. M. Gurr, S. D. Wratten, W. E. Snyder, & D. M. Y. Read, pp. 139–165. Chichester, UK: Wiley-Blackwell.

Wagner, D. L., Peacock, J. W., Carther, J. L., & Talley, S. E. (1996). Field assessment of *Bacillus thuringiensis* on nontarget Lepidoptera. *Environmental Entomology*, **25**, 1444–1454.

Wang, Z.-Y., He, K.-L., Zhang, F., Lu, X., & Babendreier, D. (2014). Mass rearing and release of *Trichogramma* for biological control of insect pests of corn in China. *Biological Control*, **68**, 136–144.

Wapshere, A. J. (1989). A testing sequence for reducing rejection of potential biological control agents for weeds. *Annals of Applied Biology*, **114**, 515–526.

Webster, F. M. (1909). Investigations of *Toxoptera graminum* and its parasites. *Annals of the Entomological Society of America*, **2**, 67–87.

Weller, D. M., Raaijmakers, J. M., McSpadden Gardener, B. B., & Thomashow, L. S. (2002). Microbial populations responsible for specific soil suppressiveness to plant pathogens. *Annual Review of Phytopathology*, **40**, 309–348.

Weseloh, R. M. (1998). Possibility for recent origin of the gypsy moth (Lepidoptera: Lymantriidae) fungal pathogen *Entomophaga maimaiga* (Zygomycetes: Entomophthorales) in North America. *Environmental Entomology*, **27**, 171–177.

Wetzstein, H. Y. & Phatak, S. C. (1987). Scanning electron microscopy of the uredinial stage of *Puccinia canaliculata* on yellow nutsedge, *Cyperus esculentus* (Cyperaceae). *American Journal of Botany*, **74**, 100–106.

Whipps, J. M. & Davies, K. G. (2000). Success in biological control of plant pathogens and nematodes by microorganisms. In *Biological Control: Measures of Success*, ed. G. Gurr & S. Wratten, pp. 231–269. Dordrecht, Netherlands: Kluwer Academic Publishers.

White, G. (1997). Population ecology and biological control of weeds. In *Biological Control of Weeds: Theory and Practical Application*, ed. M. Julien & G. White (ACIAR Monograph No. 49), pp. 39–45. Canberra: Australian Centre for International Agricultural Research.

Wijnands, F. G. & Kroonen-Backbier, B. M. A. (1993). Management of farming systems to reduce pesticide inputs: The integrated approach. In *Modern Crop Protection: Developments and Perspectives*, ed. J. C. Zadoks, pp. 227–234. Wageningen, Netherlands: Wageningen Pers.

Williams, T., Arredondo-Bernal, H. C., & Rodriguez-del-Bosque, L. A. (2013). Biological pest control in Mexico. *Annual Review Entomology*, **58**, 119–140.

Williamson, M. & Fitter, A. (1996). The varying success of invaders. *Ecology* **77**, 1661–1666.

Winkler, K., Wäckers, F., Bukovinszkine-Kissa, F., & van Lenteren, J. (2006). Sugar resources are vital for *Diadegma semiclausum* fecundity under field conditions. *Basic and Applied Ecology*, **7**, 133–140.

Winston, R. L., Schwarzländer, M., Hinz, H. L., Day, M. D., Cock, M. J. W., & Julien, M. H. (eds.) (2014). *Biological Control of Weeds: A World Catalogue of Agents and Their Target Weeds*,

5th edn. Morgantown, WV: US Department of Agriculture, Forest Service, Forest Health Technology Enterprise Team.

Wood, A. R. (2012). *Uromycladium tepperianum* (a gall-forming rust fungus) causes a sustained epidemic on the weed *Acacia saligna* in South Africa. *Australasian Plant Pathology*, **41**, 255–261.

Yandoc-Ables, C. B., Rosskopf, E. N., & Charudattan, R. (2006). Plant pathogens at work: Progress and possibilities for weed biocontrol. Part 2: Improving weed control efficiency. Retrieved from www.apsnet.org/publications/apsnetfeatures/Pages/WeedBiocontrolPart2.aspx.

Yandoc-Ables, C. B., Rosskopf, E. N., & Charudattan, R. (2007). Plant pathogens at work: Progress and possibilities for weed biocontrol: classical versus bioherbicidal approaches. *Plant Health Progress*. DOI: 10.1094/PHP-2007-0822-01-RV.

Yaninek, J. S. & Hanna, R. (2003). Cassava green mite in Africa – A unique example of successful classical biological control of a mite pest on a continental scale. In *Biological Control in IPM Systems in Africa*, ed. P. Neuenschwander, J. Langewald & C. Borgemeister, pp. 61–76. Wallingford, UK: CABI Publishing.

Zelazny, B., Lolong, A., & Pattang, B. (1992). *Oryctes rhinoceros* (Coleoptera, Scarabaeidae) populations suppressed by a baculovirus. *Journal of Invertebrate Pathology*, **59**, 61–68.

Index

Printed in the United States
by Baker & Taylor Publisher Services